MICROBIAL TECHNOLOGY
CONCEPTS AND APPLICATIONS

MICROBIAL TECHNOLOGY
CONCEPTS AND APPLICATIONS

R. Puvanakrishnan

Former Director Grade Scientist and Head
Department of Biotechnology
Central Leather Research Institute
Chennai, Tamilnadu

S. Sivasubramanian

Research Scientist
King Institute of Preventive Medicine
Chennai, Tamilnadu

T. Hemalatha

Research Associate
Department of Biotechnology
Central Leather Research Institute
Chennai, Tamilnadu

MJP PUBLISHERS

CHENNAI   NEW DELHI   TIRUNELVELI

Cataloguing-in-Publication Data

Puvanakrishnan, R. (1946 –).
  Microbial Technology Concepts and Applications /
by R. Puvanakrishnan, S. Sivasubramanian and
T. Hemalatha. Chennai : MJP Publishers, 2012

MJP Publishers

© Publishers, 2012   New No. 5, Muthu Kalathy Street
All rights reserved   Triplicane
Printed and bound in India   Chennai 600 005

          Publisher : J.C. Pillai
   Managing Editor : C. Sajeesh Kumar
     Project Editor : P. Parvath Radha
Acquisitions Editor : C. Janarthanan
     Editorial Team : B. Ramalakshmi, V.R. Padma, N. Yamunadevi,
                          S. Jeevasruthi, B. Annalakshmi, C. Devi
          CIP Data : Prof. K. Hariharan, Librarian
                          RKM Vivekananda College, Chennai.

Prof. Dr. Asit Baran Mandal
FASc., FRSC (UK)
DIRECTOR

CSIR-Central Leather Research Institute
(Council of Scientific & Industrial Research)
ADYAR, CHENNAI - 600 020. INDIA
Phone :Off - 24910846, 24910897 Res - 24421703, 24915822
Fax : 24912150, 24911589
e-mail : directorclri@gmail.com; clrim@vsnl.com
abmandal@clri.res.in; abmandal@hotmail.com

CSIR-CLRI

FOREWORD

Biotechnology is one of the frontier technologies, which has the potential in providing substantial benefits to society in a wide range of sectors such as agriculture, medicine, health, leather and environmental protection. The advancements in biotechnology have resulted in the emergence of a new area viz. microbial technology.

In the leather industry, microbial technology has a tremendous role to play in (i) the production of hydrolytic enzymes such as protease, lipase and amylase from bacteria and fungi, (ii) microbial control in leather processing and (iii) the application of microbes in the treatment of tannery effluents.

The information available on the use of microbes and microbial enzymes in leather processing is scanty and scattered. This book is a sincere attempt to create an awareness about the applications of microbial technology and microbial enzyme technology in leather and other industries. The first part of the book deals with the concepts of microbial technology while the second part delineates the scale up by submerged as well as solid state fermentation and downstream processing. The third part talks about the applications of enzymes in medicine and industry. In addition, the application of microbes and microbial enzymes in leather industry is illustrated in great detail.

I take this opportunity to congratulate my senior colleague, Dr. R. Puvanakrishnan for his vision in bringing out this book and my appreciations go to the young researchers Dr. S. Sivasubramanian and Dr. T. Hemalatha for providing valuable support in this stupendous task. My sincere thanks are due to M/s. MJP Publishers for this nice and important publication, which will be of immense value to the students of biochemistry, biotechnology, microbiology, leather technology, tanners and technicians in leather industry and enzyme manufacturers.

I offer my best wishes to the authors and the publishers.

Prof. Dr. A.B.Mandal 09/04/2012

PREFACE

The term 'Biotechnology' has gained currency only in the last century even though the fundamental ideas were established very much earlier. Humans have employed microorganisms very early for making wine, vinegar, curd, etc. Some of these processes have become an inseparable part of the domestic food preparations of every home and as such we may not even refer to them as biotechnology. These processes are based on the natural capabilities of the respective microorganisms and they are known as 'traditional biotechnology'.

Humans have continued their quest for (i) enhancing the natural capabilities of microorganisms for innovation and development of a wide range of novel process technologies and (ii) expanding the array of microorganisms with far-reaching potentials. The recent development of recombinant DNA technology which modifies microorganisms to produce highly potent products has been one of the landmarks of biotechnology. For example, the human insulin gene is transformed and expressed in *E. coli*. The insulin produced by these genetically engineered microbes is used in the management of diabetes. These and other similar examples constitute the 'modern biotechnology'.

The rapid strides in the growth of biotechnology have resulted in the emergence of a new branch of science, viz., microbial technology. Bacteria are utilized for detoxification of industrial effluents, in combating oil spills, for treatment of sewage and for biogas production. Microbes are employed for the extraction of metals from low-grade ores, termed as "microbial mining". The above-mentioned examples are only a few of a multitude of utilization of microbes in industrial applications.

In addition, there is another branch of specialization, viz., 'microbial enzyme technology' which is an extension of "microbial technology". Microorganisms, specifically bacteria and fungi, produce enzymes such as proteases, lipases and amylases for use in leather, textile, detergent, dairy and other industries. Enzymes have found uses in the various pretanning processes of leather manufacture such as soaking, dehairing, bating and degreasing. There is a vast scope for wider adaptation of microbial enzyme technology in leather and other industries.

With India's policy of laying special emphasis on the value-added export of finished leathers and leather products, demands for high-quality auxiliaries and potent enzyme products have increased. This calls for the standardization of latest fermentation techniques in the manufacture of enzymatic formulations.

In 1988, R.Puvanakrishnan and Susil C.Dhar published a monograph entitled "Enzyme technology in beamhouse practice" which dealt mostly with the applications of enzymes in pretanning processes of leather manufacture. This book was welcomed by students, teachers and technologists and received positive criticisms from the editors of various journals. Two decades have already passed and there is an absolute need for revising the book with emphasis on latest developments along with suggestive ideas for future research. The target audience for this new book will be the undergraduate students of biotechnology, microbiology and biochemistry, students of leather technology, technicians in leather industry and enzyme manufacturers. The authors earnestly hope that this new and updated book would be welcomed by tanners, technologists and students.

We place on record our sincere appreciation to all those who have helped us in this stupendous task of book writing. Our heartfelt thanks go to Dr A.B.Mandal, Director, CLRI, Chennai, for his kind support. Dr.T.Uma Maheswari, Research Associate, Department of Biotechnology, CLRI, merits special thanks from the authors for her unstinted support in shaping the chapters on solid-state fermentation and waste management. It would not have been possible to bring the book to the present shape without the support of our friends from the academia and also our faculty who critically reviewed different chapters as mentioned below.

Dr.P. Sankaranarayanan   Chapter 11
Former Professor, Department of Biochemistry
Madras Medical College, Chennai

Dr.M.Balakrishnan   Chapters 1, 13, 14, 16, 19 and 21
Former Head, Planning, Monitoring and Evaluation
CLRI, Chennai

Dr.B.Lokanadam   Chapters 14, 15, 16,17,18 and 19
Former Deputy Director and Head
Department of Biophysics
CLRI, Chennai

Mr.K.Parthasarathy   Chapters 14, 16 and 19
Former Head, Tannery Division
CLRI, Chennai

Dr.G.Jayamathi   Chapters 2, 3, 4, 6 and 11
Professor and Head, Department of Biochemistry
Meenakshi Academy for Higher Education and Research
Chennai

Dr.Meignana Lakshmi   Chapter 8 and 10
Professor and Head
Department of Biotechnology, School of Bioengineering
S.R.M. University, Kattangulathur

Dr.Rajesh   Chapter 8
Professor and Head
Department of Bioprocessing, School of Bioengineerin
S.R.M. University, Kattangulathur

Professor R. Selvaraj Department of Microbiology, Saveetha Dental College Saveetha University, Chennai	Chapter 2, 6 and 7
Professor B. Ramesh Head, Department of Biotechnology Sankara Arts and Science College, Kanchipuram	Chapter 5
Ms. J. Margaret Marie Assistant Professor, Department of Chemistry Women's Christian College, Chennai	Chapter 6 and 10
Ms.N.S. Kaviyarasi Department of Biochemistry Bangalore City College, Bangalore	Chapters 3 and 4
Dr. J.Geraldine Sandana Mala Pool Officer, Department of Biotechnology, CLRI, Chennai	Chapters 9 and 10
Dr.K.Anbukkarasi Senior Research Fellow National Dairy Research Institute, Haryana	Chapter 7

The authors thank Ms. S. Saraswathi for her excellent documentation support. The first author (R.Puvanakrishnan) gratefully acknowledges the Emeritus Scientist project awarded by CSIR, New Delhi.

When we talked to M/s. MJP Publishers and mooted the idea of publishing this book, they accepted our suggestion immediately and our heartfelt thanks go to them. We take this opportunity to thank Mr. J. C. Pillai, Publisher, Mr. C. Sajeesh Kumar, Managing Editor, Ms. Parvath Radha, Project Editor, Ms. B. Ramalakshmi and Ms. V. R. Padma, Assistant Editors, and the design editors, Ms. N. Yamuna Devi, Ms. S. Jeevasruthi, Ms. S. Annalaksmi and Ms. C. Devi. We are pleased to work with this enterprising team.

The authors thank wholeheartedly their family members for their kind cooperation and support. Any suggestions or positive criticisms are welcome and may be sent to puvanakrishnan@yahoo.com.

R. Puvanakrishnan
S. Sivasubramanian
T. Hemalatha

CONTENTS

15. ENZYMES IN SOAKING

ABBREVIATIONS

2-AP	2-Aminopurine
2-D PAGE	Two-dimensional polyacrylamide gel electrophoresis
5-BU	5-Bromouracil
6-APA	6-aminopenicillanic acid
ACP	Acid phosphatase
ADDS	Advanced drug delivery systems
AIDS	Acquired immuno deficiency syndrome
AIWPS	Advanced integrated wastewater pond system
ALP	Alkaline phosphatase
ALT	Alanine aminotransferase
AMG	Amyloglucosidase
AMP	Adenosine monophosphate
APTS	3-aminopropyltriethoxysilane
APS	Ammonium persulphate
ASD	Axillary seborrheic dermatitis
ASP	Activated sludge process
AST	Aspartate aminotransferase
ATP	Adenosine triphosphate
ATPS	Aqueous two-phase system
AV	Acne vulgaris
BAC	Bacterial artificial chromosomes
BAEE	N-Benzoyl-L-arginine ethyl ester
BCA	Bicinchoninic acid
BCID	Bromo-4-chloro-indolyl phosphate
BIT	1,2-benzisothiazolin-3-one

BOD	Biological oxygen demand
BSA	Bovine serum albumin
BSE	Bovine spongiform encephalopathy
BTEE	N-Benzoyl-L-tyrosine ethyl ester
CAACO	Chemo autotrophic activated carbon oxidation
CBB	Coomassie brilliant blue
cDNA	Complementary DNA
CDPR	Carbon dioxide production rate
CETP	Common effluent treatment plant
CF	Coagulation-flocculation
CGT	Cylcodextrin glucosyl transferases
CJD	Cruzefelt jacob disease
CK	Creatine kinase
CLRI	Central leather research institute
CMC	Carboxymethyl cellulose
CMP	Cytidine monophosphate
COD	Chemical oxygen demand
COX	Cyclooxygenase
CPCB	Central pollution control board
CPG	Controlled porous glass
CTP	Cytidine triphosphate
DEAE-cellulose	Diethylaminoethyl-cellulose
DFP	Diisopropyl fluorophosphate
DLMNP	Drug loaded magnetic nanoparticles
DNA	Deoxyribonucleic acid
DO	Dissolved oxygen
dsDNA/RNA	Double stranded DNA/RNA
EC	Enzyme commission
ECL	Enhanced chemi luminescence
ECM	Extracellular matrix

EDC	1-ethyl-3-(dimethyl-aminopropyl) carbodiimide. hydrochloride
EDTA	Ethylene diamine tetra acetic acid
EES	Ethylethane sulphonate
EF	Elongation factor
EMBR	Enzyme immobilized membrane bioreactors
EMS	Ethylmethane sulphonate
ES complex	Enzyme-substrate complex
ESR	Electron spin resonance
FAD	Flavin adenine dinucleotide
FALGPA	N-(3-(2-Furoyl) acryloyl)-Leu-Gly-Pro-Ala
G6PD	Glucose-6-phosphate dehydrogenase
GA	Genetic algorithms
GAG	Glycosaminoglycans
GC	Gas chromatography
GCF	Gingival cervicular fluid
GGT	Gamma glutamyl transferase
GMO	Genetically modified organism
GMP	Guanosine monophosphate
GRAS	Generally regarded as safe
GTF	Glucosyl transferase
GTP	Guanosine triphosphate
HA	Hyaluronic acid
HIC	Hydrophobic interaction chromatography
HIV	Human immuno deficiency virus
HPA	Hide powder azure
HPLC	High performance liquid chromatography
IC	Immobilized cellulose
IEF	Isoelectric focusing
IPTG	Isopropyl β-D-1-thiogalactopyranoside

LC	Liquid chromatography
LDH	Lactate dehydrogenase
LPS	Lipopolysaccharide
MAC	Mammalian artificial chromosomes
MB	Muscle brain
MBR	Membrane bio reactor
MBT	Methylene bis(thiocyanate)
MCCC	Multiple contact counter current
MI	Myocardial infarction
MINAS	Minimum acceptable standards
MMS	Methylmethane sulphonate
MT	Metallothioneins
NAD	Nicotinamide adenine dinucleotide
NADP	Nicotinamide adenine dinucleotide phosphate
NMR	Nuclear magnetic resonance
NRV	Non return valves
NTG	N-Methyl-N′-nitro-N-nitrosoguanidine
N-VP	N-vinyl-2-pyrrollidone
OUR	Oxygen uptake rate
PAFC	Poly aluminium ferric chloride
PAS	Periodic acid schiff
PBR	Packed bed reactor
PCMC	p-chloro meta cresol
PCP	Pentachlorophenol
PCR	Polymerase chain reaction
PDAB	p-dimethylamino benzaldehyde
PEG	Polyethylene glycol
PG	Proteoglycans
PRPP	Phosphoribosyl-α-pyrophosphate

PRR	Proton relaxation rate
PSA	Prostate specific antigen
PSO	Particle swam optimization
PVA	Polyvinyl alcohol
RNA	Ribonucleic acid
RPC	Reverse phase chromatography
RQ	Respiratory quotient
SAMe	S-adenosyl methionine
SBR	Sequencing batch reactor
SCFE	Supercritical fluid extraction
SCP	Single cell protein
SDH	Succinate dehydrogenase
SDS	Sodium dodecyl sulphate
SDS-PAGE	Sodium dodecyl sulphate-polyacrylamide gel electrophoresis
SEC	Size exclusion chromatography
SEM	Scanning electron microscopy
SGPT	Serum glutamic pyruvic transaminase
SK	Streptokinase
SMBS	Sodium metabisulphite
SmF	Submerged fermentation
ss DNA/RNA	Single stranded DNA/RNA
SSF	Solid state fermentation
STR	Stirred tank reactor
TAG	Triacylglycerol
TCA	Trichloroacetic acid
TCMBT	2-(cyanomethylthio)benzothiazole
TDS	Total dissolved solids
TEM	Transmission electron microscopy

TEMED	N,N,N',N'-tetramethylethylenediamine
TLC	Thin layer chromatography
TNO	The Netherlands co-operation
TOC	Total organic carbon
tPA	Tissue plasminogen activator
TPCK	L-1(p-toluenesulfonyl) amido-2-phenylethyl chloromethyl ketone
TRAP	Tartrate-resistant acid phosphatase
TSS	Total suspended solids
TTP	Thymidine triphosphate
UASB	Upflow anaerobic sludge blanket
UMP	Uridine monophosphate
UNIDO	United nations industrial development organization
USFDA	United states food and drug administration
UV	Ultraviolet
VOC	Volatile organic compounds
YAC	Yeast artificial chromosome
YPG agar	Yeast peptone glucose agar

PART I

MICROBIAL TECHNOLOGY-CONCEPTS

- Introduction
- Fundamentals of Microbiology
- Proteins—An Overview
- Enzymes—General Perspective
- Immobilized Enzymes and Microbial Whole Cell Technology
- Nucleic Acids—Structure and Function
- Genetic Engineering

1

INTRODUCTION

The early twentieth century was undoubtedly the age of chemistry and physics, but the twenty-first century has emerged as the age of biotechnology. Biotechnology is an interdisciplinary field and is, in essence, the use of biological knowledge for producing highly potent/profitable biological products. Biotechnology has in its fold many areas including biology, microbiology, biochemistry, molecular biology, genetics, chemistry, chemical engineering and bioinformatics. Originally, when biological mechanisms were not clearly understood, biotechnology was limited to its practice as a skill in fermentation.

The modern era of biotechnology offers advancements in various fields such as agriculture, medicine, industry, nutrition, environment, etc. One example of modern biotechnology is genetic engineering. Genetic engineering is the process of transferring individual genes from one organism to another or modifying the genes in an organism to remove or add a desired trait or characteristic. The new biotechnology revolution started in the 1980s when scientists tried to alter precisely the genetic constitution of living organisms and this had a profound impact on almost all areas of traditional biotechnology. Thus, genetic engineering could be the answer for the production of potent enzymes and biological products.

Many present-day biotechnological processes have their origins in ancient fermentation such as the brewing of beer and the manufacture of bread, cheese and wine. However, it was the discovery of antibiotics and their large-scale production that created the greatest advances in fermentation technology and biochemical engineering. Since then, a phenomenal development in this technology has been observed not only in the production of antibiotics but also in other useful biochemical products such as enzymes, hormones, vaccines, etc.

The rapid strides in the growth of biotechnology have resulted in the emergence of a new branch of science, viz., "microbial technology." Many important areas are covered under this branch, and include the following.

   i. Treatment of industrial and domestic effluents using a consortium of microorganisms

  ii. Mineral extraction through "bioleaching"

 iii. Production of antibiotics such as penicillin, streptomycin, erythromycin, etc. by fungi and bacteria

  iv. Production of single cell proteins (SCP) from bacteria, yeasts, fungi or algae for human consumption as well as animal feed

   v. Degradation of petroleum and management of oil spills

An offshoot of microbial technology is the emergence of "microbial enzyme technology" which has excellent industrial applications. Some examples are as follows:

   i. production of enzymes such as protease, lipase and amylase from bacteria and fungi for use in leather, detergent, and textile industries, etc.,

  ii. production of ethanol, lactic acid, glycerine, citric acid, acetone, etc. mainly from bacteria and

 iii. immobilization of enzymes and microbial whole cells for the production of various biologicals

Some of the objectives of this book are the following

   i. To discuss the fundamental aspects of microorganisms, proteins, enzymes, immobilized enzymes, nucleic acids and genetic engineering to provide an understanding of the principles and applications of microbial technology.

  ii. To traverse the path of microorganisms starting from its isolation, media standardization, scale-up by submerged as well as solid-state fermentation, downstream processing, application of microbes in effluent treatment, microbial control in the curing process before leather manufacture and use of microbial enzymes in soaking, dehairing, bating and degreasing processes of leather manufacture.

 iii. To bring out the uses of microbes and microbial enzymes in medicine and industry.

This book will have three major parts.

For a deep understanding of the characteristics of the microbes and microbial enzymes as well as their effective manipulation, a basic knowledge of microorganisms, proteins, enzymes, immobilized enzymes, nucleic acids and genetic engineering is essential and these aspects are covered in Part-I.

Chapter 2 deals with the general organization and specific features of microorganisms, viz., prokaryotes and eukaryotes, classification, characteristics and structure of archaea, bacteria, fungi, algae, protozoa and viruses. Enzyme is generally a protein and the properties and characteristics of protein should be properly understood for developing potent enzymatic products. Hence, the basics of amino acids, structure, classification, denaturation and renaturation and estimation of protein are dealt with in detail in Chapter 3. Chapter 4 details the classification, specificity, factors affecting enzyme activity, isoenzymes, Michaelis–Menten kinetics and inhibition of enzymes.

Chapter 5 dwells on immobilized enzyme technology as well as immobilized whole cell technology. The advantages and disadvantages of immobilized enzymes, characteristics of different supports, methods of immobilization, applications of immobilized enzymes, immobilization of microbial whole cells and their applications are brought out in great detail.

Microorganisms isolated from nature usually produce biologically active molecules in low quantities. Low productivity of metabolite results in high manufacturing cost per unit of the product. Manufacture of commercially important metabolites directly from the microorganisms isolated from nature is often not viable. Hence, improvement of microbial strains is very essential to get a higher yield and product output. To achieve this, basic knowledge on nucleic acids and recombinant DNA technology is necessary. Chapter 6 speaks at length on the components of nucleic acids, structure of RNA and DNA, replication, transcription, translation and classification and induction of mutation. Chapter 7 describes the important developments in genetic engineering, tools and steps in rDNA technology and applications of rDNA technology.

Part-II illustrates the methods for the isolation of microorganisms, media standardization, scale-up by submerged culture as well as solid-state fermentation and downstream processing. Part-II gives a clear idea as to how a product is developed from a microorganism at different stages. Chapters 8, 9 and 10 portray the above aspects.

Part-III talks about the applications of microbes and microbial enzymes in medicine and industry. The application of enzymes in medicine and industry (other than leather) is discussed in Chapters 11 and 12. Once the general applications of enzymes are delineated, the next focus will be on the application of microbial technology in leather industry.

The most important application of microbial technology is in leather industry. Discussion in this area will help the reader to understand the importance of microbial control, use of microbial enzymes in the pretanning processes of leather manufacture and the application of microbes in the treatment of effluents from leather industry (Chapter 13). Accordingly, Chapter 14 discusses the microbial control methodologies that are used in the curing process. Microbial proteases have an enviro-friendly application in the dehairing process of leather manufacture. In addition, they are employed in the soaking and bating processes. Chapters 15, 16 and 17 discuss the uses of microbial proteases in soaking, dehairing and bating respectively and show that enzymatic processes are more enviro-friendly than the conventional chemical processes. Chapter 18 elaborates the methods for the degreasing of skins, and stresses the importance of using lipolytic enzymes in place of conventional solvents and surfactants.

A consortium of microorganisms has an important application in the treatment of effluent from leather industry to remove different pollutants. Chapter 19 discusses the various procedures followed in effluent treatment and specific strategies for using microbes.

Whenever an enzyme preparation is used in the industry, the success of an enzyme application could be measured only by a proper assay procedure. Chapter 20 discusses the various assay procedures for evaluating enzymes such as proteases, lipases, amylases and collagenases and for determination of protein content.

Chapter 21 tries to give a glimpse of what is in store for the future of microbial technology in the next decade. A glossary is added wherein all the important terms used in the areas of microbial technology and microbial enzyme technology are defined. This will help the student to understand the terminology before reading the text. A list of abbreviations used in this book is also provided along with an index. Books for further reading are suggested at the end of each chapter in Part-I and Part-II. Detailed lists of references are provided for all the chapters in Part-III and this will help the students to pursue research in the applied areas of microbial technology.

2
FUNDAMENTALS OF MICROBIOLOGY

INTRODUCTION

Microorganisms are indispensable components of our ecosystem. They make possible the cycles of carbon, oxygen, nitrogen, and sulphur that take place in terrestrial and aquatic ecosystems. Microorganisms are very tiny organisms—but the science of microbiology is quite vast.

Some algae, fungi and two bacteria (*Thiomargarita* and *Eupulopiscium*) are larger and visible to the naked eye. Hence, it is suggested that microbiology should be defined not in terms of the size. At present, there is a general agreement to include five major groups as microorganisms, viz., virology, bacteriology, mycology, phycology and protozoology (studies on viruses, bacteria, fungi, algae and protozoa respectively). These studies are concerned with their form, structure, reproduction, physiology, metabolism and classification.

Most of the microorganisms are unicellular in which all the life processes are performed by a single cell. In unicellular microbes, the following characteristics are common:

  i. the ability to reproduce
 ii. the ability to ingest or assimilate food substances and metabolize them for energy and growth
iii. the ability to excrete waste products
 iv. the ability to react to changes in their environment, sometimes called irritability
  v. susceptibility to mutation.

In microbiology, we encounter a unique group of microorganisms which may represent the borderline of life. These are the infectious agents called viruses (also called living chemicals), which are simpler in structure and composition than single cells. Viruses provide an exciting challenge and opportunity to gain a better understanding of the nature of complex organic substances that may bridge the gap between the living and nonliving world.

PROKARYOTES AND EUKARYOTES

All living organisms can be sorted into two major groups—prokaryotes and eukaryotes—based on the fundamental structure of their cells. The distinguishing features of these two groups are summarized in Table 2.1.

Prokaryotes

Prokaryotes are small, relatively simple organisms that lack a true membrane-enclosed nucleus. The DNA is not contained within a membrane but is coiled up in a region of the cytoplasm called the nucleoid. The cytoplasm possesses simple internal structure, lacking membrane-bound organelles. Additionally, the DNA is less structured compared to eukaryotes and is in the form of a single loop. Most prokaryotes are made up of just a single cell (unicellular) but there are a few that are made of collections of cells (multicellular). Prokaryotes include two groups, the Bacteria and the Archaea.

Eukaryotes

Eukaryotes are organisms made up of cells that possess a membrane-enclosed nucleus (that holds genetic material) as well as specialized membrane-enclosed organelles. The linear form of DNA is condensed with histoprotein and organized into chromosomes. Eukaryotes may be either multicellular or single-celled organisms and they include algae, fungi, protozoa, plants and animals.

Table 2.1 Difference between prokaryotic and eukaryotic cells

Prokaryotes	Eukaryotes
Include bacteria, cyanobacteria and archaea.	Include algae, fungi, protozoa, plants and animals.
Usually the size ranges between 0.05 μm to 1.0 μm in diameter.	Greater than 5 μm in width or diameter.
Always unicellular.	Often multicellular.
Cell division is by binary fission.	Cell division is by mitosis or meiosis.
Reproduction is always asexual.	Reproduction is sexual or asexual.
Genetic material floats freely in the cytoplasm, and is called nucleoid.	Genetic material resides within the nucleus, mitochondria and chloroplasts.

(Contd.)

Table 2.1 (Continued)

Prokaryotes	Eukaryotes
DNA is circular without bound proteins and is not bound by nuclear membrane; single chromosome.	DNA is linear, bound by nuclear membrane and associated with histone proteins to form chromatin; more than one chromosome.
Nucleolus absent; functionally related genes may be clustered.	Nucleolus present; functionally related genes not clustered.
Zygote nature is merozygotic (partial diploid).	Zygote is diploid.
Mesosome is present.	Mesosome is absent.
Smaller ribosomes (70S) distributed in cytoplasm.	Larger ribosomes (80S) arrayed on membranes as in endoplasmic reticulum; 70S in mitochondria and chloroplasts.
Cytoplasmic streaming is absent.	Cytoplasmic streaming is present.
Absence of membrane-bound organelles.	Presence of membrane-bound organelles such as mitochondria, chloroplasts, Golgi bodies, endoplasmic reticulum.
Cytoplasmic membranes generally do not contain sterols; contain respiratory and (in some) photosynthetic machinery.	Sterols are present in cytoplasmic membranes; do not carry out respiration and photosynthesis.
Cell wall is made up of peptidoglycan (murein or mucopeptide).	Absence of peptidoglycan.
Motility is by rigid rotating flagella, simple fibril, made of flagellin.	Motility is by waving cilia or flagella, multifibrilled with "9+2" microtubules, made of tubulin.
Use a wide variety of metabolic pathways, particularly that of anaerobic energy-yielding reactions; some fix nitrogen gas; some accumulate poly-β-hydroxybutyrate as reserve material.	Use some common metabolic pathways such as glycolysis, anaerobic respiration, Krebs cycle and oxidative phosphorylation.
G+C% (Guanine + cytosine) ratio is 28 to 73.	G+C% ratio is about 40.

MAJOR GROUPS OF MICROORGANISMS

The major groups of microorganisms include bacteria, fungi, algae, protozoa and viruses. Their important characteristics are summarized in Table 2.2.

Table 2.2 Characteristics of major groups of microorganisms

Groups	Size range	Important characteristics
Viruses	0.015–0.2 μm	Acellular, extremely small infectious agents which are visible only under electron microscope, cannot be grown in synthetic media, all are obligate parasites, require living cells within which they are reproduced.
Bacteria	0.2 by 100 μm	Prokaryotic, unicellular, most abundant and ubiquitous, simple internal structure, grow on artificial laboratory media; reproduction asexual, characteristically by simple cell division.
Algae	1.0 μm to many feet	Eukaryotic, typically autotrophic, unicellular and multicellular; mostly occur in aquatic environments, contain chlorophyll, reproduction by asexual and sexual processes.
Protozoa	2.0–10.0 μm	Eukaryotic, unicellular, mostly abundant in moist habitat, some cultivated in laboratory much like bacteria; some are intracellular parasites; reproduction by asexual and sexual processes.
Fungi — yeasts	5.0–10.0 μm	Eukaryotic, unicellular, mostly present in sugar-rich environments, laboratory cultivation much like that of bacteria; reproduction by asexual cell division, budding or sexual processes.
Fungi— moulds	2.0–10.0 μm	Eukaryotic, multicellular with many distinctive structural features; cultivated in laboratory much like bacteria; some produce toxins, reproduction by asexual and sexual processes.

1 μm = 10^{-6} metre

HISTORY OF MICROBIOLOGY

Lucretius, a Roman philosopher (about 98–55 BC), and Girolamo Fracastoro (1478–1553), a physician, suggested that disease was caused by invisible living organisms even before the microbes were physically seen. But, the first person to observe and describe microorganisms accurately was the amateur microscopist Antoni van Leeuwenhoek in 1664. This was the beginning of microbiology. A chronological arrangement of important events in the history of microbiology is given in Table 2.3.

Table 2.3 Important events in microbiology

Year	Investigator	Contribution
1664	Antoni van Leeuwenhoek	First to observe, record and report microbes, describes it as animalcules
1765–1776	Lazzaro Spallanzani	First to disprove spontaneous generation
1786	Müller	First classification of bacteria
1798	Edward Jenner	Discovers vaccination for smallpox
1847–1850	Ignaz Philip Semmelweiss	Introduces use of antiseptics
1822–1895	Louis Pasteur	Discovers lactic acid and alcoholic fermentation, proposes germ theory of disease, develops immunization techniques (anthrax and rabies vaccine). Successfully disproves spontaneous generation theory
1867	Joseph Lister	Father of antiseptic surgery, develops aseptic techniques; isolates bacteria in pure culture
1820–1893	John Tyndall	Develops tyndallization
1843–1910	Robert Koch	Develops pure culture technique, Koch's postulates, discovers causative agents of anthrax and tuberculosis, first to use gelatin as solidifying agent in culture media
1854–1915	Paul Ehrlich	Develops modern concept of chemotherapy and chemotherapeutic agents
1880	Laveran	Discovers *Plasmodium*, cause of malaria
1882	Elie Metchnikoff	Discovers phagocytosis
1884	Hans Christian Gram	Develops Gram stain to differentiate bacteria

(Contd.)

Table 2.3 (Continued)

Year	Investigator	Contribution
1866–1925	August Paul von Wassermann	Introduces complement fixation tes for syphilis
1887	Richard Petri	Develops Petri dish
1887–1890	Sergei Winogradsky	Studies sulphur and nitrifin; bacteria (chemolithotrophs)
1889–1901	Martinus Beijerinck	Isolates root nodule bacteria, concep of virus, enrichment cultur technique
1902	Karl Landsteiner	Discovers human blood group
1915–1917	Fredrick W. Twort and Felix H. d'Herelle	Independently discover bacteriophages
1928	Frederick Griffith	Discovers *Pneumococcus* transformation
1921–1929	Alexander Fleming	Discovers lysozyme and penicillin
1933	Ruska	Develops first transmission electro microscope
1941	George Beadle and Edward Tatum	"One gene–one enzyme" hypothesi
1943	Max Delbruck and Salvador Luria	Discover inheritance of genetic characters in bacteria
1944	Oswald Avery, Colin Macleod, Maclyn McCarty	Confirm DNA is genetic material
1944	Selman Waksman and Albert Schatz	Discover streptomycin
1946	Edward Tatum and Joshua Lederberg	Discover bacterial conjugation
1951	Barbara McClintock	Discover transposable elements
1952	Joshua Lederberg and Norton Zinder	Bacterial transduction
1953	James Watson, Francis Crick, Maurice Wilkins and Rosalind Franklin	Discover structure of DNA to be double helix
1960	Francois Jacob, David Perrin, Carmon Sanchez and Jacques Monad	Propose the concept of operon

(Contd.)

Table 2.3 (Continued)

Year	Investigator	Contribution
1967	Thomas Brock	Discovers bacteria growing in boiling hot springs
1969	Howard Temin, David Baltimore, Renato Dulbecco	Discover retroviruses
1969	Thomas Brock and Hudson Freeze	Isolate *Thermus aquaticus*, source of *Taq* DNA polymerase
1977	Fred Sanger, Steven Niklen and Alan Coulson	Develop methods of DNA sequencing
1980	Gerd Binnig and Heinrich Rohrer	Invent scanning tunnelling microscope
1982	Karl Stetter	Isolates first prokaryote with temperature optimum >100°C
1983	Luc Montagnier	Discovers HIV
1995	Criag Venter and Hamilton Smith	Complete sequence of a bacterial genome
1997	Heidi Schulz	Discovers *Thiomargarita namibiensis,* the largest known bacterium
1999	The Institute for Genomic Research (TIGR) and others	Over 100 microbial genomes sequenced and further work is in progress

MICROSCOPIC EXAMINATION OF MICROORGANISMS

The microscope is a powerful characteristic tool of the microbiology laboratory as it enables the visualization of microbes. The magnification it provides reveals the secrets of microbial cell structure which are invisible to the naked eye. The magnifications attainable by microscopes range from ×100 to ×400,000. In addition, different kinds of microscopes are available and many techniques have been developed by which specimens of microorganisms can be prepared for examination. Each type of microscopy and each method of preparing specimens for examination may be suitable for demonstration of specific morphological features.

Microscopes are of two categories, light (or optical) and electron, depending upon the principle on which magnification is based. Light microscopy in which magnification is obtained by a system of optical lenses using light waves, includes (i) bright-field (ii) dark-field (iii) fluorescence and (iv) phase-contrast. Modern microscopes are all compound microscopes. The electron microscope, as the name suggests, uses a beam of electrons in the place of light waves to produce the image, and the specimens can be examined by either transmission (TEM) or scanning electron microscopy (SEM). In general, light microscopes are used to observe intact cells under low magnification whereas electron microscopes are used to study the internal structures and the details of cell surfaces. A comparison of different types of microscopy is given in Table 2.4.

Table 2.4   Comparison of different types of microscopy

Type of microscopy	Maximum useful magnification	Appearance of specimen	Useful applications
Bright-field	1000–2000	Specimens stained or unstained; bacteria generally stained and forms dark or coloured image under bright background	For gross morphological features of bacteria, yeast, mould, algae and protozoa
Dark-field	1000–2000	Generally living, unstained cells appear bright or lighted in dark background	For microorganisms that exhibit some characteristic morphological feature in the living state and in fluid suspension
Fluorescence	1000–2000	Bright and coloured; stained with only fluorochrome dye	Essential tool in medical microbiology and microbial ecology where fluorescent dye fixed to organism reveals the organism's identity
Phase-contrast	1000–2000	Varying degree of darkness, unpigmented living cells that are not clearly visible in the bright field microscope are observed	Especially useful for studying microbial motility, determining the shape of living cells, and detecting bacterial components such as endospores and inclusion bodies, widely used in studying eukaryotic cells

(Contd.)

Table 2.4 (Continued)

Type of microscopy	Maximum useful magnification	Appearance of specimen	Useful applications
Differential Interference Contrast (DIC)	1000–2000	Live unstained cells; similar to the phase-contrast microscope	Structures such as cell walls, endospores, granules, vacuoles, and eukaryotic nuclei are clearly visible
Electron (Transmission and Scanning)	2,00,000–4,00,000	Viewed on fluorescent screen, cells can be observed after special specimen preparation and staining	Examination of viruses, the ultrastructure of microbial cells and details of cell surfaces
Confocal Scanning, Laser (CSLM)	1000–2000	Fluorescently stained specimens	Specimens with a depth of 1 μm or less can be observed, high-resolution, three-dimensional images of cell structures and complex specimens such as biofilms can be observed
Scanning probe (scanning-tunneling and atomic force)	>1,00,000	Surface map can be displayed on a computer screen or plotted on paper. The resolution is so great that individual atoms are observed easily	Used to directly view DNA, useful in studying biological molecules, to study the behaviour of living bacteria and other cells

ARCHAEA

It is a group of unicellular microorganisms very similar to bacteria. In the past, it was classified under bacteria and named as archaebacteria. But now, archaea is reclassified into a separate domain as the organisms of this group have many characteristic differences from other forms of organisms.

Classification

The archaea is divided into the phyla Euryarchaeota and Crenarchaeota. The euryarchaeotes include seven classes (Methanobacteria, Methanococci, Halobacteria, Thermoplasmata, Thermococci, Archaeglobi and Methanopyri). The methanogens, extreme halophiles, sulphate reducers, and many extreme thermophiles with sulphur-dependent metabolism are located in the Euryarchaeota. The Crenarchaeotes are thought to resemble the ancestors of the archaea, and almost all the well-characterized species are thermophiles or hyperthermophiles. Only one class, Thermoprotei, comes under the phylum Crenarchaeota. Two new phyla, Nanoarchaeota and Korarchaeota, have also been proposed based on 16S rRNA gene sequencing.

Characteristics of Archaea

The Archaea are quite diverse, both in morphology and physiology. They are either gram-positive or gram-negative and may be spherical, rod-shaped, spiral, lobed, plate-shaped, irregularly shaped, or pleomorphic. Some are single cells, whereas others form filaments or aggregates. They range in diameter from 0.1 to over 15 μm, and some filaments can grow up to 200 μm in length. Multiplication may be by binary fission, budding, fragmentation, or other mechanisms. Nutritionally, they range from chemolithoautotrophs to chemoorganotrophs. Most are hyperthermophiles that can grow above 100°C, and a few are mesophiles. Recently, archaea have been discovered in cold environments. A few are symbionts in animal digestive systems.

Structure of Archaea

Cell wall   The cell wall structure and chemistry of archaea differ from that of the bacteria. Gram-negative archaea lack the outer membrane and complex peptidoglycan network of gram-negative bacteria. They usually have a surface layer of protein or glycoprotein subunits. None have the muramic acid and D-amino acids characteristic of bacterial peptidoglycan. Gram-positive archaea can have a variety of complex polymers in their walls. Cell wall of some methanogens consists of pseudomurein, a peptidoglycan-like polymer that has L-amino acids in its cross links, N-acetyltalosaminuronic acid instead of N-acetylmuramic acid, and β $(1 \rightarrow 3)$ glycosidic bonds instead of β $(1 \rightarrow 4)$ bonds.

Lipids and membranes   The nature of membrane lipids is the most distinguishing feature of archaea. They differ from both bacteria and eukaryotes in having branched-chain hydrocarbons attached to glycerol by ether links rather than fatty acids connected by ester links (Figure 2.1).

Sometimes, two glycerol groups are linked to form an extremely long tetraether. Cells can adjust the overall length of the tetraethers by cyclizing the chains to form pentacyclic rings, and biphytanyl chains may contain from one to four cyclopentyl rings. The membrane lipids are nonpolar lipids, which usually are derivatives of squalene that yield membranes of different rigidity and thickness. Polar lipids (phospholipids, sulpholipids, and glycolipids) are also present in archaeal membranes. Archaeal membranes may contain a mix of diethers, tetraethers, and other lipids based on their need for stability, e.g., the membranes of extreme thermophiles are almost completely tetraether monolayers.

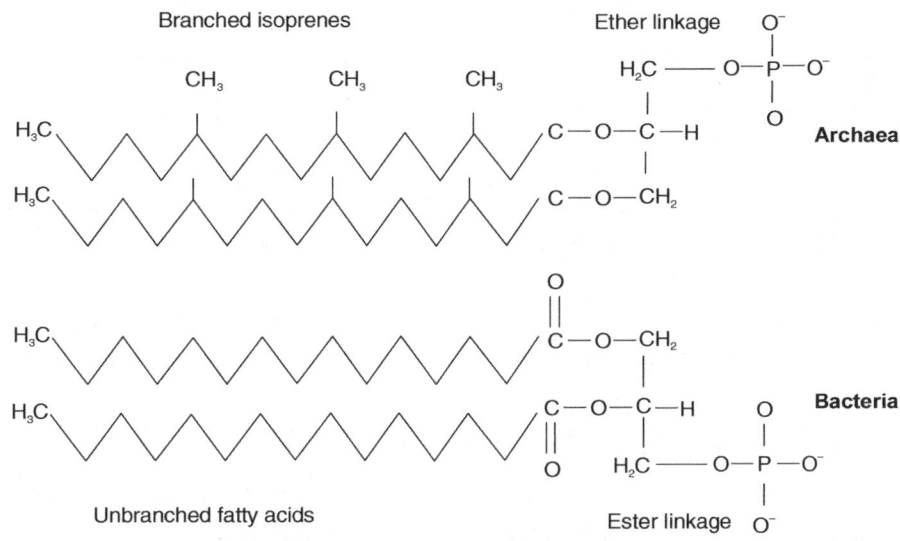

Figure 2.1 Structural difference between archaeal and bacterial membrane lipids

Archaeal Genetics

The archaeal chromosomes are single, closed DNA circles similar to that of bacteria but, the genomes of some archaeons are significantly smaller than the bacteria. The variation in G+C content is great i.e., about 21 to 68%. Archaea have few plasmids. Archaeal mRNA and promoters are similar to that of bacteria rather than eukaryotes. Archaea have polygenic mRNA but there is no evidence for mRNA splicing. The archaeal initiator tRNA carries methionine similar to eukaryotic initiator tRNA, but the TΨC arm of archaeal tRNA lacks thymine and contains pseudouridine or 1-methylpseudouridine. Despite some similarities, there are also many differences in archaea from other organisms. The 70S type archaeal ribosomes have quite variable shapes and sometimes differ from that of both bacterial and eukaryotic ribosomes. But, they resemble eukaryotic ribosomes in their sensitivity to anisomycin, insensitivity to chloramphenicol and kanamycin and their elongation factor-2 (EF-2) reacts with diphtheria toxin as that of the eukaryotic EF-2. Unlike prokaryotes, some archaea are found to have histone proteins that bind with DNA to form nucleosome-like structures. Finally, archaeal DNA-dependent RNA polymerases resemble the eukaryotic enzymes and not the bacterial RNA polymerase.

BACTERIA

Size, Shape and Arrangement

The shape of a bacterium is governed by its rigid cell wall. Bacteria occur in three main shapes *viz.,* spherical, rodlike and spiral.

i. Spherical bacteria are called cocci (singular: coccus). Based on the plane of division, the cells may occur in pairs (diplococci), in groups of four (tetracocci), in irregular grapelike clumps (staphylococci), long chains of cocci (streptococci) or in a cubical arrangement of eight or more (sarcinae).

ii. Rodlike bacteria are called bacilli (singular: bacillus). They generally occur singly, but may occasionally be found in pairs (diplobacilli) or chains (streptobacilli). Bacilli differ considerably in their length-to-width ratio, the coccobacilli being so short and wide that they resemble cocci. The shape of the terminal ends of the bacilli often varies between species and may be flat, rounded, cigar-shaped, or bifurcated. A few rod-shaped bacteria are curved to form distinctive commas called vibrios.

iii. Spiral-shaped bacilli are called spirilla if rigid (singular: spirillum) and spirochaetes if flexible. Short incomplete spirals are called vibrios or comma bacteria.

Although most bacterial species have cells that are of a fairly constant and characteristic shape, some assume variety of shapes e.g., oval- to pear-shaped. *Hyphomicrobium* produces a bud at the end of a long hypha, *Gallionella* produce nonliving stalks, Actinomycetes form characteristic long multinucleate hyphae or filaments, Walsby's square bacteria is shaped like flat, square to rectangular boxes and some bacteria are variable in shape and lack a single, characteristic form, and are called pleomorphic (Figure 2.2).

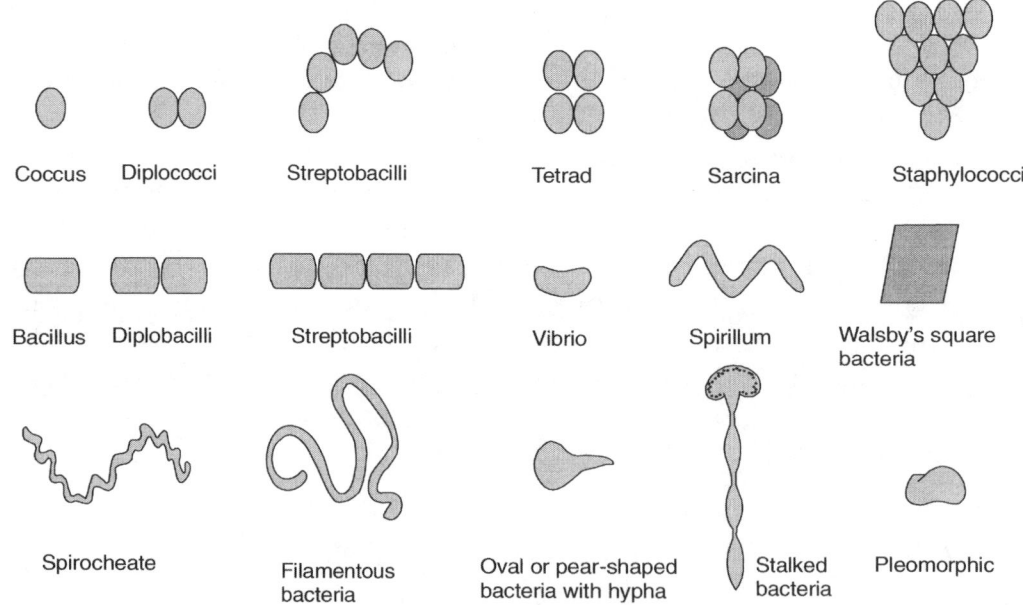

Figure 2.2 Different shapes of bacteria

Bacteria vary in size as much as in shape. Majority of the bacteria are approximately 0.5–1.0 μm in diameter. But, there are some exceptions; the smallest are under the genus

Mycoplasma of about 0.3 μm in diameter. Recently, even smaller Nanobacteria ranging between 0.2 μm to less than 0.05 μm in diameter have been reported. A huge bacterium, *Eupulopiscium fishelsoni* grows as large as 600 by 80 μm and more recently, an even larger bacterium, *Thiomargarita namibiensis* (100 to 750 μm in diameter), has been discovered in ocean sediments.

Structure

A variety of structures is found in prokaryotic cells. Prokaryotic cells are bounded by a chemically complexed cell wall. The periplasmic space separates the plasma membrane from the cell wall. The plasma membrane can be invaginated to form simple internal membranous structures. In addition, many cells are surrounded by external structures. A typical bacterial cell is depicted in Figure 2.3.

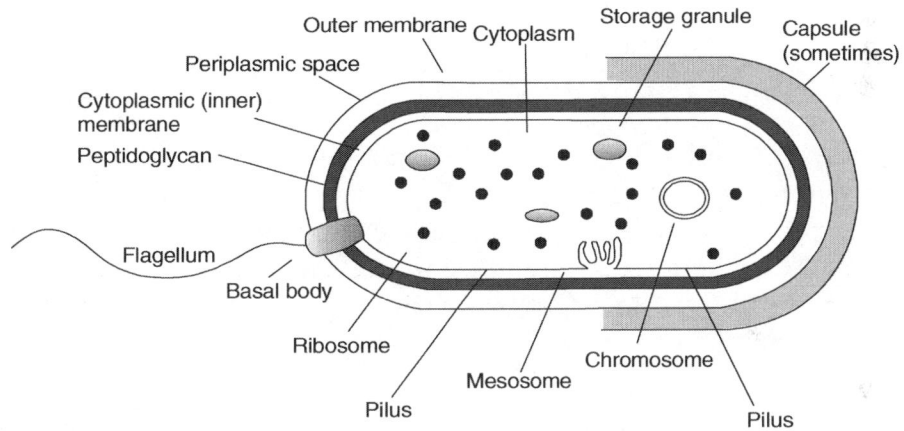

Figure 2.3   Structure of bacterial cell

Cell wall   The cell wall is a very rigid structure that gives shape to the cell. Its main function is to protect the cell wall from osmotic lysis since most bacteria live in hypotonic environments, protect the cell from toxic substances, have components that contribute to their pathogenicity and it is the site of action of several antibiotics. The shape of the bacterial cell wall is largely determined by peptidoglycan (called murein), an insoluble, porous, cross-linked polymer of enormous strength and rigidity. Peptidoglycan is found only in prokaryotes, and is found surrounding the cytoplasmic membrane. The polymer contains two sugar derivatives, *N*-acetylglucosamine and *N*-acetylmuramic acid and several different amino acids (L-alanine, D-alanine, D-glutamic acid, meso-diaminopimelic acid, L-lysine, L-ornithine, or L-diaminobutyric acid). The chains of peptidoglycan subunits are joined by cross links between the peptides. The peptide interbridge is responsible for retaining their shape and integrity.

Cell wall of Gram-positive bacteria   Gram-positive bacteria have a greater amount of peptidoglycan in their cell walls than gram-negative bacteria. The cell wall consists of a single 20–80 nm thick homogeneous peptidoglycan or murein layer lying outside the plasma membrane, i.e., 50% dry weight of the cell wall. Because of the thicker peptidoglycan layer, the walls of gram-positive cells are stronger than those of gram-negative bacteria (Figure 2.4).

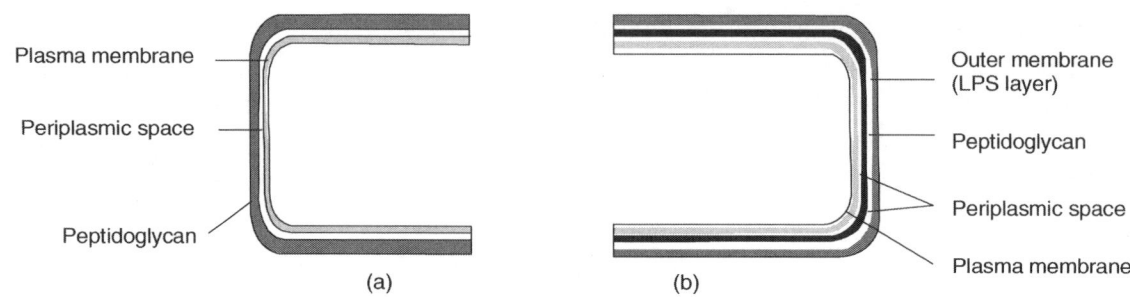

Figure 2.4   Schematic representation of cell wall of a) gram-positive and b) gram-negative bacteria

The gram-positive cell walls usually contain large amounts of negatively charged teichoic acids, polymers of glycerol or ribitol joined by phosphate groups. The teichoic acids are connected to either the peptidoglycan or to plasma membrane lipids (called lipoteichoic acids). Teichoic acids extending to the surface of the peptidoglycan give the gram-positive cell wall its negative charge. These molecules are important in maintaining the structure of the wall. Teichoic acids are not present in gram-negative bacteria.

Cell wall of Gram-negative bacteria   The cell wall of gram-negative bacteria is more complex than that of gram-positive bacteria. It has a 2 to 7 nm peptidoglycan layer (5 to 10% of wall dry weight) surrounded by a 7 to 8 nm thick outer membrane. The outer membrane of gram-negative bacteria is rich in lipids. The cell wall and outer membrane are linked by a membrane protein, Braun's lipoprotein. The most unusual component of the outer membrane is its lipopolysaccharides (LPS). These large, complex molecules consist of three parts: (1) lipid A, (2) the core polysaccharide, and (3) the O side chain. The LPS is important for the avoidance of host defences, contributes to the negative charge since the core polysaccharide contains charged sugars and phosphate, helps stabilize membrane structure, and can act as an endotoxin as lipid A is toxic. The function of the outer membrane is to serve as a protective barrier and prevent or slow down the entry of bile salts, antibiotics, and other toxic substances that might kill or injure the bacterium. It also prevents the loss of constituents like periplasmic enzymes (Figure 2.5).

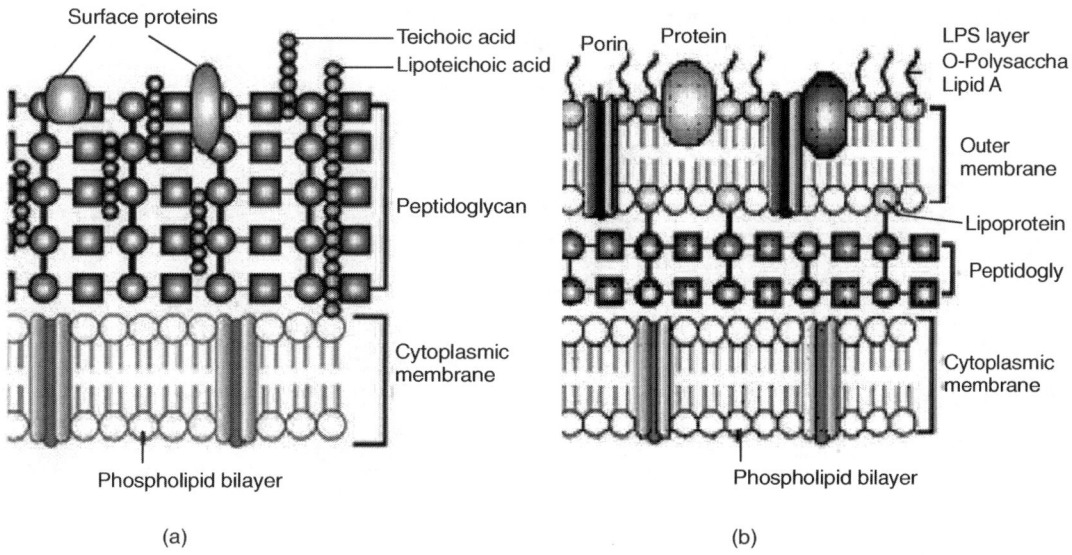

Figure 2.5   Structure of cell wall and membrane of a) gram-positive bacteria b) gram-negative bacteria

External Structures

Flagella   Bacterial flagella (singular: flagellum) are threadlike, helical appendages that protrude through the cell wall and are responsible for locomotion. They are slender, rigid structures much thinner than the flagella or cilia of eukaryotes, about 20 nm in diameter and up to 15 or 20 µm long. Based on the patterns of flagellar distribution, bacteria can be classified as monotrichous (single flagellum located at one end), amphitrichous (single flagellum at each pole), lophotrichous (cluster of flagella at one or both ends) and peritrichous (flagella are spread evenly throughout the cell) (Figure 2.6).

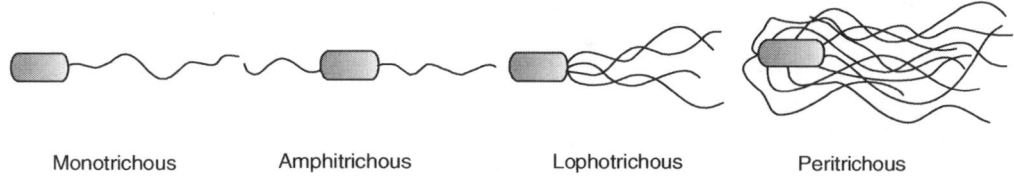

Figure 2.6   Pattern of flagellation in bacteria

Fimbriae and pili   Many gram-negative bacteria have short, fine, hairlike appendages that are thinner than flagella and are called fimbriae. A cell may have about 1,000 fimbriae that seem to be slender tubes composed of helically arranged protein subunits. They are about 3 to 10 nm in diameter and up to several micrometres long. Their function is to attach bacteria to solid surfaces and are not involved in motility.

Sex pili are similar fine appendages, about 1 to 10 per cell. Pili are larger than fimbriae i.e. around 9 to 10 nm in diameter. They are genetically determined by sex factors or conjugative plasmids and are required for bacterial mating. Some bacteriophages (viruses that infect bacteria) attach specifically to receptors on sex pili at the start of their reproductive cycle.

Capsules   Some bacterial cells are surrounded by a viscous substance forming a covering layer or envelope around the cell wall. If this layer is well-organized and cannot be easily washed off, it is termed as capsule. If the layer is a zone of diffuse unorganized material which can be washed off easily, it is called a slime layer. Other surface viscous layers, glycocalyx and S-layer, are also seen on the surface of some bacterial cells. They are usually composed of polysaccharides and sometimes made of proteins or glycoproteins. They help bacteria resist phagocytosis, protect bacteria against desiccation, exclude bacterial viruses and most hydrophobic toxic materials, aid bacterial attachment and presumably also aid in their motility.

Sheaths   Some species of bacteria particularly those from freshwater and marine environments, which grow as long filaments or trichromes, have their exterior covered by a hollow tubelike layer known as a sheath, e.g., *Leptothrix discophora* and *Sphaerotilus natans*. Its function is to provide protection to the bacteria in aquatic environments. The ability of the sheath to precipitate metallic compounds provides the bacteria with inorganic nutrients. The sheath also serves as the inert barrier to the external environment which helps the bacteria survive over a wide range of temperature and pH.

Prosthecae and stalks   A group of gram-negative bacteria possess semirigid extensions of the cell wall and cytoplasmic membrane. These cellular appendages are neither pili nor flagella and are called prosthecae, e.g., *Caulobacter*. Prosthecae aid in attachment of the cells and increase surface area for the uptake of nutrients in nutrient-poor aquatic habitats. Stalks are nonliving ribbonlike or tubular appendages produced by cells and extending from it.

Internal Structures

Immediately beneath the cell wall is the cytoplasmic membrane. Cell membranes are very thin structures about 5 to 10 nm thick and are composed primarily of phospholipids (about 20 to 30%) and proteins (about 60 to 70%). The phospholipids form a bilayer in which integral proteins (70–80%) are held tight and these proteins can be removed only by destruction of the membrane by treatment with detergents. The peripheral proteins are only loosely attached (20–30%) and can be removed by mild treatments such as osmotic shock. Bacterial membranes possess pentacyclic sterol-like molecules called hopanoids, which are responsible for stability of membranes. The lipid matrix of the membrane has fluidity and it appears to be essential for various membrane functions and is dependent on factors such as temperature and the proportion of unsaturated fatty acids to saturated fatty acids present in the phospholipids. The plasma membrane also serves as a selectively permeable barrier. It contains special receptor molecules that help bacteria to detect and respond to chemicals in their surrounding and it also serves as the location of a variety of metabolic processes like respiration, photosynthesis, synthesis of lipids and cell wall constituents, and probably chromosome segregation. It is clear that the plasma membrane is essential for the survival of microorganisms.

Mesosomes   Although bacteria lack complex membranous structures, they do have simpler internal membrane structures. Many bacteria especially gram-positive bacteria possess membrane invaginations in the form of systems of convoluted tubules and vesicles termed mesosomes. They are thought to be involved in cell wall formation during division, DNA replication and distribution to daughter cells. In contrast, peripheral mesosomes seem to be involved in export of exocellular enzymes such as penicillinase.

Cytoplasm   The cell material bounded by the cytoplasmic membrane may be divided into (1) the cytoplasmic area, which is granular in appearance and rich in the macromolecular RNA-protein bodies known as ribosomes, and is the site of protein synthesis, (2) the chromatinic area, rich in DNA and (3) the fluid portion with dissolved substances. The ribosomes of prokaryotes have a sedimentation coefficient of 70 Svedberg units (70S) and are composed of two subunits, a 50S and 30S subunit. A variety of inclusion bodies is present in the cytoplasmic matrix. They are made up of organic or inorganic material and are used for storage and for reducing osmotic pressure by tying up molecules in particulate form. Some inclusion bodies are not bound by a membrane (polyphosphate granules, cyanophycin granules, and some glycogen granules) and a few are enclosed by single-layered membrane, about 2.0 to 4.0 nm thick (poly-β-hydroxybutyrate granules, some glycogen and sulphur granules, carboxysomes and gas vacuoles). The membranes of inclusion bodies vary in composition, some are protein in nature, whereas others contain lipid.

Nuclear material   In contrast to eukaryotic cells, bacterial cells contain neither a distinct membrane-enclosed nucleus nor a mitotic apparatus. The prokaryotic chromosome is located in an irregularly shaped region called the nucleoid. It contains a single circle of double-stranded deoxyribonucleic acid (DNA), but some have a linear DNA chromosome. Recently, it has been discovered that some bacteria such as *Vibrio cholerae* have more than one chromosome.

Spores and cysts   Certain species of gram-positive bacteria produce special, dormant structures called spores, either within the cell (endospores) or external to the cell (exospores). These structures are extremely resistant to environmental stresses such as heat, radiation, chemical disinfectants and desiccation. During unfavourable environmental conditions, the metabolic processes of the cell are slowed down and the cell ceases all activities like feeding and locomotion, and this helps the organism to survive. Another type of resting, thick-walled structures formed under harsh conditions are called cysts. When the encysted microbe reaches an environment favourable to its growth and survival, the cyst wall breaks down by a process known as excystation.

Reproduction

The most common and important mode of reproduction in bacteria is transverse binary fission, in which a single cell divides after developing a transverse septum.

Budding   Some bacteria reproduce by budding, a process in which a small protuberance (bud) develops at one end of the cell; this enlarges and eventually develops into a new cell which separates from the parent.

Fragmentation   Bacteria that produce extensive filamentous growth such as *Nocardia* species reproduce by fragmentation of the filaments into small bacillary or coccoid cells, each of which gives rise to new growth.

Formation of conidiospores or sporangiospores   Species of the genus *Streptomyces* and related bacteria produce many spores per organism by developing crosswalls (septation) at the hyphal tips; each spore gives rise to a new organism.

Bacterial Recombination

Genetic recombination is the formation of a new genotype by reassortment of genes following an exchange of genetic material between two different chromosomes which have similar genes at corresponding sites. In bacteria, genetic recombination results from three types of gene transfer (Figure 2.7) which are as follows:

  i. *Transformation*   Transfer of cell-free or naked DNA from one cell to another.

 ii. *Transduction*   Transfer of genes from one cell to another by a bacteriophage

iii. *Conjugation*   Transfer of genes between cells that are in physical contact (by conjugation tube) with one another.

Plasmids

In addition to the normal chromosomal DNA, bacteria possess circular, extrachromosomal, self-replicating genetic elements called plasmids. Episomes are also extrachromosomal genetic elements which are capable of either replicating autonomously or integrating into the bacterial DNA chromosome. Plasmids are not required for host growth and reproduction, but they carry genes that give the bacterial host a selective advantage.

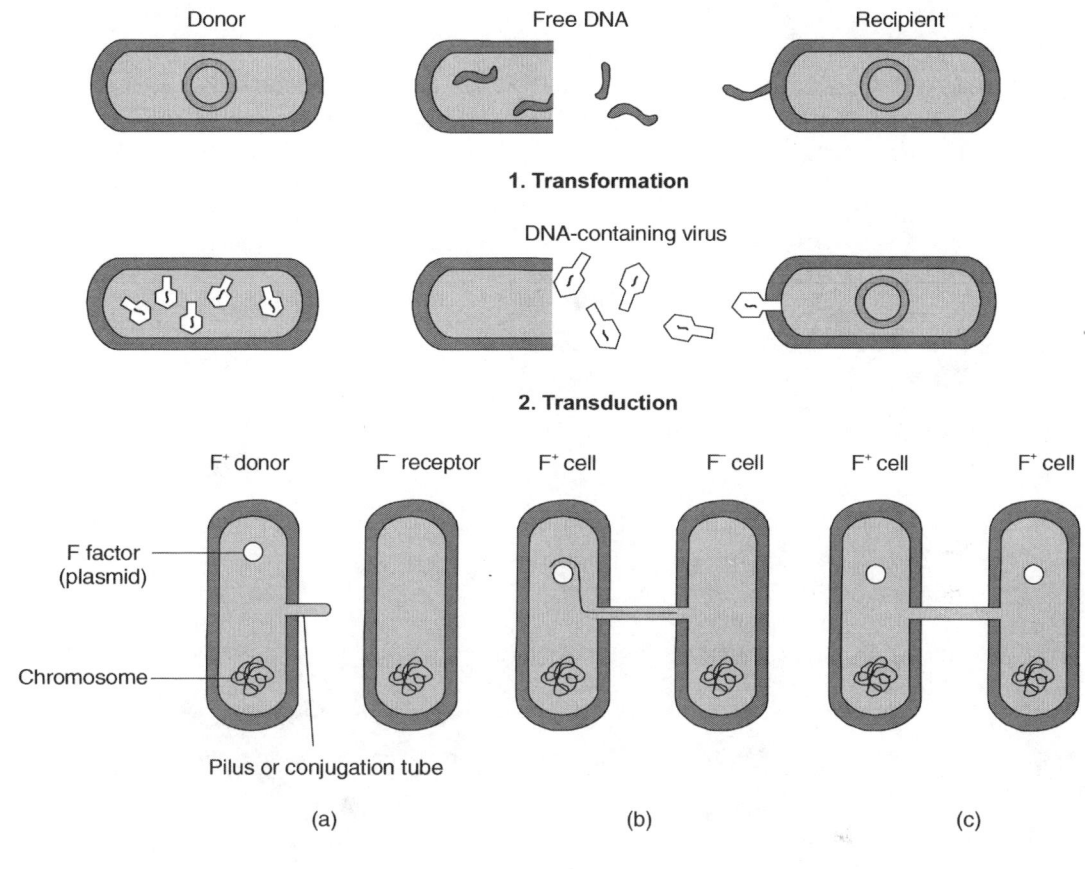

Figure 2.7 Recombination in bacteria

FUNGI

The fungi (singular: fungus) are a group of eukaryotic organisms that are of great practical and scientific interest to microbiologists. Fungi comprise moulds and yeasts. Moulds are filamentous and multicellular, whereas yeasts are usually unicellular. They generally reproduce both sexually and asexually. They are primarily terrestrial organisms and a few are freshwater or marine. Many are pathogenic, and infect plants and animals. They can also form beneficial relationships with other organisms. The phylogenetic classification of true fungi is depicted in Figure 2.8.

Characteristics of Fungi

Fungi are eukaryotic, spore-bearing, heterotrophic organisms with absorptive nutrition which lacks chlorophyll. The thallus (plural: thalli) or body of a fungus may consist of a single cell as in the yeasts; more typically, the thallus consists of filaments, 5 to 10 μm across, which are commonly branched. The yeast cell or mould filament is surrounded by a true cell wall, the exception being the slime moulds, which have a thallus consisting of a naked amoeboid mass of protoplasm. The fungal colony may be a mass of yeast cells or a filamentous mat of mould.

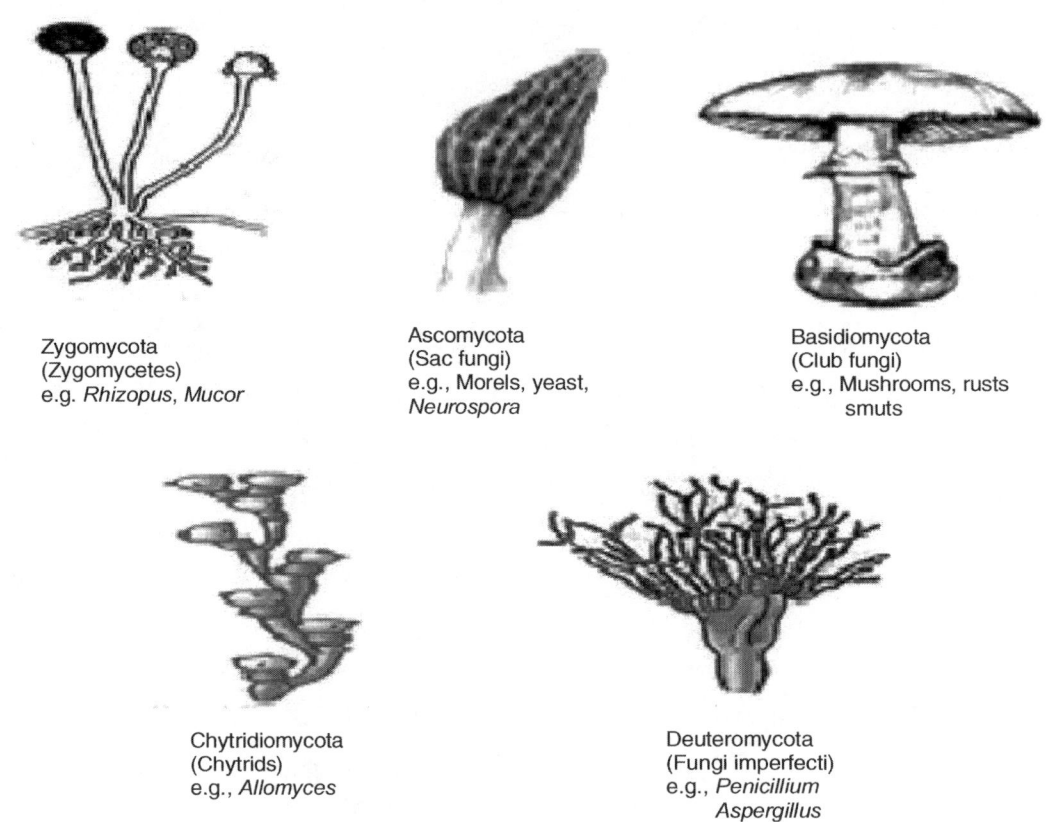

Zygomycota
(Zygomycetes)
e.g. *Rhizopus, Mucor*

Ascomycota
(Sac fungi)
e.g., Morels, yeast,
Neurospora

Basidiomycota
(Club fungi)
e.g., Mushrooms, rusts
smuts

Chytridiomycota
(Chytrids)
e.g., *Allomyces*

Deuteromycota
(Fungi imperfecti)
e.g., *Penicillium*
Aspergillus

Figure 2.8 Classification of fungi

Many fungi, especially pathogenic, are dimorphic, i.e., they have two forms. Dimorphic fungi can change from the yeast form (Y) in the animal to the mould or mycelial form (M) in the external environment in response to changes in various environmental factors. This shift is called the YM shift. In plant-associated fungi, the opposite type of dimorphism exists: the mycelial form occurs in the plant and the yeast form in the external environment.

Morphology

In general, yeast cells are larger than most bacteria and vary considerably in size, ranging from 1 to 5 μm in width and from 5 to 30 μm or more in length. They are commonly egg-shaped but some are elongated and some spherical. Yeasts neither have flagella nor other organelles of locomotion but possess most of other eukaryotic organelles. They are unicellular and have a single nucleus.

A mould consists of long, branched, threadlike filaments of cells called hyphae that form tangled mycelium. The thallus of a mould consists essentially of two parts: the mycelium (plural: mycelia) and the spores (resistant, resting or dormant cells). The mycelium is a complex of several filaments called hyphae (singular: hypha). Each hypha is about 5 to 10 μm wide, in contrast to a bacterial cell, which is usually 1 μm in diameter. Hyphae are composed of an outer tubelike wall surrounding a cavity, the lumen, which is filled or lined by protoplasm. Between the protoplasm and the wall is the plasmalemma, a double-layered membrane which surrounds the protoplasm. The hyphal wall consists of microfibrils composed for the most part of hemicellulose or chitin; true cellulose occurs only in the walls of lower fungi. Wall matrix material in which the microfibrils are embedded consists of proteins, lipids and other substances.

Growth of hypha is distal, near the tip. The major region of elongation takes place in the region just behind the tip. Hyphae occur in three forms. (i) Nonseptate or coenocytic where hyphae have no septa. (ii) Septate with uninucleate cells (iii) Septate with multinucleate cells, where each cell has more than one nucleus in each compartment (Figure 2.9).

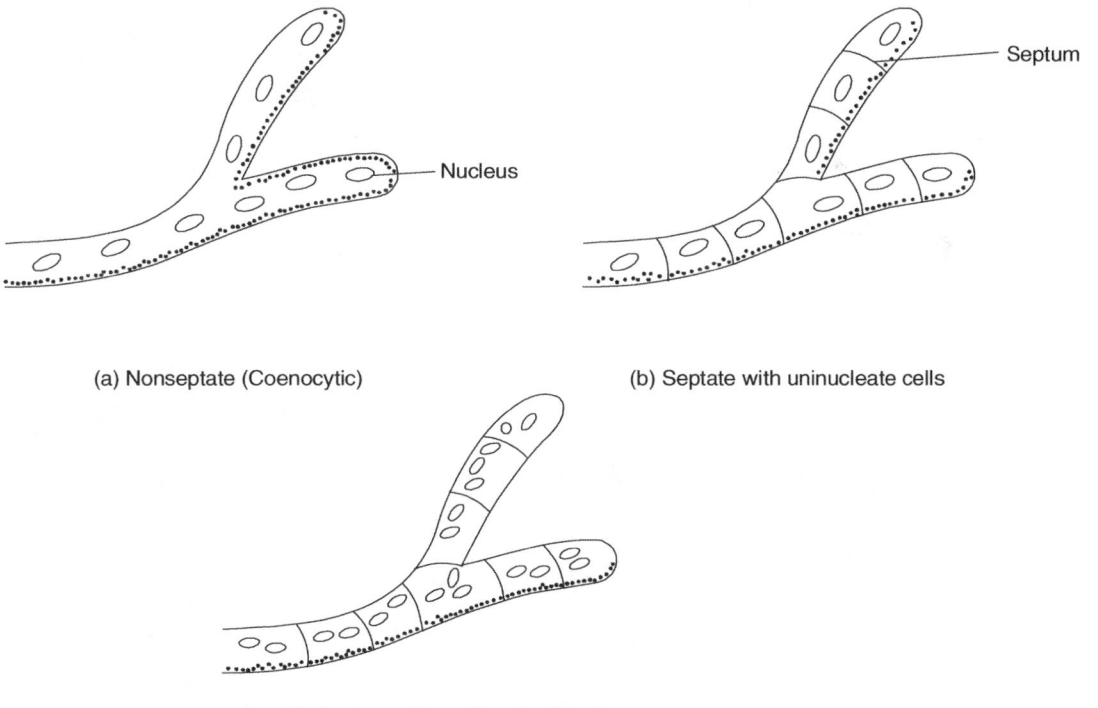

(a) Nonseptate (Coenocytic)

(b) Septate with uninucleate cells

(C) Septate with multinucleate cells

Figure 2.9 Types of hyphae

Reproduction

Fungi reproduce naturally by a variety of means. Asexual reproduction (also called somatic or vegetative reproduction) does not involve the union of nuclei, sex cells or sex organs. It may be accomplished by (1) fission of somatic cells yielding two similar daughter cells (2) budding of somatic cells or spores, each bud a small outgrowth of the parent cell developing into a new individual (3) fragmentation or disjointing of the hyphal cells, each fragment becoming a new organism or (4) spore formation. Asexual spores include sporangiospores, conidiospores, arthrospores, chlamydospores and blastospores.

Sexual reproduction is carried out by fusion of the compatible nuclei of two parent cells. The process of sexual reproduction begins with the joining of two cells and fusion of their protoplasts (plasmogamy) thus enabling the two haploid nuclei of two mating types to fuse together (karyogamy) to form a diploid nucleus. This is followed by meiosis, which again reduces the number of chromosomes to the haploid number. The various methods of sexual reproduction are gametic copulation, gamete-gametangial copulation, gametangial copulation, somatic copulation and spermatization. Sexual spores which occur less frequently include ascospores, basidiospores, zygospores, and oospores.

Water and Slime Moulds

The slime moulds and water moulds resemble fungi only in appearance and life-style. They are phylogenetically distinct in their cellular organization, reproduction and life cycles. Slime moulds are of two types, acellular and cellular. Members of Oomycota division are known as oomycetes or water moulds and consist of finely branched filaments called hyphae.

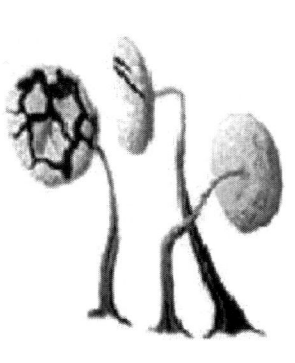
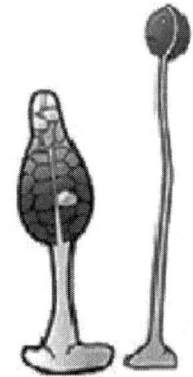
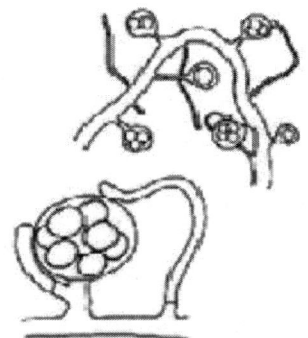

Myxomycota
(acellular slime mould)
e.g., *Physarum*

Acarisomycota
(cellular slime mould)
e.g., *Dictyostelium*

Oomycota
(water mould/egg fungi)
e.g., *Saprolegnia*

Figure 2.10   Classification of slime mould and water mould

Water moulds resemble true fungi only in appearance. They lack cell wall, and feeding is by phagocytosis. They are important decomposers in aquatic ecosystems. Some are saprophytic and a few are plant pathogens. The phylogenetic classification of slime and water moulds are depicted in Figure 2.10.

ALGAE

Algae (singular: alga) are ubiquitous unicellular eukaryotic microorganisms that have chlorophyll and carry out photosynthesis. They differ from other photosynthetic eukaryotes in lacking a well-organized vascular conducting system and in having very simple reproductive structures. Algae are of great general interest to biologists, because single algal cells are complete organisms, capable of photosynthesis and the synthesis of a multitude of other compounds which constitute the cell. Algae commonly occur in fresh, marine, or brackish water in which they may be suspended (planktonic) or attached and living on the bottom (benthic). A few algae live at the water–atmosphere interface and are termed neustonic. The important properties used for algal classification include: (1) cell wall chemistry and morphology (2) form in which food or assimilatory products of photosynthesis are stored (3) chlorophyll and accessory pigments (4) number of flagella and their location in motile cells; (5) morphology of the cells and thallus (6) habitat (7) reproductive structures and (8) life history patterns. Algae are classified into 7 divisions based on the above-mentioned properties (Figure 2.11.).

Characteristics

Algae have a wide range of sizes and shapes. They vary from a simple single cell to the more complex multicellular forms like giant kelps. Algae are unicellular, colonial, filamentous, membranous and bladelike, or tubular. The vegetative body of algae is called the thallus. Algal cells are eukaryotic. In most species, the cell wall is thin and rigid. The motile algae such as *Euglena* have flexible cell membranes, called periplasts. The cell walls of many algae are surrounded by a flexible, gelatinous outer matrix secreted through the cell wall, similar to bacterial capsule. Algae contain a discrete nucleus. Other inclusions are starch grains, oil droplets, and vacuoles. Chlorophyll and other pigments are found in membrane-bound organelles known as chloroplasts. The pyrenoid which is a dense proteinaceous area that is associated with synthesis and storage of starch may be present in the chloroplasts. Mitochondrial structure varies greatly in the algae. Mitochondrial cristae may be discoid, lamellar or tubular. Algae can be either autotrophic or heterotrophic. Most are photoautotrophic; they require only light and CO_2 as their principal sources of energy and carbon. Chemoheterotrophic algae require external organic compounds as carbon and energy sources.

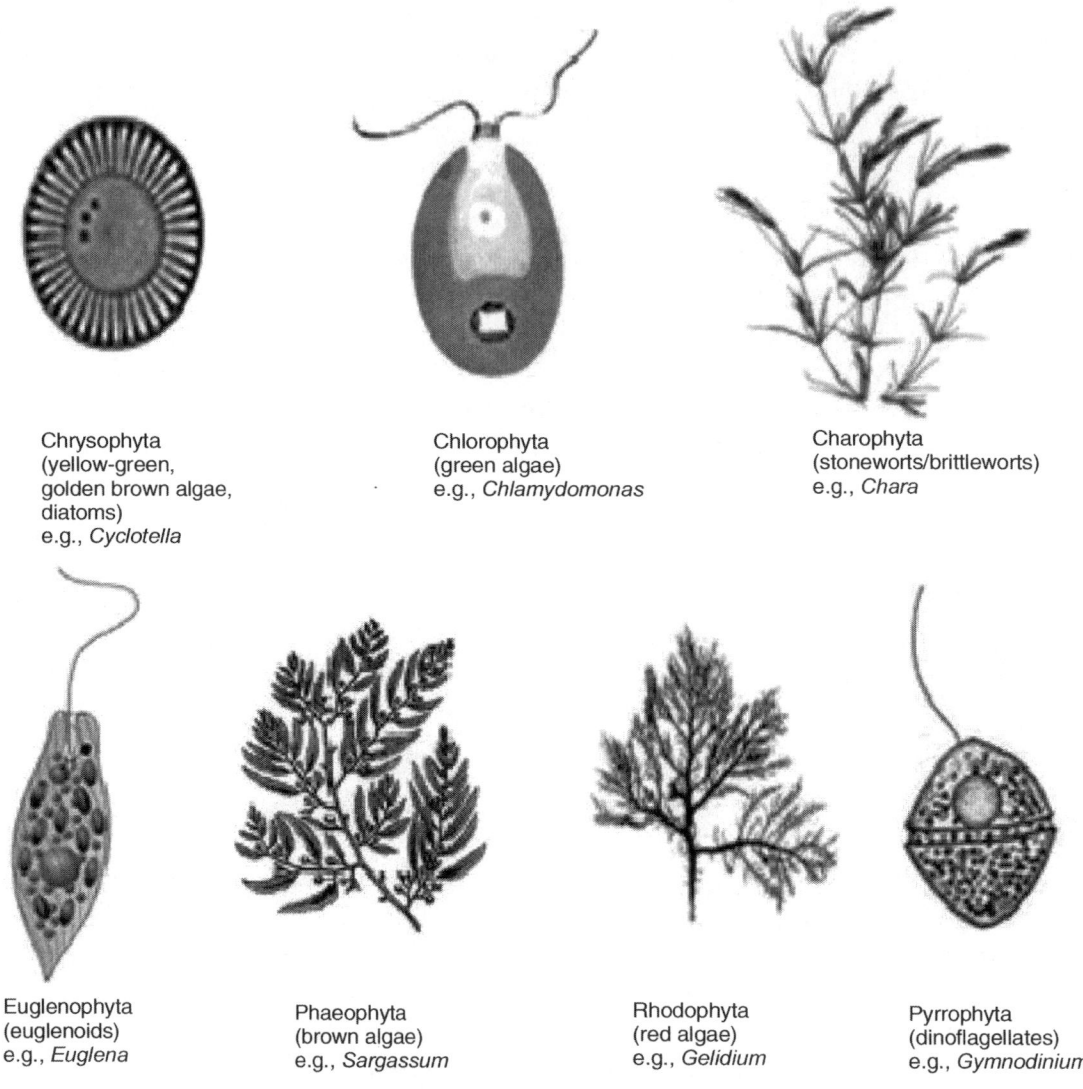

Chrysophyta
(yellow-green,
golden brown algae,
diatoms)
e.g., *Cyclotella*

Chlorophyta
(green algae)
e.g., *Chlamydomonas*

Charophyta
(stoneworts/brittleworts)
e.g., *Chara*

Euglenophyta
(euglenoids)
e.g., *Euglena*

Phaeophyta
(brown algae)
e.g., *Sargassum*

Rhodophyta
(red algae)
e.g., *Gelidium*

Pyrrophyta
(dinoflagellates)
e.g., *Gymnodinium*

Figure 2.11   Classification of algae

Algal Pigments

Chlorophyll   There are five chlorophylls: *a, b, c, d* and *e*. Chlorophyll *a* is present in all algae, as it is in all photosynthetic organisms other than anoxygenic photosynthetic bacteria.

Carotenoids   There are two kinds of carotenoids, carotenes and xanthophylls. Carotenes are linear, unsaturated hydrocarbons and xanthophylls are oxygenated derivatives of these.

Biloproteins (Phycobilins)   These are water-soluble pigments, whereas chlorophylls and carotenoids are lipid-soluble. Phycocyanin and phycoerythrin are the two kinds of phycobilins.

Motility

The motile algae, also called the swimming algae, have flagella occurring singly, in pairs or in clusters at the anterior or posterior ends of the cell. A small red or orange body, the eyespot is often present near the anterior end of motile algae.

Reproduction

Algae may reproduce either asexually or sexually. Asexual reproductive processes in algae include the purely vegetative type of cell division by which bacteria reproduce. Asexual unicellular spores are of two types, viz., zoospores (motile spores) and aplanospores (nonmotile spores). All forms of sexual reproduction are found among the algae. In these processes, there is a fusion of sex cells called gametes, to form a union in which blending of nuclear material occurs before new generations are formed. The union of gametes forms a zygote. If the gametes are identical, i.e., if there is no visible sex differentiation, the fusion process is isogamous. If the two gametes are unlike, differing in size (male and female), the process is heterogamous. As we proceed to the higher algae, the sexual cells become more characteristically male and female. The ovum (female egg cell) is large and nonmotile, and the male gamete (sperm cell) is small and actively motile. This type of sexual process is termed oogamy. Exclusively male or exclusively female thalli also exist. Although these thalli may look alike, they are of opposite sex types, since one produces male gametes and the other ova. Such algae are called unisexual or dioecious. Algae in which gametes from the same individual can unite are said to be bisexual or monoecious.

PROTOZOA

Protozoa (singular: protozoan), from the Greek *protos* and *zoon* meaning "first animal", are unicellular eukaryotic protists. They may be distinguished from other eukaryotic protists by their ability to move at some stage of their life cycle and by their lack of cell walls. Protozoa are predominantly microscopic in size. They grow widely in moist habitat, some are terrestrial protozoa and few are parasitic in plants and animals. They are aerobic or anaerobic. But, anaerobes lack mitochondria and cytochromes and possess incomplete TCA (tricarboxylic acid) cycle. Most protozoa are chemoheterotrophic. Two types of heterotrophic nutrition are found in the protozoa, which are holozoic (phagocytosis) and saprozoic (pinocytosis).

Morphology

The size and shape of these organisms show considerable variation. Certain common ciliates reach 2,000 µm or 2 mm. *Amoeba proteus* measures 600 µm or more. Like all eukaryotic cells, the protozoan cell also consists of cytoplasm, separated from the surrounding medium by a special cell envelope and the nucleus or nuclei (Figure 2.12).

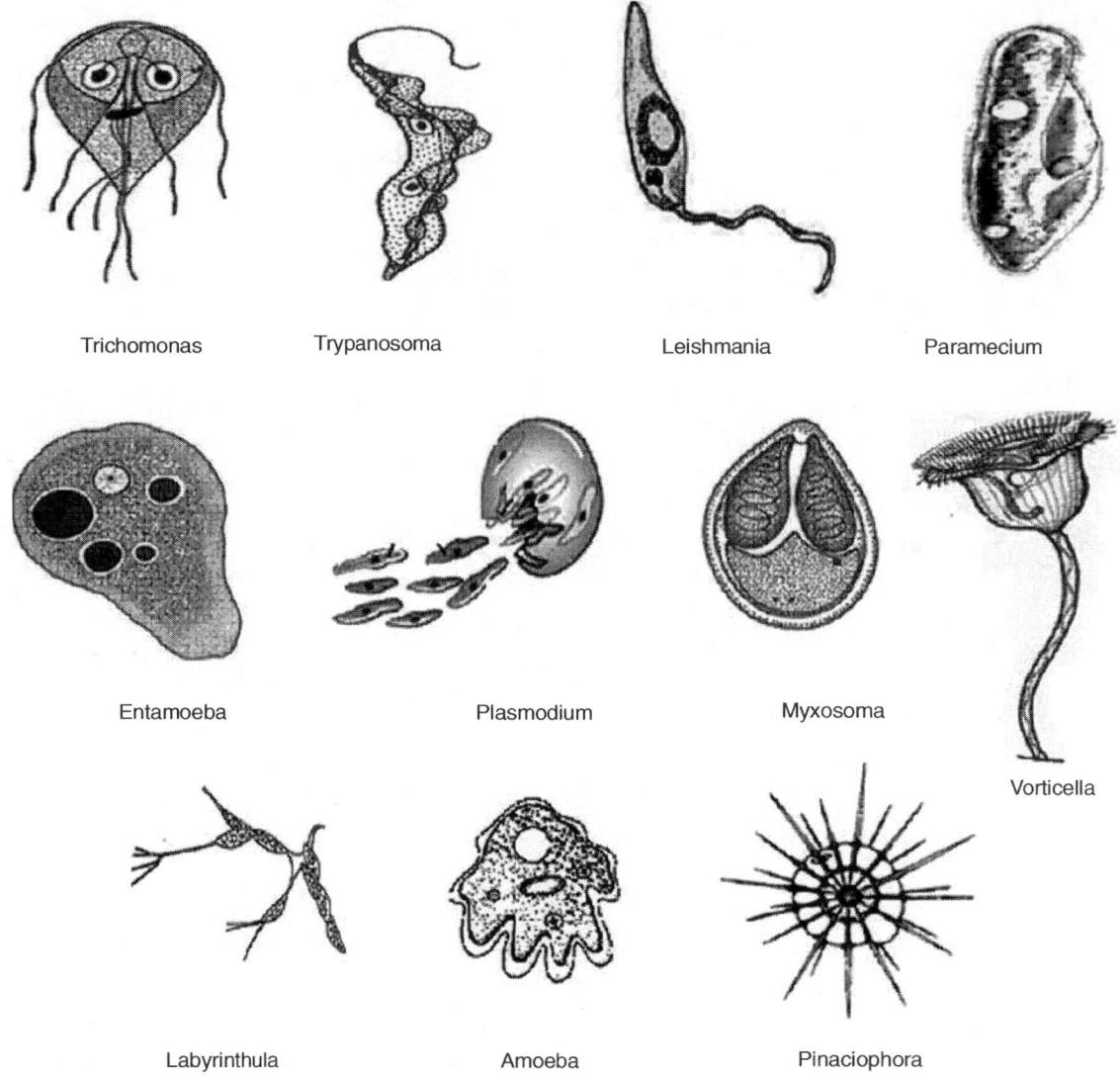

Trichomonas   Trypanosoma   Leishmania   Paramecium

Entamoeba   Plasmodium   Myxosoma   Vorticella

Labyrinthula   Amoeba   Pinaciophora

Figure 2.12   Different types of cytoplasm in protozoa

The cytoplasm is a more or less homogeneous substance consisting of globular protein molecules loosely linked together to form a three-dimensional molecular framework. Embedded within it are the various structures that give protozoan cells their characteristic appearance. In the majority of protozoa, the cytoplasm is differentiated into the ectoplasm and the endoplasm. The ectoplasm is more gel-like which gives rigidity to the cell and the endoplasm is more voluminous and fluid and contains most organelles. A thin layer supporting the cell membrane which protects and helps in retaining the shape, especially during locomotion, is known as pellicle. The cytoplasm consists of endoplasmic reticulum,

Golgi complexes, ribosomes, mitochondria, kinetosomes, food vacuoles, contractile vacuoles, secretory vacuoles, phagocytic vacuoles and nuclei.

Nucleus   The protozoan cell has one eukaryotic nucleus; however, many protozoa have two or more nuclei throughout the greater part of their life cycle. The protozoan nuclei are of various forms, sizes, and structures. They have two distinct types of nuclei *viz.*, a macronucleus and one or more micronuclei. The macronucleus is typically larger and associated with trophic activities and regeneration processes. The micronucleus is diploid and is involved in both genetic recombination during reproduction and the regeneration of the macronucleus. The essential structural elements of the nucleus are the chromosomes, the nucleolar substance, the nuclear membrane, and the karyoplasm (nucleoplasm). It has been shown that the number of chromosomes is constant for a particular species of protozoan.

Plasmalemma and other cell coverings   The cytoplasm with its various structures is separated from the external environment by a cell unit membrane (plasmalemma). The plasmalemma not only provides protection but also controls exchange of substances (semipermeable). It is the site of perception of chemical and mechanical stimuli as well as of establishment of contact with other cells (cell sensitivity to external factors). Although all protozoa possess a cell membrane, many protozoa have compound coverings of membranes modified for protection, support and movement. Such combinations of membranes are referred to as pellicle.

Feeding structures   Food-gathering structures in the protozoa are diverse and range from the pseudopodia of amoebae through the tentacular feeding tubes of suctorians to the well-developed "mouths" of many ciliates.

Cysts   Many protozoa form resistant structures called cysts at certain times of their life cycle. Cysts are able to protect the organism against adverse environmental conditions such as desiccation, low nutrient supply and even anaerobiosis. They also serve as sites for nuclear reorganization and cell division (reproductive cysts) and as a means of transfer between hosts in parasitic species. Cysts are excysted into vegetative form and this process is triggered by a return of favourable environmental conditions, e.g., cysts of parasitic species excyst and their vegetative form is called the trophozoite.

Locomotor organelles   Protozoa may move by three types of specialized locomotory organelles: pseudopodia, flagella and cilia. In addition, a few nonmotile protozoa can carry out a gliding movement by body flexion.

Reproduction

As a general rule, protozoa multiply by asexual reproduction but some protozoa reproduce by sexual means. Asexual reproduction occurs by simple cell division, which can be equal or unequal and the daughter cells are of equal or unequal sizes, respectively. If two daughter cells are formed, then the process is called binary fission; if many daughter cells are formed, it is multiple fission. Budding is a variation of unequal cell division.

Various types of sexual reproduction have been observed among protozoa. Sexual fusion of two gametes (syngamy or gametogamy) occurs in various groups of protozoa. The most common method of sexual reproduction, conjugation, is a temporary union of two individuals for the purpose of exchanging nuclear material found exclusively in ciliates. After exchange of nuclei, the conjugants separate and each of them gives rise to its respective progeny by fission or budding. However, some ciliates show "total conjugation" with complete fusion of the two organisms.

Regeneration

The capacity to regenerate lost parts is characteristic of all protozoa, from simple forms to those with highly complex structures. When a protozoan is cut into two, the nucleated portion regenerates but the anucleated portion degenerates. In general, the nucleus is necessary for regeneration. In ciliates, the macronucleus alone (or even just a portion of it) is sufficient for this process.

VIRUS

Characteristics

Viruses are infectious agents with fairly simple, acellular organization that can only be seen using the electron microscope. They are 10 to 100 times smaller than most bacteria, with an approximate size range of 20 to 300 nm. Thus, they pass through the pores of filters which do not permit the passage of most bacteria.

Viruses are obligate parasites and are incapable of growth in artificial media. They can grow only in animal or plant cells or in microorganisms. Viruses possess only one type of nucleic acid, either DNA or RNA, enclosed mainly in a protein coat or carbohydrate/lipid layers. They reproduce only in living cells; and use the cell's biosynthetic machinery to direct the synthesis of specialized particles (virions) which contain the viral genomes and transfer them efficiently to other cells. Thus, viruses are referred to as obligate intracellular parasites.

Bacteriophages, (bacterial viruses), like all viruses, are composed of a nucleic acid core surrounded by a protein coat. Bacterial viruses occur in different shapes, although many have a tail through which they inoculate the host cell with viral nucleic acid. There are two main types of bacterial viruses: lytic or virulent and temperate or avirulent.

Morphology and Structure

Virions range in size from about 10 to 300 or 400 nm in diameter. The electron microscope has made it possible to determine the structural characteristics of bacterial viruses. All phages have a nuclei acid core covered by a protein coat, or capsid. The capsid is made up of morphological subunits called capsomeres. Capsids have icosahedral, helical or binal symmetry. The capsomeres consist of a number of protein subunits or molecules called protomers. Bacterial viruses are grouped based on the following characteristics: nature of the host, nucleic acid characteristics, virion structure and composition, mode of reproduction and immunologic properties.

Genetic Material of Viruses

Viruses contain double-stranded (ds) or single-stranded (ss) DNA or RNA as their genetic material, which may be of one of the following forms.

* ds DNA with each end covalently sealed
* ds DNA with covalently linked terminal protein
* ds circular DNA   * ds DNA
* ds RNA   * ssDNA
* ss circular DNA   * ss RNA

The classification of viruses based on their nucleic acids is illustrated in Figure 2.13.

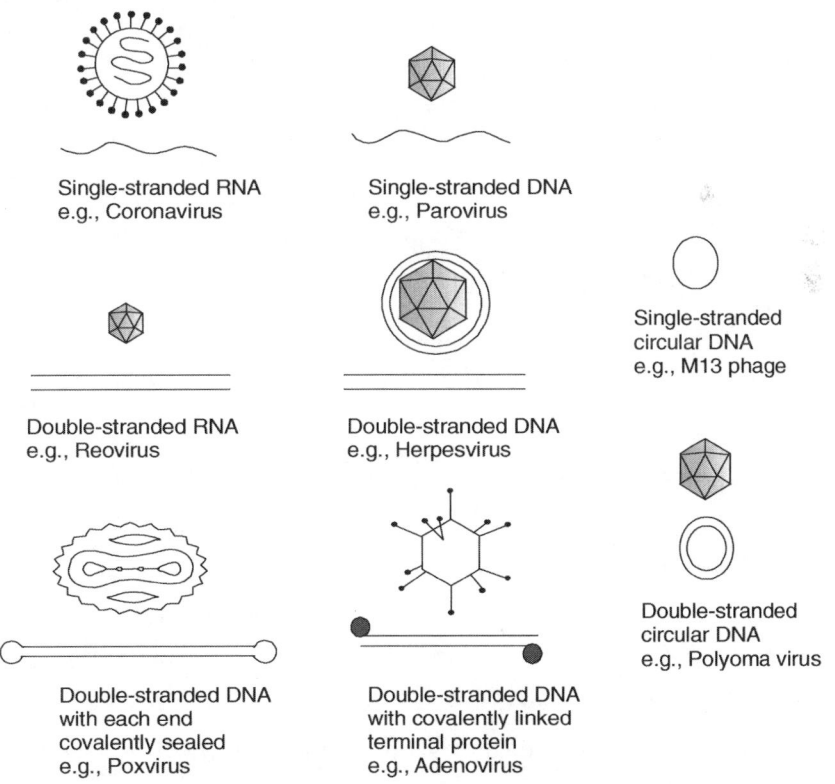

Single-stranded RNA
e.g., Coronavirus

Single-stranded DNA
e.g., Parovirus

Single-stranded
circular DNA
e.g., M13 phage

Double-stranded RNA
e.g., Reovirus

Double-stranded DNA
e.g., Herpesvirus

Double-stranded
circular DNA
e.g., Polyoma virus

Double-stranded DNA
with each end
covalently sealed
e.g., Poxvirus

Double-stranded DNA
with covalently linked
terminal protein
e.g., Adenovirus

Figure 2.13   Classification of virus based on nucleic acids

The type of genetic material found in a particular virus depends on the nature and function of the specific virus. The genetic material is not typically exposed but covered by a protein coat. All four types of genetic material are found in animal viruses. Plant viruses most often have single-stranded RNA genomes. Phages usually contain dsDNA although they have ssDNA, dsRNA and ssRNA. The viral genome can consist of a very small number of genes or up to hundreds of genes depending on the type of virus. The genome is typically organized as a long molecule that is usually straight or circular.

Virulent Bacteriophages and the Lytic Cycle

Viruses that kill their infected host cell are called virulent. The DNA in these types of viruses reproduces through the lytic cycle. When these viruses reproduce, they break open, or lyse their host cells, resulting in the destruction of the host. The whole cycle can be completed in 20–30 min. depending on a variety of factors such as temperature, growth phase and host cell physiology. Phage reproduction is much faster than typical bacterial reproduction; so, entire colonies can be destroyed very quickly (Figure 2.14). They form plaques after lysis of bacterial colonies on the medium lawn of Petri plates.

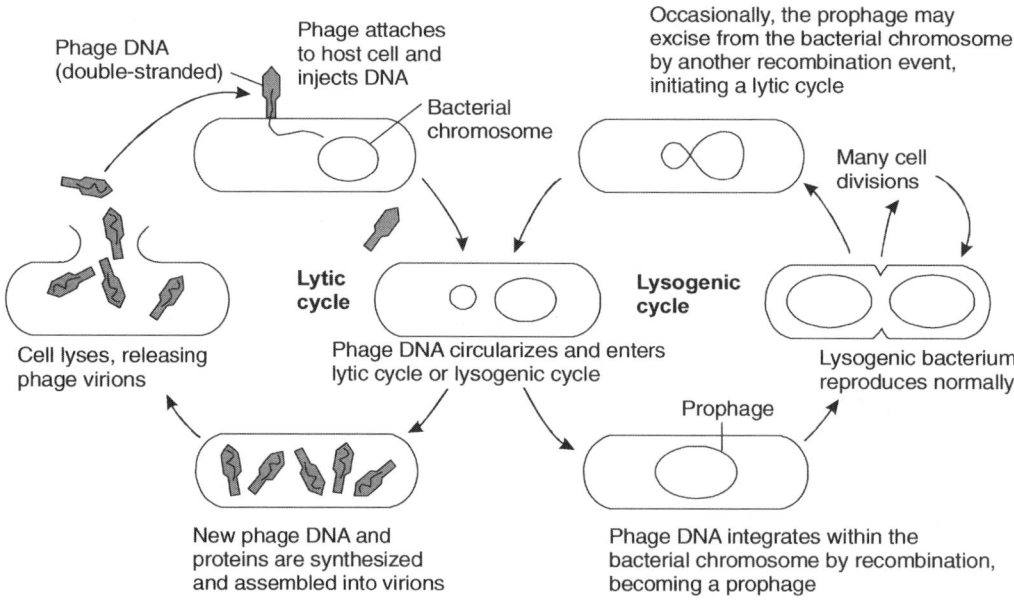

Figure 2.14   Steps in lytic and lysogenic cycle

Temperate Viruses and the Lysogenic Cycle

Temperate viruses are those that reproduce without killing their host cell. Typically, they reproduce in two ways: through the lytic cycle and the lysogenic cycle. In the lysogenic cycle, the phage's DNA recombines with the bacterial chromosome. Once it has inserted itself, it is known as a prophage. A host cell that carries a prophage has the potential to lyse, thus it is called a lysogenic cell. Figure 2.14 illustrates both the lytic and lysogenic cycles of a bacteriophage.

Virus Replication

Viruses are intracellular obligate parasites which mean that they cannot replicate or express their genes without the help of a living cell. A virion (single virus particle) is inert and it lacks components to replicate and reproduce. When a virus infects a cell, it marshals the cell's ribosomes, enzymes and much of the cellular machinery to replicate. Viral replication produces many progeny, which when complete, leave the host cell to infect other cells.

Once a virus infects its host, it utilizes the host cell machinery to produce the viral progeny components. The assembly of these components into viral capsule is a nonenzymatic spontaneous process. Viruses typically can only infect a limited number of hosts (also known as host range). The "lock and key" mechanism is the most common explanation for this range. Certain proteins on the virus particle must fit certain glycoprotein receptor sites on the particular host's cell surface.

The basic process of viral infection and virus replication occurs in 6 main steps.

1. Adsorption—virus binds to the host cell via receptor.
2. Penetration—virus injects its genome into host cell.
3. Viral genome replication—viral genome replicates using the host's cellular machinery.
4. Assembly—viral components and enzymes are produced and begin to assemble.
5. Maturation—viral components assemble and viruses fully develop.
6. Release—newly produced viruses are expelled from the host cell.

Viruses of Plants, Animals, Fungi, Algae and Insects

Animal and plant viruses vary greatly in shape and size, but they do not have the tadpole characteristic of bacteriophages. Virions range in size from 20 to 350 nm and represent the smallest and simplest infectious agents. Since most viruses measure less than 150 nm, they are beyond the limit of resolution of the light microscope and they are visible only under the electron microscope. These viruses are important from agriculture and veterinary point of view as they cause devastating diseases. Few examples are foot-and-mouth disease virus, avian influenza virus, tobacco mosaic virus, etc.

Most of the fungal viruses or mycoviruses contain double-stranded RNA and have isometric capsids which are approximately 25 to 50 nm in diameter. Many appear to be latent viruses. In algae, few viruses have been isolated and they have dsDNA genomes and polyhedral capsids. But, a virus of the green alga *Uronema gigas* resembles bacteriophages in having a tail.

Insects are infected by at least seven virus families and the three most important are the Baculoviridae, Reoviridae, and Iridoviridae. They reproduce or even use them as the primary host. The Iridoviridae are icosahedral viruses with linear double-stranded DNA genome and capsids made of lipids. The reoviruses are icosahedral with double shells and have double-stranded RNA genomes. The baculoviruses are rod-shaped, enveloped viruses of helical symmetry and with double-stranded DNA. Inclusion bodies are formed within the infected cells during insect virus infection. Granular protein inclusions are formed by granulosis viruses, usually in the cytoplasm. Nuclear polyhedrosis and cytoplasmic polyhedrosis virus infections produce polyhedral inclusion bodies in the nucleus or the cytoplasm of affected cells. Although all three types of viruses generate inclusion bodies, cytoplasmic polyhedrosis viruses are formed by reoviruses and nuclear polyhedrosis viruses and granulosis viruses are formed by baculoviruses.

Viroids and Prions

Some infectious agents of plant diseases are simpler than some viruses. The group of infectious agents called viroids is circular, single-stranded RNAs, about 250 to 370 nucleotides long. Examples of infection are potato spindle-tuber disease, exocortis disease of citrus trees, and chrysanthemum stunt disease. Proteinaceous infectious particles different from both viruses and viroids cause diseases in livestock and humans and are called prions. Examples are bovine spongiform encephalopathy (BSE) and Creutzfeldt–Jakob disease (CJD).

Retroviruses

Retroviruses are responsible for causing some forms of leukemia (a type of cancer) in humans, and the virus that causes AIDS (HIV) is a retrovirus. Despite this, retroviruses are also useful tools for protein engineering, and are used for introducing new genes into a host cell. These types of virus integrate into host DNA at random sites; however, the locations for gene insertions cannot (as yet) be controlled.

Cultivation of Viruses

Viruses cannot be cultured in artificial synthetic media, as they are unable to reproduce independently of living cells. Animal viruses are cultivated in host animals or embryonated eggs and most recently in tissue culture on monolayers of animal cells. Bacterial viruses or bacteriophages are cultivated in either broth or agar cultures of young, actively growing bacterial cells. Plant viruses are cultivated in plant tissue cultures, cultures of separated cells or cultures of protoplasts and also can be grown in whole plants.

BIBLIOGRAPHY

Prescott, L.M., Harley, J.P. and Klein, D.A. (2002). *Microbiology*, 5th edn. The McGraw-Hill Companies, Columbus, USA.

Madigan, M. and Martinko, J. (2005). *Brock Biology of microorganisms*, 11th edn. Prentice Hall, Indiana, USA.

Pelczar, M.J. (1998). *Microbiology*, 5th edn. The McGraw-Hill Companies, Columbus, USA.

3
PROTEINS—AN OVERVIEW

INTRODUCTION

Proteins are the most versatile and abundant macromolecules present in all living cells. They are formed by condensing the α-amino group of one amino acid (imino group in the case of proline) with the α-carboxyl group of another, with the concomitant loss of a molecule of water and the formation of a peptide bond. They function as catalysts, provide immune protection, transmit nerve impulses, control growth and differentiation, etc.

Peptide bond

Proteins are linear polymers built of monomer units called amino acids. They contain a wide range of functional groups and can interact with one another and with other biological molecules to form complex assemblies. The progressive condensation of many molecules of amino acids gives rise to an unbranched polypeptide chain. By convention, the N-terminal amino acid is taken as the beginning of the chain and the C-terminal amino acid as the end of the chain (proteins are biosynthesized in this direction). Polypeptide chains contain between 20 and 2000 amino acid residues and hence, have a relative molecular mass ranging between 2000 and 2,00,000. Many proteins have a relative molecular mass in the range of 20000 to 100000. The distinction between a large peptide and a small protein is not clear. Generally, chains of amino acids containing fewer than 50 residues are referred to as peptides and those with more than 50 are referred to as proteins. Most proteins contain many hundreds of amino acids (ribonuclease is an extremely small protein with only 103 amino acid residues) and many biologically active peptides contain 20 or fewer amino acids, e.g. oxytocin (9 amino acid residues), vasopressin (9), enkephalins (5), gastrin (17), somatostatin (14) and luteinizing hormone (10).

AMINO ACIDS

Amino acids are the basic building blocks of proteins and they play an important role as intermediates in metabolism. There are 20 different amino acids found in a living cell. The first to be discovered was asparagine, in 1806. The last of the 20, threonine, was not identified until 1938. From these building blocks, different organisms construct widely diverse products as enzymes, hormones, antibodies, transporters, muscle fibres, feathers, milk proteins, antibiotics and innumerable other substances having distinct biological activities. The chemical properties of the amino acids of a protein determine the biological activity of the protein. All amino acids have the following basic skeleton.

$$R - \underset{\underset{NH_2}{|}}{\overset{\overset{H}{|}}{C}} - COOH$$

All amino acids, except glycine, have a central chiral carbon atom with 4 attachments, viz., a single hydrogen atom, a COOH (carboxyl) group, an NH_2 (amino) group and a side chain (R) which differs for each of the 20 amino acids. Amino acid, determined by the presence of the amino group and carboxyl group, dissociates in water to give COO^- and H^+ ions and it is acidic because of the presence of H^+ ions in the solution. The 20 amino acids are defined by the nature of the side chain R. The chemical properties of the amino acids of a protein determine the biological activity of the protein. The 20 different amino acids are listed in Table 3.1 and their structures are depicted in Figure 3.1.

Among the 20 different amino acids, glycine is the simplest amino acid with a single hydrogen atom as its side group. Tyrosine, tryptophan and phenylalanine are aromatic amino acids, i.e., they contain an aromatic ring as a part of their side chain, while methionine and cysteine contain sulphur in their side chain. The amino acid residues in protein molecules are exclusively L-stereoisomers. D-amino acid residues have been found only in a few, generally small, peptides including some peptides of bacterial cell wall and certain peptide antibiotics. Cells are able to specifically synthesize the L-isomers of amino acids because the active sites of enzymes are asymmetric, causing the reactions they catalyse to be stereospecific. When amino acids are dissolved in water at neutral pH, they act as either acids or bases. This property is referred to as zwitterion or dipolar ion. Amino acids vary in their acid–base properties and have characteristic titration curves and isoelectric point/pH.

The amino acids are grouped into five types based on their polarity, i.e., their tendency to interact with water at biological pH. The polarity of R groups may vary from nonpolar and hydrophobic to polar and hydrophilic. The five types include:

i. Nonpolar and hydrophobic aliphatic R groups which include alanine, valine, leucine, isoleucine, glycine, methionine and proline.

ii. Relatively nonpolar and hydrophobic aromatic R groups which include phenylalanine, tyrosine and tryptophan.

iii. Polar uncharged R groups that are more hydrophilic than nonpolar groups and include amino acids serine, threonine, cysteine, asparagine and glutamine.

Table 3.1 Different amino acids

Amino acid	Three-letter symbol	One-letter symbol
Alanine	Ala	A
Arginine	Arg	R
Asparagine	Asn	N
Aspartic acid	Asp	D
Cysteine	Cys	C
Glutamic acid	Glu	E
Glutamine	Gln	Q
Glycine	Gly	G
Histidine	His	H
Isoleucine	Ile	I
Leucine	Leu	L
Lysine	Lys	K
Methionine	Met	M
Phenylalanine	Phe	F
Proline	Pro	P
Serine	Ser	S
Threonine	Thr	T
Tryptophan	Trp	W
Tyrosine	Tyr	Y
Valine	Val	V

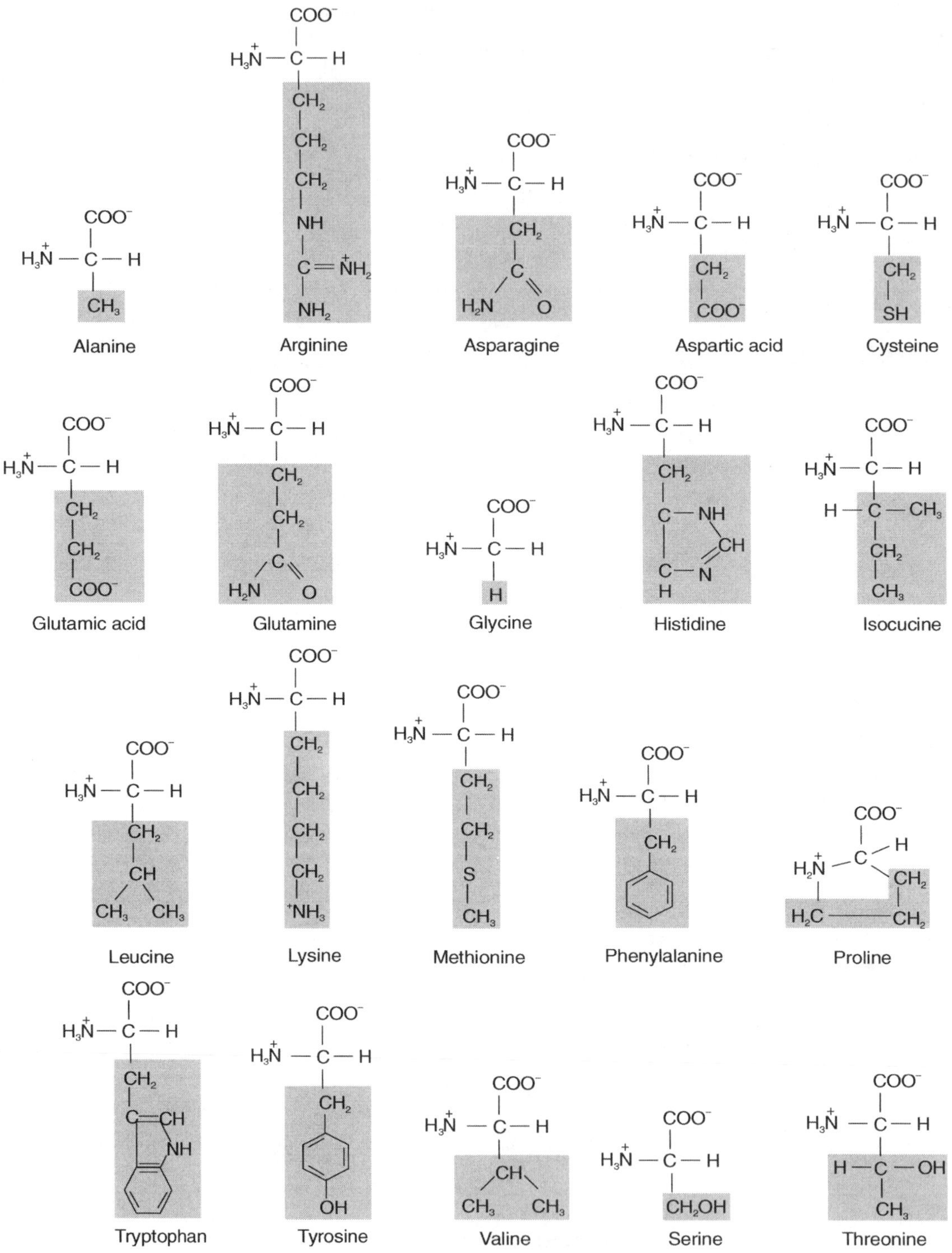

Figure 3.1 Structure of amino acids

iv. Polar charged R groups that are positively charged, hydrophilic basic group amino acids, and include lysine, arginine and histidine.

v. Polar negatively charged R groups which include acidic amino acids such as aspartic acid and glutamic acid.

In addition to the 20 common amino acids, a few uncommon amino acids are present due to the modification of common residues. They are 4-hydroxyproline, a derivative of proline, 5-hydroxylysine, derived from lysine, 6-*N*-methyllysine, a constituent of myosin, γ-carboxyglutamic acid found in prothrombin, desmosine, a derivative of four lysine residues, and selenocysteine derived from serine. Some 300 additional amino acids have been found in cells. They have a variety of functions but are not constituents of proteins.

Essential and Nonessential Amino Acids

Humans can produce 10 of the 20 amino acids. The others must be supplied in the food. Failure to obtain enough of even 1 of the 10 essential amino acids that we cannot make, results in degradation of the body's proteins. Unlike fat and starch, the human body does not store excess amino acids for later use; the amino acids must be contained in the food every day.

Nonessential amino acids   The 10 amino acids that are synthesized in the living system are alanine, asparagine, aspartic acid, cysteine, glutamic acid, glutamine, glycine, proline, serine and tyrosine. Tyrosine is produced from phenylalanine; so, if the diet is deficient in phenylalanine, tyrosine will be required as well.

Essential amino acids   They are not synthesized in the human system and have to be supplemented in the diet. A simple formula, viz., MATTVILPHLY, helps one to remember the essential amino acids: **M**ethionine, **A**rginine, **T**hreonine, **T**ryptophan, **V**aline, **I**soleucine, **L**eucine, **P**henylalanine, **H**istidine and **L**ysine. Arginine is required in the diet for children, but not for adults.

PEPTIDES AND PROTEINS

Two amino acid molecules can be covalently joined through a substituted amide linkage (peptide bond) to yield a dipeptide. Peptide bond is formed by removal of the elements of water from the α-carboxyl group of one amino acid and the α-amino group of another. Three amino acids can be joined by two peptide bonds to form a tripeptide; similarly, amino acids can be linked to form tetrapeptides, pentapeptides, and so on. A few amino acids are joined to form oligopeptide. When many amino acids are joined, the structure is called polypeptide, which is otherwise called protein. Naturally occurring peptides range in length from two to many thousands of amino acid residues. Like free amino acids, peptides have characteristic titration curves and a characteristic isoelectric pH at which they do not move in an electric field. This property can be used to separate proteins and this method is known as isoelectric focusing (For more details, *see* Chapter 10).

PROTEIN STRUCTURE

The spatial arrangement of atoms in a protein is called its conformation. Any conformation of a protein can be achieved without breaking covalent bonds. Proteins in any of their functional, folded conformations are called native proteins. Each protein possesses a characteristic three-dimensional shape, its conformation. There are four separate levels of structure and organization of proteins (Figure 3.2) which are as follows: i) Primary, ii) Secondary, iii) Tertiary and iv) Quaternary.

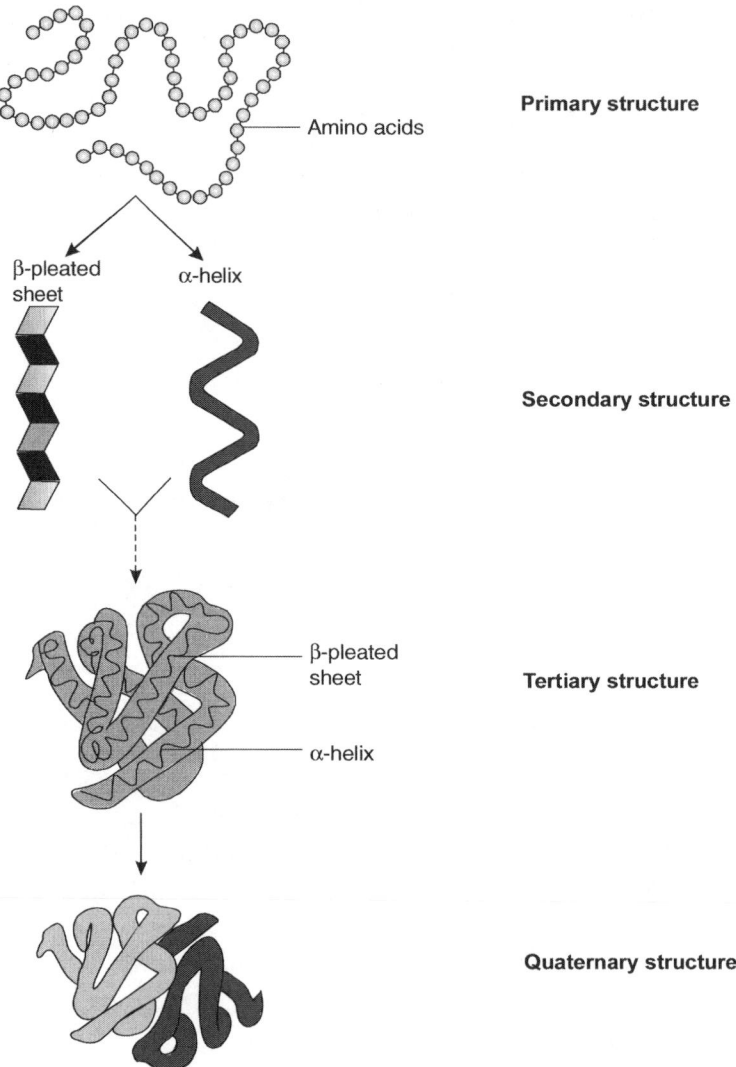

Figure 3.2   Structure of protein

Primary Structure

The primary structure of a protein defines the sequence of the amino acid residues and it is dictated by the base sequence of the corresponding gene(s). Indirectly, the primary structure also defines the amino acid composition (which of the possible 20 amino acids are actually present) and content (the relative proportions of the amino acids present).

Secondary Structure

Secondary structure defines the localized folding of some part of the polypeptide chain due to hydrogen bonding. Polypeptide chains can be folded into regular patterns such as the α-helix and β-pleated sheet. Some uncommon folding structures such as β-turns and Ω-loop are also identified. The α-helix is a coiled structure stabilized by intra-chain hydrogen bonds whereas β-sheets are stabilized by hydrogen bonding between antiparallel polypeptide strands. The polypeptide chains can change direction by making reverse turns (also known as the β-turn or hairpin bend) and loops (omega loop). Turns and loops invariably lie on the surfaces of proteins and thus often participate in interactions between proteins and other molecules. The distribution of α-helices, β-strands, and turns along a protein chain is often referred to as its secondary structure. A typical feature of α-helical structure is that it contains more number of nonpolar amino acid residues. Membrane-bound proteins possess α-helical structure and they are engaged in hydrophobic interactions with cell membrane. Some of the 20 amino acids found in proteins, including proline, isoleucine, tryptophan and asparagine, disrupt α-helical structures. Some proteins have up to 70% secondary structure but others have none.

Tertiary Structure

Tertiary structure defines the overall folding of a polypeptide chain i.e. water-soluble proteins fold into compact structures with nonpolar cores. It is stabilized by electrostatic attractions between oppositely charged ionic groups ($-NH_3^+$, COO⁻), weak van der Waals forces, hydrogen bonding, hydrophobic interactions and, in some proteins, by disulphide ($-S-S-$) bridges formed by the oxidation of spatially adjacent sulphydryl groups ($-SH$) of cysteine residues. The three-dimensional folding of polypeptide chains is such that the interior consists predominantly of non-polar, hydrophobic amino acid residues such as valine, leucine and phenylalanine. The polar, ionized, hydrophilic residues are found on the outside of the molecule, where they are compatible with the aqueous environment. However, some proteins also have hydrophobic residues on their outside and the presence of these residues is important in the processes of ammonium sulphate fractionation and hydrophobic interaction chromatography.

Quaternary Structure

Polypeptide chains can be assembled into multi-subunit structures and the spatial arrangement of subunits and the nature of their interactions are referred to as Quaternary structure. It is restricted to oligomeric proteins, which consist of the association of two or more polypeptide chains held together by electrostatic attractions, hydrogen bonding, van der Waals forces and occasionally disulphide bridges. Thus, disulphide bridges may exist within a given polypeptide chain (intra-chain) or linking different chains (inter-chain). An individual polypeptide chain in an oligomeric protein is referred to as a subunit. The subunits in a protein may be identical or different, e.g., haemoglobin consists of two α- and two β-chains, and lactate dehydrogenase consists of four (virtually) identical chains.

PROTEIN CLASSIFICATION

Because of the complexity of protein molecules and their diversity of function, it is very difficult to classify them in a single, well-defined fashion. There are alternative methods of classification, which are as follows:

    i. According to structure—globular, fibrous and intermediate proteins
   ii. According to composition—simple and conjugated
  iii. According to function—structural proteins, enzymes, antibodies, hormones, etc.

Protein Classification According to Structure

According to this classification, proteins may be globular, fibrous or intermediate proteins.

Globular proteins   These are approximately spherical in shape, are generally water-soluble and may contain a mixture of α-helix, β-pleated sheet and random structures. Globular proteins include enzymes, transport proteins and immunoglobulins.

Fibrous proteins   These are structural proteins, generally insoluble in water, consisting of long cable-like structures built entirely of either helical or sheet arrangements. Examples include hair keratin, silk fibroin, elastin, reticulin and collagen. The native state of a protein is its biologically active form.

Intermediate type proteins   These are fibrous, but soluble, e.g., fibrinogen.

Protein Classification According to Composition

This includes the simple and conjugated proteins.

Simple proteins   Only amino acids form their structure.

Conjugated proteins   Complex compounds consisting of globular proteins and tightly bound non-protein material; the non-protein material is called prosthetic group. Proteins based on prosthetic groups are listed in Table 3.2.

Table 3.2 Prosthetic group of conjugated proteins

Name	Prosthetic group	Location
Phosphoprotein	Phosphoric acid	Casein of milk, vitellin of egg yolk
Glycoprotein	Carbohydrate	Membrane structure and cell-surface receptors
Nucleoprotein	Nucleic acid	Component of viruses, chromosomes (e.g., histone proteins)
Chromoprotein	Pigment	Haemoglobin—haem (iron-containing pigment), phytochrome (plant pigment)
Lipoprotein	Lipid	Membrane structure. Lipid transported in blood as lipoprotein (e.g., Chylomicrons)
Flavoprotein	FAD (Flavine adenine dinucleotide)	Important in electron transport chain in respiration
Metalloprotein	Metals like Ca, Mo, Zn	Nitrate reductase (containing Mo), the enzyme in plants which converts nitrate to nitrite

Protein Classification According to Function

The types of proteins classified based on their function are listed in Table 3.3.

Table 3.3 Protein classification according to function

Type	Examples	Occurrence/function
Structural	Collagen	Component of connective tissue, bone, tendon, cartilage
Enzymes	Trypsin	Pancreatic enzyme; catalytic hydrolysis of proteins
Hormones	Insulin and glucagon	Help to regulate glucose metabolism
Respiratory pigment	Haemoglobin	Transports oxygen in vertebrate blood
Transport	Serum albumin	Transport of fatty acids and lipids in blood
Protective	Antibodies	Form complexes with foreign proteins
Contractile	Myosin	Moving filaments in myofibrils of muscle
Storage	Casein	Milk protein
Toxins	Diphtheria toxin	Toxin made by diphtheria bacteria

PROTEIN DENATURATION AND RENATURATION

The process of protein denaturation results in the loss of biological activity, decrease in aqueous solubility and increased susceptibility to proteolytic degradation. It can be brought about by heat and by treatment with reagents such as acids and alkalis, detergents, organic solvents and heavy-metal cations such as mercury and lead. It is associated with the loss of organized (tertiary) three-dimensional structure and exposure to the aqueous environment of numerous hydrophobic groups previously located within the folded structure.

Sometimes, a protein will spontaneously refold into its original structure after denaturation, provided the conditions are suitable. This is called renaturation.

PROTEIN PURIFICATION

The purification of a protein from a cell and tissue homogenate typically containing 10000–20000 different proteins seems a frightening task. However, in practice, on average, only four different fractionation steps are needed to purify a given protein. The reason for purifying a protein is normally to provide material for structural or functional studies and the final degree of purity required depends on the purpose for which the protein will be used. Theoretically, a protein is pure when a sample contains only a single protein species, although in practice it is more or less impossible to achieve 100% purity. Many studies on proteins can be carried out on samples that contain as much as 5–10% or more contamination with other proteins because unavoidably there will be loss of proteins in each purification step.

The degree of protein purity required depends on the purpose for which the protein is needed, e.g., a 90% pure protein is sufficient for amino acid sequence determination studies as long as the sequence is analysed quantitatively to ensure that the deduced sequence does not arise from a contaminant protein. Immunization of a rodent to provide spleen cells for monoclonal antibody production can be carried out with a sample that is considerably less than 50% pure. For studies on enzyme kinetics, a relatively impure sample can be used provided it does not contain any competing activities. On the other hand, for raising a monospecific polyclonal antibody in an animal, it is necessary to have a highly purified protein as antigen; otherwise, immunogenic contaminating proteins will give rise to additional antibodies. Equally, proteins that are used for therapeutic purpose must be extremely pure to satisfy regulatory requirements.

DETERMINATION OF PROTEIN CONCENTRATION

The need to determine protein concentration in solution is a routine requirement during protein purification. The only truly accurate method for determining protein concentration is to acid-hydrolyse a portion of the sample and then carry out amino acid analysis on the hydrolysate. However, this is relatively time-consuming, particularly if multiple samples are to be analysed. But, in practice, pure accuracy is not necessary; hence, other quicker methods that give a reasonably accurate assessment of protein concentration of a solution are used. Most of these are colorimetric methods, where a portion of the protein solution is reacted with a reagent that produces a coloured product. This coloured product is then measured spectrophotometrically and the amount of colour is related to the amount of protein present by appropriate calibration. However, none of these methods is absolute, since the development of colour is often at least partly dependent on the amino acid composition of the protein(s). The presence of prosthetic groups (e.g., carbohydrate) also influences colorimetric assays. Many workers prepare a standard calibration curve using bovine serum albumin (BSA) because of its low cost, high purity and ready availability. However, it should be understood that, since the amino acid composition of BSA will differ from the composition of the sample being tested, any concentration values deduced from the calibration graph can only be approximate.

Ultraviolet Absorption

The aromatic amino acid residues tyrosine and tryptophan in a protein exhibit an absorption maximum at a wavelength of 280 nm. However, for most proteins, the extinction coefficient lies in the range 0.4–1.5; a solution with a complex mixture of proteins with an absorbance of 1.0 at 280 nm (A_{280}), using a 1 cm path length, has a protein concentration of approximately 1 mgcm^{-3}. The method is relatively sensitive, being able to measure protein concentrations as low as 10 μgcm^{-3}. Unlike colorimetric methods, UV absorption is non-destructive, i.e., after measurement, the sample in the cuvette can be recovered and used for further study. However, the method is subject to interference by the presence of other compounds that absorb at 280 nm. Nucleic acids fall into this category having an absorbance as much as 10 times that of protein at this wavelength. Hence, the presence of only a small percentage of nucleic acid can greatly influence the absorbance at this wavelength. However, if the absorbances (A) at 280 and 260 nm wavelengths are measured, it is possible to apply a correction factor:

$$\text{Protein (mg cm}^{-3}) = 1.55\, A_{280} - 0.76\, A_{260}$$

The great advantage of this protein assay is that it can be measured continuously, e.g. in chromatographic column effluents. Even greater sensitivity can be obtained by measuring the absorbance of ultraviolet light by peptide bonds at 210 nm.

Lowry Method

Lowry method (also called Folin–Ciocalteau) has been the most commonly used one for determining protein concentration. Lowry method is reasonably sensitive, detecting up to 10 μgcm^{-3} of protein, and the sensitivity is moderately constant from one protein to another. (For more information, refer Chapter 20.)

Bicinchoninic Acid Method

This method is similar to the Lowry method in that it also depends on the conversion of Cu^{2+} to Cu^+ under alkaline conditions. Cu^+ is then detected by reaction with bicinchoninic acid (BCA) to give an intense purple colour with an absorbance maximum at 562 nm. This method is more sensitive than the Lowry method, being able to detect up to 0.5 µg protein cm^{-3}. It is generally more tolerant to the presence of compounds that interfere with the Lowry assay and hence this method is becoming increasingly popular.

Bradford Method

This method relies on the binding of the dye Coomassie brilliant blue to protein. Although it is sensitive down to 20 µg protein cm^{-3}, it is only a relative method, as the amount of dye binding appears to vary with the content of the basic amino acids arginine and lysine in the protein. This makes the choice of a standard difficult. In addition, many proteins will not dissolve properly in the acidic reaction medium. (For more information, refer Chapter 20.)

Kjeldahl Method

This is a general chemical method for determining the nitrogen content of any compound. It is frequently used for analysing complex solid samples and microbiological samples for protein content and not used for the analysis of purified proteins or for monitoring column fractions. (For more information, refer Chapter 20.)

DETECTION OF GLYCOPROTEINS

Glycoproteins have traditionally been detected on protein gels by the use of Periodic Acid–Schiff (PAS) stain. This allows mixture of glycoproteins to be distinguished. However, PAS stain is not very sensitive and often gives very weak, red-pink bands, difficult to observe on a gel. A far more sensitive method used nowadays is to blot the gel and use lectins to detect the glycoproteins. Lectins are protein molecules that bind carbohydrates, and different lectins have been found that have different specificities for different types of carbohydrates, e.g., certain lectins recognize mannose, fucose, or terminal glucosamine of the carbohydrate side-chains of glycoproteins.

The sample to be analysed is run on a number of tracks of sodium dodecyl sulphate-polyacrylamide gel electrophoresis (SDS-PAGE). Coloured bands appear at the point where the lectins bind if each blotted track is incubated with a different lectin, washed, incubated with a horseradish peroxidase-linked antibody to the lectin, and then peroxidase substrate added. In this way, by testing a protein sample against a series of lectins, it is possible not only to determine that a protein is a glycoprotein, but also to obtain information about the type of glycosylation.

DENSITOMETRY AND TRANSILLUMINATORS

Quantitative analysis, i.e., measurements of the relative amounts of different proteins in a sample can be achieved by scanning densitometry. A number of commercial scanning densitometers are available, and these work by passing the stained gel track over a beam of light (laser) and measuring the transmitted light. A graphic presentation of protein zones (peaks of absorbance) against migration distance is produced, and peak areas can be calculated to obtain quantitative data. However, such data must be interpreted with caution because there is only a limited range of protein concentrations over which there is a linear relationship between absorbance and concentration. Also, equal amounts of different proteins do not always stain equally with a given stain; so, any data comparing the relative amounts of protein can only be semiquantitative. An alternative and much cheaper way of obtaining such data is to cut out the stained bands of interest, elute the dye by shaking overnight in a known volume of 50% pyridine, and then measure spectrophotometrically the amount of dye released. Nowadays, gel documentation systems are used, which are replacing scanning densitometers. Such benchtop systems comprise a computer-linked video imaging unit attached to a small "darkroom" unit that is fitted with a white or ultraviolet light (transilluminator). Gel images can be stored on the computer, enhanced accordingly and printed as required on a thermal printer, thus eliminating the need for wet developing in a purpose-built darkroom, as is the case for traditional photography.

ELECTROELUTION

Although gel electrophoresis is used generally as an analytical tool, it can be utilized to separate proteins in a gel to achieve protein purification. Protein bands can be cut out of protein blots and sequence data obtained by placing the blot in a protein sequencer. Cutting out the stained protein bands from the protein gel and the recovery of the protein by electrophoresis is known as electroelution. A number of different designs of electroelution cells are commercially available, but the easiest method is to seal the gel piece in buffer in a dialysis sac and place the sac in buffer between two electrodes. Protein will electrophorese out of the gel piece towards the appropriate electrode but will be retained by the dialysis sac. After electroelution, the current is reversed for a few seconds to drive off any protein that has adsorbed to the wall of the dialysis sac, and then the protein solution within the sac is recovered.

WESTERN BLOTTING

Polyacrylamide gel electrophoresis (PAGE) is used to achieve fractionation of a protein mixture during the electrophoresis process. It is possible to make use of this fractionation to examine further individual separated proteins. In western blotting, the first step is to transfer or blot the pattern of separated proteins from the gel onto a sheet of nitrocellulose paper.

The method is known as protein blotting or western blotting by analogy with Southern blotting, the equivalent method used to recover DNA samples from an agarose gel. Transfer of the proteins from the gel to nitrocellulose is achieved by a technique known as electroblotting (Figure 3.3).

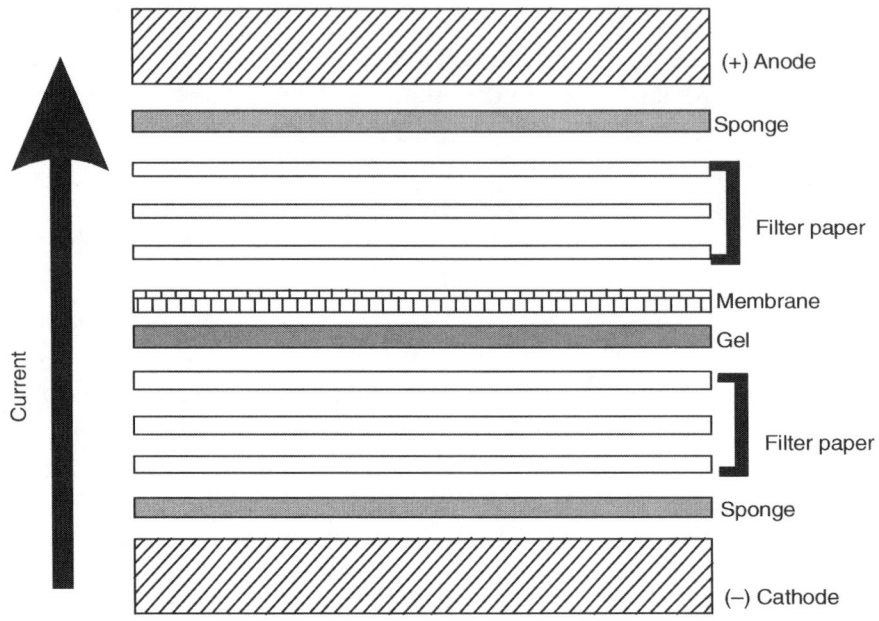

Figure 3.3 Diagrammatic representation of electroblotting

The gel to be blotted is placed on top of a filter paper saturated in buffer. The membrane is then placed on top of the gel, followed by filter papers. This sandwich is supported between two sponge pads. The whole sandwich is then placed between parallel electrodes in a buffer reservoir and electric current is passed. The sandwich must be placed in such a way that the membrane must be between the gel and anode for SDS-polyacrylamide gels, because all the proteins in SDS-PAGE carry negative charge.

In electroblotting, a sandwich of gel and nitrocellulose is compressed in a cassette and immersed in buffer, between two parallel electrodes. A current is passed at right angles to the gel, which causes the separated proteins to electrophorese out of the gel and into the nitrocellulose sheet. The nitrocellulose with its transferred protein is referred to as a blot. Once transferred onto nitrocellulose, the separated proteins can be examined further. This involves probing the blot, usually using an antibody to detect a specific protein. The blot is first incubated in a protein solution, for example 10% (w/v) bovine serum albumin, or 5% (w/v) non-fat dried milk (the so-called blotto technique), which will block all remaining hydrophobic binding sites on the nitrocellulose sheet. The blot is then incubated in a dilution of an antiserum (primary antibody) directed against the protein of interest. This IgG molecule will bind to the blot if it detects its antigen, thus identifying the protein of interest. In order to visualize this interaction, the blot is incubated further in a solution of a secondary antibody, which is directed against the IgG of the species that provided the primary antibody. For example, if the primary antibody was raised in a rabbit, then the secondary antibody would be anti-rabbit IgG. This secondary antibody is appropriately labelled so that the interaction of the secondary antibody with the primary antibody can be visualized on the blot. Anti-species IgG molecules are readily available commercially, with a choice of different labels attached.

One of the most common detection methods is to use an enzyme-linked secondary antibody. In this case, following treatment with enzyme-labelled secondary antibody, the blot is incubated in enzyme-substrate solution, when the enzyme converts the substrate into an insoluble coloured product that is precipitated onto the nitrocellulose. The presence of a coloured band therefore indicates the position of the protein of interest. By careful comparisons of the blot with a stained gel of the same sample, the protein of interest can be identified (Figure 3.4). The enzyme used in enzyme-linked antibodies is usually either alkaline phosphatase, which converts colourless 5-bromo-4-chloro-indolylphosphate (BCIP) substrate into a blue product, or horseradish peroxidase, which, with H_2O_2 as a substrate, oxidizes either 3-amino-9-ethylcarbazole into an insoluble brown product, or 4-chloro-1-naphthol into an insoluble blue product.

An alternative approach to the detection of horseradish peroxidase is to use the method of enhanced chemiluminescence (ECL). In the presence of hydrogen peroxide and the chemiluminescent substrate, luminol, horseradish peroxidase oxidizes the luminol, with concomitant production of light, the intensity of which is increased 1000-fold by the presence of a chemical enhancer. The light emission can be detected by exposing the blot to a photographic film. Corresponding ECL substrates are available for use with alkaline-phosphatase-labelled antibodies. The principle behind the use of enzyme-linked antibodies to detect antigens in blots is highly analogous to that used in enzyme-linked immunosorbent assays.

First, the primary antibody binds to the protein of interest that is present on the blot (membrane). Second, the enzyme-linked secondary antibody detects the primary antibody. Third, addition of enzyme substrate results in coloured product deposited at the site of protein of interest on the blot.

Although enzymes are commonly used as markers for secondary antibodies, other markers can also be used. These include:

[125]I-labelled secondary antibody Binding to the blot is detected by autoradiography.

Fluorescein-labelled secondary antibody This fluorescent label is detected by exposing the blot to ultraviolet light.

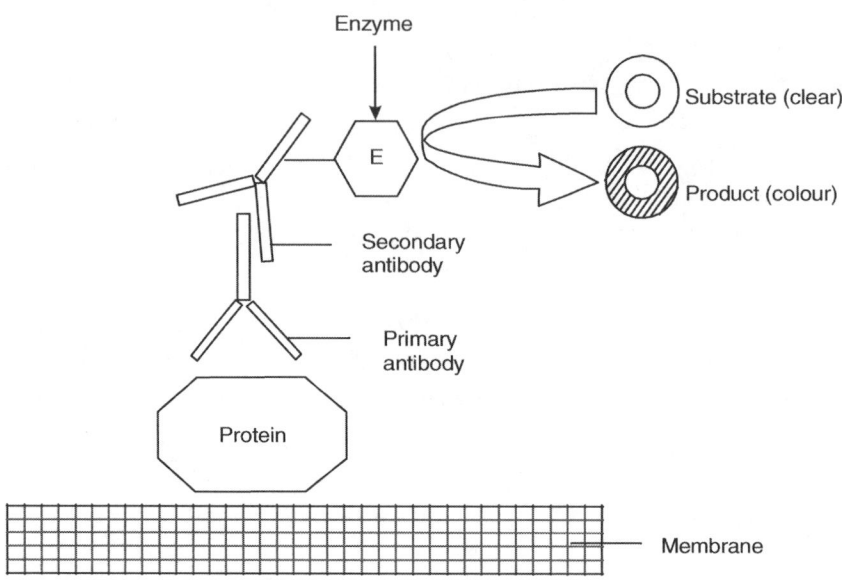

Figure 3.4 Enzyme-linked secondary antibodies in immunodetection of protein blots

[125]I-labelled protein A   Protein A is purified from *Staphylococcus aureus* and specifically binds to the Fc region of IgG molecules. ^{125}I-labelled protein A is therefore used instead of a secondary antibody, and binding to the blot is detected by autoradiography.

Gold-labelled secondary antibodies   Secondary antibodies (anti-species IgG) coated with minute gold particles are commercially available. These are directly visible as a red colour when they bind to the primary antibody on the blot.

Biotinylated secondary antibodies   Biotin is a small molecular weight vitamin that binds strongly to the egg protein avidin ($K_d = 10^{-15}$ M). The blot is incubated with biotinylated secondary antibody, then incubated further with enzyme-conjugated avidin. Since multiple biotin molecules can be linked to a single antibody molecule, many enzyme-linked avidin molecules can bind to a single biotinylated antibody molecule, thus providing an enhancement of the signal. The enzyme used is usually alkaline phosphatase or horseradish peroxidase.

Quantum dots   These are engineered semiconductor nanoparticles, with diameter of the order of 2–10 nm, which fluoresce when exposed to UV light. Quantum dot nanocrystals comprise a semiconductor core of CdSe (Cadmium selenide) surrounded by a shell of ZnS. This crystal is then coated with an organic molecular layer that provides water solubility and conjugation sites for biomolecules. Typically, therefore, secondary antibodies will be bound to a quantum dot, and the position of binding of the secondary antibody on the blot identified by exposing the blot to UV light.

In addition to the use of labelled antibodies or proteins, other probes are sometimes used. For example, radioactively labelled DNA can be used to detect DNA-binding proteins on a blot. The blot is first incubated in a solution of radiolabelled DNA, and then washed, and an autoradiograph of the blot made. The presence of radioactive bands, detected on the autoradiograph, identifies the positions of the DNA-binding protein on the blot.

BIBLIOGRAPHY

Jeremy, M. Berg, John, L. Tymoczko and Lubert Stryer. (2007). *Biochemistry*, 6th edition. WH Freeman and Company.

David, L. Nelson and Michael, M. Cox. (2008). *Lehninger Principles of Biochemistry*, 5th edition. WH Freeman and Company.

4
ENZYMES—GENERAL PERSPECTIVE

Enzymes are catalysts of biological origin and they are characterized by their extraordinary specificity and reactivity in living systems. Enzymes are found in all living organisms where they catalyse and regulate innumerable chemical reactions. Further, researches on the biochemistry of living organisms have demonstrated the importance of enzymes, and enzymology has become one of the essential sciences of the present day.

PROPERTIES OF ENZYMES

Enzymes possess the following major properties:

    i. They are biocatalysts.

    ii. All are globular proteins.

    iii. Being proteins, they are coded by DNA.

    iv. Their presence does not alter the nature of properties of the end products of the reaction.

    v. They are very efficient. In other words, a very small amount of the catalyst brings about a change in a large amount of substrate.

    vi. They are highly specific.

    vii. The catalysed reaction is reversible and some ATP-dependent reactions are irreversible.

    viii. Their activity is affected by pH, temperature, substrate concentration and time.

    ix. Enzymes lower the activation energy of the reaction they catalyse.

    x. Enzymes possess active sites where the reaction takes place. The sites have specific shapes.

CLASSIFICATION OF ENZYMES

In recent years, the rapid growth in the science of enzymology and the great increase in the number of enzymes have given rise to many difficulties in terminology. In practice, the naming of enzymes by individual workers had proven far from satisfactory. In many cases, the same enzyme became known by several different names, while there were cases in which the same name was given to different enzymes. Many of the names conveyed no idea of the nature of the reactions catalysed.

In view of this state of affairs, an Enzyme Commission (EC) was appointed by the International Union of Biochemistry, and its report published in 1964 and updated in 1972, 1978, 1984 and 1992 forms the basis of the presently accepted system. The Enzyme Commission has given a rational classification of enzymes based on reaction types and reaction mechanisms. The chemical reaction catalysed is the specific property which distinguished one enzyme from another and it is logical to use it as the basis for the classification and naming of enzymes. The enzymes are divided into groups on the basis of the type of reaction catalysed, and this, together with the name of the substrate, provides a basis for naming individual enzymes.

Difference Between Trivial and Systematic Nomenclature

It was recommended by the Enzyme Commission that there should be two nomenclatures for enzymes, one working or trivial and the other one systematic.

The trivial name will be (i) sufficiently short for general use (ii) not necessarily very exact or systematic and (iii) in many cases, the name already in current use, e.g., trypsin, pepsin and chymotrypsin.

The systematic name of an enzyme will (i) be formed in accordance with definite rules (ii) identify the enzyme precisely (iii) show the action of the enzymes as exactly as possible and (iv) include the name of the substrate. According to systematic nomenclature, any enzyme is indicated by a code number, unique for each enzyme. This number contains four elements, separated by points and arranged on these principles.

All enzymes are classified into six major categories.

1. Oxidoreductases

These enzymes catalyse the removal of H_2 (dehydrogenases) or addition of O_2 (oxygenases). The groups involved in oxidoreduction are:

$$>CHOH, \quad >CH-CH<, \quad -CHNH_2, etc.$$

e.g., L-lactate: NAD^+ oxidoreductase (E.C. 1.1.1.27) (trivial name: lactate dehydrogenase) catalyses the oxidation of lactate to pyruvate in presence of NAD^+ which is shown as follows.

$$H_3C - CH - COO^- + NAD^+ \rightleftharpoons H_3C - C - COO^- + NADH + H^+$$

$$\underset{OH}{|} \qquad\qquad \underset{O}{\|}$$

(S) Lactate   Pyruvate

It is the alcohol group of lactate which is involved in the reaction and this is indicated in the above equation.

2. Transferases

These enzymes catalyse reactions of the type:

$$AX + B \rightleftharpoons BX + A$$

and they catalyse transfer of groups which are not free during reaction from one substrate to another.

Examples of transferred groups:$- CH_3, - CH_2OH, -NH_2$

e.g., ATP: D-hexose-6-phosphotransferase (E.C. 2.7.1.1) (trivial name: hexokinase) catalyses:

$$C_5H_9O_5 \cdot CH_2OH + ATP \rightleftharpoons C_5H_9O_5 \cdot CH_2OPO_3^{2-} + ADP$$

D–Hexose   D-Hexose-6-phosphate

This enzyme will transfer phosphate to a variety of D-hexoses.

3. Hydrolases

These enzymes catalyse the splitting of compounds by addition of water across various bonds, e.g., peptides, glycoside, ester.

e.g., Orthophosphoric monoester phosphohydrolase (E.C.3.1.3.1) (trivial name: alkaline phosphatase) catalyses:

$$\underset{O}{\overset{O^-}{R - O - \underset{\|}{P} - O^-}} + H_2O \rightleftharpoons R - OH + \underset{O}{\overset{O^-}{HO - \underset{\|}{P} - O^-}}$$

Organic phosphate   Inorganic phosphate

Alkaline phosphatases are non-specific and act on a variety of substrates at alkaline pH.

Of the hydrolases, proteases are mainly used in the leather industry. Again, proteases are divided into three groups depending upon the origin, viz., plant, animal and microbial sources.

4. Lyases

These enzymes catalyse the non-hydrolytic addition or removal of groups which are free during reaction. Examples of cleaved bonds: $C-C$ in decarboxylation, $C-O$ in removal of water, $C-N$ and $C-S$.

e.g., L-histidine carboxy-lyase (E.C.4.1.1.22) (trivial name: histidine decarboxylase) catalyses:

Histidine   Histamine

5. Isomerases

These enzymes catalyse different types of isomerization, e.g., *cis-trans*, keto-enol, racemization or epimerization.

e.g., Alanine racemase (E.C.5.1.1.1) catalyses:

$$\text{L-}alanine \rightleftharpoons \text{D-}alanine$$
$$\underset{\text{racemase}}{\text{Alanine}}$$

6. Ligases

These enzymes catalyse the condensation of two molecules coupled with the breakdown of a pyrophosphate bond from a nucleotide triphosphate.

The reactions are represented as:

$$X + Y + ATP \rightleftharpoons X - Y + ADP + P_i$$

or

$$X + Y + ATP \rightleftharpoons X - Y + AMP + PP_i$$

e.g., L-glutamate: ammonia ligase (E.C.6.3.1.2) (trivial name: glutamine synthetase) catalyses the conversion of L-glutamate to L-glutamine

L-Glutamate   L-Glutamine

SPECIFICITY

One of the remarkable properties of enzymes is that they differ from chemical catalysts most strikingly in their specificity. Enzyme specificity is tested by utilizing a number of compounds in which the structure is varied systematically. Enzymes are highly specific both in the reaction catalysed and in their choice of reactants, which are called 'substrates'.

Absolute and Relative Specificity

If an enzyme acts only on one or two substrates, it is absolutely specific. A typical example of absolute specificity is the action of urease on urea to form ammonia and carbon dioxide. Another type of specificity known as relative specificity is seen among the peptidases that catalyse the hydrolysis of peptide bonds in protein. Even though all peptidases catalyse the hydrolysis of peptide bonds, their specificities are different since each can recognize and hydrolyse peptide bonds in different positions in a protein. Chymotrypsin recognizes and hydrolyses only those peptide bonds next to phenylalanine and tyrosine. Trypsin acts only on peptide bonds next to basic amino acids and its action is seen below:

Substrate Specificity

Trypsin acts only at peptide, amide or ester bonds involving arginine or lysine residues. α-benzoyl-L-argininamide or α-benzoyl-L-lysinamide serves as a model substrate. The specificity for reaction of the cationic group of the substrate with an aspartyl group of the enzyme is so exact that deamination will not occur with compounds in which the cationic group is displaced by one CH_2 group from the amide bond. Neither the shorter chain of α-benzoyl L-ornithinamide nor the longer chain of α-benzoyl L-homoargininamide provides a suitable structure for reaction with trypsin as shown in Figure 4.1.

The specificity of some proteases showing their major sites of action are given in Table 4.1.

Active Site

The active site of an enzyme is the region that binds the substrate (and the prosthetic group, if any) and contributes the residues that directly participate in the making and breaking of bonds. These residues are called catalytic groups. The active site is a three-dimensional entity. The active site of an enzyme is not a point, a line, or even a plane. It is an intricate three-dimensional form made up of groups that come from different parts of the linear amino acid sequence, e.g., in lysozyme, the important groups in the active site are contributed by residues numbered 35, 52, 62, 63, 101 in the linear sequence of 129 amino acids.

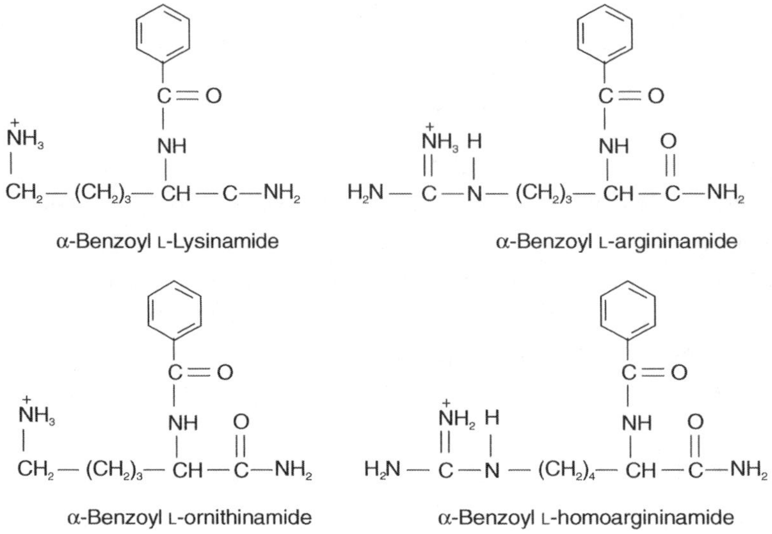

Figure 4.1   Specificity of trypsin

Table 4.1   Specificity of some proteases showing their major sites of action

Enzyme	Source	Major site of action
Trypsin	Pancreas	Arginine (Arg), Lysine (Lys)
Chymotrypsin	Pancreas	Tryptophan (Trp), Phenylalanine(Phe), Tyrosine (Tyr)
Pepsin	Gastric mucosa	Trp, Phe, Tyr, Methionine (Met), Leucine (Leu)
Papain	Papaya fruit	Arg, Lys, Glycine (Gly)

The specificity of binding depends precisely on the defined arrangement of atoms in an active site. A substrate must have a matching shape to fit into the site. Emil Fischer's "lock and key" model suggests that an enzyme like a lock, can only be fitted by a substrate (key) of precisely the correct shape (The substrate and enzyme active sites have complementary shapes). This view is modified in the induced fit model of Koshland where an enzyme changes its conformation and binds the substrate in such a way as to make the binding more complete (The enzyme active site forms a complementary shape to the substrate after binding) (Figure 4.2a and b). The substrate binds to the enzyme because the arrangement of charged groups on the enzyme exactly compliments the charged groups on the substrate, or the hydrophobic parts of the substrate exactly fit hydrophobic grooves or pockets in the enzyme.

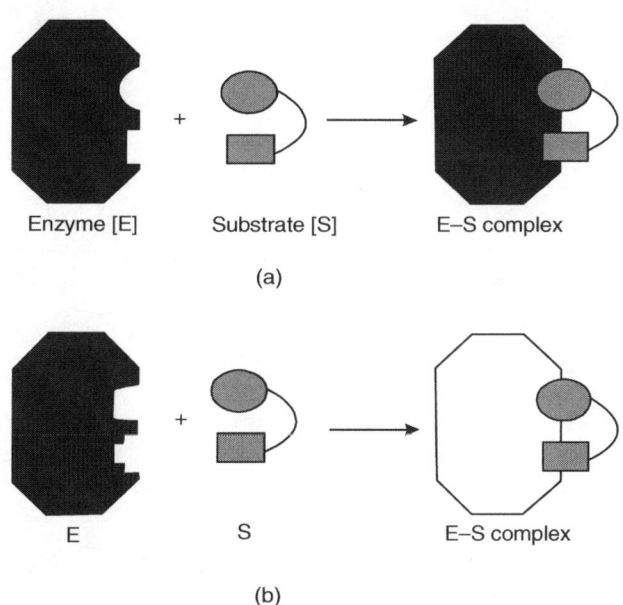

Figure 4.2 Schematic view of (a) Lock and key model and (b) Induced fit model

Chymotrypsin is irreversibly inactivated by reaction with certain phosphorus derivatives, e.g., Diisopropyl fluorophosphate (DFP). A derivative is formed with the specific serine residue (No.195) of the enzyme which is essential for the activity. There are about 28 serine residues but only one combines with DFP. This shows that only a small well-defined part of chymotrypsin is responsible for its activity and the active serine (No.195) forms a part of the 'active site' of the enzyme. Considerable evidence also implicates a histidine residue (No.57) in the catalytic activity of chymotrypsin. Treatment of chymotrypsin with L-1(*p*-toluenesulphonyl) amido-2-phenylethyl chloromethyl ketone (TPCK) results in loss of activity and formation of the imidazole derivative substituted at the third position of histidine residue (No.57).

From these examples, it is seen that serine (No.195) and histidine (No.57) form part of the 'active site' of chymotrypsin. The reagents commonly used to detect specific groups and the groups attacked are shown in Table 4.2.

Table 4.2   Compounds commonly used to detect specific groups on the active site

Compound formula	Compound	Main reactive groups
	Acetic anhydride	Lys (—NH$_2$), Ser(—OH), Tyr (—OH)
I—CH$_2$—COOH	Iodoacetic acid	Cys (—SH), His (—NH)
ClHg—⟨benzene⟩—COOH	p-chloromercury benzoate	Cys (—SH)
O$_2$N—⟨benzene⟩—F, NO$_2$	2,4 - dinitrofluorobenzene	Lys (—NH$_2$), Tyr (—OH), His(—NH)
F—P(OCH(CH$_3$)$_2$)$_2$, O	Diisopropylfluorophosphate (DFP)	Ser(—OH) on active centre of chymotrypsin
	L-1(p-toluenesulphonyl) amido-2-phenylethyl chloromethylketone (TPCK)	His (—NH) on active centre of chymotrypsin

Isoenzymes

Isoenzymes are multiple forms of an enzyme with the same substrate specificity. The multiple forms are due to the genetic differences in their primary structures. They catalyse the same chemical reaction but differ in some of their physico-chemical properties. They can be separated using the techniques of electrophoresis and chromatography. The physico-chemical properties of these enzymes often affect their catalytic activity and they also differ in such properties as their K_m, sensitivity to heat and effect of inhibitors. Lactate dehydrogenase (LDH) is a typical example of an isoenzyme which catalyses the reaction:

$$H_3C—CH—COO^- + NAD^+ \rightleftharpoons H_3C—C—COO^- + NADH + H^+$$
$$\quad\quad | \quad\quad\quad\quad\quad\quad\quad\quad\quad\quad ||$$
$$\quad\quad OH \quad\quad\quad\quad\quad\quad\quad\quad\quad\quad O$$

(S) Lactate   Pyruvate

The enzyme, as found in many species, is a tetramer of molecular weight about 1, 40,000 daltons. Although each subunit has a molecular weight of about 35,000 daltons, two types of different amino acid composition are found within each species. They are the M-form, which predominates in skeletal muscle and the H-form, which is a predominant subunit in the heart. The two types of subunits are produced by separate genes. Each monomer is catalytically inactive but it can combine with others of the same or different type to produce the active tetrameric enzymes. Five isoenzymes of LDH can exist, viz., H_4 (LDH_1), H_3M (LDH_2), H_2M_2 (LDH_3), HM_3 (LDH_4), and M_4 (LDH_5). Although they catalyse the same reaction, they do so with different characteristics. The properties of H_3M, H_2M_2 and HM_3 are intermediate between those of H_4 and M_4.

COENZYMES AND ACTIVATORS

In many cases, if an enzyme is mixed with its substrate under the appropriate conditions, either no catalysis occurs or there is only a slight activity. This is due to the absence of a coenzyme or activator.

Coenzymes

These are low molecular weight organic compounds which are actively involved in the catalysis. They often act as acceptors or donors of specific chemical groups. Nicotinamide adenine dinucleotide (NAD), for example, accepts and donates hydrogen atoms and is a coenzyme for many dehydrogenases. Coenzyme is the name given to soluble cofactors while the term "prosthetic group" is reserved for coenzymes that are firmly attached to the protein. Some examples of coenzymes and their involvement in enzymatic reactions are shown in Table 4.3.

Table 4.3 Coenzymes and the enzymatic reactions in which they are involved

Coenzyme	Enzyme
Nicotinamide adenine dinucleotide (NAD^+) and nicotinamide adenine dinucleotide phosphate ($NADP^+$)	Alcohol dehydrogenase
Flavin adenine dinucleotide (FAD)	Glucose oxidase
Coenzyme A (CoA. SH)	Pyruvate dehydrogenase
Pyridoxal phosphate	Aspartate transaminase
Biotin	Acetyl-CoA carboxylase
Tetrahydrofolate	Formate tetrahydrofolate ligase
Coenzyme B_{12}	Methyl malonyl-CoA mutase
Adenosine triphosphate	Pyruvate kinase

Activators

Non-specific substances which are necessary for the activity of an enzyme, or which activate a precursor of an enzyme are often called activators. In most of the enzymatic reactions, metallic ions perform this role. In metal-activated enzymatic reactions, purified enzymes may have to be activated by the addition of metal ions. It has already been shown that ternary complexes are formed between an enzyme, metal ion and substrate. The involvement of metal ions in enzymes may be investigated by nuclear magnetic resonance (NMR), electron spin resonance (ESR) and proton relaxation rate (PRR) enhancement techniques. Metalloenzymes are enzymes in which the metal is tightly bound to the enzyme structure and is retained by the enzyme on purification. There is no clear-cut distinction between metalloenzymes and metal-activated enzymes.

Examples of a few metalloenzymes or metal-activated enzymes with their activators are shown in Table 4.4.

Table 4.4 Metalloenzymes/metal-activated enzymes and their activators

Enzyme	Activator
α-amylase and trypsin	Calcium
Creatine kinase and pyruvate kinase	Magnesium
Nitric oxide reductase	Molybdenum and iron
Carboxypeptidase A and carbonic anhydrase	Zinc
Arginase and yeast phosphatase	Manganese, cobalt, nickel, iron
Tyrosinase and ascorbic acid oxidase	Copper

SYNZYMES

A relatively new approach involves synthesizing molecules mimicking the action of natural enzymes due to the incorporation of some features in their active sites. Such artificial enzymes are called as synzymes. Synzymes follow the Michaelis–Menten kinetics just as natural enzymes.

Some synzymes are derivatized proteins. An example is myoglobin which functions as oxygen carrier in muscle. When the group $[Ru(NH_3)_5]^{3+}$ is attached to three surface histidine residues of myoglobin, it starts functioning like an oxidase. It oxidizes ascorbic acid and reduces molecular oxygen. This derivatized myoglobin is almost as effective as the natural ascorbate oxidase.

A different example for the synzyme is that the starting material need not necessarily be a protein. Synzymes with chymotrypsin-like characteristics have been obtained based on cyclodextrins, consisting of 6–10 D-glucose units linked head-to-tail in a ring. Glycosidic oxygen atoms and —CH linkages point inwards, creating a hydrophobic environment which can act as a binding pocket. Catalytic activity is provided by the attachment of imidazole, hydroxyl and carboxyl groups, as seen in the active sites of serine proteases. The resulting synzyme resembles chymotrypsin in its esterase activity and it is stable. Cyclodextrins may also be linked to pyridoxyl coenzymes, producing synzymes with transaminase activity.

ABZYMES

Abzymes are antibodies capable of catalysing a chemical reaction. One of the major factors governing specificity is the stability of the enzyme-bound transition state which exists during the conversion of enzyme-bound substrate to products. A potential substrate which can form a relatively stable transition-state when bound to the enzyme will be converted to products at an appreciable rate. Transition-state analogues are stable compounds which resemble the transition-state compounds thought to be formed as part of a reaction sequence. When an analogue of the supposed transition-state of a particular reaction is injected into a mouse, the immune system will make antibodies against it and some of these may be able to catalyse the reaction in question. These abzymes may be proliferated by the techniques of monoclonal antibody production or by recombinant DNA technology. The first antibody to be commercialized is the abzyme with an aldolase activity. Aldolase abzyme could be used for the synthesis of epothilone A, an anticancer compound.

EXTREMOZYMES

It is now known that enzymes can function at temperatures as high as 140°C and as low as below freezing point. Organisms living in these conditions are described as extremophiles and the enzymes that function under these conditions are therefore described as extremozymes. Most of these extremophiles belong to archaea group of bacteria.

Extremozymes are broadly classified as follows:

i. Extremely psychrophilic enzymes, which function at temperatures approaching the freezing point of water; for instance, subtilisin S41 from the psychrophile Bacillus S41 differs from other subtilisins in several respects, suggesting structural differences between these psychrophilic enzymes and those extracted from mesophiles.

ii. Extremely halophilic enzymes function better in salt solutions. For instance, aspartate aminotransferase from *Haloferas mediterranei* is inactivated rapidly at room temperature in low salt solutions, but does not denature even at 78.5°C in 3.3 M KCl suggesting that the enzyme is more stable in higher salt concentration. A protease from *Halobacterium halobium* is also similarly affected.

iii. Extremely thermophilic enzymes function at high temperatures and several examples of these enzymes are available now. DNA polymerase from *Thermus aquaticus* and *Pyrococcus furiosus* used for PCR is one such example.

iv. Extremely barophilic enzymes have been isolated from microbes living in deep sea which are both barophils and psychrophiles.

FACTORS AFFECTING ENZYMATIC ACTIVITY

Substrate Concentration

The rate of enzyme-catalysed reactions increases with increasing substrate concentration but above a certain substrate concentration, the rate of enzyme action ceases to increase (Figure 4.3). The shape of the curve is commonly explained on the basis of catalytically active sites on the enzyme, which react with the substrate. Maximum velocity V_{max}, of the reaction is reached when all the sites are occupied by the substrate molecules.

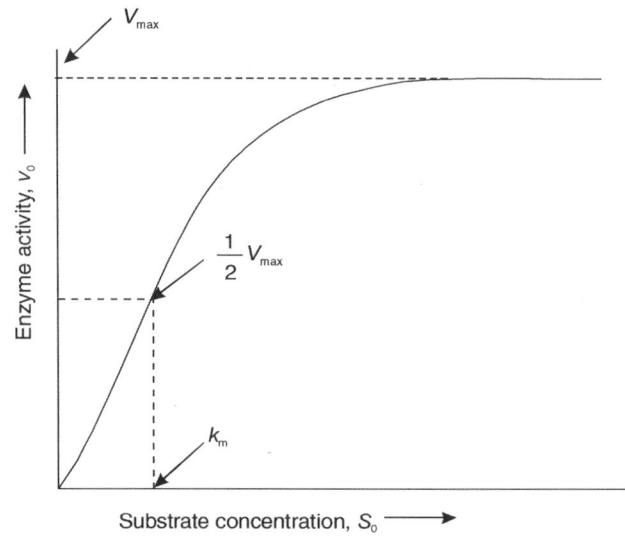

Figure 4.3   Effect of substrate concentration on enzyme activity

Michaelis–Menten equation   L.Michaelis and M.L.Menten in 1913, postulated the existence of an enzyme–substrate complex as the basis for a theoretical analysis of enzymatic reactions. Let us consider a single substrate reaction where there is just one substrate-binding site per enzyme.

$$\underset{E}{\text{Enzyme}} + \underset{S}{\text{Substrate}} \underset{k_{-1}}{\overset{k_1}{\rightleftharpoons}} \underset{ES}{\text{Enzyme – Substrate complex}} \underset{k_2}{\rightarrow} \text{Enzyme} + \text{Product}$$

The Michaelis–Menten assumption was that the equilibrium between enzyme, substrate and enzyme–substrate complex was instantly set up and maintained, and the breakdown of the enzyme–substrate complex to products was negligible to disturb this equilibrium. Briggs and Haldane extended this idea and derived an equation assuming steady-state conditions, i.e., the rate of breakdown of the complex is the same as the rate of formation during the period of the measurement. Using this assumption, the equation is written as

$$k_1[E][S] = k_{-1}[ES] + k_2[ES] = [ES](k_{-1} + k_2)$$

Separating the constants from the variables,

$$\frac{[E][S]}{[ES]} = \frac{[k_{-1} + k_2]}{k_1} = K_m$$

where K_m is another constant.

The total concentration of enzyme present $[E_0]$ must be the sum of the concentration of free enzyme $[E]$ and the concentration of bound enzyme $[ES]$.

Substituting $[E] = [E_0] - [ES]$ in the above equation,

$$\frac{([E_0] - [ES][S])}{[ES]} = K_m$$

$$K_m[ES] = ([E_0] - [ES])[S]$$

$$= [E_0][S] - [ES][S]$$

$$[ES][S] + K_m[ES] = [E_0][S]$$

$$[ES]([S] + K_m) = [E_0][S]$$

$$[ES] = \frac{[E_0][S]}{[S] + K_m}$$

The term $[ES]$ governs the rate of formation of products (the overall rate of reaction) according to the relationship:

$$v_0 = k_2[ES]$$

If this is substituted in the above equation,

$$\frac{v_0}{k_2} = \frac{[E_0][S]}{[S] + K_m}$$

$$v_0 = \frac{k_2[E_0][S]}{[S] + K_m}$$

Moreover, when the substrate concentration is very high, all the enzyme is present as the enzyme–substrate complex and the maximum velocity, V_{max}, is reached. Under these conditions,

$$V_{max} = k_2[E_0]$$

Hence,

$$v_0 = \frac{V_{max}[S]}{[S] + K_m}$$

Finally, since the substrate concentration, $[S_0]$, is usually greater than the enzyme concentration,

$$v_0 = \frac{V_{max}[S_0]}{[S_0] + K_m} \text{ at constant} [E_0]$$

This equation has retained the name Michaelis–Menten equation and K_m is called Michaelis constant.

A graph of v_0 against $[S_0]$ will have the form of rectangular hyperbola consistent with experimental findings for many enzyme-catalysed reactions. V_{max}, the maximum initial velocity at a particular $[E_0]$, can be obtained from the graph as shown in Figure 4.3. K_m can also be obtained from the graph.

When, $v_0 = \frac{1}{2} V_{max}$ and when this is substituted in Michaelis–Menten equation,

$$\frac{V_{max}}{2} = \frac{V_{max}[S_0]}{[S_0] + K_m}$$
$$V_{max}([S_0] + K_m) = 2(V_{max})[S_0]$$
$$K_m = [S_0]$$

Therefore, K_m is the substrate concentration at which an enzyme reaches half the maximal velocity.

Significance of Michaelis–Menten equation The Michaelis–Menten equation, as modified by Briggs and Haldane, is applicable to many enzyme-catalysed reactions and the constants V_{max} and K_m can be determined using this equation. V_{max} varies with the total concentration of the enzyme present, but K_m is independent of enzyme concentration and is characteristic of the system being investigated. It can, thus, be used to identify a particular enzyme. K_m value of an enzyme varies widely. This means that the substrate concentration needed to saturate the enzyme is different for different enzymes. An enzyme with a large K_m value is one that is saturated at low substrate concentration. For the same enzyme, K_m value changes when different substrates are used. In this context, a large K_m value indicates that the affinity of an enzyme for a particular substrate is low and a small K_m value indicates an inverse relationship. K_m values of some enzymes are given in Table 4.5 and for most enzymes, K_m value lies between 10^{-1} and 10^{-6} M.

Table 4.5 K_m values of some typical enzymes

Enzyme	Substrate	K_m (M)
Chymotrypsin	Acetyl-L-tryptophanamide	5×10^{-3}
Lysozyme	Hexa -N-acetylglucosamine	6×10^{-6}
β-galactosidase	Lactose	4×10^{-3}
Trypsin	Benzoyl-L-arginine ethyl ester	5×10^{-5}
Pepsin	Acetyl-L-phenylalanyl diiodotyrosine	7.5×10^{-5}
Papain	Benzoyl-L-arginine ethyl ester	1.89×10^{-3}
β-Amylase	Amylose	7.3×10^{-5}

Lineweaver–Burk plot   The graph of the Michaelis–Menten equation showing enzyme activity against substrate concentration is not entirely satisfactory for the determination of V_{max} and K_m (Figure 4.3). Unless there are at least three consistent points on the plateau of the curve at different [S] values, an accurate value of V_{max} and hence of K_m cannot be obtained. The graph, being a curve, cannot be accurately extrapolated upwards from non-saturating values of [S].

Lineweaver and Burk overcame this problem without making further assumptions. The Michaelis–Menten equation was simply inverted as shown below:

$$v_0 = \frac{V_{max}[S_0]}{[S_0] + K_m}$$

$$\frac{1}{v_0} = \frac{[S_0] + K_m}{V_{max}[S_0]}$$

$$= \frac{[S_0]}{V_{max}[S_0]} + \frac{K_m}{V_{max}[S_0]}$$

$$\therefore \frac{1}{v_0} = \frac{K_m}{V_{max}} \frac{1}{S_0} + \frac{1}{V_{max}} \text{(Lineweaver – Burk equation)}$$

This is of the form $y = mx + c$, which is the equation for straight line graphs. A plot of y against x has a slope m and intercepts c on the Y-axis. A plot of $1/v_0$ against $1/[S_0]$ (Lineweaver–Burk plot) for systems obeying the Michaelis–Menten equation is shown in Figure 4.4. The graph being linear can be extrapolated even if no experiment has been performed at a saturating substrate concentration and from the extrapolated graph, the values of K_m and V_{max} can be determined.

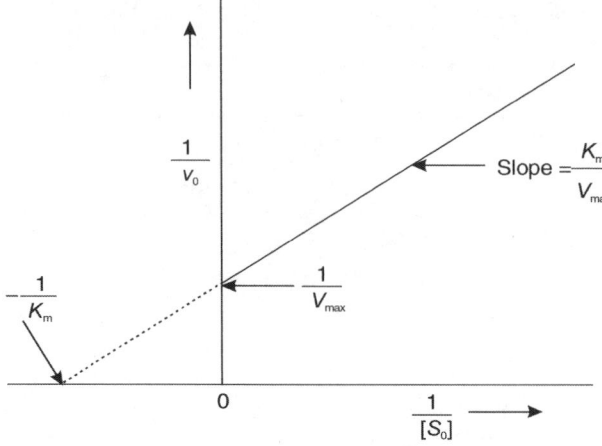

Figure 4.4 Lineweaver–Burk plot

Enzyme Concentration

The relation between enzyme concentration and its activity is demonstrated by maintaining substrate concentration, pH and temperature constant while the concentration of the enzyme solution is varied.

Three different conditions are expressed in Figure 4.5. 'A' indicates normal response showing that the activity varies linearly with enzyme concentration. 'B' shows that some activator is present in the enzyme preparation and hence the reaction does not proceed in a linear way. 'C' shows a condition of substrate exhaustion where there is no increase in reaction velocity once the substrate is depleted.

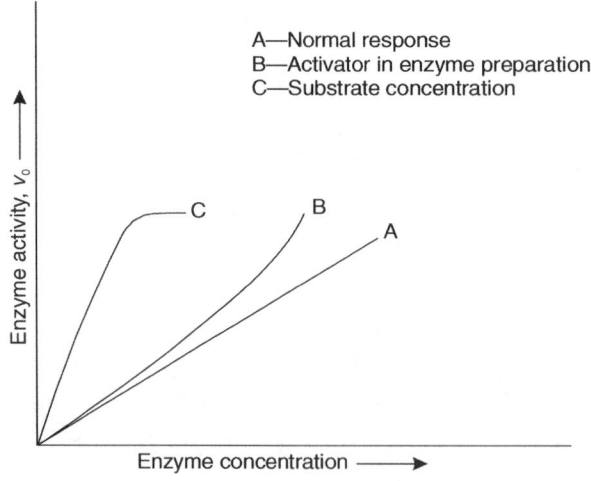

Figure 4.5 Effect of enzyme concentration on activity

Temperature

Enzyme-catalysed reactions are similar to other reactions in that the rate is increased by increasing temperatures—up to a point as shown in Figure 4.6. As the temperature increases beyond 45–50°C, the rate decreases. This decrease is caused by thermal denaturation of the enzyme protein or the inactivation of a thermolabile component in the enzyme system. When thermal energy becomes great enough to cause the rupture of a few bonds, the neighbouring bonds are weakened and the whole molecule becomes denatured. The optimum temperature of enzymes under physiological conditions is close to 40°C. The maximum velocity of enzyme reaction is obtained around the optimum temperature.

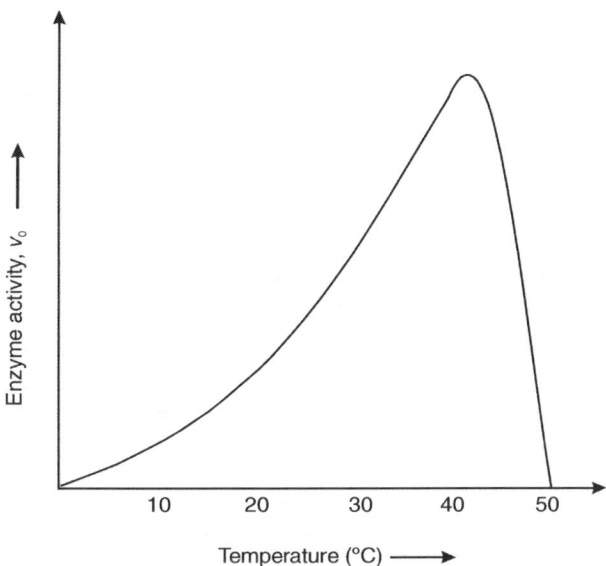

Figure 4.6　Effect of temperature on enzyme activity

pH

When the enzymatic activity is plotted against pH, a bell-shaped curve is obtained. This indicates a marked dependence of an enzyme on the pH of the reaction mixture with enzyme activity decreasing rapidly on either side of the optimum pH (Figure 4.7). The pH optimum of an enzyme is dependent upon a number of experimental parameters including temperature, nature of substrate, concentration of buffer, ionic strength of medium and purity of enzyme preparation.

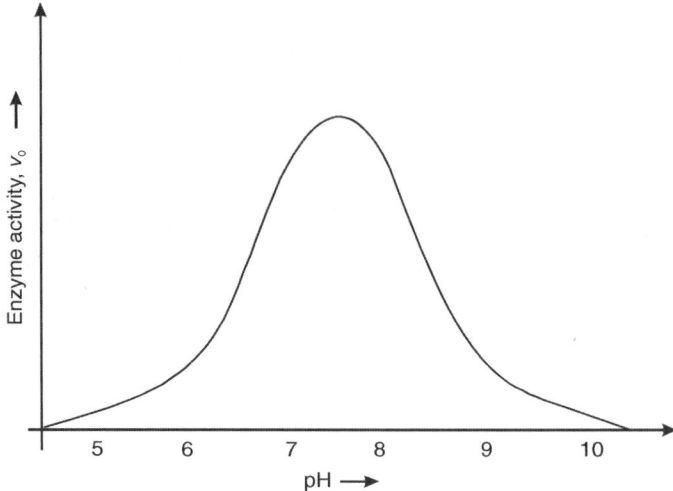

Figure 4.7 Effect of pH on enzyme activity

INHIBITORS

An enzyme inhibitor is a chemical that combines with an enzyme and reduces or eliminates its catalytic activity. The decrease in catalytic activity means that the inhibitor has affected the active site of the enzyme directly, by damaging or physically clogging up the active site. Indirectly, it may change the three-dimensional shape of the entire protein. Some inhibitors act against only one enzyme, while others inactivate a wide spectrum of enzymes. There are two types of inhibitors: (i) Irreversible (ii) Reversible. Irreversible inhibitors cannot be removed from an enzyme by dialysis. Reversible inhibitors bind to an enzyme in a reversible fashion and can be removed by dialysis. Simple dilution can restore full enzymic activity. Reversibility refers to whether or not the enzyme–inhibitor complex can dissociate to regenerate the enzyme.

Irreversible Inhibition

An irreversible inhibitor binds to the active site of the enzyme by an irreversible reaction:

$$E + I \rightarrow EI$$

and hence cannot subsequently dissociate from it. A covalent bond is usually formed between inhibitor and enzyme. The inhibitor may act by preventing the enzyme–substrate binding or it may destroy some component of the catalytic site. Irreversible inhibitors effectively reduce the activity of the enzyme present.

Many irreversible inhibitors attack—SH groups (in cysteine side chains) which are often found at the active sites of enzymes. Important examples are alkylating agents such as iodoacetic acid and iodoacetamide which form covalent linkages with the essential—SH groups.

$$E-SH + I - CH_2COOH \rightarrow E-S-CH_2COOH + HI$$

Enzyme   Iodoacetic acid

Another well-known group is the organophosphorous compounds which react with essential—OH groups (in serine side chains) of some enzymes. An example is diisopropylfluorophosphate (DFP). It is a nerve poison and inactivates acetylcholinesterase, an enzyme essential in nerve function.

Irreversible inhibitors are useful in the investigation of the active site of an enzyme. The inhibitor will remain firmly bound to one of the amino acids in the active site of the enzyme and thus act as a marker for easy identification.

Reversible Inhibition

Some of the major types of reversible inhibition are discussed below:

(a) Competitive inhibition (b) Non-competitive inhibition (c) Uncompetitive inhibition (d) Substrate inhibition and (e) Allosteric inhibition.

Competitive Inhibition

When a compound competes with a substrate for the active site on the enzyme molecule and thereby reduces the catalytic activity of the enzyme, it is known as competitive inhibitor. These compounds are usually structural analogues of the substrates.

Let us consider a typical enzymatic reaction.

$$E + S \overset{1}{\rightleftharpoons} ES \overset{2}{\rightarrow} E + P$$

Two conditions exist here. If the inhibitor interferes in the first step, it combines with the active centre of the enzyme and brings about a different orientation in the enzyme. When the inhibitor interferes in the second step, it combines with [ES] complex, retarding its dissociation into product and enzyme. This is possible by changing the conformation of the enzyme in the [ES] complex.

An important factor in competitive inhibition is the concentration of substrate and inhibitor with respect to each other. If $S : I$ ratio known as affinity ratio is $50 : 1$, then I is able to alter the reaction velocity. If this ratio is > 50, the substrate nullifies the effect of I. Since a higher concentration of substrate will resume the enzymatic activity, competitive inhibition is reversible. At very high substrate concentration, molecules of substrate will outnumber molecules of inhibitor and the effect of the inhibitor will be negligible. Hence, V_{max} for the reaction is unchanged. As the active sites are directly involved, the K_m of the enzyme is altered, i.e., increased by the inhibitor. The Lineweaver–Burk plot showing the effect of competitive inhibition is depicted in Figure 4.8.

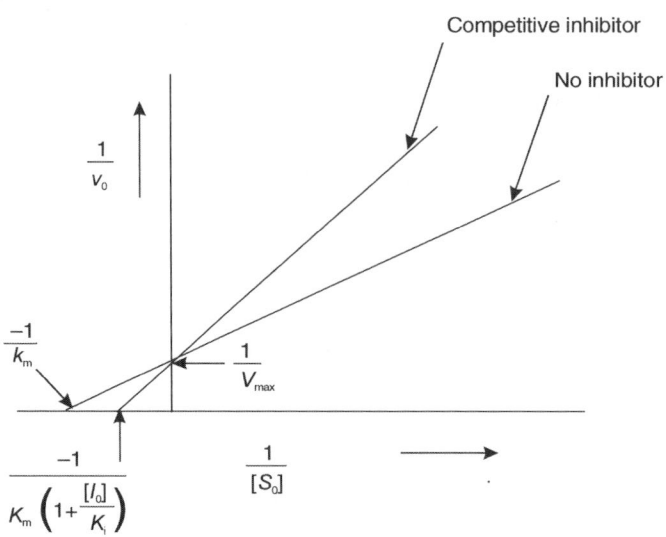

Figure 4.8 Lineweaver–Burk plot showing the effect of competitive inhibition

A typical example of competitive inhibition is cited below. Succinate dehydrogenase (SDH) readily oxidizes succinate to fumarate in the presence of a suitable hydrogen acceptor (A). When inhibitor (I) is present, the rate of reaction is dependent upon the concentration of enzyme, substrate and inhibitor. The structural similarity between the substrate and inhibitor helps the binding of I with E to effect inhibition. In the above reaction, malonate serves as the strong inhibitor

Both malonate and succinate can combine with the active site of the enzyme but no product is formed in the presence of malonate. So, malonate inhibits SDH. Another example of competitive inhibition is the inhibition of folic acid synthesis by sulphanilamide in many microorganisms.

Sulphanilamide, a structural analogue of *p*-aminobenzoic acid, will block folic acid synthesis and the resulting deficiency of this essential vitamin is fatal to the organism.

Non-competitive Inhibition

As the name implies, no competition occurs between substrate and inhibitor. In this type, the inhibitor combines rather strongly with a site other than the active site on the enzyme surface (Figure 4.9). Non-competitive inhibitor lowers the maximum velocity (V_{max}) attainable with a given amount of enzyme but does not affect the K_m value. Since *I* and *S* may combine at different sites, formation of both *EI* and *EIS* is possible. As *EIS* may break to form the products at a slower rate than *ES*, the reaction is slowed down but not stopped. The reactions are shown below:

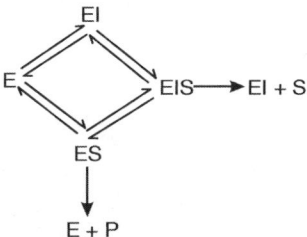

When $1/v_0$ is plotted against $1/[S_0]$ at various concentrations of inhibitor, the results, as shown in Figure 4.9, are obtained. There has been no significant alteration of the conformation of active site when *I* is bound.

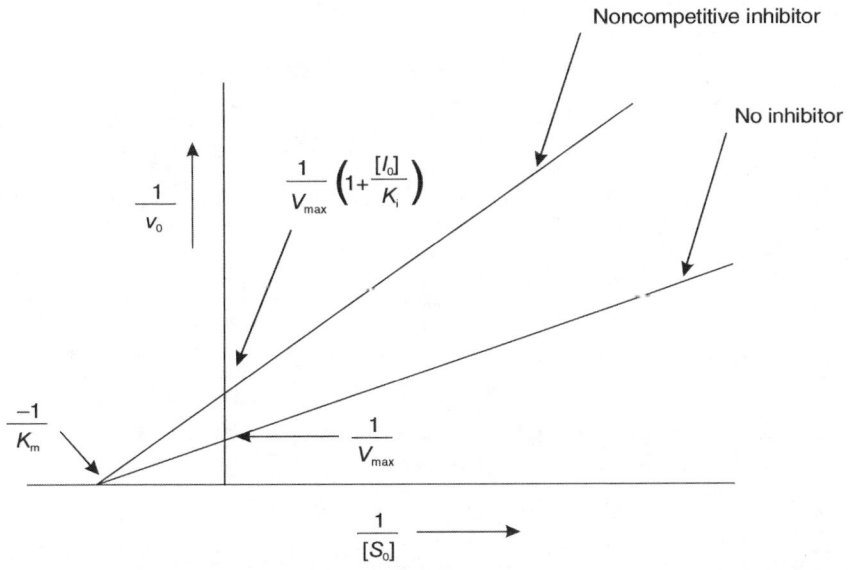

Figure 4.9 Lineweaver–Burk plot showing the effect of simple linear non-competitive inhibition

Uncompetitive Inhibition

Uncompetitive inhibitors bind only to the enzyme–substrate complex and not to the free enzyme. Substrate-binding could bring about a change in the conformation of the enzyme and reveal an inhibitor-binding site. The inhibitor does not compete with the substrate for the same binding site and hence, the inhibition cannot be overcome by increasing the substrate concentration. Uncompetitive inhibition may be represented as below:

$$E + S \rightleftharpoons ES \longrightarrow E + P$$

$$-I \updownarrow +I$$

$$ESI$$

As shown in Figure 4.10, both K_m and V_{max} are altered.

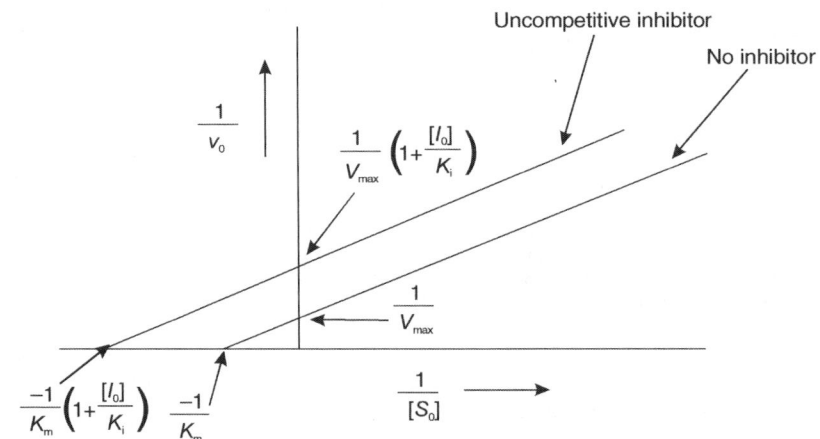

Figure 4.10 Lineweaver–Burk plot showing the effect of uncompetitive inhibition

The diagrammatic representation of competitive, non-competitive and uncompetitive inhibition is shown in Figure 4.11.

Substrate Inhibition

A characteristic of enzyme-catalysed reactions is that for a given enzyme concentration, the initial reaction velocity increases with increasing initial substrate concentration to a limiting value, V_{max}. At still higher substrate concentrations, the initial velocity is sometimes found to be less than the maximum value. It could be that the substrate, at very high concentrations, can inhibit its own conversion to product.

One possible mechanism for substrate inhibition at high substrate concentration could be explained with relevance to succinate dehydrogenase. For a reaction to take place, both carboxyl groups of the substrate have to bind to the enzyme.

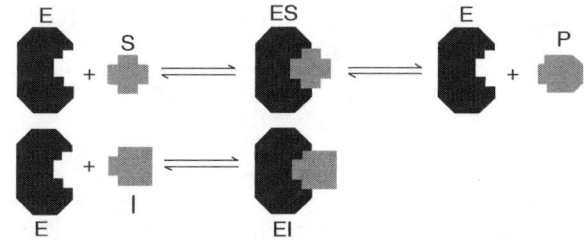

(a) Competitive inhibition, I binding to same site as S

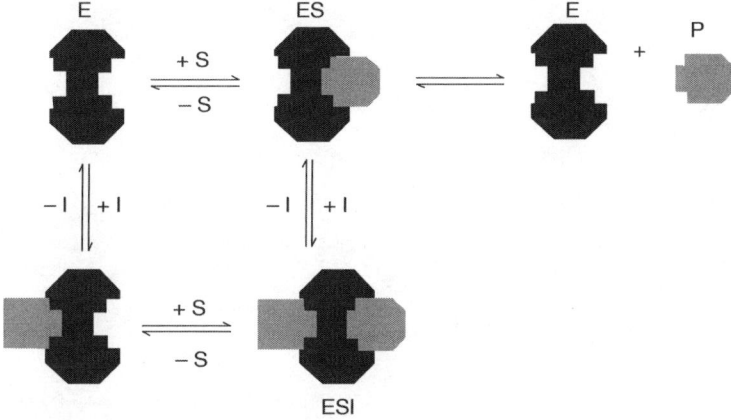

(b) Simple linear non-competitive inhibition

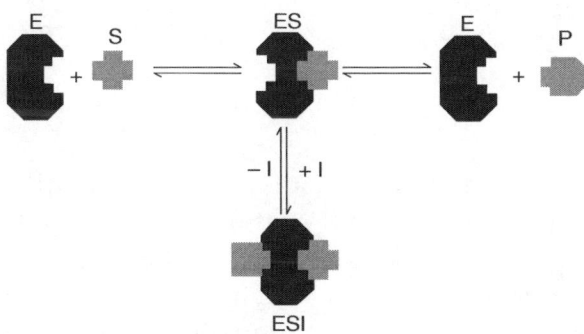

(c) Uncompetitive inhibition

Figure 4.11 Diagrammatic representation of competitive, non-competitive and uncompetitive inhibition (E—Enzyme, I—Inhibitor, S—Substrate and P—Product)

At very high substrate concentrations, there is an increased possibility of carboxyl groups from two separate substrate molecules binding to the same enzyme. In this particular condition, a reaction cannot take place until one of them dissociates away. The mechanism of substrate inhibition could be explained as shown in Figure 4.12.

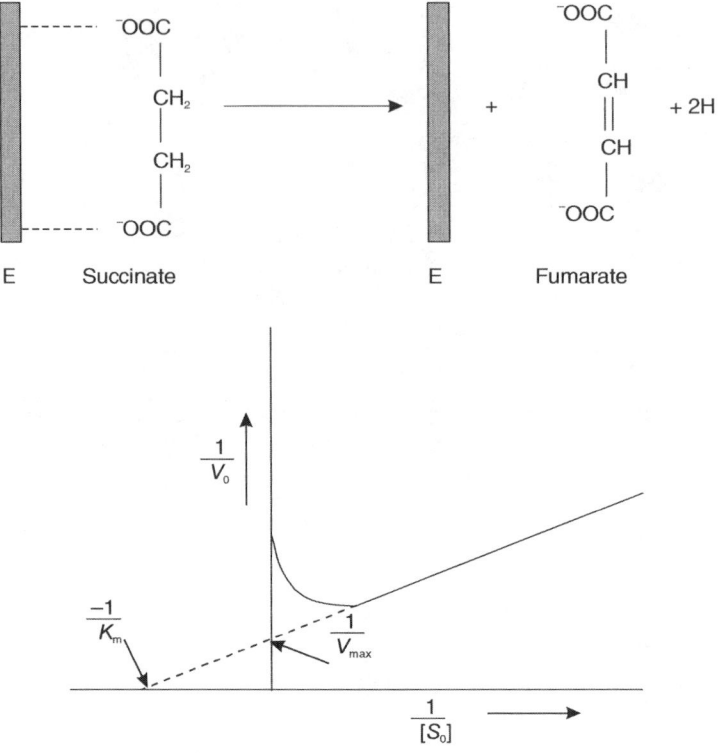

Figure 4.12 Lineweaver–Burk plot showing the effect of substrate inhibition

Substrate inhibition occurs when a molecule of substrate binds to one site on the enzyme and then another molecule of substrate binds to a separate site on the enzyme to form a dead-end complex.

Allosteric Inhibition

An allosteric inhibitor binds to the enzyme at a site other than the substrate-binding site. The term "allosteric inhibition" is used when the inhibitor, instead of forming a complex with the enzyme, influences conformational changes which may alter the binding characteristics of the enzyme for the substrate or the subsequent reaction characteristics or both. The Michaelis–Menten plot becomes more sigmoidal (S-shaped) which means that the rate of reaction is reduced at low substrate concentration (Figure 4.13).

If the binding characteristics alone are affected, V_{max} will usually remain unchanged and the inhibition pattern could be regarded as competitive. Similarly, other forms of allosteric inhibition, where V_{max} is altered, could be regarded as giving non-competitive or mixed inhibition depending on whether K_m is changed or not. However in most cases, the Michaelis–Menten equation is not obeyed in the presence of allosteric inhibitors. A typical case of allosteric inhibition is the inhibition of aspartate transcarbamoylase by cytidine triphosphate (CTP).

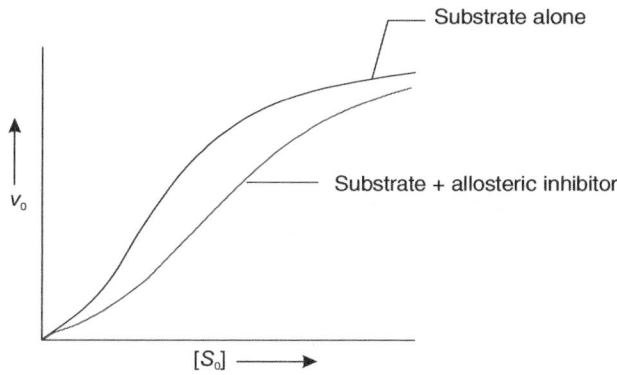

Figure 4.13 Effect of allosteric inhibitor on the binding of a substrate

Applications of Enzyme Inhibitors

Mechanism of enzyme inhibition is important in that (i) it provides an understanding about the possible ways by which metabolic acitivity could be controlled *in vivo,* (ii) it gives an insight into the mechanism by which enzymes promote their catalytic activity and (iii) it allows specific inhibitors to be synthesized and used as therapeutic agents to block key metabolic pathways underlying clinical conditions. Some of the examples of enzyme inhibitors acting as therapeutic agents are shown in Table 4.6.

Table 4.6 Enzyme inhibitors used as therapeutic agents

Inhibitor	Enzyme	Application
Aspirin	Cyclooxygenase (COX-1 and COX-2)	Inhibition of prostaglandin and thromboxane synthesis
Mevinolin	Hydroxymethyl glutaryl - CoA reductase	Inhibition of cholesterol synthesis
Allopurinol	Xanthine oxidase	Treatment of gout
Disulfiram	Aldehyde dehydrogenase	Treatment of alcoholism
Sulphonamides Trimethoprim	Dihydrofolate synthase	Treatment of bacterial infections
Methotrexate	Dihydrofolate synthase	Anticancer effect
Celastatin	Dehydropeptidase-I	Enhancement of antibacterial action
Ritonavir, Saquinavir	HIV protease	HIV therapy

BIBLIOGRAPHY

David, L. Nelson and Michael, M.Cox. (2009). *Lehninger Principles of Biochemistry,* 5th edn. WH Freeman and Company Inc., New York.

Palmer, T. (2004). *Enzymes.* East-West Press, New Delhi.

Berg, J.M., Tymoczko, J.L. and Stryer, L. (2005). *Biochemistry.* WH Freeman and Company and Sumanas Inc., New York.

5

IMMOBILIZED ENZYMES AND MICROBIAL WHOLE CELL TECHNOLOGY

INTRODUCTION

The world market for industrial enzymes, which was worth US $2.3 billion by the end of 2007, is expected to reach US $2.7 billion by 2012, with a compound annual growth rate of 4% (Chellappan *et al.*, 2011). Despite the fundamental biological significance, enzymes have not yet found their rightful place in industry due to many factors. To use expensive enzymes only once with the substrate has many disadvantages: (i) cost of enzymes (ii) the need to remove the enzyme from the product and (iii) concomitant large reactors with high capital and energy cost.

To eliminate the inherent disadvantages in enzymes and to obtain highly active and stable catalysts having well-defined specificity, a new classical approach, viz., the immobilization of enzymes has been made. The term "Immobilized Enzyme" was first suggested at a meeting on "Enzyme Engineering" held in Henniker, New Hampshire, USA, in August 1971. During that conference, a classification of immobilized enzymes was proposed, as shown below.

Classification of Immobilized Enzymes

"Immobilization" of an enzyme means the physical confinement or localization of enzyme molecules during a "continuous" catalytic process (Zaborsky, 1973). The localization of enzyme molecules can be brought about by diverse means such as their covalent attachment to water-insoluble functional polymers, their adsorption onto water-insoluble organic or inorganic supports, their entrapment within gel matrices or semi-permeable micro-capsules and their containment within semi-permeable membrane-dependent devices. Immobilized enzymes may be recovered from the reaction mixture and used over and over again, thus improving the economy of the process.

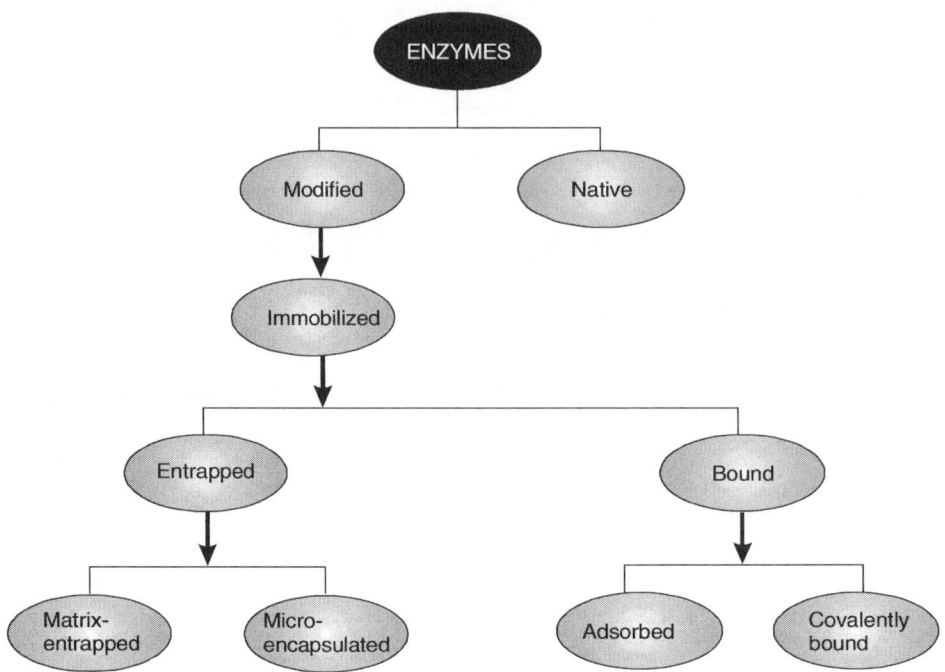

The earliest attempts on immobilization of enzymes were made by Grubhofer and Schleith (1953), in which carboxypeptidase, diastase, pepsin and ribonuclease were immobilized on diazotized polyaminostyrene resin. In the 1960s, extensive studies were carried out at the Weizmann Institute of Science, Rehovot, Israel, on new immobilization techniques and on the enzymatic, physical and chemical properties of immobilized enzymes.

Advantages of Immobilized Enzymes

There are many advantages of immobilized enzymes.

1. Conventionally, enzymes are used only once in commercial processes before being either inactivated or removed from the product. The substitution of immobilized enzymes would permit reuse of the enzymes, since recovery is easy and processing can be accomplished on a continuous basis, reducing enzyme and labour costs.

2. Operational advantages, viz., (a) possibility of batch or continuous operational modes, (b) rapid termination of reactions, (c) controlled product formation, (d) greater variety of engineering designs for continuous processes and (e) reduction in effluent disposal problems.

3. The binding with a carrier may change the optimum pH of the enzyme and this may permit an enzyme to act efficiently at a pH not optimal for that enzyme. This may have applicability in a variety of areas where the enzyme cannot be used in its native free form.

4. Immobilized enzymes can operate in the presence of relatively high concentrations of organic solvents. This will be of value when the substrate has low solubility in water, but high solubility in organic solvents.

5. Immobilized enzymes have increased storage and temperature stabilities leading to wide applications.

Disadvantages of Immobilized Enzymes

i. Initial cost of the immobilization process is high.
ii. Possible loss of enzyme activity on immobilization cannot be avoided.
iii. Initial investment should be made for plant construction.
iv. Technically, the process is more complex.
v. More skilled supervision of process is required.
vi. Unique sanitation and toxicology problems are observed.
vii. Protein contamination in the product needs to be prevented.
viii. Microbial contaminations especially in continuous systems is a major problem, and needs to be prevented.
ix. In the continuous immobilization process, the loss of enzyme activity occurs during every cycle. This results in reduction in the productivity of the process.

SUPPORTS/CARRIERS

The most important contributing component to the performance of an immobilized enzyme system is the support. Support or carrier is an inorganic or organic material to which the enzyme is bound.

The ideal support material must meet the following criteria:

i. Wide chemical functionality to permit reaction with a variety of groups of proteins.
ii. Low cost to permit an economically feasible operation.
iii. Resistance to microbial attack as well as changes in pH.
iv. Retention of physical shape following changes in pressure, temperature, solvent composition, etc.
v. Non-eroding and chemically non-reactive with the solvent and should not break when used in suspension.
vi. Devoid of toxicity.
vii. The ability to form different shapes for applying the immobilized biocatalyst to various types of reactors.
viii. High affinity for enzyme and low affinity for substrate and product.

The supports are commonly divided into two categories, viz., organic and inorganic.

Organic Supports

Amino acid polymers, polysaccharides, poly (methylmethacrylate), nylon, polystyrene, chitin and collagen are some of the organic supports used widely.

Amino acid polymers   Excellent review has been published on the use of polymer materials for enzyme immobilization and their applications in bioreactors (Fang *et al.*, 2011).

Amino acid copolymers are among the earliest supports for enzyme immobilization. Cellulase has been immobilized on poly-L-glutamic acid (Takeuchi and Makino, 1987) and this immobilized cellulase (IC) is water-soluble in neutral and alkaline solutions, where IC has the activity. IC can be made insoluble by lowering the pH so that it can be recovered from the reaction mixture with its activity.

Polysaccharides   Among polysaccharide supports, cellulose derivatives, viz., DEAE-cellulose, CM-cellulose, sepharose, phenoxyacetyl cellulose, aminobenzoyl cellulose, etc., have been widely used as carriers for immobilization of enzymes (Simionescu *et al.*, 1987). The hydroxyl groups on celluloses are made to react with the non-functional groups on enzymes through suitable coupling agents.

Poly(methylmethacrylate), nylon and polystyrene Macroporous poly(methylmethacrylate) resins and macroporous polystyrene resins with varied particle sizes have been employed for *Candida antarctica* B lipase adsorption (Chen *et al.*, 2007a; Chen *et al.*, 2007b). Immobilized enzyme derivatives prepared with nylon and polystyrene can be synthesized in the form of tubes. Invertase from *Saccharomyces cerevisiae* (yeast) has been covalently bound to a macroporous polystyrene anion exchanger via benzoquinone and glutaraldehyde (Mansfeld and Schellenberger, 1987). Even though nylon and polystyrene possess similar mechanical properties, the extreme hydrophobic nature of polystyrene might limit the usefulness of this support. Nylon, non-porous in nature, has been employed advantageously in interface with blood systems. Nylon does not respond to the vigorous clotting reaction that is readily noted with high-surface-area inorganic carriers such as porous glass (Messing, 1978). Nylon, being relatively hydrophilic in nature, makes it more suitable for enzyme immobilization, and commercially, available nylon tubes have been used for the immobilization of invertase (Onyezili, 1987).

Chitin   Native chitin contains approximately equal amounts of calcium carbonate and chitin. The calcium carbonate can be removed by treatment with acid, giving a rigid porous material that is relatively resistant to both chemical and microbial attack. Krill chitin, which is more porous than chitins from other sources, has been modified with carbon disulphide (CS_2) and used as a support for the immobilization of diastase, β-amylase, α-amylase and amyloglucosidase (Synowiecki *et al.*, 1987). The dithiocarbamino groups formed after reaction of CS_2 with the amino group of chitin can participate in binding the enzyme without denaturation and loss in activity.

Collagen Collagen as an enzyme support offers a number of potential advantages, which are as follows: (i) it is a natural material with good mechanical and hydrodynamic properties (ii) it offers a low resistance to substrate diffusion (iii) it has an open internal structure with many potential binding sites for enzyme attachment (iv) it is inexpensive and available in abundance and (v) it can be processed into films, membranes and other forms that retain their structural identity.

There is an organizational regularity in the packing of collagen molecules into microfibrils. This results in holes spaced at regular and repeating intervals along the length of the microfibrils. It is believed that these regions (holes) are the primary sites for enzyme binding. Immobilization of enzymes on collagen involves the formation of a network of non-covalent bonds such as salt linkages, hydrogen bonds and van der Waals interactions acting together between collagen and the enzyme.

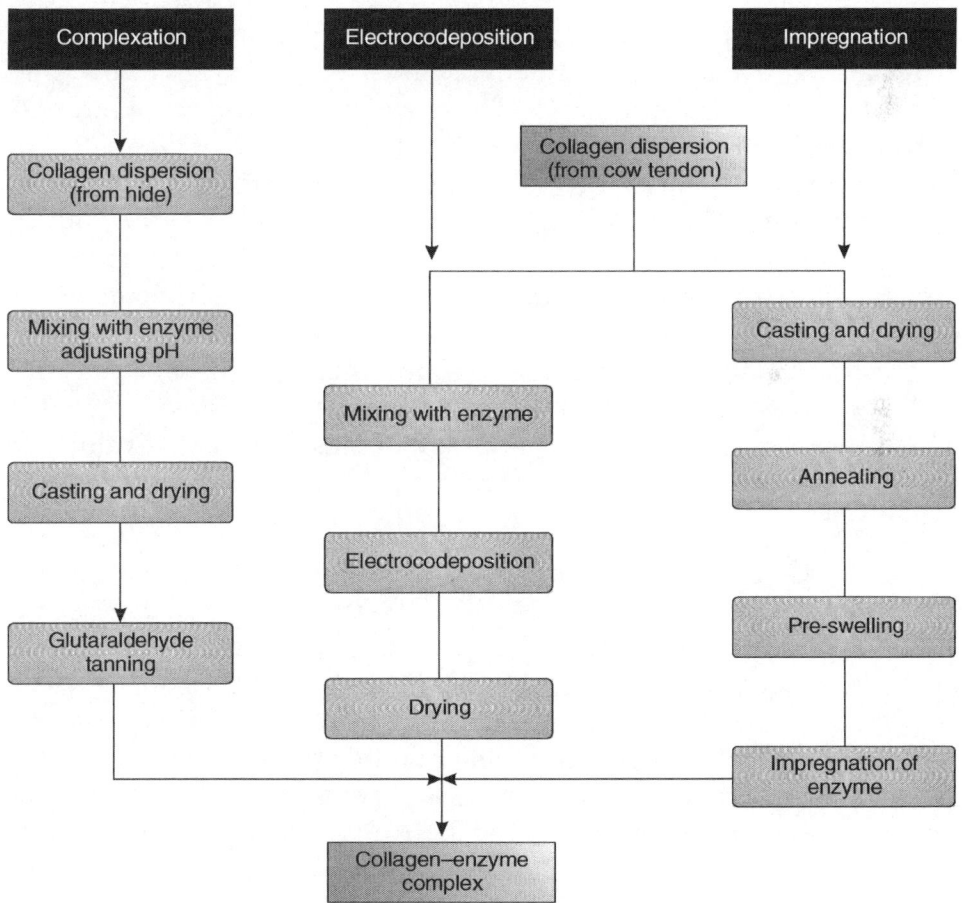

Figure 5.1 Methods of enzyme immobilization using collagen as carrier

Collagen–enzyme complexes have been prepared by three different procedures (Figure 5.1) which are as follows:

 i. impregnating a pre-swollen membrane with enzyme
 ii. macromolecular complexation and
 iii. electrocodeposition

Inorganic Supports

Porous glass, silica and alumina impregnated with nickel oxide, porous ceramics of various types, sand and carbon are some of the inorganic materials to which enzymes have been bound by adsorption as well as by covalent coupling.

Porous glass   Porous glass commonly used for immobilization has a pore diameter in the range of 200–1500 Å and particle size in the range of 20–80 mesh. There are certain advantages in the use of porous materials like porous glass and porous ceramics. The porous materials have relatively high surface areas per unit weight or unit volume which make them ideal for large reactors. The enzyme is bonded on an internal surface which is protected from the turbulent and harsh external environment. The major disadvantage of porous carriers is the internal surface available for bonding enzymes. When an internal surface is involved for bonding the enzyme, not only does the coupling/cross linking agent require access to the surface but also the much larger enzyme and perhaps substrate molecule must undergo these diffusional constraints. In addition, when a carrier has a broad pore size distribution, only a limited number of pores will be large enough to accommodate both enzyme and substrate. Therefore, only a small portion of the total surface area will be utilized (Messing, 1978). Porous glass has been reported to be partly soluble, and breaks down at pH value near 7.0 or above. This problem can be overcome by coating the glass with zirconium oxide, an insoluble material.

Ceramic carriers   Ceramic carriers are less expensive than porous glass and are more insoluble. Ceramic carriers composed of SiO_2, TiO_2 and Al_2O_3, have been used for immobilization. Coupling is achieved by first preparing the silanized ceramics followed by treating with glutaraldehyde and finally adding the enzyme. Bentonite has been utilized for the coupling of an alkaline protease from *Bacillus subtilis* by this method (Sreenivasalu *et al.*, 1984). Commercial fungal cellulose preparations from *Trichoderma viride* and *Aspergillus niger* have been immobilized on ceramics such as alumina and titania using titanium chloride and on their silanized derivatives using glutaraldehyde (Shimizu and Ishihara, 1987).

Carbon   Many enzymes have been immobilized on colloidal carbon, granular carbon and activated carbon.

The use of activated carbon for enzyme immobilization has several advantages. (i) Its pore volume is suitable for enzyme immobilization. (ii) The cost incurred is low (iii) activated carbon-immobilized enzymes can be fluidized or suspended in a slurry reactor more easily than enzymes supported on porous glass. (iv) Activated carbon possesses mechanical strength comparable to porous glass materials. (v) Enzyme loadings achieved on activated carbon are similar to those obtained on porous glass. The demerits associated with the system include the following (i) changes in ionic strength may cause desorption; and (ii) support may be subjected to microbial or proteolytic attack.

Sand Sand is another silica support which has been used to immobilize urease and alcohol dehydrogenase (Brotherton *et al.*, 1976), hemicellulases (Puls *et al.*, 1977), trypsin (Puvanakrishnan and Bose, 1980) and pepsin (Puvanakrishnan and Bose, 1984).

The advantages of utilizing a non-porous material like sand as a support are: (i) the cost is very low (ii) the enzyme is immobilized on the external surface of the carrier, eliminating the diffusional constraints with respect to the substrate and (iii) this may be suitable for large-volume applications with low concentrations of substrate even though it has a lower, effective surface area and enzyme-loading capacity than porous glass (Olson and Korus, 1977). The enzyme-loading problem may be partially overcome by using fine particles.

The disadvantage of this carrier is that the surface area is low due to its non-porous nature. Hence, the available surface area for the attachment of the enzyme is limited and there is a need to optimize the concentration of enzyme to be immobilized with respect to the size of the sand particles.

Disadvantages of Organic Supports

There are certain general limitations associated with the use of organic carriers viz. (i) many organic carriers change configuration under differing conditions, thus creating changes in flow rates if used in columns (ii) they are susceptible to attack by microorganisms or enzymes (iii) many enzymes immobilized on organic carriers have poor stability under operating conditions (iv) many organic matrices are of extremely small particle size and gelatinous in nature.

Advantages of Inorganic Supports

Inorganic carriers have many advantages over organic polymers, which include the following (i) inorganic materials are not susceptible to pH and solvent conditions and hence they will not change size or configuration during usage (ii) they are not susceptible to microbial attack (iii) inorganic materials can be easily shaped, permitting a wide variety of configurations (iv) enzymes coupled to inorganic materials appear to have greater operational stability.

METHODS FOR ENZYME IMMOBILIZATION

The methods can be classified into three basic categories (Figure 5.2), and Table 5.1 summarizes the three methods.

Table 5.1 Characteristics of different immobilization systems

Characteristics	Carrier-binding method			Intermolecular cross linking	Entrapment
	Adsorption	Ionic binding	Covalent binding		
Preparation	Simple	Simple	Difficult	Difficult	Difficult
Enzyme activity	Low	High	High	Moderate	High
Binding force	Weak	Moderate	Strong	Strong	Strong
Substrate specificity	Unchangeable	Unchangeable	Changeable	Changeable	Unchangeable
General applicability	Low	Moderate	Moderate	Low	High
Cost of immobilization	Low	Low	High	Moderate	Low

Carrier-binding Method

When enzymes are bound to water-insoluble carriers, much care is required in the selection of carriers as well as binding techniques. The carrier binding method can be further divided into three categories depending on the mode of binding of the enzyme, viz., adsorption, ionic binding and covalent binding.

Adsorption method   This method is based on the physical adsorption of enzyme protein on the surface of water-insoluble carriers. The commonly employed adsorbents are: alumina, anion exchange resins, calcium carbonate, activated carbon, clays, collagen, diatomaceous earth, kaolinite, bentonite, hydroxyapatite, porous glass, chitin, etc.

The adsorption of an enzyme onto a water-insoluble material is dependent on experimental variables such as pH, nature of the solvent, ionic strength, concentration of protein as well as adsorbent. This type of adsorption of enzymes onto water insoluble matrices can be attributed either to an ion exchange mechanism, or to simple physical adsorption at the external surface of a particle, or to "physico-chemical bonds" created by hydrophobic interactions, van der Waals forces, etc.

It has been found that enzymes can be readily immobilized by using an adsorbent containing tannin as a ligand. The adsorbent is prepared by the reaction of tannin previously activated by CNBr or epichlorohydrin with aminohexyl cellulose. The tannin-aminohexyl cellulose can be used as an adsorbent for enzyme immobilization. Hydrophobic forces play an important role in this immobilization. For continuous processing, it is highly desirable to have a strongly formed association, since weak binding leads to eventual depletion of the enzyme, with possible contamination of the product.

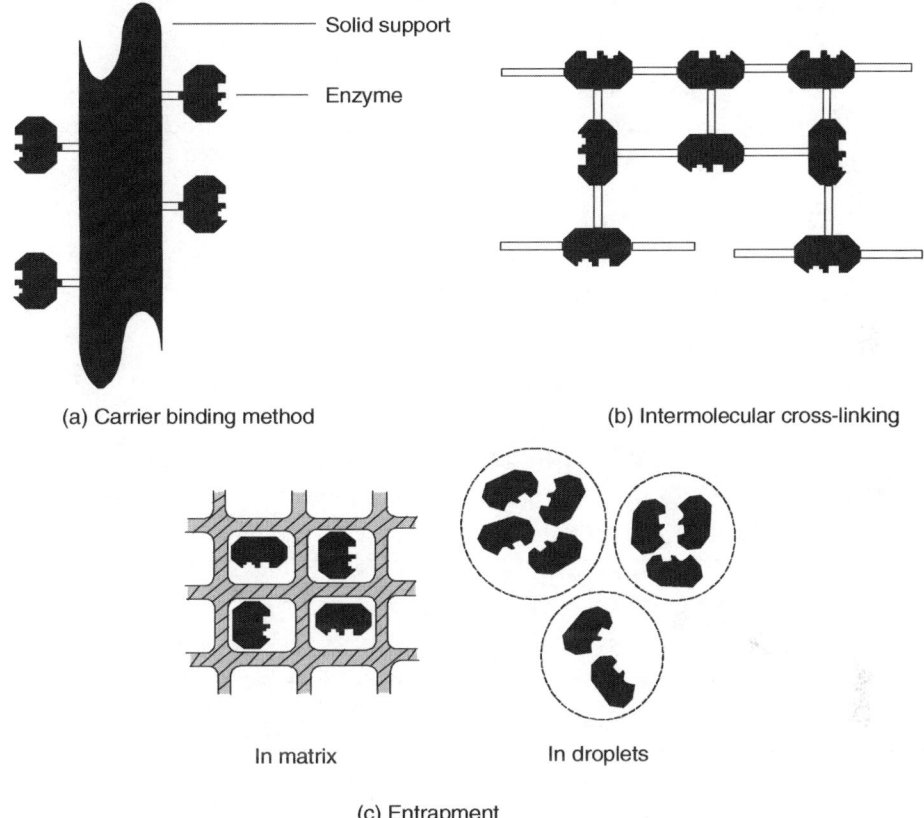

Figure 5.2   Methods of enzyme immobilization

Polypropylene powder obtained with metallocene catalysts (PP_{met}) and commercial polypropylene particles obtained with Ziegler-Natta catalysts (PP_{ZN}) show relatively high activity and stabilities due to the porous surface and smaller particle size of the carrier (Foresti and Ferreira, 2007).

The advantages of adsorption methods for immobilizing enzymes are simplicity, greater choice of differently charged and shaped carriers and the possibility of achieving simultaneous purification and immobilization. Regeneration of carrier is also possible. No conformational change of the enzyme protein or destruction of its active centre is observed. On the other hand, the disadvantages are that the adsorbed enzyme may leak out from the carrier during the process since the binding force between the enzyme protein and carrier is weak. Optimal conditions for adsorption, i.e., pH, ionic strength and temperature must be determined to achieve and maintain good adsorption and activity.

Ionic binding method   This method is based on the ionic binding of enzyme protein to water-insoluble carriers containing ion exchange residues. In some cases, not only ionic binding but also physical adsorption may be involved in the binding.

Advantages of this technique are that the binding of enzyme to the carrier can be easily carried out and its preparative conditions are mild in comparison with those necessary for the covalent binding method. Therefore, ionic binding method may not cause any change in the conformation as well as in the reactive sites of the enzyme protein.

The binding forces between enzyme and carrier are stronger than adsorption method but are weaker than covalent binding. As the binding forces between enzyme protein and carriers are weaker than in covalent binding, leakage of enzyme from the carrier may occur in substrate solutions of high ionic strength or upon variations of pH.

Covalent binding method   The covalent attachment of water-soluble enzyme molecules via non-functional amino acid residues to water-insoluble functionalized supports is the most prevalent method for immobilizing enzymes. A large number of chemical reactions have been used for the covalent binding of enzymes by way of their non-functional groups to inorganic carriers such as ceramics, glass, iron, zirconium and titanium, to natural polymers such as sepharose and cellulose, and to synthetic polymers such as nylon, polyacrylamide and other vinyl polymers and copolymers possessing reactive chemical groups. The groups that take part in the covalent binding of enzymes to carrier are (i) α- or ε-amino group (ii) α- or β- or γ- carboxyl group (iii) sulphhydryl group of cysteine (iv) hydroxyl group of serine and threonine (v) imidazole group of histidine and (iv) phenolic group of tyrosine. In coupling reactions, these groups react with carriers containing reactive groups such as diazonium, acid azide, isocyanate and halides.

The major advantage of the method is that the reaction is not affected by pH, ionic strength of the medium or substrate concentration. The enzyme is strongly bound, and is thus unlikely to be lost in the process. The selection of conditions for immobilization by covalent binding is more difficult than in the cases of physical adsorption and ionic binding. The reaction conditions required for covalent binding are relatively complicated and not particularly mild. Therefore, in some cases, covalent binding alters the conformational structure and active centre of the enzymes, resulting in major loss of activity. One of the ways to avoid active site involvement is to use a coupling technique which does not react with the functional group present at the active site. Besides, the other potential problem is the difficulty in removing the enzyme from the carrier. Hence, this method is better suited to expensive enzymes whose stability is significantly improved by covalent binding to the carrier.

Some of the methods for covalent binding of enzymes to supports are shown in Table 5.2.

Table 5.2 Different methods for covalent binding of enzymes to supports (El Nashar, 2010).

Reaction	Support-Enzyme linkage
Diazotization	SUPPORT— $N{=}N$ —Enzyme
Alkylation and arylation	SUPPORT—CH_2—NH—ENZYME
Alkylation and arylation	SUPPORT—CH_2—S—ENZYME
Schiff's base formation	SUPPORT—$CH{=}N$—ENZYME
Amide bond formation	SUPPORT—CO—NH—ENZYME
Amidation reaction	SUPPORT—CNH—NH—ENZYME
Thio-disulphide interchange	SUPPORT—S—S—ENZYME
Carrier binding with bifunctional reagents	SUPPORT—$O(CH_2)_2$ $N{=}CH$ $(CH_2)_3$ $CH{=}N$—ENZYME

The covalent binding method can be further classified into (a) diazo (b) peptide and (c) carrier linking methods according to the mode of linkage.

(a) Diazo method   In this method, carriers containing aromatic amino groups are diazotized with nitrous acid and then the diazonium derivatives are reacted with enzyme proteins. Aromatic amino derivatives of the polymers of amino acids, polyacrylamide, polystyrene, copolymers of ethylene-maleic anhydride and porous glass are commonly employed as carriers for this method. The groups on enzyme proteins participating in diazo coupling include free amino group, the imidazole group of histidine and the phenolic group of tyrosine.

Much of the work on covalent bonding has involved the use of the reagent, 3-aminopropyltriethoxysilane (APTS). This has proved to be the most successful reagent for generating a covalent bridge between the inorganic surface (oxide or hydroxyl groups) and the enzyme. Interest in amino alkyl controlled porous glass (CPG) stems from the pioneering work of Weetall and his collaborators, who first introduced amino alkyl CPG as an enzyme support and described methods of preparing suitable derivatives for enzyme immobilization (Weetall and Filbert, 1974). Once the covalent link between the inorganic support and the amino propyl groups is formed, the aminoalkylated glass can be reacted with enzyme directly. However, the aminoalkylated glass is activated by a suitable procedure before reacting with enzyme.

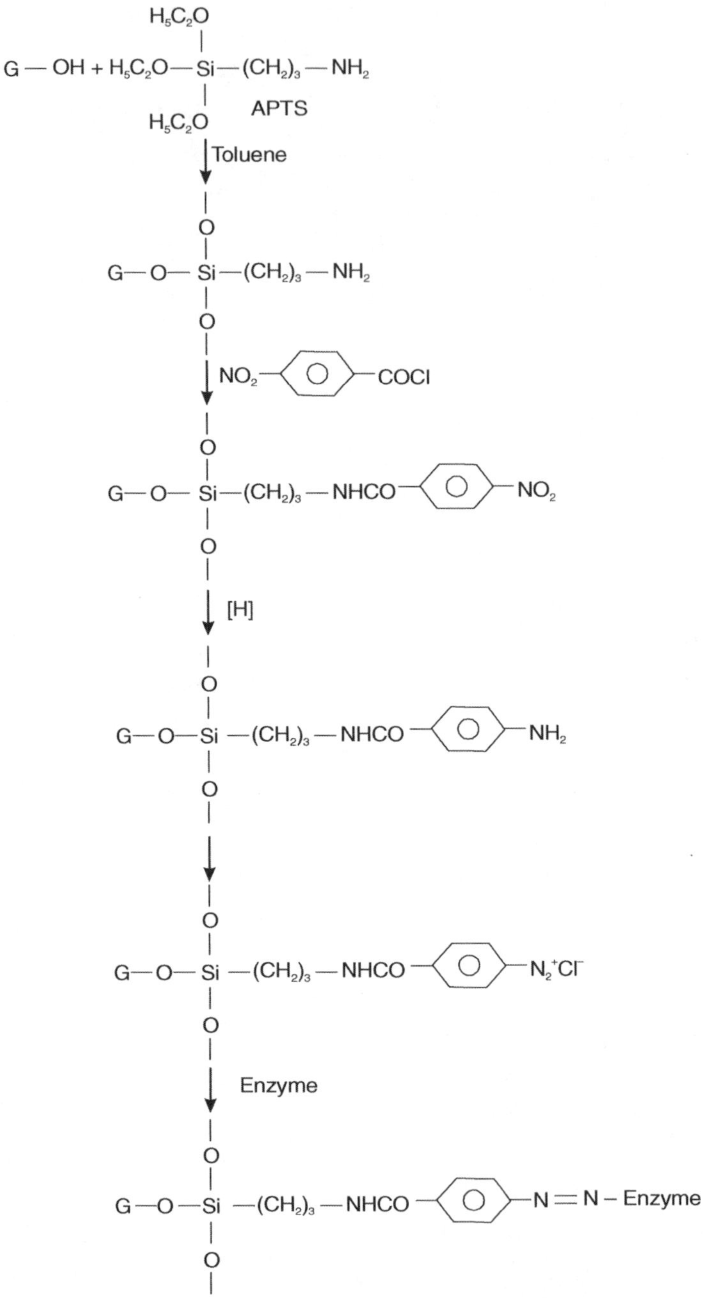

Figure 5.3 Diazo method of enzyme immobilization

Amino alkyl groups are incorporated into glass by refluxing porous glass with 3-aminopropyltriethoxysilane in toluene. The resulting aminoalkylated porous glass is reacted with *p*-nitrobenzoyl chloride and then reduced to give the *p*-aminobenzoyl derivative. This is diazotized to the diazonium derivative and coupled with the enzyme as shown in Figure 5.3. Phospholipase A$_2$ has been coupled by this method (Adamich *et al.*, 1978).

(b) Peptide binding method This method is based on the formation of peptide bonds between the enzyme protein and a water-insoluble carrier as shown below.

i. Water-insoluble carriers containing carboxyl groups are converted to reactive derivatives such as acid azide, chloride, isocyanate, etc. These derivatives then react with free amino groups in the enzyme to form peptide bonds.

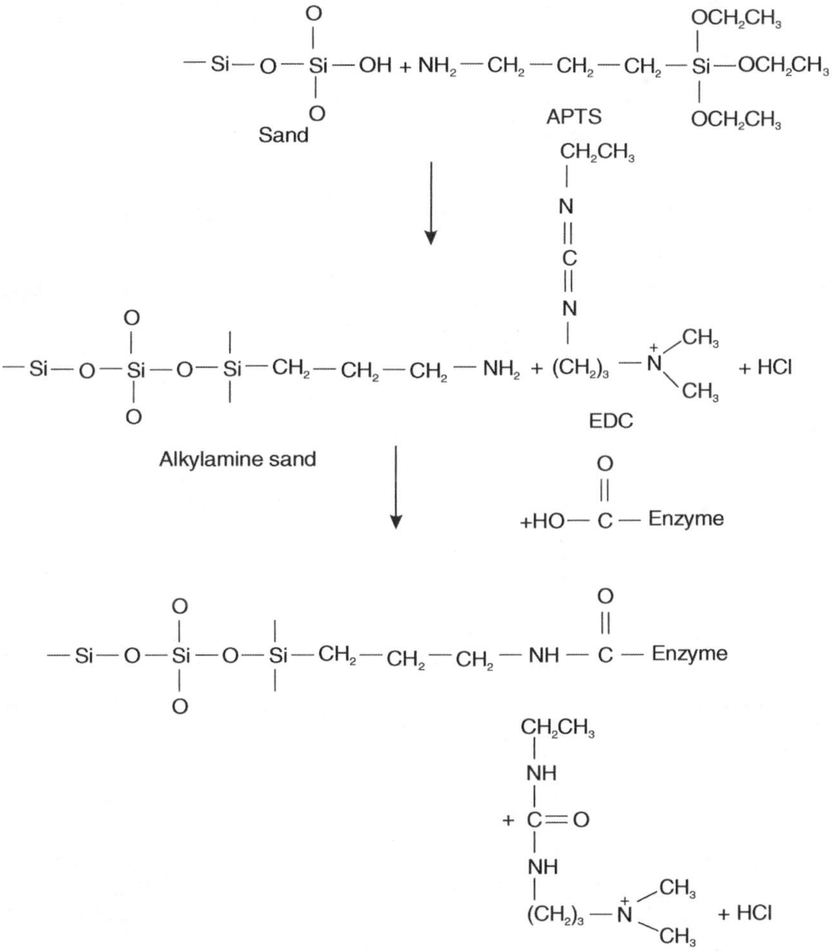

Figure 5.4 Peptide binding method of enzyme immobilization

ii. By using condensing reagents such as carbodiimide, peptide bonds are formed between free carboxyl or amino groups in the enzyme and amino or carboxyl groups in the carrier. Pepsin has been coupled to alkylamine sand through the amino groups present on the alkylamine sand surface by this method as shown in Figure 5.4 (Puvanakrishnan and Bose, 1984). Ye *et al.* (2005) have prepared ultrafiltration hollow fibre membranes from poly(acrylonitrile-co-maleic acid) for *Candida rugosa* lipase immobilization with 1-ethyl 3-(dimethyl-aminopropyl) carbodiimide hydrochloride (EDC) / N-hydroxyl succinamide as coupling agent.

(c) Carrier cross-linking method   This method is based on the formation of cross-links between the amino group of carrier and the amino group of enzyme protein using bifunctional reagents. Glutaraldehyde is more commonly employed as the bifunctional reagent and many enzymes have been immobilized by the formation of Schiff bases between the amino groups of carriers and enzyme protein. As carriers possessing amino groups, AE-cellulose, DEAE-cellulose, partially deacetylated chitin, aminoalkyl derivatives of sand, porous glass and ceramics have been used. Trypsin (Puvanakrishnan *et al.*, 1980) has been immobilized on alkylamine sand as shown in Figure 5.5.

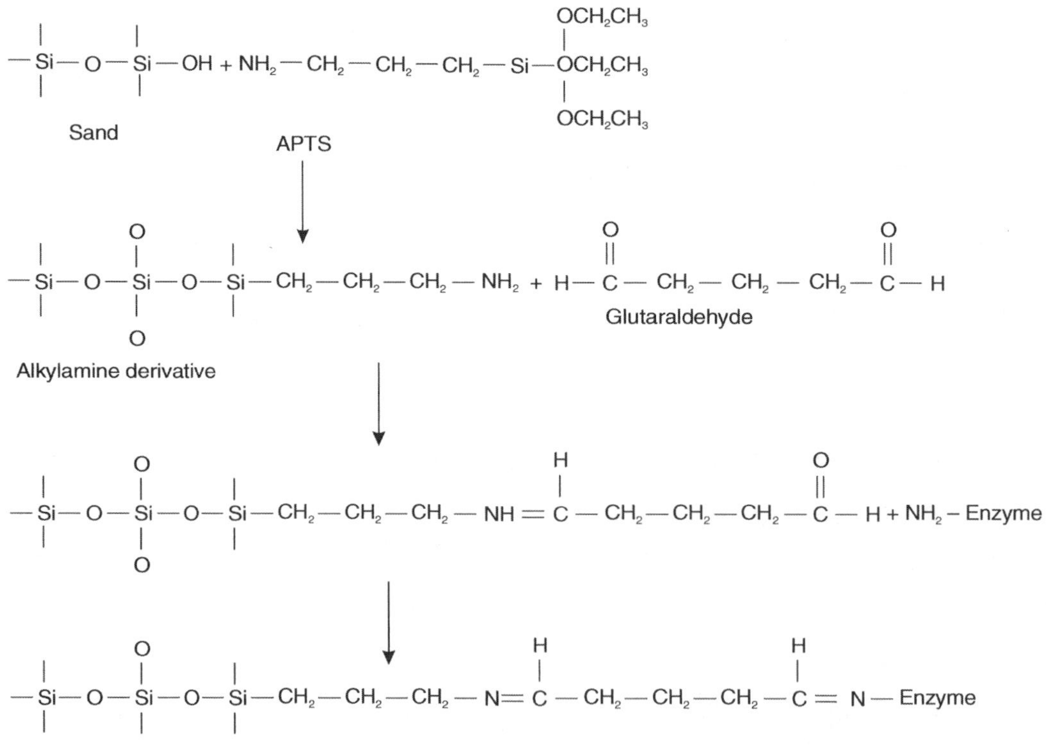

Figure 5.5   Carrier cross-linking method of enzyme immobilization

Studies have been carried out on the immobilization of lipase from *Candida rugosa*. Three different supports are used, two with amino groups (Sepabeads EC-EA and Sepabeads

EC-HA) differing in spacer length (two and six carbons, respectively) and one with epoxy group (Sepabeads EC-EP). Lipase immobilization is carried out by two conventional methods (via epoxy groups and via glutaraldehyde) and with periodate method for modification of lipase. The results of activity assays have shown that lipase has retained 94.8% or 87.6% of activity after immobilization via epoxy groups or with periodate method respectively while glutaraldehyde method is inferior with only 12.7% of retention (Prlainovic *et al.*, 2011).

Cyclodextrin glucosyl transferases (CGTases) represent one of the most important groups of microbial amylolytic enzymes. These enzymes catalyse the formation of cyclodextrins from starch and related $\alpha(1 \rightarrow 4)$-linked glucose polymers via a transglycosylation reaction. Cyclodextrins are used as complexing agents in food, pharmaceutical and cosmetic industries. CGTase from *Thermoanaerobacter* sp. is covalently attached to Eupergit C, which consists of macroporous beads with a diameter of 100–250 μm made by copolymerization of N,N-methylene-bis-(methacrylamide), glycidyl methacrylate, allyl glycidyl ether and methacrylamide. Different immobilization parameters are optimized and the maximum yield of bound protein is around 80% (8.1 mg/g support) (Martin *et al.*, 2003).

The covalent attachment offers numerous advantages:

1. The coupling of the enzymes to a functional support is easy to conduct.
2. The solid-supported catalysts are easy to handle in manipulative operations. Easy removal of the catalyst allows quick termination of a reaction at a particular stage and permits selective or partial transformations to be achieved.
3. Various physical forms of the support material, e.g., flat sheet, powder, large sized particle, fibre, etc., can be employed.
4. The immobilization is dependent on the formation of covalent bonds between the enzyme and support. The binding force between enzyme and carrier is strong, and leakage of the enzyme does not occur even in the presence of salt solutions of high ionic strength.

The disadvantages of covalent attachment are the following.

1. The selection of conditions for immobilization by covalent binding is more difficult than in the cases of physical adsorption and ionic binding.
2. Certain enzymes are extremely sensitive to changes in pH, ionic strength, etc., and the conditions necessary for their successful immobilization may completely abolish activity.
3. A great disadvantage of covalently coupled enzymes is their low catalytic efficiency on high molecular weight substrates, caused mainly by steric repulsion of the macromolecules.

Intermolecular Cross-linking

This method is based on the formation of chemical bonds as in the covalent binding method, but water-insoluble carriers are not used in this method. The immobilization of enzymes is performed by the formation of intermolecular cross-linkages between the enzyme molecules by means of bi- or multifunctional reagents. Common multifunctional reagents that have been employed for the preparation of immobilized enzymes by this method are glutaraldehyde, hexamethylene diisocyanate, 1, 5-difluoro-2, 4-dinitrobenzene and diazobenzidine. The functional groups of enzyme proteins participating in the reactions include the α-amino group at the amino terminus, the ε-amino group of lysine, the phenolic group of tyrosine, the sulphydryl group of cysteine and the imidazole group of histidine.

The correct choice of reaction conditions is one of the prerequisites in the study of the chemical modification of proteins with multifunctional reagents for producing immobilized enzymes. The concentrations of the enzyme and the reagent, the pH and ionic strength of the solution, etc. must be chosen so carefully as to favour "intermolecular" crosslinking between enzyme molecules. A second important consideration is the degree of "intramolecular" cross-linking. Low concentrations of the enzyme and reagent enhance the probability of functional groups of a reagent molecule to react with amino acid residues only of the same enzyme molecule and not of different molecules. The control of "intra" versus "intermolecular" cross linking is experimentally very difficult to achieve and immobilization of an enzyme with multifunctional reagents will involve both types of crosslinks.

Three different methods are employed for preparing immobilized enzymes by intermolecular cross-linking. Enzyme immobilization is carried out with (i) multifunctional reagent alone or (ii) multifunctional reagent in the presence of a second protein (co-protein) or (iii) multifunctional reagent after adsorption of the enzyme on water-insoluble surface-active support. The pros and cons of carrier-free versus carrier-bound immobilized enzymes and of each type of carrier-free enzyme have been reviewed critically (Cao *et al.*, 2003).

Preparation of immobilized enzyme using a multifunctional agent is simple and gives a derivative which is almost a pure protein. These enzyme gel derivatives can be dispersed readily in aqueous solutions. The drawbacks in this approach are the need for rigid controls of pH, concentrations, etc., the need for a large amount of protein or co-protein, the unavoidable inactivation of the enzyme caused by the chemical modification and the gelatinous nature of these enzyme derivatives making them unsuitable for use in columns.

Penicillin G acylase (penicillin amidohydrolase) is immobilized in a simple and effective way by physical aggregation of the enzyme using a precipitant (ammonium sulphate or tert-butyl alcohol or polyethylene glycol [PEG]) followed by chemical cross-linking to form insoluble cross-linked enzyme aggregates (Cao *et al.*, 2000).

Entrapment

This method is based on confining enzymes in the lattice of a polymer matrix (matrix entrapment) or enclosing the enzymes in a semi-permeable polymer membrane which is known as microencapsulation.

Lattice type   This method involves entrapping enzymes physically within the interstitial spaces of a cross-linked water-insoluble polymer. Various polymers such as alginate, carrageenan, cellulose triacetate, gelatin, polyacrylamide, polyvinyl alcohol and polyacrylic acid have been used for enzyme immobilization by this technique. The form and nature of lattice vary. Pore size of lattice should be adjusted to prevent the loss of enzyme from the lattice due to excessive diffusion especially when using agar and carrageenan lattices since they have large pore sizes, i.e., < 10 µ.

Entrapment of an enzyme is generally performed by polymerizing an aqueous solution containing acrylamide and the enzyme in the presence of N,N´-methylene-bis-acrylamide as a cross-linking agent. Initiation of polymerizing reaction is done by potassium persulphate or riboflavin and the reaction is accelerated by β-dimethylaminopropionitrile, N,N,N´,N´-tetramethylethylene diamine or alum. Bernfeld and Wan (1963) used this technique first to entrap trypsin, papain, amylase etc., in a gel lattice of polyacrylamide.

The advantages of lattice entrapment for immobilizing enzymes are as follows:

i.   Only small amounts of enzyme are needed.

ii. No chemical modification of the enzyme is expected and consequently, the enzyme's intrinsic properties are not changed.

iii. This method permits the preparation of immobilized enzyme derivatives of widely different physical forms.

Disadvantages of the method also exist and include the following:

i.   There is a chance of chemical and thermal inactivation of enzymes taking place during the gel formation.

ii. Leakage of low molecular weight enzyme from the lattice or an enzyme from within the crosslinked polymeric network is possible.

iii. Lattice-entrapped enzymes show very little activity towards macromolecular substrates.

Microcapsule type   Encapsulations within natural polymer-based hydrogel beads have gained much attention due to their excellent biocompatibility. So far, polysaccharide hydrogel beads such as agarose beads, alginate beads, chitosan beads, κ-carrageenan beads and nanogels of cholesterol-bearing pullulan have been used as supports (Jegannathan *et al.*, 2009; Sawada and Akiyoshi, 2010).

The immobilization of enzymes by entrapping the molecules within permanent semipermeable microcapsules has been first reported by Chang (1971). Extremely well-controlled conditions are required for the preparation of enzyme microcapsules. The method of encapsulation is cheap and simple but its effectiveness largely depends on the stability of enzyme although the catalyst is very effectively retained within the capsule. Three different procedures are available for microencapsulation of enzymes: (i) interfacial polymerization method (ii) liquid drying method and (iii) phase separation method.

The advantages of microencapsulation are the following. (i) This method provides an extremely large surface area for contact of substrate and catalyst and (ii) It allows the simultaneous immobilization of many enzymes in a single step.

The disadvantages of this method are the high protein concentration necessary for microcapsule formation, occasional inactivation of enzymes and the possibility of enzyme incorporation into the membrane wall and the restriction of substrates to low molecular weight substances.

REACTORS FOR IMMOBILIZED ENZYMES

A vessel employed to carry out the desired conversion using an enzyme is called enzyme reactor. Several types of reactors are available and the choice of a reactor type depends on the form of the enzyme (free or immobilized) to be used, kinetics of reaction and the scale of operation.

The different types of reactors (Figure 5.6) used for enzyme-mediated conversions are as follows: (i) stirred tank reactor (ii) membrane reactor (iii) continuous flow reactor, e.g., packed bed reactor, continuous-flow stirred tank reactor and fluidized bed reactor. The reactors may be operated either in batch or continuous mode.

The principles developed for general heterogeneous catalysis in synthetic chemistry result in well-known reactor configurations. Differences between enzyme catalysis and other systems result from the nature of the biocatalyst and reaction medium. Soft particles containing the biocatalyst, such as alginate beads, may limit the pressure drop in fixed-bed reactors.

The decision on specific reactor design will be based on a careful analysis of the kinetic properties of the reaction system. If the enzyme shows a strong substrate-surplus inhibition, a continuously operated reactor with complete back mixing working at high conversion is advantageous. A reaction with strong product inhibition may utilize a batch reactor or a plug flow reactor to achieve higher volume and catalyst-specific productivities. An extractive bioreactor may be used if substrates and products show different solubilities. By using this reactor configuration, the destabilizing effect of organic solvents may also be overcome, because the enzyme is separated from the organic phase, which is used to extract the insoluble product. The aqueous phase containing the enzyme will be saturated until the maximum solubility of the substrate is reached.

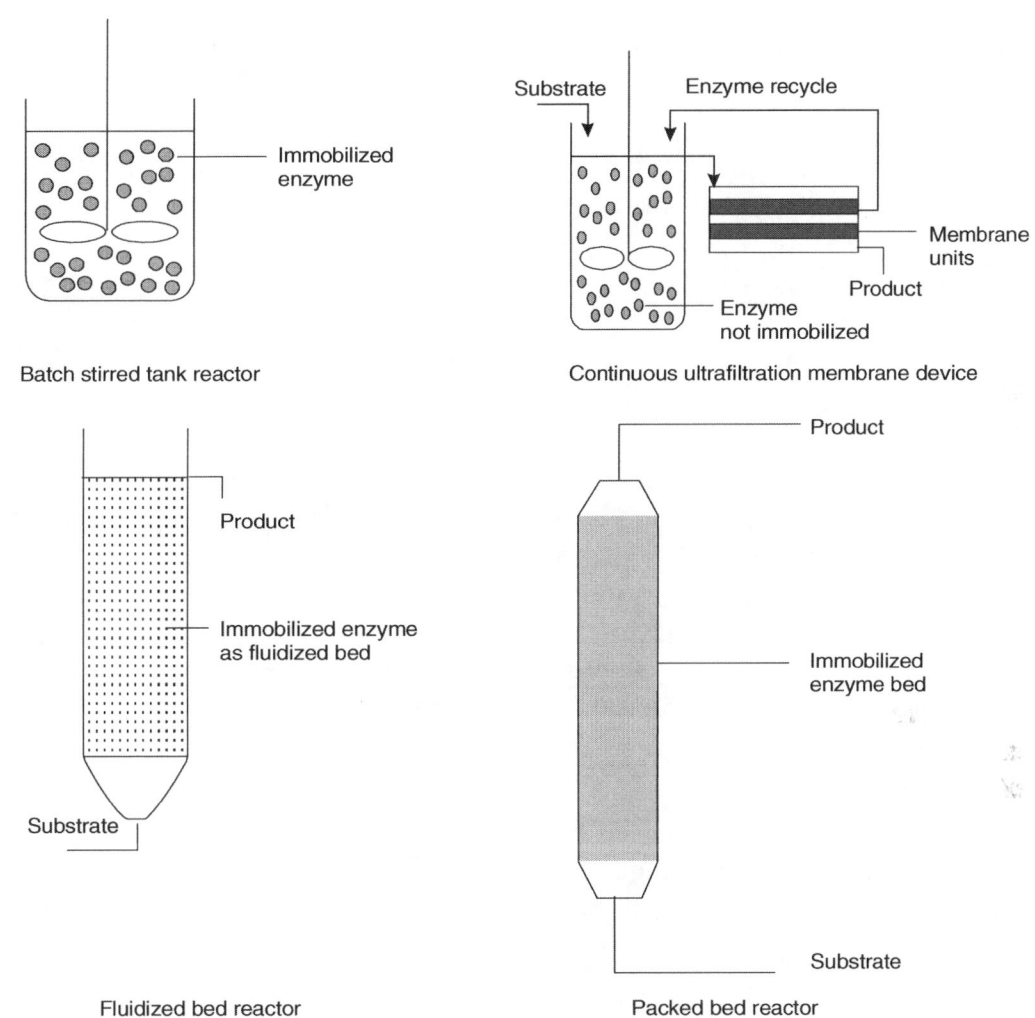

Figure 5.6 Types of reactors

When it comes to the application of immobilized enzymes, bioreactor is one of the important areas. Compared to fluidized-bed reactors, enzyme-immobilized membrane bioreactors (EMBR) stand out as a specific mode for combining product separation process with enzymatic catalysis in continuous processes. Until now, EMBR have been widely used in fine chemical synthesis, food industries and waste water treatment (Mazzei *et al.*, 2009; Le-Clech, 2010).

PROPERTIES OF IMMOBILIZED ENZYMES

It is important to understand the changes in physical and chemical properties which an enzyme would be expected to undergo upon immobilization. Changes have been observed in the stability of enzymes and in their kinetic properties because of the microenvironment imposed upon them by the supporting matrix and by the products of their own action.

Stability

The stability of the enzymes might be expected to either decrease or increase on immobilization, depending on whether the carrier provides a microenvironment capable of denaturing the enzyme protein or of stabilizing it. Inactivation due to autodigestion of proteolytic enzymes could be reduced by isolating the enzyme molecules by immobilizing them on a matrix. It has been found that enzymes coupled to inorganic carriers are generally more stable than those attached to organic polymers when stored at 4° or 23°C. Stability to denaturing agents may also be better upon immobilization.

Kinetic Properties

Changes in activity of enzymes due to the actual process of immobilization have been studied critically. There is usually a decrease in the specific activity of an enzyme upon immobilization, and this can be attributed to denaturation of the enzyme protein caused by the coupling process. Once an enzyme has been immobilized, it finds itself in a microenvironment that may be drastically different from that existing in free solution. The new microenvironment may be a result of the physical and chemical character of the support matrix alone, or it may result from interactions of the matrix with substrates or products involved in the enzymatic reaction.

Catalytic properties of immobilized enzymes can often be altered favourably to allow operation under broader or more rigorous reaction conditions, e.g., immobilized glucose isomerase can be used continuously for over 100 h at temperatures between 60° and 65°C.

The diffusion of substrate from the bulk solution to the microenvironment of an immobilized enzyme can limit the rate of the enzyme reaction. The rate at which substrate passes over the insoluble particle affects the thickness of the diffusion film, which in turn determines the concentration of substrate in the vicinity of the enzyme and hence the rate of reaction.

The effect of the molecular weight of the substrate can also be significant. Diffusion of large molecules will obviously be limited by steric interactions with the matrix, and this is reflected in the fact that the relative activity of bound enzymes towards high molecular weight substrates has been generally found to be lower than that towards low molecular weight substrates. This, however, may be an advantage in some cases, since the immobilized enzymes may be protected from attack by large inhibitor molecules.

APPLICATION OF IMMOBILIZED ENZYMES

Soluble enzymes have found a wide range of industrial applications in recent times. Since immobilized enzymes have certain special characteristics with inherent advantages associated with their use, considerable interest has been focussed on the study of their applications in various fields such as medicine, food processing, chemical processes, analytical chemistry, leather industry, etc. The future potential for immobilized enzymes lies in novel applications and the development of new products rather than as an alternative to existing processes using non-immobilized enzymes.

1. Medicine

Soetan *et al.* (2010) have extensively reviewed the biochemical and biotechnological applications of immobilized enzymes and drugs.

i. *Drug delivery systems*   Advanced drug delivery systems (ADDS) have found applications in many biomedical fields. Adoption of different types of membranes in ADDS has made it possible to release a drug in an optimal fashion according to the nature of a disease. Examples of drug delivery systems include glucose-sensitive insulin and drug-loaded magnetic particles.

ii. *Glucose-sensitive insulin*   Drug delivery systems in which a drug is liberated in response to a chemical signal (e.g., insulin release in response to rising glucose concentration) can be achieved using this system. The exposure of a glucose-sensitive insulin releasing system results in the oxidation of glucose to gluconic acid and thus a decrease in the pH, protonation and shrinking of the polymer, leading to an increased release of insulin. The polymer swells in size at normal body pH (pH=7.4) and closes the gates. It shrinks at low pH (pH=4) when the blood glucose level increases thus opening from the nanoparticles (Sona, 2010).

iii. *Drug-loaded magnetic nanoparticles*   Nanotechnology offers the methods to send the drugs to targeted sites. The drug is released in a controlled manner reducing side effects due to lower dosage and minimizing drug degradation by using pathways other than gastrointestinal. Magnetic nanoparticles are recently applied in various fields such as MRI imaging, water treatment, hyperthermia and drug delivery systems. Drug-loaded magnetic nanoparticles (DLMNP) have several advantages such as: small particle size, large surface area, magnetic response, biocompatibility and nanotoxicity. DLMNP are administered through injection and directed using external magnets to the right organ. Only small dosage is required resulting in fewer side effects.

Yu *et al.* (2008) have reported a novel *in vivo* strategy for combined cancer imaging and therapy by employing thermally cross-linked superparamagnetic iron oxide nanoparticles as a drug delivery carrier. Kettering *et al.* (2009) have used magnetic iron nanoparticles with cisplatin (drug used to treat various types of cancer) adsorbed in them for drug release in magnetic healing treatments for cancer.

iv. *Leukemia therapy*   Immobilized asparaginase in the form of microcapsules has been used for leukemia therapy. When physiological saline, native asparaginase and asparaginase microcapsules have been injected into lymphosarcoma-affected mice, the implanted tumours first appear after 9 days in the groups administered physiological saline, after 14 days in those administered native asparaginase and only after 60 days in the groups given asparaginase microcapsules. This clearly indicates the potentials of asparaginase microcapsules in suppressing the growth of lymphosarcoma for a longer period.

v. *Immobilized urease for urea removal*   The composite system for an artificial kidney consists of an extracorporeal shunt containing the immobilized urease and various types of adsorbents. In operation, the blood containing high concentrations of urea, creatine, uric acid, etc. enters the extracorporeal shunt containing the immobilized urease and the various types of adsorbents. The urease converts urea to carbon dioxide, and ammonia and ammonium ions along with other charged contaminants are then adsorbed.

2. Food Processing

Immobilized enzymes are of great value in the processing and analysis of food samples. The extent of lactose hydrolysis, whey processing, skimmed milk and cheese production, etc. have been greatly enhanced by using respective enzymes such as lactases, rennin and lipases in immobilized forms (Khan and Alzohairy, 2010). In Japan, the fermentation industry has improved its processing efficiency for amino acids through the use of immobilized aminoacylase. Aminoacylase and glucose isomerase have been demonstrated to be techno-economically feasible. Immobilization of glucoamylase, lactase, protease and flavour-modifying enzymes has received significant attention for food processing (Carpio *et al.*, 2000).

i. *Prevention of chill haze*   To prevent the chill haze (cloudiness), which develops on storage of beer due to the presence of tannin proteins, native papain is mainly used. When the native proteolytic enzyme remains in beer for a long time, an undesirable excess proteolysis occurs. Thus, utilization of papain cross-linked with glutaraldehyde and collagen–papain membrane have been recommended for controlled treatment to prevent chill haze in beer.

ii. *High-fructose corn syrup*   D-glucose is only 70% as sweet as sucrose and is comparatively less soluble in water. Fructose, on the other hand, is 30% more soluble than sucrose and is twice as soluble as glucose at lower temperatures. Therefore, glucose syrup is treated with glucose isomerase to produce high-fructose syrup. 42% fructose syrup is good enough for many uses; but for soft drinks, a 55% fructose syrup is required since high fructose content gives the syrup enhanced sweetening properties with low calorific values. Glucose isomerase is obtained from *Actinoplanes missouriensis, Bacillus coagulans* and *Streptomyces* spp. It is remarkably thermostable and acts at very high substrate concentrations. It is used in the immobilized state obtained by cross-linking with glutaraldehyde; in some cases, a protein diluent may be added. The process is continuous using a packed bed reactor.

iii. *Lactose removal from milk*　About 5% lactose is present in milk food and this must be removed before feeding babies deficient in intestinal lactase. Lactase is used to remove lactose from milk and whey since many people are sensitive to lactose. Yeast (*Kluveromyces lactis*) lactase is immobilized in cellulose triacetate fibres and used in batch stirred tank reactor (STR) employing conditions such as temperature of 5°C and pH in the range of 6.4–6.8 for processing of milk and sweet whey. But, fungal lactase (from *Aspergillus niger*, having optimum pH 3.0–4.0) is immobilized on porous silica and is used in a packed bed reactor (PBR) for treatment of acid whey. The limitations of the technique are (i) product inhibition (by galactose) (ii) production of unwanted oligosaccharides and (iii) microbial contamination.

3. Production of L-amino Acids

Mould aminoacylase has been employed for the industrial production of several L-amino acids. A drawback in this procedure is that in order to isolate L-amino acids from the enzyme reaction mixture, it is necessary to remove the enzyme by pH alteration or by heat treatment. Thus, even if enzyme activity remains in the reaction mixture, the enzyme has to be discarded as there is no suitable procedure for isolating the active enzyme from the mixture. To overcome these disadvantages and to improve this enzymatic method, extensive studies have been carried out for the continuous optical resolution of DL-amino acids using a column packed with immobilized aminoacylase. Aminoacylase bound to DEAE-sephadex has been chosen as the most advantageous enzyme preparation for the industrial production of L-amino acids.

4. Production of Antibiotics

6-aminopenicillanic acid (6-APA) is used as an important intermediary for the synthesis of semi-synthetic penicillin and has been produced industrially from natural penicillin G (phenylacetyl-6-APA) or penicillin V (phenoxyacetyl-6-APA) by the action of penicillin amidase. Both penicillins contain a nucleus of 6-APA and a side chain. The antibiotic activity of the penicillin molecule is governed by the side chain and, when removed and replaced with another, can profoundly alter the antibiotic spectrum and other properties. Penicillin amidase extracted from *E. coli* has been immobilized by covalent binding to DEAE-cellulose activated with 2,4-dichloro-6-carboxymethylamine-S-triazine and used in a column for the production of 6-APA. The immobilized enzyme can be used for continuous operation without significant loss of activity at 37°C for 11 weeks. An advantage of this method is that proteins and other impurities causing allergic reaction are not present in the 6-APA obtained.

Many pharmaceutical companies now operate immobilized enzyme processes for the production of 6-APA on an industrial scale.

Enzyme-based routes are acknowledged as an environment-friendly approach, avoiding organic solvents and working at room temperature. Among different alternatives, the kinetically controlled synthesis, using immobilized penicillin G acylase in aqueous environment with the simultaneous crystallization of the product is the most promising one (Giordano *et al.*, 2006).

5. Bioremediation

There are more than 1,00,000 commercially available dyes with over 7×10^5 tons of dyestuff produced annually and used extensively in textile, dyeing and printing industry (Akhtar *et al.*, 2005a). It is estimated that about 10–15% of the dyes are discharged into the industrial effluents. The discharge of waste water containing high concentration of reactive dyes is a major challenge from the treatment point of view. Recently, peroxidase from bitter gourd (*Momordica charantia*) immobilized on ConA-sephadex has been found highly effective in decolorizing reactive textile dyes compared to its soluble counterpart as the immobilized enzyme loses only 50% activity even after ten cycles of usage (Akhtar *et al.*, 2005b).

6. Production of Biodiesel

Biodiesel has gained importance for its ability to replace fossil fuels which are likely to run out within a century. Biodiesel fuel does not produce sulphur dioxide, halogens and carbon monoxide and the evolution of soot particles is also minimized (Iso *et al.*, 2001). Immobilized enzymes could be employed in biodiesel production with the aim of reducing the production cost by reusing the enzyme (Jegannathan *et al.*, 2006). Most of the researchers have used lipases from various sources for biodiesel production, and various supports such as ceramics, kaolinites, silica and zeolites have been used for lipase immobilization (Yagiz *et al.*, 2007).

7. Analytical Applications

i. *Biosensors*   Enzyme electrodes are a new type of detector or biosensor designed for the potentiometric or amperometric assay of substrates such as urea, amino acids, glucose, alcohol, lactic acid and several elements and ions. An enzyme electrode is basically a combination of an enzyme and an electronic sensing device. In design, the electrode is composed of a given electrochemical sensor in close contact with a thin permeable enzyme membrane capable of reacting specifically with the given substrates. The embedded enzymes in the membrane produce O_2, hydrogen ions, ammonium ions, CO_2 or other small molecules depending on the enzymatic reactions occurring, which are readily detected by the specific sensor. The magnitude of the response determines the concentration of the substrate.

Enzyme-based electrodes represent a major application of immobilized enzymes in medicine (Khan and Alzohairy, 2010). Biosensors used in clinical applications possess advantages such as reliability, sensitivity, accuracy, ease of handling and low cost compared to conventional detection methods. These characteristics in combination with the unique properties of an enzyme render an enzyme-based biosensor ideal for biomedical applications (D'Orazio, 2003). Nanotechnology is also increasingly used in the design of biologically optimized and optically enhanced surfaces for Surface Plasmon Resonance Biosensors (Hoa *et al.*, 2007). The first enzyme electrode has been prepared by entrapping glucose oxidase in a gelatinous polymer coating over the surface of a polarographic oxygen electrode. When the electrode is placed in contact with a glucose solution, glucose and O_2 diffuse into the gelatinous layer around the electrode where the diffusing glucose is oxidized, producing gluconic acid.

ii. *Analysis of urea and glucose* Immobilized enzymes have also been used successfully for analytical purposes in columns. A column of glucose oxidase has been used for the analysis of glucose (Weibel *et al.*, 1973). Urease and glucose oxidase covalently attached to partially hydrolysed nylon powder have been utilized for automated analyses of urea and glucose respectively (Inman and Hornby, 1972).

8. Leather Industry

A good deal of work has already been carried out on the use of soluble enzymes in the pretanning operations for the manufacture of different types of leather. But, not much work on the use of immobilized enzymes in leather processing has been reported. An immobilized pancreatic enzyme product has been developed for use in bating and dehairing of skins and hides (Puvanakrishnan *et al.*, 1980b). A new bating process for the manufacture of glace kid (Puvanakrishnan *et al.*, 1980a) and a novel method of enzymatic dehairing of sheep skins (Puvanakrishnan *et al.*, 1981) have been developed by utilizing the immobilized enzyme product. Use of immobilized enzyme product in dehairing and bating will be discussed in the subsequent Chapters 16 and 17.

IMMOBILIZATION OF MICROBIAL WHOLE CELLS

Immobilization of intracellular enzymes can be carried out only when the enzymes are extracted from the cells. But, the extracted intracellular enzymes become unstable in many cases. On the other hand, if microbial cells are immobilized directly without the extraction of the enzymes, the cells can be used as a solid catalyst avoiding extraction procedures. Moreover, if microbial cells having multienzyme systems are immobilized, fermentative methods involving multienzyme reactions may be replaced with continuous reactions involving immobilized cells. In these immobilized cells, the cells are in growing, resting and autolysed states. In some cases, the immobilized microbial cells are dead but the enzyme activities remain. This immobilization procedure involves certain preconditions: (i) the microorganism should not contain interfacing enzymes catalysing side reactions (ii) interfering enzymes, if any, should be easily inactivated by simple methods such as heat treatment or pH alteration and (iii) the substrate and product should pass easily through the microbial cell membrane.

Three main methods are available for the immobilization of microbial cells.

Carrier binding method   This is based on linking microbial cells directly to water-insoluble carriers. This method is not found to be advantageous because of the leakage of the enzyme through autolysis during the enzyme reaction.

Covalent binding method   The covalent binding method is one of the most frequently used techniques for the immobilization of enzymes. However, this method is not so widely used for cells because of the toxicity of the agents and the difficulty of finding the proper conditions of immobilization. Navarro and Durand (1977) have immobilized *Saccharomyces carlsbergensis* by covalent binding to porous silica beads that have been treated with 3-aminopropyltriethoxysilane and activated with glutaraldehyde. *Serratia marcescens, Saccharomyces cerevisiae* and *Saccharomyces amurcea* are also bound covalently to borosilicate glass or zirconia ceramics by using an isocyanate coupling agent (Messing *et al.*, 1979).

Cross-linking method   Microbial cells are immobilized by cross-linking among the cells with bi- or multifunctional reagents. *E. coli* cells having high aspartase activity have been cross-linked with glutaraldehyde by this method.

Entrapping method   Extensive work has been carried out for entrapping microbial cells directly into polymer matrices. Commonly employed matrices are polyacrylamide, alginate, carrageenan and anionic polyurethanes.

Application of Immobilized Microbial Whole Cells

The greatest potential of immobilized cell systems lies in replacing complex fermentations such as secondary metabolites, water analysis and waste treatment, continuous malting processes, nitrogen fixation, synthesis of steroids and other valuable medical products. There are many industrial applications for immobilized microbial whole cells.

Production of aspartic acid   Production of aspartic acid from fumaric acid using *E. coli* cells immobilized in polyacrylamide gel is the best described immobilized cell process. Since 1973, this method has been used to produce 1700 kg of aspartic acid per day per 1 m^3 column. Aspartic acid, which is used in medicines and as a food acidulant, has previously been produced industrially by fermentation using soluble aspartase.

Production of ethanol   Many types of natural, synthetic and inorganic support materials such as alginate beads (Najafpour *et al.*, 2004), sorghum bagasse (Yu *et al.*, 2007), zeolite (Shindo *et al.*, 2001), grape skins (Mallouchos *et al.*, 2002), glass beads (Arasaratnam and Balasubramaniam, 1998), Chrysotile (Joekes *et al.*, 1998) and agar (Lebeau, 1997) have been used for ethanol production from microbial whole cells. Chibata and Tosa (1980) have extensively investigated the production of ethanol by κ-carrageenan-entrapped *S. carlsbergensis* and *S. cerevisiae* cells. When the gels containing about 3.5×10^6 *S. carlsbergensis* per ml of gel are incubated in a nutrient medium at 30°C for 60 h, the number of living cells reaches a constant level of about 5.4×10^9 cells per ml of gel. Under these conditions, the immobilized cells produce continuously about 100 mg per ml (12.7% by volume) of ethanol from 200 mg per ml of glucose at a retention time of 2.5 h (Chibata and Tosa, 1980). In another study, *S. cerevisiae* cells entrapped in κ-carrageenan have been shown to produce about 110 mg per ml of ethanol at a retention time of 2.8 h (Wada *et al.*, 1981).

Cross-linked graft copolymers of carboxymethyl cellulose (CMC) with N-vinyl-2-pyrrolidone (N-VP) is used as a new support material for whole cell immobilization. *S. cerevisiae* is immobilized using entrapment method in the graft copolymers of CMC with N-VP for ethanol fermentation. Immobilized yeast is reused four times without activity loss. Ethanol production is shown to increase to 59.3 g/L from 46.4 g/L when percentage of N-VP in the graft copolymer is increased. This method for immobilization of *S. cerevisiae* has potential for production of ethanol (Gokgoz and Yigitoglu, 2011).

Digestion of raffinose in sugar beet juice   Raffinose present in sugar beet juice is digested with raffinase activity and this leads to a 3% more sugar recovery and reduction in cost of disposal of molasses. Raffinase activity is provided by immobilized cells of the mould *Mortierella vinacea* var. *raffinoseutilizer*. The process is carried out in batch mode using stirred tank reactors. This enzyme may be used to remove raffinose and stachyose present in soybean milk as these sugars cause flatulence.

Production of acrylamide   Acrylamide, used to produce several economically useful polymers, is obtained by addition of a water molecule to acrylonitrile as follows:

$$CH_2 = CHCN + H_2O \longrightarrow CH_2 = CHCONH_2$$

This reaction is achieved by nitrile hydratase activity obtained from *Rhodococcus* sp. The bacterial cells are immobilized in polyacrylamide gel and used at 10°C and pH 8.0–8.5 in semibatch process keeping the substrate concentration below 3%. About 4000 tons of acrylamide is produced every year using this process.

Production of tannase   The ability of immobilized cell cultures of *A. niger* FELT FT3 to produce extracellular tannase has been investigated. Production of tannase is increased by entrapping the fungus in nylon scouring mesh cubes compared to free cells. Using optimized parameters of six scouring mesh cubes and inoculum size of 1×10^6 spores/ml, tannase production of 3.98 U/ml is obtained from the immobilized cells compared to free cells (2.81 U/ml) and the increment is about 41.64%. The immobilized cultures exhibit significant tannase production stability for two repeated runs (Darah *et al.*, 2011).

Lipase production Different matrices, viz., agar, alginate and polyacrylamide have been examined for the immobilization of whole cells of *Ralstonia pickettii* for production of lipase. When different concentrations of polyacrylamide are tried for immobilization of *R. pickettii* whole cells, 15% polyacrylamide blocks show a retention activity of 66 % (25 U/ml per min) when compared to that of the free cells (40 U/ml/min). Optimal immobilized whole cell concentration (polyacrylamide blocks) for lipase production is 20%, and polyacrylamide blocks are reused three times effectively for lipase production. Among the three matrices examined (agar, alginate and polyacrylamide), polyacrylamide has given the best performance (Hemachander *et al.*, 2001).

BIBLIOGRAPHY

Adamich, M., Voss, H.F. and Dennis, E.A. (1978). "Cobra venom phospholipase A2 immobilized to porous glass beads." *Arch. Biochem. Biophys.* 189, 417–423.

Akhtar, S., Khan, A.A. and Husain, Q. (2005a). "Partially purified bitter gourd (*Momordica charantia*) peroxidase catalyzed decolorization of textile and other industrially important dyes." *Bioresour. Technol.* 96, 1804–1811.

Akhtar, S., Khan, A.A. and Husain, Q. (2005b). "Potential of immobilized bitter gourd (*Momordica charantia*) peroxidases in the decolorization and removal of textile dyes from polluted wastewater and dyeing effluent." *Chemosphere.* 60, 291–301.

Arasaratnam, V. and Balasubramaniam, K. (1998). "The use of monochloroacetic acid for improved ethanol production by immobilized *Saccharomyces cerevisiae.*" *World J. Microbiol. Biotechnol.* 14, 107–111.

Bernfeld, P. and Wan, J. (1963). "Antigens and enzymes made insoluble by entrapping them into lattices of synthetic polymers." *Science.* 142, 678–679.

Bornscheuer, U.T. (2003). "Immobilizing enzymes: how to create more suitable biocatalysts." *Angen. Chem. Int. Ed. Engl.* 42, 3336–3337.

Brotherton, J.E., Emery, A, and Rodwell, V.W. (1976). "Characterization of sand as a support for immobilized enzymes." *Biotechnol. Bioeng.* 18, 527–543.

Cao, L., Langen, L.V. and Sheldon, R.A. (2003). "Immobilised enzymes: carrier-bound or carrier-free?" *Curr. Opin. Biotechnol.* 14, 387–394.

Cao, L., Rantwijk, F.V. and Sheldon, R.A. (2000). "Cross-linked enzyme aggregates: a simple and effective method for the immobilization of penicillin acylase". *Org. Lett.* 2, 1361–1364.

Carpio, C., Gonzalez, P., Ruales, J. and Batista-Viera, F. (2000). "Bone-bound enzymes for food industry application." *Food Chem.* 68, 403–409.

Chang, T.M.S. (1971). "The *in vivo* effects of semipermeable microcapsules containing L-asparaginase on 6C3HED lymphosarcoma." *Nature.* 229, 117.

Chen, B., Miller, E.M., Miller, L., Maikner, J.J. and Gross, R.A. (2007a). "Effects of macroporous resin size on *Candida antarctica* lipase B Adsorption, Fraction of active molecules, and catalytic activity for polyester synthesis." *Langmuir.* 23, 1381–1387.

Chen, B., Miller, E.M. and Gross, R.A. (2007b). "Effects of porous polystyrene resin parameters on *Candida antarctica* Lipase B adsorption, distribution, and polyester synthesis activity." *Langmuir.* 23, 6467–6474.

Chibata, I. and Tosa, T. (1980). "Immobilized microbial cells and their applications." *Trends in Biochem. Sci.* 5, 88–90.

D'Orazio, P. (2003). "Biosensors in clinical chemistry." *Clin. Chim. Acta.* 334, 41–69.

Darah, I., Sumathi, G., Jain, K. and Lim, S.H. (2011). "Tannase enzyme production by entrapped cells of *Aspergillus niger* FETL FT3 in submerged culture system." *Bioproc. Biosystem. Eng.* 34, 795–801.

El Nashar, M.M.M. (2010). "Immobilized molecules using biomaterials and nanobiotechnology." *J. Biomat. Nanobiotechnol.* 1, 61–77.

Fang, Y., Huang, X.J, Chen, P.C. and Xu, Z.K. (2011). "Polymer materials for enzyme immobilization and their application in bioreactors." *BMB Reports.* 44, 87–95.

Foresti, M.L. and Ferreira, M.L. (2007). "Analysis of the interaction of lipases with polypropylene of different structure and polypropylene-modified glass surface." *Colloids Surf. A; Physicochem. Eng. Aspects.* 294, 147–155.

Giordano, R.C., Ribeiro, M.P. and Giordano, R.L. (2006). "Kinetics of beta-lactam antibiotics synthesis by penicillin G acylase (PGA) from the viewpoint of the industrial enzymatic reactor optimization." *Biotechnol. Adv.* 24, 27–41.

Gokgoz, M. and Yigitoglu, M. (2011). "Immobilization of *Saccharomyces cerevisiae* on to modified carboxymethylcellulose for production of ethanol." *Bioprocess Biosyst Eng.* 34, 849–857.

Grubhofer, N. and Schleith, L. (1953). "Modified ion-exchange resins as specific adsorbent." *Naturwissenschaften.* 40, 508–508.

Hemachander, C., Bose, N. and Puvanakrishnan, R. (2001). "Whole cell immobilization of *Ralstonia pickettii* for lipase production." *Process Biochem.* 36, 629–633.

Hoa, X.D., Kirk, A.G. and Tabrizian, M. (2007). "Towards integrated and sensitive surface plasmon resonance biosensors: a review of recent progress." *Biosen. Bioelectron.* 23, 151–160

Inman, D.J. and Hornby, W.E. (1972). "The immobilization of enzymes on nylon structures and their use in automated analysis." *Biochem.J.* 129, 255–262.

Iso, M., Chen, B., Eguchi, M., Kudo, T. and Shrestha, S. (2001). "Production of biodiesel fuel from triglycerides and alcohol using immobilized lipase." *J. Mol. Catal. B: Enz.* 16, 53–58.

Jegannathan, J., Bassi, A. and Nakhla, G. (2006). "Pre-treatment of high oil and grease pet food industrial wastewaters using immobilized lipase hydrolyzation." *J. Hazard. Mater.* 137, 121–128.

Jegannathan, K.R., Chan, E.S. and Ravindra, P. (2009). "Physical and stability characteristics of *Burkholderia cepacia* lipase encapsulated in κ-carrageenan." *J. Mol. Catal. B; Enzym.* 58, 78–83.

Joekes, I., Moran, P.J.S., Rodrigues, J.A.R., Wenchausen, R., Tonella, E. and Cassiola, F. (1998). "Characterization of *Saccharomyces cerevisiae* immobilized onto chrysotile for ethanol production." *J. Chem. Technol. Biotechnol.* 73, 54–58.

Kettering, M., Zorn, H., Bremer-Streck, S., Behring, H., Zeisberg, M., Bergeman, C., Hergt, T., Halbhuber, J., Kaiser, A. and Hilger, I. (2009). "Characterization of iron oxide nanoparticles adsorbed with cisplatin for biomedical applications." *Phys. Med. Biol.* 54, 5109–5121.

Khan, A.A. and Alzohairy, M.A. (2010). "Recent advances and applications of immobilized enzyme technologies: A Review Res." *J. Biol. Sci.* 5, 565–575.

Lebeau, T., Jouenne, T and Junter, G.A. (1997). "Fermentation of D-xylose by free and immobilized *Saccharomyces cerevisiae* cultures." *Biotechnol. Lett.* 19, 615–618.

Le-Clech, P. (2010). "Membrane bioreactors and their uses in wastewater treatments." *Appl. Microbiol. Biotechnol.* 88, 1253–1260.

Mallouchos, A., Reppa, P., Aggelis, G., Kanellaki, M., Koutinas, A.A. and Komaitis, M. (2002). "Grape skins as a natural support for yeast immobilization." *Biotechnol. Lett.* 24, 1331–1335.

Mansfeld, J. and Schellenberger, A. (1987). "Invertase immobilized on macroporous polystyrene: Properties and kinetic characterization." *Biotechnol. Bioeng.* 29, 72–78.

Martin, M.T., Plon, F.J., Alcalde, M. and Ballesteros, A. (2003). "Immobilization on Eupergit C of cyclodextrin glucosyltransferase (CGTase) and properties of the immobilized biocatalyst." *J. Molecul. Catalysis B: Enzymatic.* 21, 299–308.

Mazzei, R., Giorno, L., Piacentini, E., Mazzuca, S. and Drioli, E. (2009). "Kinetic study of a biocatalytic membrane reactor containing immobilized β-glucosidase for the hydrolysis of oleuropein." *J. Membr. Sci.* 339, 215–223.

Messing, R.A. (1978) In: *"Advances in Biochemical Engineering"*, Vol.10. Springer Verlag. pp. 60–61.

Messing, R., Oppermann, R.A. and Kolot, F.B. (1979). "Pore dimensions for accumulating biomass in immobilized microbial cells." *Am. Chem. Soc. Symp. Ser.* 106, 13–28.

Najafpour, F., Younesi, H. and Ismail, K. (2004). "Ethanol fermentation in an immobilized cell reactor using *Saccharomyces cerevisiae*." *Bioresour. Technol.* 92, 251–260.

Navarro, J.M. and Durand, G. (1977). "Modification of yeast metabolism by immobilization onto porous glass." *Eur. J. Appl. Microbiol.* 4, 243–254.

Olson, A.C. and Korus, R.A. (1977). In: R.L.Ory and A.J. St. Angelo (Eds.). *Enzymes in Food and Beverages processing.* Amer. Chem. Soc. pp.100–131.

Onyezili, F.N. (1987). "Glutaraldehyde activation step in enzyme immobilization on nylon." *Biotechnol. Bioeng.* 29, 399–402.

Prlainovic, N.Z., Knezevix-Jugovic, Z.D., Mijin, D.Z. and Bezbradica, D.I. (2011). "Immobilization of lipase from *Candida rugosa* on Sepabeads(®): the effect of lipase oxidation by periodates." *Bioprocess Biosyst. Eng.* 34, 803–810.

Puls, J., Sinner, M. and Dietrichs, H.H. (1977). "Hydrolysis of hemicelluloses by immnobilised enzymes." *Trans. Tech. Sect., Can. Pulp Paper Assoc.* 3, 64–72.

Puvanakrishnan, R. and Bose, S.M. and Dhar, S.C (1980b). "A new process for the bating of skins and hides by the use of immobilized pancreatic enzyme product for the manufacture of different types of leather." Indian Patent No.151089 (1980).

Puvanakrishnan, R. and Bose, S.M. (1980). "Studies on the immobilization of trypsin on sand." *Biotechnol. Bioeng.* 22, 919–928.

Puvanakrishnan, R. and Bose, S.M. (1984). "Immobilization of pepsin on sand: Preparation, characterization and application." *Indian J. Biochem. Biophys.* 21, 323.

Puvanakrishnan, R. and Bose, S.M. and Dhar, S.C. (1981). "Comparative studies on enzymatic unhairing using immobilized pancreatic enzyme product and CLRI enzyme depliant "M" in the manufacture of grain garment leather." *Leath. Sci.* 28, 32.

Puvanakrishnan, R., Bose, S.M. and Dhar, S.C. (1980a). "Comparative studies on bating using immobilized pancreatic enzyme product and pancreatin bate in the manufacture of glace kid." *Leath.Sci.* 27, 81.

Sreeja, C., Jasmin, C., Soorej, M., Kishore, B.A., Elyas, K.K., Bhat, S.G. and Chandrasekaran, M. (2011). "Characterization of an extracellular alkaline serine protease from marine *Engyodontium album* BTMFS10." *J. Ind. Microbiol. Biotechnol.,* 38, 743–752.

Sawada, S.I. and Akiyoshi, K. (2010). "Nano-encapsulation of lipase by self assembled nanogels: Induction of high enzyme activity and thermal stabilization." *Macromol. Biosci.* 10, 353–358.

Soetan, K., Aiyelaugbe, O. and Blaiya, C. (2010). "A review of the biochemical, biotechnological and other applications of enzymes." *African J. Biotechnol.* 9, 382–393.

Shimizu, K. and Ishihara, M. (1987). "Immobilization of cellulolytic and hemicellulolytic enzymes on inorganic supports." *Biotechnol. Bioeng.* 29, 236–241.

Shindo, S., Takata, S., Taguchi, H. and Yoshimura, N. (2001). "Development of novel carrier using natural zeolite and continuous ethanol fermentation with immobilized *Saccharomyces cerevisiae* in a bioreactor." *Biotechnol. Lett.* 23, 2001–2004.

Simionescu, C., Popa, M.I. and Dumitriu, S. (1987). "Bioactive polymers XXX. Immobilization of invertase on the diazonium salt of 4-aminobenzoylcellulose." *Biotechnol. Bioeng.* 29, 361–365.

Sona, P. (2010). "Nanoparticulate drug delivery systems for the treatment of Diabetes." *Digest Journal of Nanomaterials and Biostructures.* 5, 411–418.

Sreenivasulu, S., Dhar, S.C. and Puvanakrishnan, R. (1984). "Bentonite: A novel support for the immobilization of alkaline proteinase." *Leath. Sci.* 31, 335–341.

Synowiecki, J., Sikorska-Siondalska, A. and El-Bedawey, A.E. (1987). "Adsorption of enzymes on krill chitin modified with carbon disulfide." *Biotechnol. Bioeng.* 29, 352–354.

Takeuchi, T. and Makino. (1987). "Cellulase immobilized on poly-L-glutamic acid." *Biotechnol. Bioeng.* 29, 160–164.

Wada, M., Kato, J. and Chibata, I. (1981). "Continuous production of ethanol in high concentration using immobilized growing yeast cells." *Eur. J. Appl. Microbiol. Biotechnol.* 11, 67–71.

Weetall, H.H. and Filbert, A.M. (1974). "Porous glass for affinity chromatography applications." *Meth. Enzymol.* 34, 59–72.

Weibel, M.K., Dritschilo, W., Bright, H.J. and Humphrey, A.E. (1973). "Immobilized enzymes: a prototype apparatus for oxidase enzymes in chemical analysis utilizing covalently bound glucose oxidase." *Anal. Biochem.* 52, 402–414.

Yagiz, F., Kazan., D and Akin, A.L.N. (2007). "Biodiesel production from waste oils by using lipase immobilized on hydrotalcite and zeolites." *Chem. Eng. J.* 134, 262–267.

Ye, P., Xu, Z.K., Wang, Z.G., Wu, J., Deng, H.T. and Seta, P. (2005). "Comparison of hydrolytic activities in aqueous and organic media for lipases immobilized on poly(acrylonitrile-co-maleic acid) ultrafiltration hollow fiber membrane." *J. Mol. Catal. B: Enzym.* 32, 115–121.

Yu, J., Zhang, X. and Tan, T. (2007). "A novel immobilization method of *Saccharomyces cerevisiae* to sorghum bagasse for ethanol production." *J. Biotechnol.* 129, 415–420.

Yu, M., Jeorg, Y., Park, J., Park, S., Kim, J., Min, J., Kim, K. and Jon, S. (2008). "Drug-loaded superparamagnetic iron oxide nanoparticles for combined cancer imaging and therapy in vivo." *Angewandte Chemice Inter National Edition.* 47, 5362–5365.

Zaborsky, O.R. (1973). *Immobilized Enzymes.* CRC Press, Cleveland, Ohio, pp.1–2.

6

NUCLEIC ACIDS—STRUCTURE AND FUNCTION

Nucleic acids, viz. deoxyribonucleic acid (DNA) and ribonucleic acid (RNA), are biomolecules essential for life. They have quite different, although related, functions within the cell. Nucleic acids account for 0.5–1 percent of the dry weight of a cell. DNA and RNA are macromolecular structures composed of regular repeating units called nucleotides. The two biomolecules differ in the chemical structure of their component nucleotides. DNA serves as the carrier of genetic information from one generation to the next, for almost all organisms (except a few viruses). RNA plays a major role in the biosynthesis of proteins.

COMPONENTS OF NUCLEIC ACIDS

Nucleic acids are composed of three types of molecules—pentose sugar, nitrogenous bases (purines and pyrimidines) and phosphoric acid.

Pentose Sugar

Two types of pentoses are found in nucleic acids. The sugar present in RNA is L-ribose, while DNA contains 2´-deoxy-D-ribose. The difference lies in the presence/absence of the hydroxyl group at position 2 of the sugar ring as illustrated in Figure 6.1. The oxygen atom present at the second carbon of D-ribose is missing in 2´-deoxy-D-ribose and thus the name.

Organic Bases

The organic bases found in nucleic acids are heterocyclic compounds containing nitrogen in their rings; they are also called nitrogenous bases. Nitrogenous bases are grouped into two classes based on their chemical structure. They are the purines [adenine (A) and guanine (G)] and pyrimidines [thymine (T), cytosine (C) and uracil (U)] (Figure 6.1). Pyrimidines have a six-member ring, while purines have fused five and six member rings. DNA contains four different bases, viz., A, G, C and T, while RNA contains the same bases, except for thymine (T) which is replaced by uracil (U). The only difference between uracil and thymine is the presence of a methyl substituent at carbon-5 in thymine.

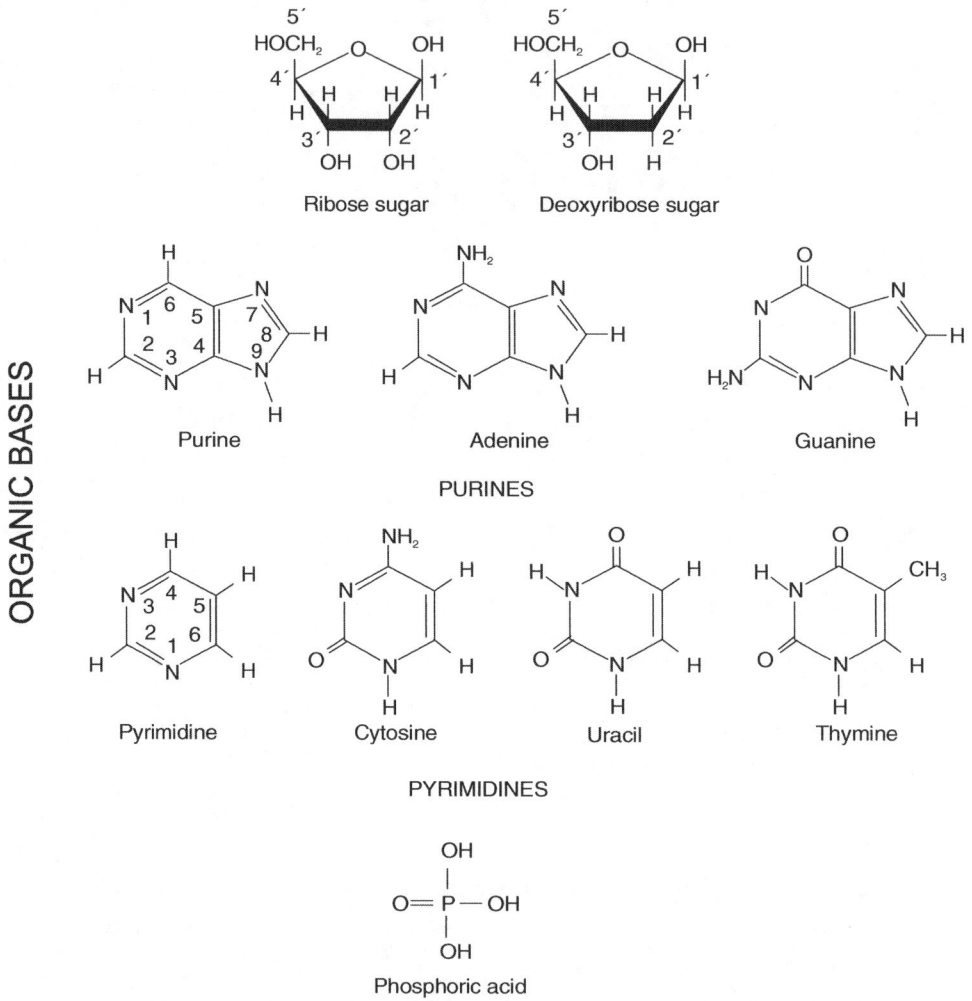

Figure 6.1   Components of nucleic acids

Phosphoric Acid

Phosphoric acid (H_3PO_4) (Figure 6.1) has three reactive (—OH) groups, of which two are involved in forming the sugar–phosphate backbone of DNA. A phosphate moiety binds to the 5′ carbon of one and the 3′ carbon of the neighbouring pentose molecule of DNA to produce the phosphodiester linkage. When DNA or RNA is broken into its constituent nucleotides, the cleavage may take place on either side of the phosphodiester bond.

NUCLEOSIDES

In nucleic acids, organic bases are linked with the pentose molecules. The linkage between ribose and the different organic bases yields ribosides, while those between deoxyribose and the organic bases yield deoxyribosides.

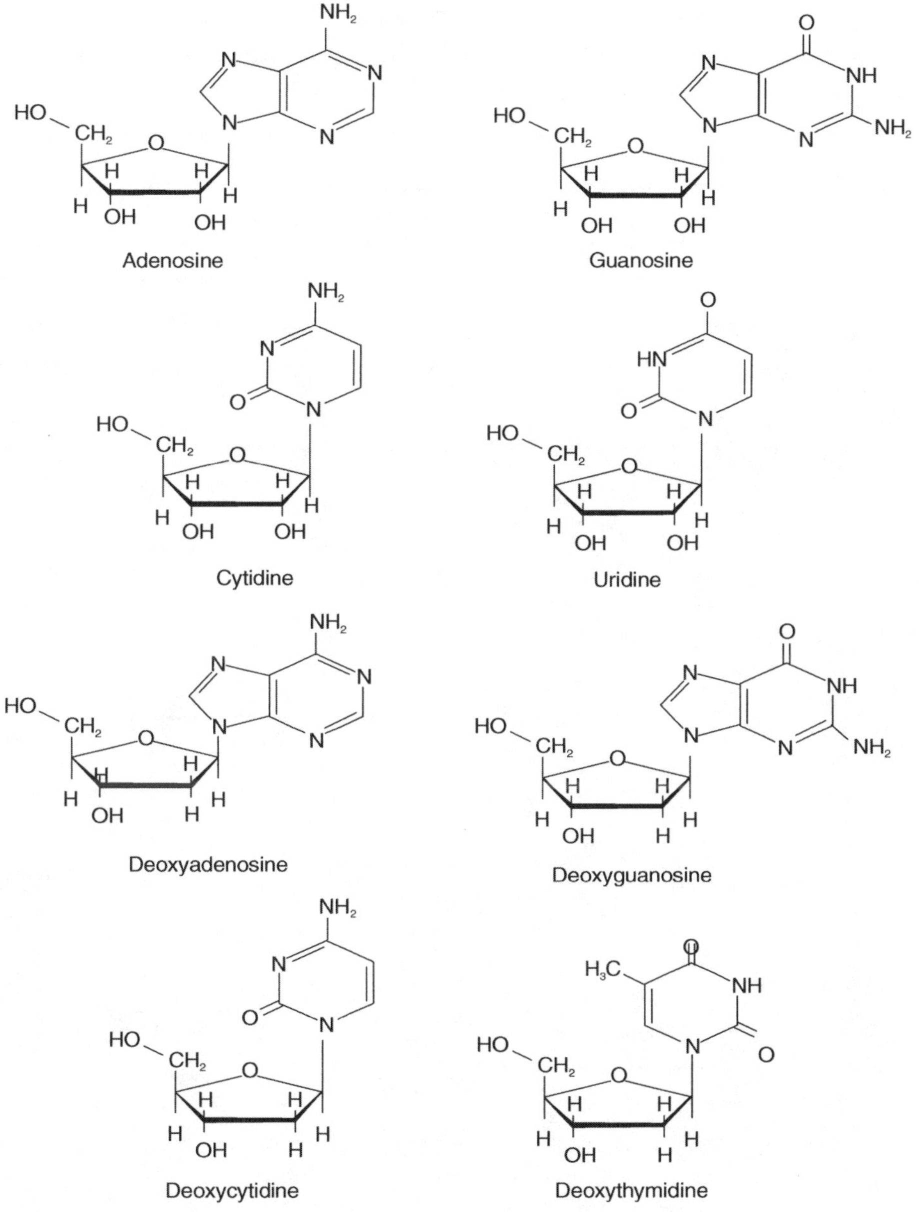

Figure 6.2   Nucleosides

The four common ribosides are called adenosine, guanosine, cytidine and uridine. Similarly, the four common deoxyribosides are deoxyadenosine, deoxyguanosine, deoxycytidine and deoxythymidine or simply called thymidine (Figure 6.2).

NUCLEOTIDES

A nucleotide is formed when a phosphate group is attached to either the 3'C or the 5'C of the pentose molecule of a nucleoside. The nucleotides produced by the different ribosides are together known as ribonucleotides or ribotides. Similarly, the nucleotides formed by the four deoxyribosides are called deoxyribonucleotides or deoxyribotides (Figure 6.3).

Ribonucleotides	Deoxyribonucleotides
Adenosine monophosphate (AMP)	Deoxyadenosine monophosphate (dAMP)
Guanosine monophosphate (GMP)	Deoxyguanosine monophosphate (dGMP)
Cytidine monophosphate (CMP)	Deoxycytidine monophosphate (dCMP)
Uridine monophosphate (UMP)	Deoxythymidine monophosphate (dTMP)

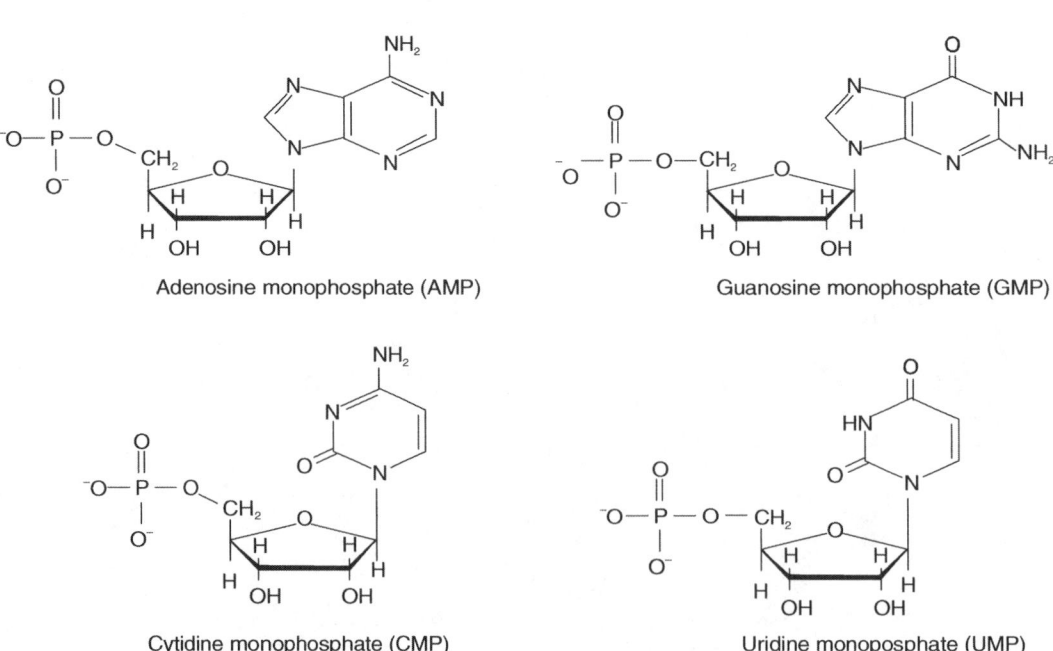

Adenosine monophosphate (AMP)

Guanosine monophosphate (GMP)

Cytidine monophosphate (CMP)

Uridine monoposphate (UMP)

Figure 6.3   Nucleotides (Continued)

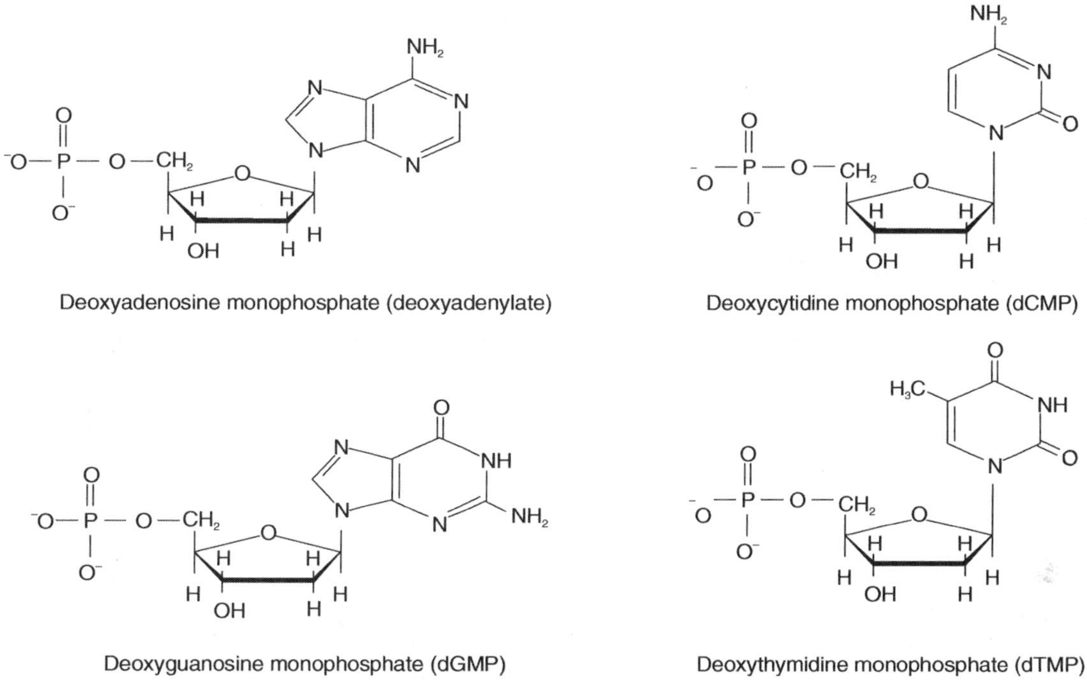

Figure 6.3 Nucleotides

PRIMARY STRUCTURE OF DNA

DNA molecule is double-stranded. Each of the two strands of a DNA molecule is a multimer of deoxyribonucleotides known as a polynucleotide. The two polynucleotide chains in the double helix are associated through hydrogen bonding between the bases. The neighbouring nucleotides in a polynucleotide are joined with each other by phosphodiester linkage, which generates the primary structure of DNA. Thus, a polynucleotide chain consists of several pentose and phosphate molecules linked with each other by phosphodiester linkage. Each pentose has any one of the four organic bases attached to its 1′C; the base is not directly involved in the formation of sugar-phosphate backbone of a polynucleotide.

DNA DOUBLE HELIX

Watson and Crick used chemical analyses and X-ray crystallographic studies of Wilkins and Franklin to build the models of DNA molecule (Figure 6.4).

In 1953, they proposed the double-helix model of DNA and the main features of this model are summarized below:

1. DNA molecule is made up of two polydeoxyribonucleotide strands.

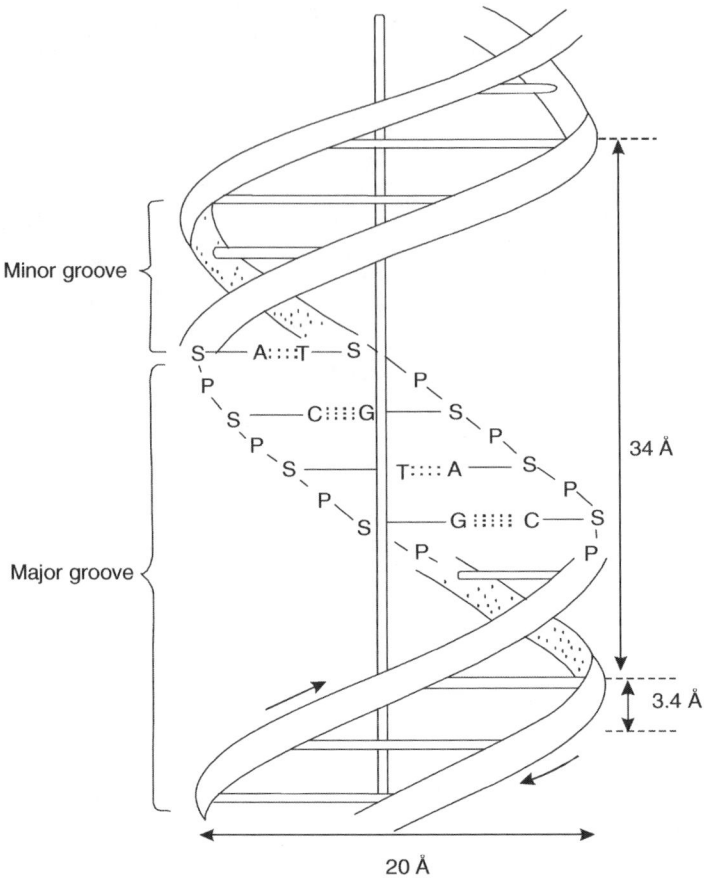

Figure 6.4   Watson–Crick model of DNA (A—Adenine; T—Thymine; C—Cytosine; G—Guanine; P—Phosphate; S—Sugar)

2. The two strands of the DNA molecule run in opposite directions (antiparallel), i.e., one strand runs in $5' \rightarrow 3'$ direction, while its partner runs in $3' \rightarrow 5'$. This antiparallel orientation of the two strands is essential for formation of hydrogen bonds between the pairs of DNA bases.

3. Each polydeoxyribonucleotide strand is composed of many deoxyribonucleosides joined by phosphodiester linkage between their sugar units (Figure 6.5).

The sequence or order of the nucleotides defines the primary structure of DNA and RNA. The nucleotides of the polymer are linked by phosphodiester bonds connecting through the oxygen on the $5'$ carbon of one to the oxygen on the $3'$ carbon of another. According to complementary base-pairing, adenine pairs with thymine and guanine pairs with cytosine in DNA.

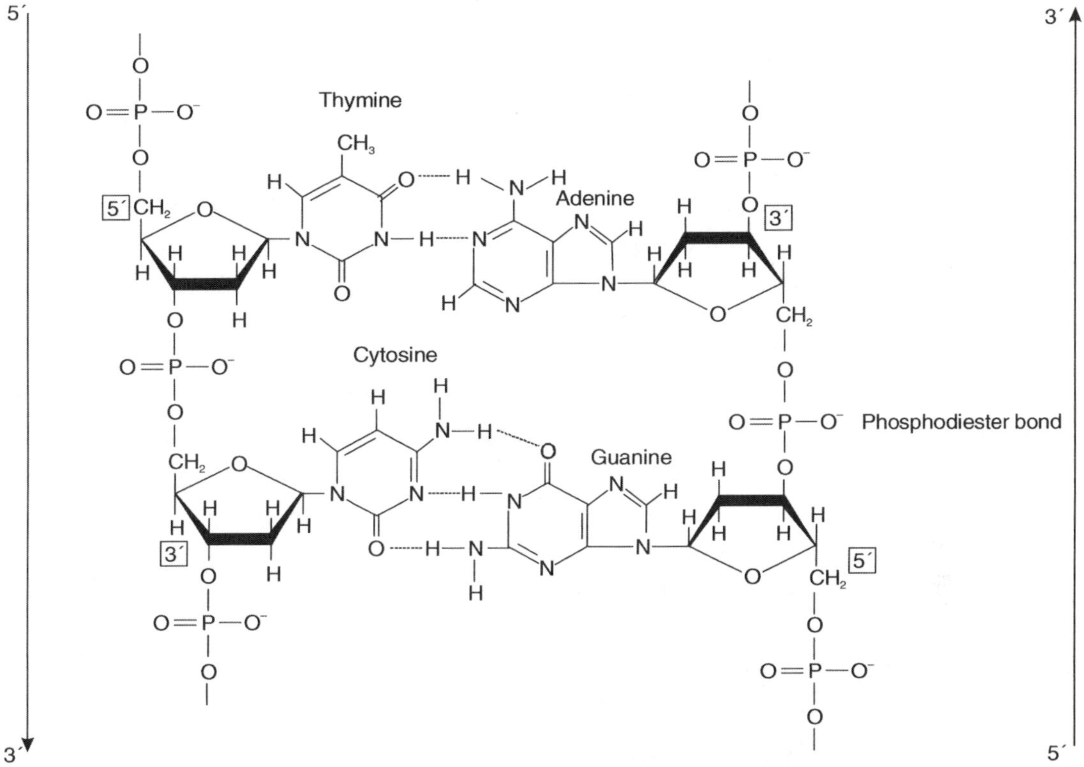

Figure 6.5 Phosphodiester bonds between nucleotides and complementary base-pairing

4. The base sequence of the two strands of a DNA molecule show the following universal relationship: (i) wherever adenine occurs in one strand, thymine is present in the corresponding positions of the other strand and vice versa (ii) similarly, the sites at which guanine is present in one strand are occupied by cytosine in the second strand and vice versa. Due to this relationship, the two strands of a DNA molecule are called complementary strands. Further, this accounts for the equality of A to T and G to C in DNA.

5. **Complementary base-pairing**   Figure 6.5 shows that in their usual forms, A can pair specifically only with T, while G can hydrogen-bond specifically only with C. This is described as base-pairing and the paired (G with C or A with T) bases are complementary to each other. The adenine present in one strand of a DNA molecule is linked by two hydrogen bonds with the thymine located opposite to it in the second strand, and vice versa. Similarly, G located in one strand forms three hydrogen bonds with the C present opposite to it in the second strand and vice versa. A hydrogen bond is relatively weak. But, since a DNA molecule has many hydrogen bonds, they effectively keep the two strands of DNA molecule together. Hydrogen bonds are readily disrupted when DNA is either heated or

when the pH of a DNA suspension is raised in the alkaline range; this produces single-stranded DNA (denaturation).

6. The bases are flat structures lying in pair perpendicular to the axis of the helix. The energy released by hydrogen-bonding between the paired bases contributes to the thermodynamic stability of the duplex.

7. The two strands of a DNA molecule are coiled together in a right-handed helix forming the DNA double helix and this represents the B-form of DNA. The diameter of this helix is 20 Å, while its pitch (length of helix required to complete one turn) is 34 Å. In each DNA strand, the bases occur at a regular interval of 3.4 Å so that about 10 base pairs are present in one pitch of a DNA double helix. The base pairs in a DNA molecule are stacked between sugar–phosphate backbones of the two strands.

8. During replication, the two strands of DNA molecule uncoil and the unpaired bases in the single-stranded regions of the two strands bind with their complementary bases present in the cytoplasm nucleotides. These nucleotides are joined by phosphodiester linkages generating complementary strands of the old ones. This mechanism is called semi-conservative replication. This provides almost error-free, high-fidelity replication of the genetic material.

9. Sometimes, errors in the base-pairing may occur during replication which would account for the occurrence of mutations.

STRUCTURAL FORMS OF DNA

The B-form is found in fibres of very high (92%) relative humidity and in solutions of low ionic strength. This corresponds to the form that is thought to prevail in the living cell. A-form is found in fibres at 75% relative humidity and requires the presence of sodium, potassium or cesium as the counter ion. Instead of lying flat, the bases are tilted with regard to the helical axis and there are more base pairs per turn. The C-form occurs when DNA fibres are maintained in 66% relative humidity in the presence of lithium ions. It has fewer base pairs per turn than B-DNA. The D-form and E-form have the fewest base pairs per turn (only 8 and 7½) and are taken up only by certain DNA molecules that lack guanine. The Z-form provides the most striking contrast; it is a left-handed helix whereas all the others are right-handed. It has the most base pairs per turn, and so has the least twisted structure. The sugar–phosphate backbone follows a zigzag path along the helix, hence the name, Z-DNA.

STRUCTURE OF RNA

The primary structure of RNA is the same as that of DNA: a polynucleotide chain with sugar–phosphate backbone. RNA, like DNA, is a polynucleotide; it is produced by the phosphodiester linkages between ribonucleotides in the same manner as in the case of DNA. RNA nucleotides have ribose sugar which participates in the formation of sugar phosphate backbone of RNA. Thymine is usually absent in RNA, and uracil is found in its place. Usually, RNA is single-stranded, but double-stranded RNA is also found. Since RNA is single–stranded, the base composition is not restricted to the overall G to C and A to U equalities. However, base–pairing can take place within and between RNA molecules. In most organisms, RNA is involved in biosynthesis of proteins, but in some viruses, it serves as the genetic material.

TYPES OF RNA

There are three types of RNA: messenger RNA (mRNA), transfer RNA (tRNA) and ribosomal RNA (rRNA).

mRNA

The key features of mRNA are:

i. It is single-stranded. It is formed by transcription of complementary sections of DNA.

ii. It provides information for amino acid sequence of the polypeptide.

iii. Its molecular weight varies from 25,000 to 1,000,000.

iv. After its formation, it is transported selectively from the nucleus to the ribosomes.

v. Polycistronic mRNA codes for more than one polypeptide, while monocistronic mRNA codes for single polypeptide.

tRNA

i. They act as carriers of amino acids during protein synthesis.

ii. They have a characteristic clover-leaf structure.

iii. They show considerable double-helical structure and more than 50% of bases in tRNA are paired.

iv. They are relatively small, 73–93 nucleotides long, with molecular weight 23,000 to 30,000.

v. There is at least one tRNA for each amino acid. The three nucleotides at the 3′end are CCA in all tRNA molecules. This segment is the aminoacid acceptor region. The amino acid attaches to one of the –OH group of the terminal adenine nucleotide.

vi. tRNA has three loops, dihydrouridine (DHU) loop, anticodon loop and pseudouridine (T ψ C) loop. Sometimes, an extra loop having variable number of nucleotides is also present.

vii. The anticodon loop has an anticodon (a triplet base which is complementary to the codon for the particular amino acid) which can bind to the appropriate codon on the mRNA (Figure 6.6).

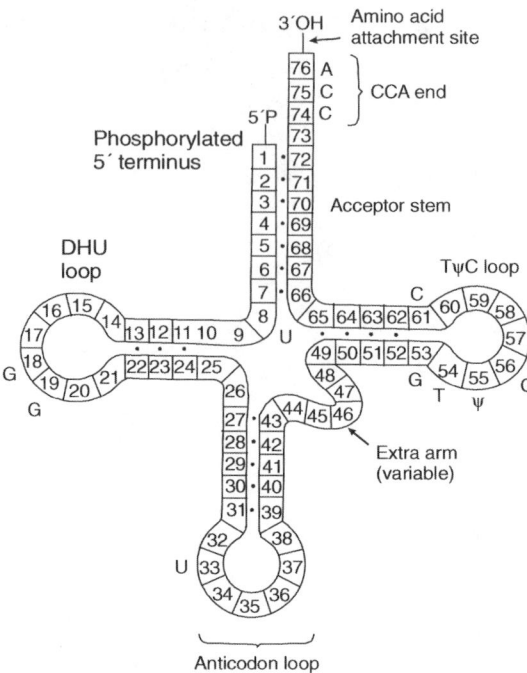

Figure 6.6 Structure of t-RNA. The folded structure of tRNA molecule gives rise to three loops viz., Dihydrouridine (DHU) loop, anticodon loop and Pseudouridine (T ψ C) loop, an extra arm and an amino acid acceptor stem.

viii. Specific enzymes enable amino acids to bind to their tRNA molecules via ester bonds. The ester bonds are formed between the —COOH group of the amino acid and the —OH at the end of the tRNA nucleic acid chain and their formation needs energy from the hydrolysis of ATP. The correct attachment of tRNA-amino acid to ribosome-bound mRNA also requires energy from the hydrolysis of ATP and occurs by anticodon–codon base-pairing.

rRNA

i. It is a structural component of the ribosomes. It makes about 80% of total cellular RNA.

ii. The molecules consist of single strands of RNA and have molecular weight up to 1,000,000.

iii. The specific function of rRNA is not fully established, but it binds to the mRNA to provide the site for protein synthesis.

A comparison between DNA and RNA is represented in Table 6.1.

Table 6.1   DNA AND RNA—A COMPARISON

Characteristic	DNA	RNA
Pentose	Deoxyribose	Ribose
Base	A, T, G, C. Uracil absent	A, U, G, C. Thymine is replaced by Uracil.
Number of strands	Generally double-stranded	Generally single-stranded
Function	Genetic material only	Generally involved in protein biosynthesis. Genetic material in some viruses.
Origin	i. Replication of pre-existing DNA ii. In case of infection by RNA virus, reverse transcription of genetic RNA	i. Genetic RNA either through transcription of DNA or through replication of RNA by RNA-dependent RNA polymerase. ii. Non-genetic RNA from transcription of DNA

REPLICATION OF DNA

The two strands of DNA double helix get separated to act as templates for the synthesis of daughter strands. After a round of replication, two daughter duplexes are formed; each one consists of one parental strand and one newly formed strand. Thus, the parental strand is conserved from one generation to the next. This mechanism is described as semiconservative replication. The evidence for semiconservative replication of DNA was first presented by Meselson and Stahl in 1958. DNA replication is catalysed by the enzyme DNA polymerase and it involves several other proteins/enzymes. Semiconservative replication of DNA (Figures 6.7a and b) occurs with very high fidelity; it leads to the perpetuation of genetic information. The generalized steps in DNA replication are as follows.

1. The first step in semiconservative replication is the separation of the two complementary strands from each other at a point in the DNA duplex called "origin" where replication begins. The features of origin points conferring this unique property are not fully understood.

2. Two enzymes, DNA gyrase and DNA helicase, bind to the origin points and induce the unwinding of DNA double helix.

3. Certain proteins, called single-strand binding proteins, bind to the single-stranded regions thus produced; as a result, these regions remain single-stranded.

4. Primase initiates the transcription of $3' \rightarrow 5'$ strand (where $3'$ end is single-stranded); it generates a 10–60 nucleotide-long primer RNA (transcribed in $5' \rightarrow 3'$ direction). The sequence of primer so formed is dictated by the DNA template, following the base-pair rules.

5. DNA polymerase can catalyse only the unidirectional extension of a nucleotide chain, i.e., it catalyses the addition of deoxynucleotides to the $3'$ OH end of the chain but not to the $5'$ phosphate end. The free $3'-OH$ of RNA primer provides the initiation point for DNA polymerase for the sequential addition of deoxyribonucleotides. In prokaryotes, e.g. *E. coli, B. subtilis,* etc., DNA polymerase III catalyses DNA replication. DNA polymerase progressively adds deoxyribonucleotides to the free $3'-OH$ of this growing polynucleotide chain so that the replication of the $3' \rightarrow 5'$ strand of the DNA molecule is continuous (growth of the new strand in $5' \rightarrow 3'$ direction).

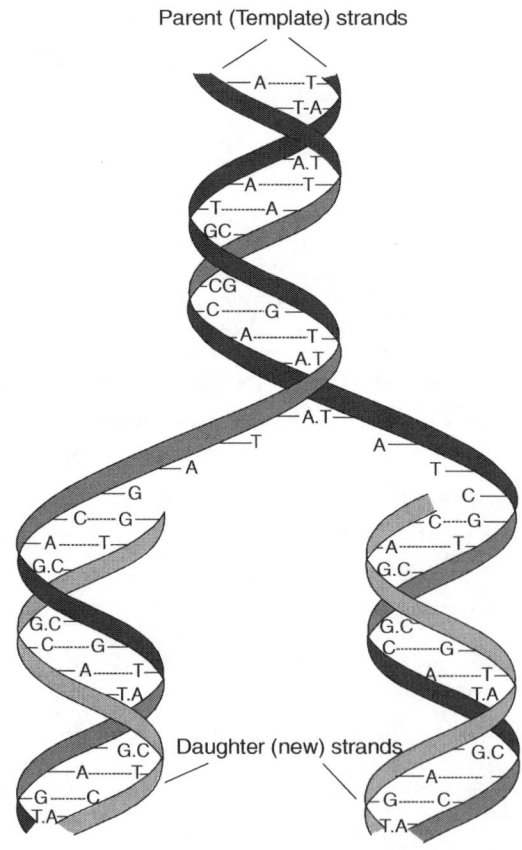

Figure 6.7a   Semiconservative replication of DNA

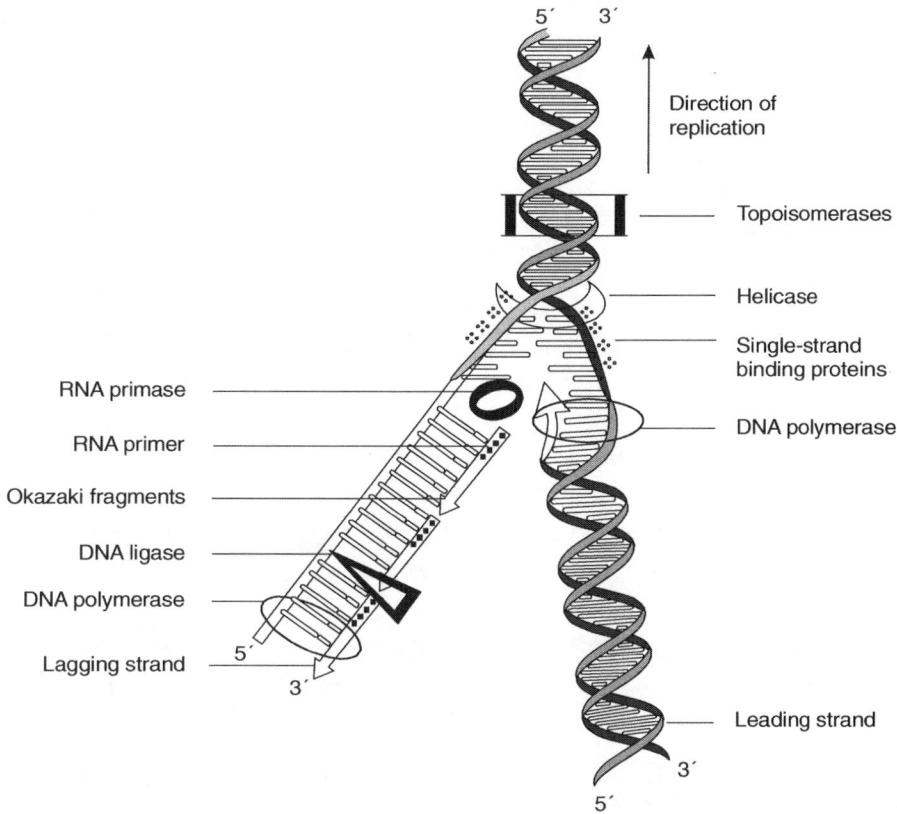

Figure 6.7b   Replication fork depicting leading and lagging strands

6. The replication of the second strand ($5'\rightarrow3'$ strand) of the DNA molecule is discontinuous. The replication of this ($5'\rightarrow3'$) strand begins somewhat later than that of the $3'\rightarrow5'$ strand. Consequently, a segment of the $5'\rightarrow3'$ stand (with reference to the origin) of a DNA molecule always replicates later than the homologous segment of the $3'\rightarrow5'$ strand.

7. When replication of the $3'\rightarrow5'$ strand has progressed for some time, primase initiates the synthesis of RNA primer on the $5'\rightarrow3'$ strand close to the replication fork (away from the origin); the RNA synthesis progresses towards the origin. The $3'-OH$ of this primer RNA provides the initiation point for DNA polymerase to catalyse the replication of the lagging strand. Obviously, the replication of lagging strand proceeds from the replication fork towards the origin, i.e., its direction is opposite to that of the leading strand (from origin towards the fork). However, the new strand is synthesized from the 5'-end to the 3'-end ($5'\rightarrow3'$) in the case of both leading and lagging strands of a DNA molecule.

8. When replication of the leading strand has proceeded further, primase again initiates the transcription of the lagging strand near the fork. The primer RNA thus formed provides the free 3′—OH for the replication of lagging strand.

9. Clearly, the replication of lagging strand generates small polynucleotide fragments called Okazaki fragments, after the biochemist who discovered them. The replication of lagging strand is discontinuous in that it has to be initiated several times, and every time, one Okazaki fragment is produced. Okazaki fragments are about 1000–2000 nucleotides long in *E. coli* while they are only 100–200 nucleotides long in eukaryotes.

10. The exonuclease activity of the DNA polymerase successively removes RNA nucleotides from the RNA primer associated with the newly synthesized DNA strands/Okazaki fragments. This enzyme also catalyses the filling of the gaps so generated in the new strands through semiconservative replication of the old strands. The Okazaki fragments (after the gap-filling by DNA polymerase) are joined together by the enzyme polynucleotide ligase which catalyses the formation of phosphodiester bonds between the immediate neighbour nucleotides of the adjacent fragments.

11. Since initiation of the discontinuous strand (the one formed from Okazaki fragments) cannot begin until the continuously formed strand has opened the helix, synthesis of the discontinuous strand lags behind that of the continuous strand. Thus, the continuous strand is referred to as leading strand and the discontinuous strand is referred to as lagging strand.

12. The above description pertains to the DNA replication progressing in one direction from the origin; the same events also occur in the opposite direction of the origin. It appears that the rate of replication in both the directions from the origin are comparable.

13. In the above model, each replication fork is engaged in synthesizing both the strands of a template DNA molecule; however, the replication of lagging strand is somewhat delayed as compared to that of the leading strand.

14. Topoisomerases are isomerase enzymes that act on the topology of DNA. During replication, DNA becomes overwound ahead of a replication fork. If left unabated, this tension would eventually bring replication to a halt. In order to overcome this topological problem, topoisomerases bind to either single-stranded or double-stranded DNA and cut the phosphate backbone of the DNA. This intermediate break allows the DNA to be untangled or unwound, and, at the end of these processes, the DNA backbone is resealed again.

15. In *E. coli*, a replication fork moves through 60,000 base pairs (10 base pairs per turn, hence, 6000 turns of the helix) per minute, while in eukaryotes, the rate is about ten times less.

THE CENTRAL DOGMA

DNA ultimately controls every aspect of cellular function, primarily through protein synthesis. In 1958, Crick proposed that the central dogma of molecular biology (Figure 6.8) which states that the information present in DNA (in the form of base sequence) is transferred to RNA by transcription and then from RNA, it is transferred to protein by translation (in the form of amino acid sequence). This information does not flow in the reverse direction, i.e., from protein to RNA to DNA. DNA molecules provide the information for their own replication. The process of reverse transcription (RNA to DNA) has been discovered in RNA tumour virus, and certain RNA viruses replicate purely through the RNA form.

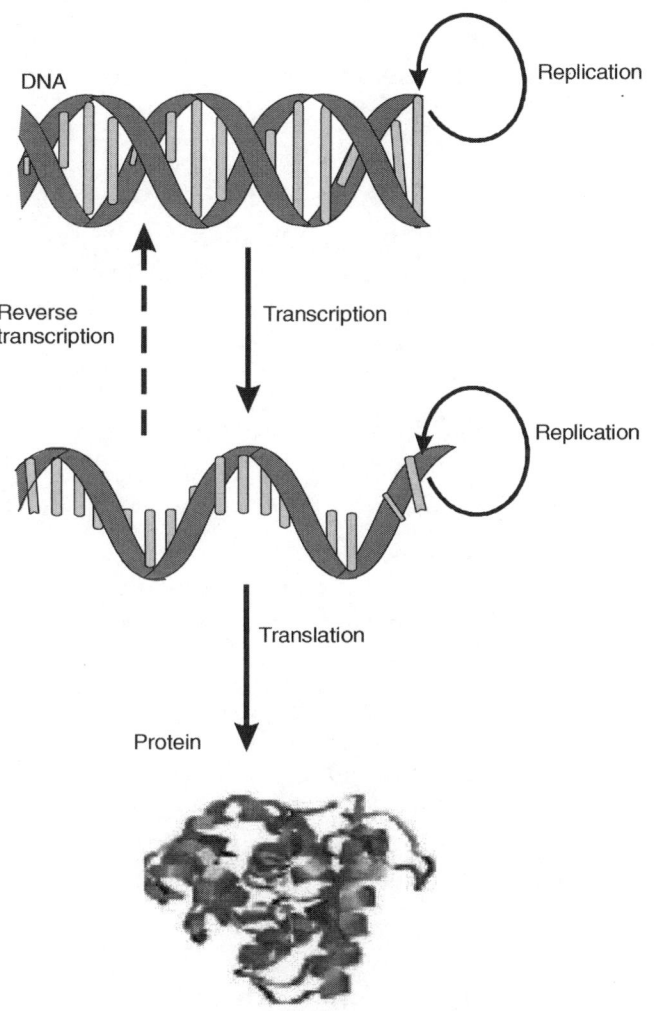

Figure 6.8   Central dogma

(The genetic information contained in DNA is passed on to protein via RNA. The process of DNA replication, transcription and translation occurs in a normal cell while the process of reverse transcription and RNA replication occurs apparently in viral infection.)

PROTEIN SYNTHESIS

The genetic information is carried by the sequence of DNA (or in some virus by RNA). This information is converted into the amino acid sequence of proteins by the process of protein synthesis. Protein synthesis consists of two separate events: transcription and translation. Transcription consists of the production of RNA copies of a strand of DNA and is catalysed by DNA-directed RNA polymerase (RNA polymerase). The RNA polymerase activity was first shown by Weiss in 1960 in rat liver. The reading of the message of mRNA and the synthesis of protein molecules specified by this message is referred to as translation; it involves ribosomes, tRNA and a number of enzymes performing specific functions.

TRANSCRIPTION

The production of RNA copies of DNA template is known as transcription. Only one strand of the DNA duplex is copied during transcription and it is called as the antisense strand, while the other strand is called as the sense strand. The new RNA molecule produced has a base sequence which is similar to that of the sense strand and it is complementary to the antisense strand. The RNA synthesis proceeds from $5' \rightarrow 3'$ direction and hence, orientation of the antisense strand (template strand) of the DNA duplex is in $3' \rightarrow 5'$ direction with respect to the direction of transcription.

Transcription is carried out by RNA polymerase. The *E. coli* RNA polymerase is capable of recognizing the appropriate promoter site for initiation of transcription and continuing the synthesis of RNA using DNA as template. The complete enzyme is called as the holoenzyme which consists of the core enzyme and sigma factor (σ). The sigma factor is involved in binding of the RNA polymerase to the promoter and thereby the initiation of transcription. Sigma factor is released from the holoenzyme when the RNA chain reaches 8–9 bases; the core enzyme then continues transcription. Thus, the important function of sigma factor is to ensure that the RNA polymerase binds firmly only to the promoter. Transcription begins with the attachment of RNA polymerase holoenzyme with the promoter region of the transcription unit and comes to an end when the core enzyme reaches the terminator site and dissociates from the DNA. It includes (i) initiation (ii) elongation and (iii) termination (Figure 6.9).

Initiation

The initiation of transcription begins with the binding of RNA polymerase with the promoter region of a transcription unit and ends with the movement of the process to the elongation phase. The σ factor of the holoenzyme is required during the phase both for the recognition of promoter sequences and for the initiation of transcription. Initially, the enzyme binds to the duplex DNA of the promoter region and induces strand separation in the stretch of DNA duplex bound to the enzyme.

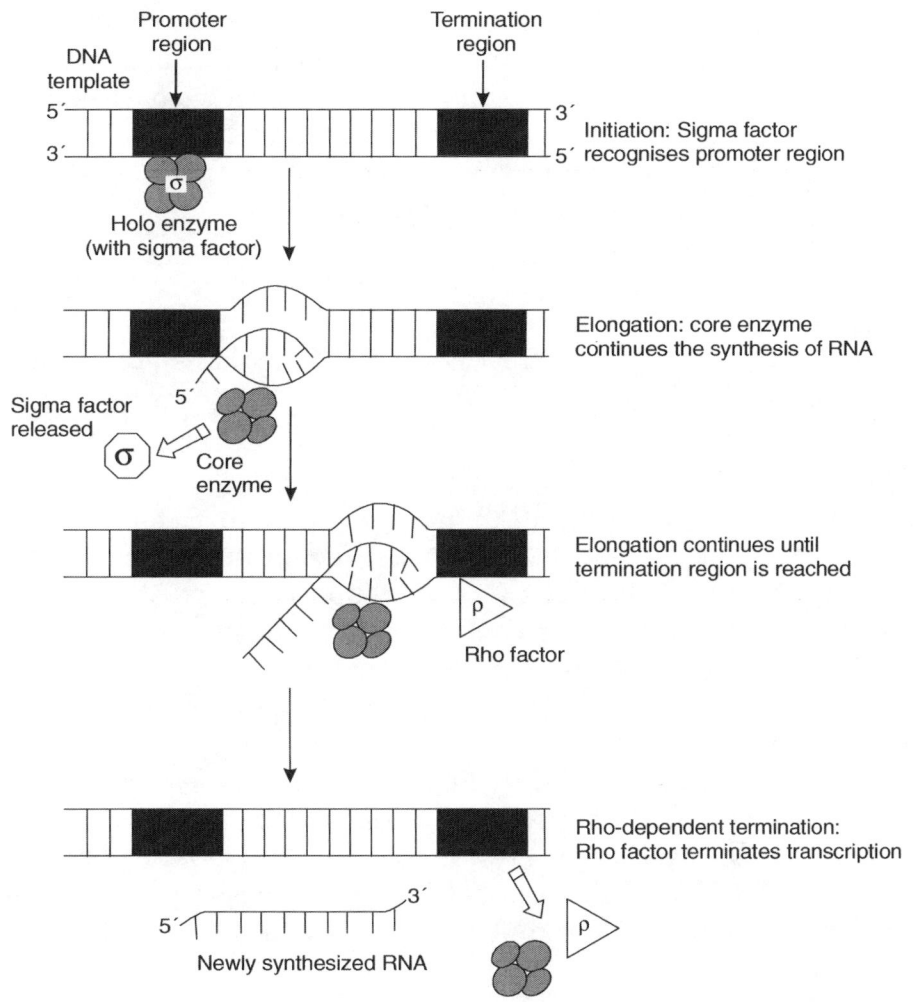

Figure 6.9 Transcription

Once the template strand (the antisense strand) becomes available, RNA polymerase begins to incorporate the ribonucleotides beginning at the start point. The nucleotides used for RNA synthesis are riboside-5´-triphosphates which have a free 3´—OH. The availability of a free 3´—OH is an absolute requirement for RNA synthesis just as it is in the case of DNA replication. Therefore, RNA synthesis also proceeds in the 5´→ 3´ direction (the orientation of the RNA strand being synthesized) and the template DNA strand must be oriented as $3' \rightarrow 5'$. Unlike DNA polymerase, RNA polymerase does not require primer polynucleotide; it starts the formation of RNA transcripts with a single ribonucleoside triphosphate. The polymerase scrutinizes the incoming nucleotides, most likely by comparing the overall geometry of the template and the incoming nucleotides and allows only the correct ribotides to occupy the site opposite a template base in the $5' \rightarrow 3'$ orientation. When the first two ribotides have been aligned, RNA polymerase catalyses the formation of phosphodiester linkage between them. In this reaction, the $3' - OH$ of the first nucleotide and the first of the three $5'$-phosphates of the second nucleotide participate; the terminal two phosphates of the second nucleotide are released as a pyrophosphate group. The dinucleotide thus formed remains associated with the template and the enzyme, while the third ribotide becomes properly aligned; a phosphodiester linkage now forms between the free $3' - OH$ of the second ribotide of the dinucleotide and the first $5'$-phosphate of the 3rd ribotide. In this manner, the nascent RNA chain keeps on growing. The RNA chain remains as RNA–DNA hybrid, with its template; the length of this hybrid molecule may be around 12 bp at any given time. As the RNA synthesis progresses, the RNA progressively separates from the RNA–DNA hybrid.

Elongation

Very early in this phase, the polymerase holoenzyme loses its σ factor and remains as the core enzyme; this phenomenon may be important in permitting the core enzyme to leave the promoter region and move on with the transcription. As the polymerase moves along the template DNA duplex, new regions become single-stranded and available for RNA synthesis. Correct nucleotides are sequentially added to the $3' - OH$ of the growing RNA chain. As the RNA chain elongates, the stretches of RNA at the $5'$-end progressively separate from the template DNA. The template DNA, therefore becomes progressively free and its two strands reassociate to form the DNA double helix. Thus, elongation involves the disruption of DNA structure so that transient single-stranded regions are progressively generated in the DNA duplex; one of the two strands supports RNA synthesis and remains for sometime as RNA–DNA hybrid; the RNA later dissociates from the DNA strand and the two strands of the template DNA reassociate to form the DNA duplex.

Termination

When the core enzyme reaches the terminator site, there is no further addition of ribotides to the RNA chain. The RNA molecule dissociates from the template DNA strand, and the core enzyme frees itself from the template DNA molecule. The DNA strands in the melted region reassociate to form normal double helix and the process of transcription catalysed by this polymerase molecule comes to an end. The released core enzyme will soon bind to a segment of DNA irrespective of the base sequence, and would begin the search for a promoter once the σ factor binds to it.

The prokaryotic termination sites have been classified into two groups, viz., (ρ) rho-independent terminators and rho-dependent terminators. Rho-independent terminators have the following two features: (i) a typical hairpin having a stem and a loop formed due to base-pairing. This secondary structure of the RNA most likely causes the RNA polymerase to pause for ~ 60 seconds. The hairpin usually contains a G-C-rich region near the base of the stem (ii) A run of U residues immediately follows the hairpin.

Rho-dependent terminators: Certain transcription units have terminators which require a polypeptide called the rho-factor for the termination of transcription.

PROCESSING OF MESSENGER RNA (mRNA)

mRNAs are rarely processed in prokaryotes, but this is a regular feature in eukaryotes. In all these cases, the genes contain sequences called exons (expressed sequences) and introns (intervening sequences). Exons are sequences that form the mature mRNA and they are represented in the proteins coded by the concerned genes. In contrast, introns are deleted during the mRNA processing and are not represented in the concerned proteins. The excision of introns and the joining together of all the exons of a gene in a proper sequence so as to yield a mature mRNA is known as splicing. Also, a poly (A) tail (3´ end) and a 5´-cap (5´-end) are added to the eukaryotic hnRNA (heterogenous nuclear RNA). About 70% of the mRNAs contain poly (A) tails and all the mRNAs of eukaryotes are capped.

In prokaryotes, translation proceeds along with transcription, but in eukaryotes, transcription is completed in the nucleus and the mRNA is translocated to the cytoplasm where translation takes place. Two other RNA molecules viz., tRNA and rRNA are important for the process of translation. So, before going ahead with the process of translation, a brief outlook on the different types of RNA and more importantly the genetic code is essential.

THE GENETIC CODE

The number and the sequence of bases in mRNA specifying an amino acid is known as codon, while the set of bases in a tRNA that base-pair with a codon of an mRNA is known as anticodon. The sequence of bases in an anticodon is exactly the opposite of that present in the codon, since the codon and anticodon segments run antiparallel to each other when they base-pair, e.g., the codon (5´) AUG(3´) has the anticodon (3´)UAC(5´). The set of all the codons that specify the 20 amino acids is termed as the genetic code or genetic language or coding dictionary. The genetic language has only four alphabets, A, U, C and G, the four bases making up polynucleotides. The triplet code or the three-letter code produces 64 $(= 4^3)$ different codons as listed in Figure 6.10.

Crick and his coworkers in 1961 concluded that a codon has most likely three bases (or a multiple of three) and that the code is nonoverlapping, degenerate and commaless. A commaless code means that all the bases in a polynucleotide are parts of codons and that no base serves as punctuation mark. In the same year, Nirenberg and Matthei reported that a synthetic polyribonucleotide, poly(U), produced a polypeptide consisting only of phenylalanine in a cell-free system. This was the first report of a specific codon (UUU) in mRNA coding for a particular amino acid. Subsequently, Nirenberg and his group and Ochoa and coworkers used synthetic polyribonucleotides and more particularly, triribonucleotides to decipher the meaning of most of the codons. The meaning of other codons and the confirmation of already known ones was provided by Hargobind Khorana and coworkers.

		Second base of codon							
		U		**C**		**A**		**G**	
U	UUU	phe	UCU	ser	UAU	tyr	UGU	cys	U
	UUC	phe	UCC	ser	UAC	tyr	UGC	cys	C
	UUA	leu	UCA	ser	UAA	STOP	UGA	STOP	A
	UUG	leu	UCG	ser	UAG	STOP	UGG	trp	G
C	CUU	leu	CCU	pro	CAU	his	CGU	arg	U
	CUC	leu	CCC	pro	CAC	his	CGC	arg	C
	CUA	leu	CCA	pro	CAA	gin	CGA	arg	A
	CUG	leu	CCG	pro	CAG	gin	CGG	arg	G
A	AUU	ile	ACU	thr	AAU	asn	AGU	ser	U
	AUC	ile	ACC	thr	AAC	asn	AGC	ser	C
	AUA	ile	ACA	thr	AAA	lys	AGA	arg	A
	AUG	met*	ACG	thr	AAG	lys	AGG	arg	G
G	GUU	val	GCU	ala	GAU	asp	GGU	gly	U
	GUC	val	GCC	ala	GAC	asp	GGC	gly	C
	GUA	val	GCA	ala	GAA	glu	GGA	gly	A
	GUG	val	GCG	ala	GAG	glu	GGG	gly	G

First base of codon (left axis) — *Third base of codon* (right axis)

*= START

Figure 6.10 The genetic code

The Coding Dictionary

It has become clear from various experiments that a sequence of three nucleotides in the mRNA (a triplet code or codon) codes for an amino acid and that the code is nonoverlapping and commaless. The meaning of all the possible 64 triplet codons has been determined beyond any doubt. Three codons, UAA, UAG and UGA are nonsense codons as they do not code for any amino acid; these codons act as polypeptide termination codons. There is a considerable degeneracy in the code; only two amino acids, methionine (AUG) and tryptophan (UGG) are coded by a single codon each. The codon AUG acts as the initiator codon; in prokaryotes, it codes for formyl methionine at the beginning of a polypeptide, while it specifies methionine if it occurs elsewhere in the mRNA.

Ambiguity, Degeneracy and Universality of the Genetic Code

There is no evidence that the genetic code is ambiguous *in vivo*. Ambiguity denotes that a single codon may code for more than one amino acid (with the exception of AUG in prokaryotes). Eighteen of the 20 amino acids are coded by more than one codon, i.e., the genetic code is degenerate. The degeneracy arises because of (i) the presence of more than one variety of tRNA (each with a different anticodon specific to one amino acid (ii) the ability of an anticodon (on tRNA) to base-pair with more than one codon of mRNA. Also, there is considerable evidence that the genetic code is essentially universal, i.e., a codon codes for the same amino acid in all the organisms. The evidence for universality comes from a variety of experiments, but with exceptions in ciliates and mitochondria.

TRANSLATION

The translation process may be divided into the following three distinct steps based on the processes, accessory factors and speed: (i) initiation (ii) elongation and (iii) termination (Figure 6.11).

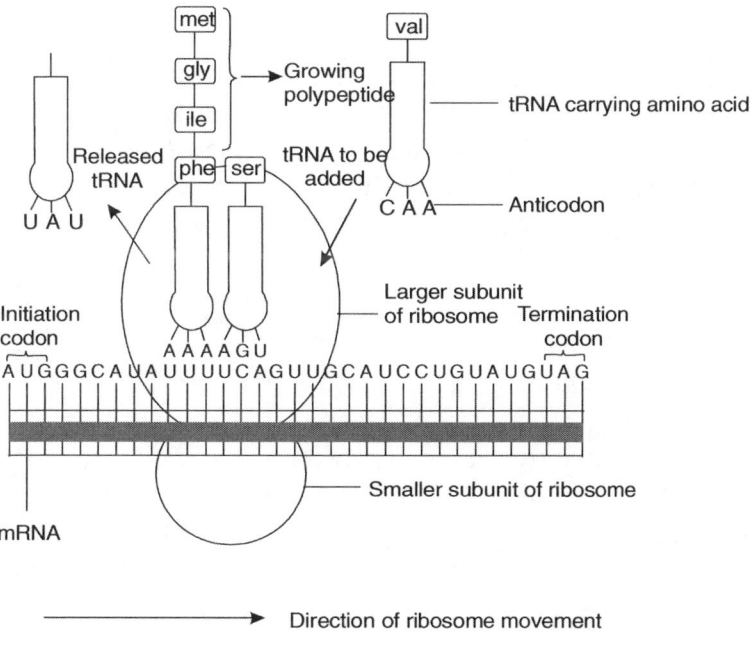

Figure 6.11 The process of translation

Initiation

Translation is initiated by the binding of the 30S subunit of ribosome (the 40S subunit in the case of eukaryotes) to the mRNA. It is followed by the binding of the first aminoacyl-tRNA, the initiator tRNA formylmethionyl-tRNA^{fMet}, to the P site; this forms the initiation complex. This is a relatively slow step in protein synthesis. In prokaryotes, the initiation complex is formed when the initiation codon, AUG of mRNA binds to the P site of the 30S subunit of a ribosome, and a tRNA^{fMet} molecule carrying formylmethionine (formylmethionyl-tRNA^{fMet}) base-pairs with the codon AUG. A formylmethionine molecule has a—CHO (formyl) group attached to the amino group of methionine. In eukaryotes, methionine (not formylmethionine) tRNA^{Met} (methionyl- tRNA^{Met}) binds to the initiation codon AUG. The formation of initiation complex requires a number of initiation factors, viz., GTP, and the factors IF1, IF2, and IF3 in *E. coli*, while in eukaryotes, at least nine different initiation factors are known. Hence, initiation comprises all the events that precede the formation of the first peptide bond.

Elongation

Elongation process includes all the reactions from the formation of the first peptide bond to the formation of the last peptide bond, the amino acids being added to the polypeptide chain one at a time. Elongation is the most rapid process of translation; in bacteria ~15 amino acids are added in one second at 37°C, while in eukaryotes, it is only ~2 amino acids/second.

Once the complete ribosome is formed at the initiation complex, the ribosome is in a ready state: it has P site occupied by the initiator aminoacyl-tRNA and its A site is free. Any aminoacyl-tRNA can enter the free A site; this entry is mediated by the elongation factor and the type of aminoacyl-tRNA entry is determined by the mRNA codon available at the A site of the ribosome.

The free A site is now occupied by another aminoacyl-tRNA specified by the mRNA codon. The amino acids carried by the tRNA molecules occupying the P and A sites adhere to the surface of the larger subunit of the ribosome; a peptide bond is formed between the two amino acids by the enzyme peptidyl transferase. The peptide bond forms when the amino acid dissociates from the tRNA occupying the P site (tRNA^{fMet} in this case); then the tRNA molecule is released from the P site. (The amino acid is attached to tRNA by its —COOH group, and the peptide bond involves the—COOH group of the amino acid attached to the tRNA on P site and the NH$_2$ group of that attached to the tRNA at the A site).

The tRNA from the A site, along with the corresponding mRNA segment, now moves to the P site making the A site free for the attachment of a new aminoacyl-tRNA molecule. In this way, the entire message of the mRNA is read sequentially; this is referred to as the polypeptide elongation.

Termination

Termination describes those events that occur after the formation of the last peptide bond and culminates with the separation of the complete ribosome from the mRNA. Termination is signalled by the three termination codons (UAG, UAA and UGA) and is mediated by certain proteins called release factors.

The signal for termination of the polypeptide resides in three chain termination or nonsense codons, UAA, UAG and UGA. Generally, no tRNA has anticodons for any of these three codons, but some suppressor mutations produce tRNA with anticodons for one or the other of these nonsense codons. Generally, a single nonsense codon is present at the termination site, but in some cases, two different nonsense codons may be situated side by side.

As the ribosome reaches the nonsense codon, three simultaneous events occur: (i) detachment of the polypeptide chain from the tRNA at P site, (ii) release of the tRNA from the P site and (iii) separation of the ribosome from the mRNA. This marks the end of the translation process associated with this ribosome particle.

MUTATIONS

The replication and distribution of genetic material is extremely accurate so that the genetic information is usually passed on from one generation to the next without alteration. But, occasionally, errors do occur both during replication and distribution of the genetic material giving rise to sudden heritable changes in the characters of organisms; such alterations are called mutations, while individuals showing these changes are known as mutants.

A mutation may arise due to a change in the base sequence of a gene; such mutations are called as gene mutations or point mutations. But, changes in chromosomal number and structure also produce heritable changes in phenotype; these are termed as chromosomal mutations.

Classification of Mutation

There are several ways in which mutation can be classified; some of the common ones are listed in Table 6.2.

Induction of Mutation

Mutations are induced by a number of agents; the agents capable of inducing mutations are called mutagens. The different mutagenic agents may be classified into two broad groups viz., (i) physical mutagens (different types of radiations) and (ii) chemical mutagens (a variety of chemical compounds) Table 6.3.

Table 6.2   Classification of mutation

Character	Types
Direction of mutation	**Forward** A mutation from the wild type allele to a mutant allele is called forward mutation. **Reverse** A mutation from mutant allele to the wild type allele is known as reverse mutation.
Cause of mutation	**Spontaneous** Occurs naturally **Induced** Originates in response to treatment with some mutagen.
Tissue of origin	**Somatic** Mutations occurring in somatic cell. **Germinal** Mutations occurring in reproductive tissues.
Effect on survival	Lethal, sublethal, subvital, vital and supervital
Type of trait affected	**Visible or morphological** mutation which alters a morphological feature of the organism so that the presence of the mutant allele is visually detected. **Biochemical** Mutations which do not produce changes in morphological traits but prevent the production of a biochemical by the organism.
Quantum of morphological effect produced	**Macromutations** Produce large-enough changes in the phenotypes. **Micromutations** Produce small morphological effects which could be easily confused with the effects produced by environmental factors.
Cytological basis	**Chromosomal** Detectable changes in either chromosome number or structure. **Gene or point mutation** Alterations in base sequence of genes. **Cytoplasmic** Changes of mitochondrial or chloroplast DNA.
Molecular basis	**Base substitution** Replacement of a single base in a DNA molecule (gene). Replacement of a purine by another purine or pyrimidine by another pyrimidine is known as transition, while in transversion, a purine is replaced by a pyrimidine or vice versa. **Deletion** One or more bases are deleted or lost from a gene **Addition** Insertion of one or more bases in a gene

(Contd.)

Table 6.2 Continued

Character	Types
Type of amino acid replacement in the polypeptide	**Missense** Replacement of a single amino acid by another amino acid in the respective polypeptide.
	Nonsense Replacement of an amino acid specifying codon by a stop codon (UAA, UGA UAG). As a result, it terminates the growth of the polypeptide chain.
	Frameshift mutation Changes all the amino acids of the concerned protein which are located beyond the point of addition/deletion of the bases.

Table 6.3 Some of the commonly used physical and chemical mutagens

Class	Mutagen
Physical mutagens	
Ionizing radiations	
1. Particulate radiations	Alpha rays (α rays), Beta rays (β rays), fast neutrons, thermal neutrons
2. Nonparticulate radiations	X-rays, gamma rays (γ rays)
Nonionizing radiations	UV rays
Chemical mutagens	
Alkylating agents	Mustard gas, nitrogen mustard, ethylmethane sulphonate (EMS), methylmethane sulphonate (MMS), Ethylethane sulphonate (EES), N-Methyl-N'-nitro-N' nitrosoguanidine (NTG)
Base analogues	5-Bromouracil (5-BU), 2-aminopurine (2-AP)
Acridine dyes	Acriflavin, proflavin, acridine orange, ethidium bromide
Deamination agents	Nitrous acid (HNO_2)
Other chemical mutagens	Hydroxylamine, sodium azide

Repair of DNA Damage

All organisms have developed the means to repair damage to their DNA, in addition to the proofreading mechanism. The most common way of dealing with mutations is by means of a method called excision repair, in which enzymes recognize and cut out the altered region of DNA and then fill in the missing bases using the other strand as template. Mismatch repair is used to repair the incorporation of a base that has escaped the proofreading system. Less frequently, the alteration in DNA is simply reversed by direct repair mechanisms, the best known of which is photoreactivation. This involves an enzyme called DNA photolyase, which breaks the bonds formed between adjacent thymine bases by UV light before replication takes place. Unusually the photolyase is dependant on visible light (> 300 nm) for activation. Another form of direct repair involving the enzyme methylguanine transferase allows the reversal of the effects of alkylating agents.

BIBLIOGRAPHY

Benjamin Lewin. (2008). *Genes IX*. Jones and Bartlett Publishers, London.

Gardner, E.J., Simmons, M.J. and Snustad, D. P. (2006). *Principles of Genetics*, 8th edition. Wiley, India.

Snyder, L. and Champness, W. (2007). *Molecular Genetics of Bacteria*, 3rd edition. ASM Press.

7

GENETIC ENGINEERING

Genetic engineering refers to the deliberate and controlled manipulation of an organism's genome. Specifically, it means transferring defined segments of genetic material from one organism to another. It uses various techniques to manipulate the genes of the organism which include recombinant DNA technology, molecular cloning, gene targeting, transformation, etc. A brief summary of the history and development of genetic engineering is given in Table 7.1.

Table 7.1   Brief summary of history and development of genetic engineering

1866	Gregor Mendel published an account of his experiments with peas, explaining the laws of genetics.
1944	DNA was recognized to carry hereditary information.
1953	The double-helix structure of DNA is described.
1960–1970	Isolation of restriction enzymes and their use to analyse DNA structure
1972-73	The first successful genetic engineering experiment (insertion of a gene from an African clawed toad into bacterial DNA), was conducted, heralding the era of recombinant DNA technology.
1974	The expression of a gene from a different species in a bacterium, for the first time
1975	DNA sequencing method was developed
1977	First complete genetic code of an organism (base sequence of a complete genome). The organism was phage ΦX174 and its genetic code is 5375 bases long
1978	Bacteria made to produce human somatostatin and human insulin from synthetic genes

(*Contd.*)

Table 7.1 (Continued)

1982	Insulin made by genetically engineered bacteria to be approved for use in Britain and USA
1981/82	"Super mouse" created by inserting human growth hormones into mouse DNA (First transgenic animal produced)
1983	First transgenic plant produced
1985	First transgenic farm animals produced (rabbit, pig and sheep)
1985	Polymerase chain reaction was discovered by Kary Mullis
1986	First controlled release of genetically engineered organisms into environment; The FDA approves the first genetically engineered hepatitis B vaccine
1989	First patented transgenic animal, the oncomouse
1990	Human genome project started. First successful gene therapy for SCID in USA; USFDA approved first genetically engineered food, chymosin, produced by genetically engineered bacteria
1991	First gene therapy trials on humans
1990-92	First transgenic cereal plant (maize and wheat); Regulations for deliberate release of genetically engineered organisms established in the USA and EU. First complete base sequence of a chromosome (yeast chromosome III)
1993	First human gene therapy trial in UK. Gene therapy for cystic fibrosis and SCID begun in UK. FDA approves genetically engineered Bovine Growth Hormone (BGH), a drug designed to increase milk production in cows
1994	Genetically engineered tomato ("Flavr Savr") marketed in USA.
1995	*Bt* potato approved as safe by EPA and marketed in USA.
1997	First cloned mammal produced from a single cell. Dolly (sheep) was developed from a single udder cell
1998	First reproducible mice clones developed. These mice were used to study cancer and the effects of ageing
2000	Completion of a rough draft of Human Genome in Human Genome Project. World's first germ-line genetically modified monkey (Rhesus) born in Oregon, USA.
2001	World's first genetically modified humans were developed. The babies were the result of ooplasmic transplantation. First human 'embryo' cloned (Dolly method)

(Contd.)

Table 7.1 (Continued)

2002   Rice genome decoded.

World's first cloned cat was developed.

2003   Glofish, the first biotech pet, hit North American market. Specially bred to detect water pollutants, the fish glowed red under black light, thanks to the addition of the natural bioluminescence gene

Complete sequence of human Y chromosome published in Nature.

2004   Korean researchers treated spinal cord injury by transplanting multipotent adult stem cells from umbilical cord blood. Researchers produced red blood cells from hematopoietic stem cells.

First human embryos created that were true clones of the sixteen women who provided the cells to make them.

2005   Researchers differentiated human blastocyte stem cells into neural stem cells and finally into spinal motor neuron cells.

2009   11 transgenic crops were grown commercially in 25 countries, the largest of which (by area grown) were the USA, Brazil, Argentina, India, Canada, China, Paraguay and South Africa.

2010   Scientists at the J. Craig Venter Institute created the first synthetic bacterial genome, and added it to a cell containing no DNA. The resulting bacterium, named Synthia, was the world's first synthetic life form.

IMPORTANT CONTRIBUTIONS IN GENETIC ENGINEERING

The potential of genetic engineering is vast and varied. Genetic engineering techniques, which have been applied in the fields of medicine and health care, agriculture, veterinary science and industries and for the control of pollution, offer great hope. Some of the important contributions of biotechnology in various fields are briefed below:

Medical Biotechnology

* Monoclonal antibodies produced by hybridoma technology are used for disease diagnosis.

* Recombinant vaccines produced by genetically engineered microbes are cleaner and safer.

* Valuable drugs like human insulin, human interferon, human and bovine growth hormones, factor VIII for hemophilia A and factor IX for hemophilia B and erythropoietin (EPO) for treating anemia are produced by genetically engineered microbes.

❋   Gene therapy is used to cure genetic diseases like Huntington's chorea and cystic fibrosis.

❋   Molecular markers are used for identification of parents/criminals.

Plant Biotechnology

❋   Embryo culture to rescue otherwise inviable hybrids, to recover haploid plants from interspecific hybrids, for micropropagation of orchids, etc.

❋   Rapid clonal multiplication through meristem culture, e.g., many fruit and forest trees.

❋   Recovery of virus and other pathogen-free stocks of clonal crops; meristem culture is generally combined with thermotherapy/cryotherapy.

❋   Rapid isolation of homozygous lines by chromosome doubling of haploids produced through anther culture/interspecific hybridization/ovary culture.

❋   Isolation of stable somaclonal variants with improved yield/yield traits/disease resistance/resistance to cold, herbicides, metal toxicity, salt and other abiotic stresses.

❋   Gene transfers for insect resistance, protection against viruses, herbicide resistance, storage protein improvement, etc.

❋   Molecular markers, e.g., RFLPs and RAPDs for linkage mapping and mapping of quantitative trait loci.

Animal Biotechnology

❋   Test tube babies in humans; involves *in vitro* fertilization and embryo transfer.

❋   Production of transgenic animals for increased milk production, growth rate, resistance to diseases, etc.

❋   *In vitro* fertilization in farm animals.

❋   The production of recombinant vaccines (subunit, recombinant DNA, synthetic peptide, anti-idiotype, deletion mutant) for use against various animal diseases that are caused by bacterial, viral or parasitic infections is a field of active research.

❋   Application of DNA/RNA probes in the diagnosis of infectious diseases, detection of genetic errors, determination of sex of embryos, analysis of pedigrees and monitoring of physiological changes induced by the introduction of new genetic material.

❋   Production of enzymes, yeast cultures, probiotics and feed pH adjusters for use in livestock farming.

Industrial Biotechnology

❋   Production of useful compounds, e.g., ethanol, lactic acid, glycerine, citric acid, gluconic acid, acetone, etc. from less useful substrates.

❋ Production of antibiotics, e.g., penicillin, streptomycin, erythromycin.

❋ Transformation of less useful and cheaper compounds into more useful and valuable ones, e.g., steroid hormones from sterols

❋ Production of enzymes, e.g., α-amylase, proteases, lipases, etc. for use in detergent, textile, leather industry, etc.

❋ Mineral extraction through leaching from low-grade ores, e.g., copper, uranium, etc.

❋ Immobilization of enzymes for their repeated industrial application.

❋ Protein/enzyme engineering to change the primary structure of existing proteins/ enzymes to make them more efficient.

❋ Production of immunotoxins by joining a natural toxin with a specific antibody.

Environmental Biotechnology

❋ Efficient sewage treatment, deodorization of human excreta.

❋ Degradation of petroleum and management of oil spills.

❋ Detoxification of wastes and industrial effluents.

❋ Biocontrol of plant diseases and insect pests by using viruses, bacteria, fungi, etc.

GENETIC VARIATION

After the discovery of the structure of DNA, there has been a spectacular unraveling of complex interactions required to express the coded chemical information of the DNA molecule into cellular and organismal expression. Changes in the DNA molecule making up the genetic component of an organism are the means by which organisms evolve and adapt themselves to new environments. The precise role of DNA is to act as a reservoir of genetic information. In nature, changes in the DNA of an organism can occur in two ways: mutation and genetic recombination.

Mutations

The high-fidelity semiconservative replication of DNA ensures transmission of genes from parents to progeny without change; this is the reason for stability of genetically controlled phenotypes over generations. However, a low frequency (10^{-4} to 10^{-7} per gene per generation) of changes occurs in genes naturally (spontaneous mutations); these mutations are the ultimate sources of all the heritable variations observed in living forms. Clearly, mutations create the variation that is exploited through selection during strain development (for more details refer Chapter 6).

Genetic Recombination

This is another source of genetic variation, which occurs regularly during the crossover event of meiotic cell division (necessary for the formation of gametes) in eukaryotes.

This phenomenon has been successfully exploited in plant and animal breeding programmes. In prokaryotes, on the other hand, recombination occurs when foreign DNA is brought into the cell during transformation or transduction or conjugation (for more details refer Chapter 2).

Limitations The natural processes of gene transfer vary appreciably in their range and specificity. In general, they are rather imprecise which makes the recovery and desired gene combination dependent on efficient screening and selection. In addition, their range in terms of the species involved is rather restricted depending on sexual compatibility (sexual reproduction) and virus–host range. These processes put serious limitations on the movement of genes across taxonomic borders. The above problems could be circumvented by genetic manipulation.

Genetic Manipulation

The manipulation of the genetic material in organisms can be achieved in three clearly definable ways—organismal, cellular and molecular.

Organismal manipulation Genetic manipulation of whole organisms has been happening naturally by sexual reproduction since time immemorial. Active control of sexual reproduction has been practised in agriculture for centuries. It is a very random process and can take a long time to achieve desired results.

Cellular manipulation Cellular manipulations of DNA have been used for over two decades, and involve cell fusion and the regeneration of whole plants from these cells. This is a semi-random or directed process in contrast to organismal manipulations, and the changes can be more readily identified. Successful biotechnological examples of these methods include monoclonal antibodies and the cloning of many important plant species.

Molecular manipulation Molecular manipulations of DNA and RNA are being performed since the past two decades and they are known as genetic engineering or recombinant DNA technology. With this technology, it is now possible to add or delete parts of the DNA molecule with high precision and the products can be easily identified. Current industrial ventures are concerned with the production of new genetically engineered organisms and of numerous compounds ranging from pharmaceuticals to commodity chemicals.

Among the three methods of genetic manipulation, the fascinating field of recombinant DNA technology is now bringing dramatic changes in biotechnology.

RECOMBINANT DNA (rDNA) TECHNOLOGY

A recombinant DNA molecule is produced by combining two or more DNA segments usually originating from different organisms. More specifically, a recombinant DNA molecule is made by inserting the desired DNA fragment into a small replicating molecule (vectors like plasmid, phage or virus) to enable its cloning in an appropriate host. Then the desired DNA fragment gets amplified along with the vector and results in a molecular clone of the inserted DNA. This is achieved by using specific enzymes for cutting the DNA (restriction enzymes) into suitable fragments and then for joining together the appropriate fragments (ligation, catalysed by ligases). In this manner, a recombinant DNA molecule may be produced which contains a gene from one organism joined to regulatory sequences from another organism; such a gene is called chimaeric gene. The purpose of rDNA technology is

1. To obtain a large number of copies of specific DNA fragments.
2. To produce and recover large quantities of the proteins expressed by the concerned gene (e.g. antibodies in tobacco plants).
3. To integrate the gene of interest into the chromosome of a target organism where it expresses itself.
4. To develop a basic understanding of the function and regulation of the gene product.
5. To identify new genes for which protein products have not been isolated (reverse genetics).
6. To correct endogenous genetic defects (e.g., sickle cell anemia).
7. To express foreign genes in disease-susceptible hosts (e.g., Disease resistance genes in agricultural crops).

Vectors are the vehicles used for inserting and obtaining multiple copies of desired DNA fragments. The most commonly used vectors are either bacterial plasmids or DNA viruses. The process of transformation includes the integration of the gene of interest into a vector (chimaeric vectors) and introduction of this vector into a suitable organism called host, which is usually a bacterium or yeast. The transformed host cells are selected and cloned. The vector present in such clones would replicate either in synchrony with or independent of the host cell; the gene present in the vector expresses itself by directing the synthesis of the concerned polypeptide; it may also not express itself. The step concerned with transformation of a suitable host with a chimaeric vector and cloning of the transformant cells is called DNA cloning or gene cloning. However, often DNA or gene cloning is taken to include both the development of chimaeric vectors as well as their cloning in a suitable host.

A clone is obtained from a single individual or cell by means of asexual progeny and this process or technique of producing a clone is called cloning. As a result, all the individuals of a clone have the same genotype which is also identical with that of the individual from which the clone was derived. Therefore, the genomes present in members of a single clone are identical; this applies to the recombinant DNA as well.

Therefore, gene or DNA cloning produces large numbers of copies of the gene/DNA being cloned. rDNA technology is used as a synonym for cloning used in the broader sense. A rather popular term used for these activities is genetic engineering.

TOOLS OF rDNA TECHNOLOGY

i. Restriction Endonucleases

Enzymes that introduce internal cuts in DNA molecules are called endonucleases. Many endonucleases cleave DNA molecules at random sites. Restriction endonucleases are a class of endonucleases that bind specifically to and cleave dsDNA at specific sites within or adjacent to particular sequences known as recognition sequence/sites. The recognition sequences are different and specific for the different restriction enzymes. The presence of restriction enzymes was postulated by W. Arber during the 1960s, while the first true restriction endonuclease was isolated in 1970. Smith, Nathans and Arber were awarded the Nobel Prize for Physiology and Medicine in 1978 for the discovery of endonucleases. Restriction endonucleases serve as the tool for cutting DNA molecules at predetermined sites, which is the basic requirement for gene cloning or recombinant DNA technology.

Table 7.2   Major classes of restriction endonucleases

Class	Abundance	Recognition site	Composition	Use in recombinant research	Examples
Type I	Less common than type II	Cut both strands at a non-specific location >1000 bp away from recognition site	Three-subunit complex: individual recognition, endonuclease, and methylase activities	Not useful	*Eco*K, *Eco*B
Type II	Most common	Cut both strands at a specific, usually palindromic, recognition site (4–8 bp)	Endonuclease and methylase are separate, single-subunit enzymes	Very useful	*Eco*RI, *Hind*III *Bam*H1
Type III	Rare	Cleavage of one strand only, 24–26 bp downstream of the 3' recognition site	Endonuclease and methylase are separate two-subunit complexes with one subunit in common	Not useful	*Eco*P1, *Eco*P15

Types of restriction endonucleases   There are three major classes of restriction endonucleases. Their grouping is based on the types of sequences recognized, the nature of the cut made in the DNA, and the enzyme structure. Type I and III restriction endonucleases are not useful for gene cloning because they cleave DNA at sites other than the recognition sites and thus cause random cleavage patterns. In contrast, type II endonucleases are widely used for mapping and reconstructing DNA *in vitro* because they recognize specific sites and cleave just at these sites (Table 7.2).

ii. Vector

A vector is a DNA molecule that has the ability to replicate in an appropriate host cell and into which the DNA fragment to be cloned (called DNA insert) is integrated for cloning. Therefore, a vector must have an origin of DNA replication, a multicloning site, a selectable marker and function in the host cell. Any self-replicating extra-chromosomal small genome (e.g. plasmid, phage and virus) can be used as vectors (Table 7.3). A good vector must

1. Possess the ability to replicate autonomously. When the objective of cloning is to obtain a large number of copies of the transgene, the vector replicon must be under relaxed control so that it can generate multiple copies of itself in a single host cell.

2. Be easy to isolate and purify.

3. Be capable of being easily introduced into the host cells, i.e., transformation into the host.

4. Have suitable marker genes that allow easy detection and/or selection of the transformed host cells.

5. Have the ability to integrate either itself or the transgene it carries into the genome of the host cell if the vector is used in gene transfer.

6. Contain unique target sites for restriction enzymes, so that DNA insert can be integrated without disrupting an essential function.

7. When expression of the DNA insert is desired, the vector should contain suitable control elements (e.g., promoter, operator and ribosome-binding sites).

Cloning and expression vectors   All vectors used for propagation of DNA inserts in a suitable host are called cloning vectors. But when a vector is designed for the expression of the protein specified by the DNA insert, it is termed as expression vector. As a rule, such vectors contain at least the regulatory sequences, i.e., promoters, operators, ribosomal binding sites, etc., having optimum function in the chosen host. It is desirable that all cloning vectors have relaxed replication control so that they can produce multiple copies per host cell. The commonly used types of vectors are discussed below.

Table 7.3 Principal features and applications of different cloning vector systems

Vector	Basis	Size of insert	Major application
Plasmid	Naturally occurring multicopy plasmids	< 10 kb	Subcloning and downstream manipulation, cDNA cloning and expression assays
Phage	Bacteriophage λ	5–20 kb	Genomic DNA cloning, cDNA cloning, and expression libraries
Cosmid	Plasmid containing a bacteriophage λ *cos* site	35–45 kb	Genomic library construction
BAC (Bacterial artificial chromosome)	*Escherichia coli* F factor plasmid	75–300 kb	Analysis of large genomes
YAC (Yeast artificial chromosome)	*Saccharomyces cerevisiae* centromere, telomere, and autonomously replicating sequence	100–1000 kb (1 Mb)	Analysis of large genomes, YAC transgenic mice
MAC (Mammalian artificial chromosome)	Mammalian centromere, telomere, and origin of replication	100 kb to > 1 Mb	Under development for use in animal biotechnology and human gene therapy

Plasmid vectors   A plasmid is a small extra-chromosomal DNA molecule that is capable of independent replication and transmission. Plasmids are circular and may exist either independent of or may become integrated into the bacterial chromosome; generally they are not essential for the host cell except under specific environments. There are several types of bacterial plasmids, but the three widely studied types are: *F* plasmids (responsible for conjugation), *R* plasmids (carry genes for resistance to antibiotics) and *Col* plasmids (code for colicins, the protein that kills sensitive *E. coli* cells; they also carry genes that provide immunity to the particular colicin). The plasmids may be either conjugative/transmissible or nonconjugative. Conjugative plasmids mediate DNA transfer through conjugation and as a result, spread rapidly among the bacterial cells of a population, e.g., *F* plasmids, many *R* plasmids and some *Col* plasmids. Nonconjugative plasmids do not mediate DNA transfer through conjugation, e.g., many *R* plasmids and most *Col* plasmids. Examples of plasmid vectors are pBR322 (Figure 7.1), pSC101, pUC7.

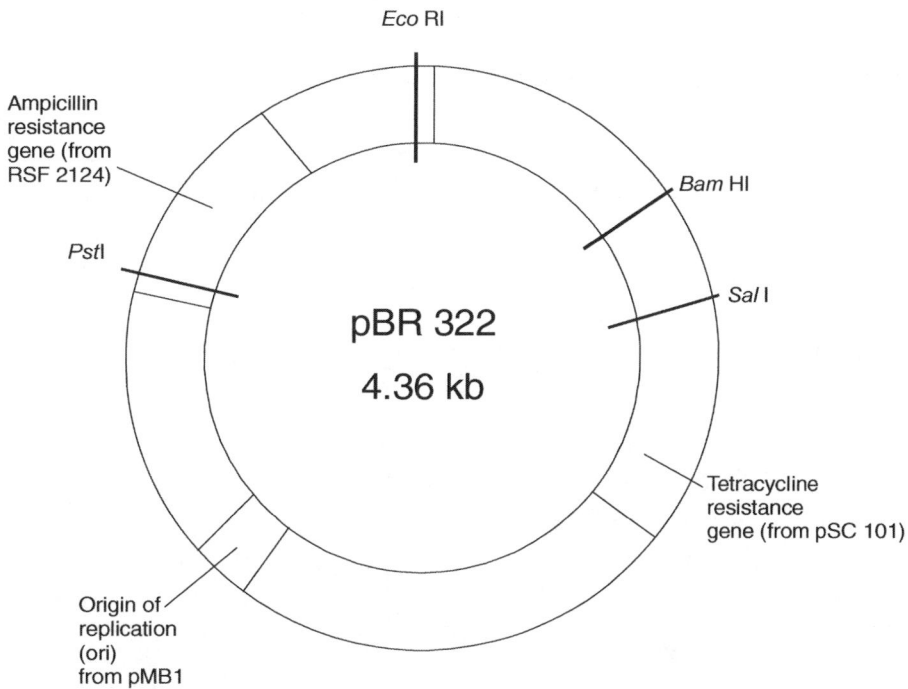

Figure 7.1 pBR322

Figure 7.1 shows a map of pBR322 depicting *ori* (origin of replication), the two resistance markers (*tet^r* and *amp^r*)that allow selection of the recombinant vectors, and cleavage sites for restriction enzymes viz., *Bam*HI and *Sal* I. (pBR322 has 30 known restriction sites, only 2 are shown for clarity).

Bacteriophage vectors Bacteriophages are viruses that attack bacteria. Several bacteriophages are used as cloning vectors, the most commonly used *E. coli* phages being λ (lambda) and M13 phages. Plasmid vectors are introduced into bacterial cells which are then cloned and selected for the recovery of recombinant vectors. In contrast, the phage vectors are directly tested on a bacterial lawn (a continuous bacterial growth on an agar plate) where each phage particle forms a plaque (a clear bacteria-free zone in the bacterial lawn). Phage λ vectors are particularly useful for preparing genomic libraries, because they can hold a larger piece of DNA than a plasmid vector. Phage vectors present two advantages over plasmid vectors: (1) they are more efficient than plasmids for cloning of large DNA fragments; the largest cloned insert size in a λ vector is just over 24 kb, while that for plasmid vectors is less than 15 kb. In addition (2) it is easier to screen a large number of phage plaques than bacterial colonies for the identification of recombinant vectors.

Cosmid vectors   Cosmids are a type of hybrid plasmids with bacterial *OriV* region, unique restriction sites, selectable markers and most importantly they carry one or more *cos* sites derived from λ phage. The *cos* sites are the sequences needed for binding and cleavage of terminase so that under appropriate conditions, they are packaged *in vitro* into empty λ phage particles. Cosmids are able to contain 37 to 52 kb of λ DNA while normal plasmids carry only 1 to 20 kb. Cosmid vectors are constructed using recombinant DNA techniques. The typical features of cosmids are as follows: (i) they can be used to clone DNA inserts of up to 45 kb, (ii) they can be packaged into λ particles which infect host cells, which is manyfold efficient than plasmid transformation, (iii) selection of recombinant vector is based on the procedure applicable to the plasmid making up the cosmid and (iv) these vectors are amplified and maintained in the same manner as the contributing plasmid.

Phasmid vectors   These vectors are shortened linear λ genomes containing DNA replication and lytic function plus the cohesive ends of the phage; their middle nonessential segment is replaced by a linearized plasmid with an intact replication module. In practice, a phasmid vector contains several tandem copies of the plasmid to make it longer than 38 kb, the minimum size needed for packaging in λ particles. During construction of the recombinant DNA, one or more copies of the plasmid are deleted from and the DNA insert is integrated into the vector; but, generally, one copy of the plasmid is retained in the recombinant vector. Phasmids, both recombinant and unaltered, are packaged in λ particles *in vitro* and used for infection of appropriate *E. coli* cells.

Shuttle vectors   These vectors have been constructed in such a way that they can replicate in cells of two distinct species; therefore they contain two origins of replication, one specific for each host species, as well as those genes necessary for their replication but not provided by the host cells. These vectors are created by recombinant techniques. Some of them can be grown in two different prokaryotic species, while others can propagate in a prokaryotic species (usually *E. coli)* and in a eukaryotic one (yeast, plants, animals). Since these vectors can be grown in one host and then moved into another without any extra manipulation they are called shuttle vectors.

Yeast vectors   The analysis of eukaryotic DNA sequence has been facilitated by cloning the genes in prokaryotes. But some functions such as glycosylation, mitosis, meiosis, etc. are absent in prokaryotes. When genes functionally related to such a function are to be analysed, those genes have to be cloned in a eukaryotic system. Yeast system offers the best possibility to study, as it is small and easy to grow and manipulate. In addition, the biochemistry and regulation of yeast is very much like that of higher eukaryotes. Thus yeast can be a very good host like *E. coli,* for studying the structure and functions of eukaryotic gene products.

Various yeast vectors have been designed, ever since the ability and utility of yeast was confirmed. All of them have three features in common.

1. All of them contain unique target sites for a number of restriction enzymes.
2. All of them can replicate in *E. coli* at high copy number.
3. They have markers that can be used to select recombinant yeasts. The tough polysaccharide wall of yeast is an effective barrier to DNA molecules. Therefore, yeast cell wall is enzymatically digested to produce spheroplasts which can take up DNA following treatment with $CaCl_2$; walls regenerate in specific media. The different vectors used in yeast may be grouped as plasmid vectors, autonomously replicating sequences (ARS) vectors, micro-chromosome vectors and yeast artificial chromosome (YAC) vectors.

Vectors for animals   Animal cells used for transformation may be tissue culture cell lines, *Xenopus* oocytes, early embryos (for obtaining transgenic animals) and even body tissues of individuals (for gene therapy/vaccination). The first animal vector has been derived from the primate papova virus and simian virus 40 (SV40). Subsequently vectors have been developed from many other viruses also, e.g., papilloma virus, adenoviruses, the Epstein–Barr herpes virus, vaccinia viruses (all for mammals) and baculoviruses (for insects).

Vectors for plants   Plant cells do not contain any plasmid. But two plasmids called *p*Ti and *p*Ri are present in the bacteria *Agrobacterium tumefaciens* and *A. rhizogenes,* respectively, provide a naturally occurring transformation system. These plasmids transfer a part of their DNA called T-DNA, into the genomes of most dicot and some monocot plants. These plasmids, especially the Ti plasmid, have been used to develop a variety of vectors. In addition, plant viruses can transfer recombinant genes into plants and can also be used effectively as vectors for the following reasons. 1) Plant viruses adsorb to and infect cells of intact plants, 2) Relatively large amounts of virus can be produced from infected plants, leading to the prospect of large amount of foreign protein being expressed from recombinant viruses, 3) Some virus infections are systematic. They are spread throughout the whole plant. In some cases intact viruses are transported through the vascular system of the plant. They are widely used as vectors, but progress has been made only with two groups of plant viruses viz., cauliflower mosaic viruses (CaMV) and Gemini viruses, both DNA viruses. The purpose of plant vectors is almost always a stable transformation ordinarily in the form of integration in plant genomes. But in the case of virus vectors, the objective is to produce large quantities of the protein encoded by DNA insert.

iii. Host

The DNA insert cloned in the vector should be transformed inside a host for multiplication. A good host should

1. be easy to transform.
2. support the replication of recombinant DNA.
3. be free from elements that interfere with replication of recombinant DNA.
4. lack active restriction enzymes.
5. not have methylases, since these enzymes would methylate the replicated recombinant DNA which as a result would become resistant to useful restriction enzymes.
6. be deficient in normal recombination function so that the DNA insert is not altered by recombination events.

Bacteria and yeasts are the hosts of choice for DNA cloning. *E. coli* supports several types of vectors, some natural, and many constructed, viz., plasmids, bacteriophages, cosmids, phasmids and shuttle vectors.

STEPS IN rDNA TECHNOLOGY

The various steps involved in cloning or recombinant DNA technology (Figure 7.2) are:

i. Identification and isolation of the desired gene or DNA fragment to be cloned.
ii. Insertion of the isolated gene in a suitable vector using restriction enzymes and ligase.
iii. Introduction of this vector into a suitable host (called transformation).
iv. Selection of the transformed host cells.
v. Multiplication/expression/integration followed by expression of the introduced gene in the host.

I. Isolation of the Desired Gene

The initial step in gene cloning is the identification and isolation of the desired gene or DNA fragment (DNA insert). The desired DNA inserts can be obtained from the following: cDNA libraries, genomic library, chemical or enzymatic synthesis and amplification through polymerase chain reaction. The transgene to be cloned is a critical step in gene cloning.

II. Integration of the DNA into the Vector

Once the DNA fragment to be cloned is identified and the appropriate vector is selected, the DNA segment has to be integrated into the vector at an appropriate site. The vector is cut open with a restriction enzyme that has a unique (single) target site located in the sequence where the DNA insert is to be integrated. The DNA insert is ligated with the help of DNA ligase (Figure 7.3).

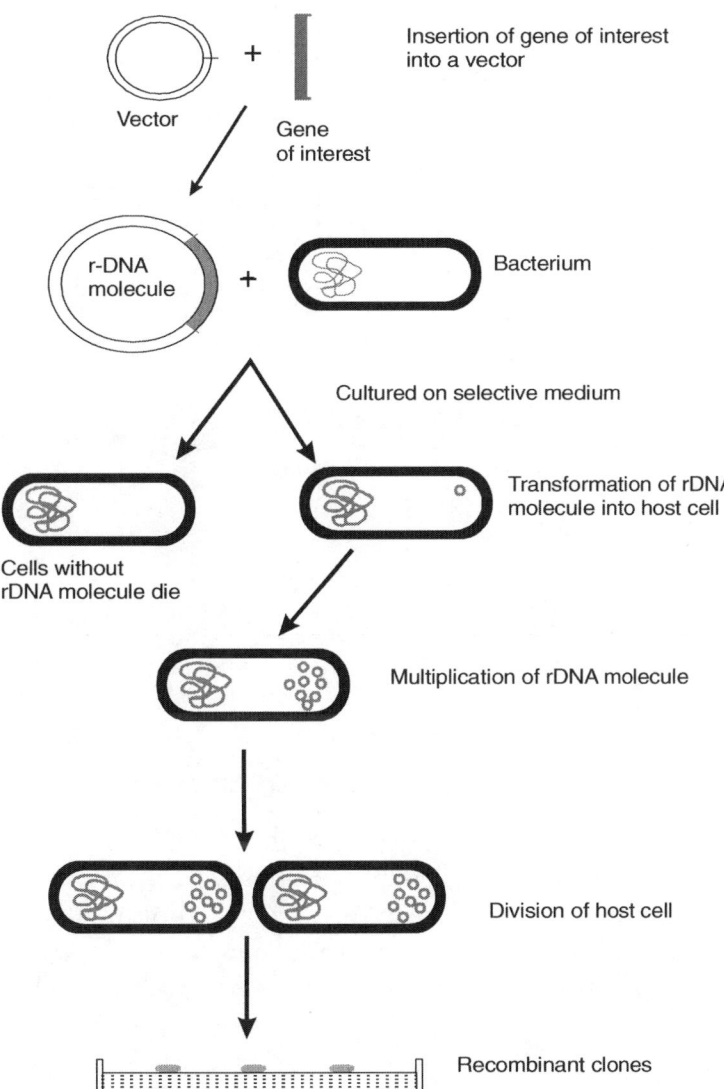

Figure 7.2 Steps in gene cloning

III. Introduction of the Vector into a Suitable Host

The recombinant vector is constructed *in vitro*; it is then generally introduced into *E. coli* to (i) select the recombinant from the unchanged vector (ii) to obtain many copies of the recombinant vector or the DNA insert (iii) to express the insert in *E. coli* itself. The various approaches for introducing recombinant vectors into bacteria especially *E. coli* are (i) increased competence of *E. coli* by CaCl$_2$ treatment (ii) infection by vectors packaged as virions.

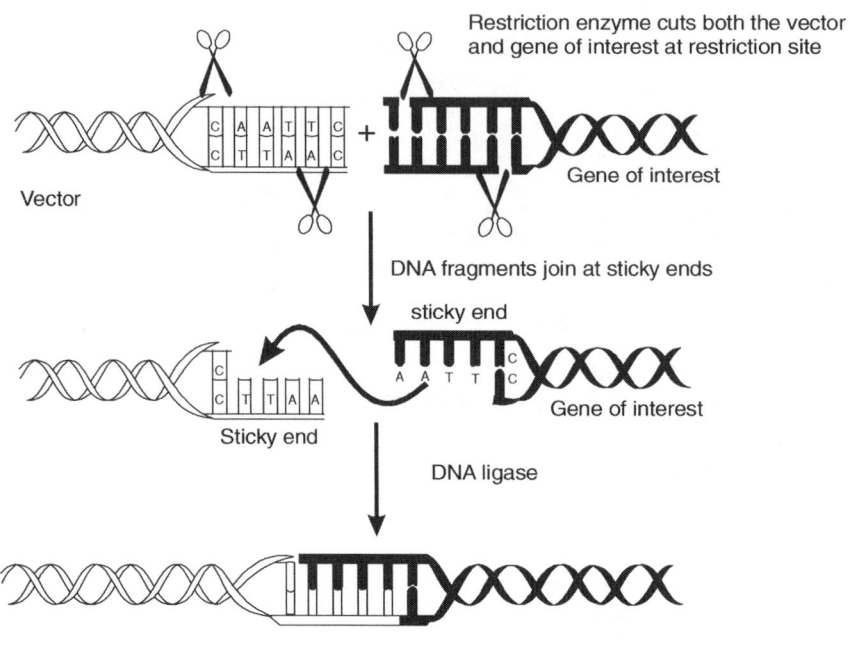

Figure 7.3   Integration of gene of interest into the vector

Amplifying recombinant DNA   The ligated recombinant DNA enters a bacterial cell by transformation. After it is in the host cell, the plasmid vector is able to replicate because plasmids normally have a replication origin. Then the donor DNA is automatically replicated along with the vector. Each recombinant plasmid that enters a cell will form multiple copies of itself in that cell. Subsequently, many cycles of cell division will take place, and the recombinant vectors will undergo more rounds of replication. The resulting colony of bacteria will contain billions of copies of the single donor DNA insert. This set of amplified copies of the single donor DNA fragment is the DNA clone.

IV. Selection of Recombinant Clones

When recombinant vector is constructed and used for transformation of *E. coli* cells, the following types of bacterial cells are obtained (i) majority of the cells are nontransformed (2) a proportion of the transformed cells contain unaltered vector, while (3) the remainder of cells have recombinant vector. The first objective of cloning experiments is to identify and isolate those small numbers of cells that contain the recombinant vector from among a very large number of nontransformed cells. The next step therefore is to identify the clone having the desired DNA insert from among the large number of clones containing the recombinant vectors. Suitable selection strategies have been devised to achieve these two critical objectives.

Reporter genes   A marker gene or reporter gene produces a phenotype which permits either an easy selection or quick identification of the cells in which it is present. Thus marker genes are either selectable or scorable. A selectable marker governs a feature which enables only such cells that possess it to survive under the selective conditions. For example, *pBR322* contains genes conferring resistance to the antibiotics, ampicillin (*amp*^{r}) and tetracycline (*tet*^{r}). *Bam*HI and *Sal*I sites are within the *tet*^{r} gene. Insertion at either of these sites yields a plasmid that is *amp*^{r} *tet*^{s}, (ampicillin-resistant, tetracycline-sensitive) because the *tet* gene is inactivated. Prior to plating, the cells are grown in a medium containing cycloserine and tetracycline. Cycloserine kills growing cells, but *tet*^{s} cells are merely inhibited, not killed by tetracycline. Thus in this growth medium *tet*^{r} cells (which grow) are killed and *tet*^{s} cells (which are inhibited) survive. Plating of cells treated in this way on agar containing ampicillin yields *amp*^{r} *tet*^{s} colonies; all these possess *pBR*322 containing a donor DNA fragment (Figure 7.4). On the other hand, scorable markers produce distinct phenotypes which allow easy identification of the cells having them from those which do not contain them. Examples of such genes are *gus* (β-galacturonidase that produces blue colour in the presence of appropriate substance), *lux* (luciferase that produces phosphorescence), etc. Scorable markers do not allow selective multiplication of the cells having them; they only enable their easy identification.

As shown in Figure 7.4, *pBR322*, *Bam*HI and *Sal*I sites are within the *Tet*^{r} gene. Insertion at either of these sites yields a plasmid that is *amp*^{r} *tet*^{s}, because the *tet* gene is inactivated. Prior to plating, the cells are grown in a medium containing cycloserine and tetracycline. Cycloserine kills growing cells, but *tet*^{s} cells are merely inhibited, not killed by tetracycline. Thus in this growth medium *Tet*^{r} cells (which grow) are killed and *Tet*^{s} cells (which are inhibited) survive. Plating of cells treated in this way on agar containing ampicillin yields *Amp*^{r} *Tet*^{s} colonies; all these possess *pBR*322 containing a donor DNA fragment.

rDNA clone identification   After obtaining a population of recombinant vectors, the next step is to identify a clone which has the DNA insert of interest. The technique used for identification has to be highly precise and extremely sensitive to allow an accurate detection of a single clone from among the thousands obtained from a cloning experiment. The various strategies used for this purpose are colony hybridization, hybrid arrested translation (HART), hybrid selection, complementation, FACS, etc.

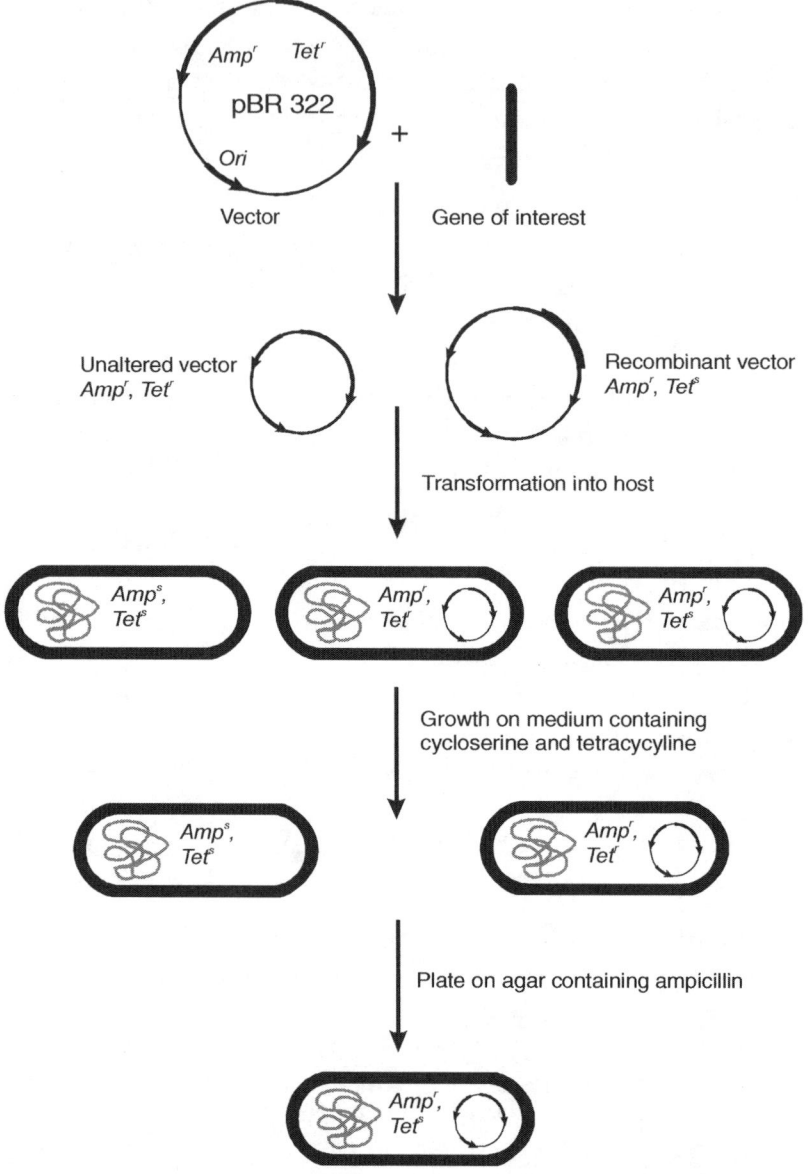

Figure 7.4   Selection of recombinant clones using antibiotic resistance gene

V.   Multiplication, Expression and Integration of the DNA Insert in Host Genome

Once the clone containing the desired DNA insert is identified, it is multiplied in *E. coli* to obtain sufficient number of copies to be used in one or more of the following ways

(i) it can be used for a structural analysis of the insert, e.g., DNA sequencing, chromosome walking, etc. (ii) it may be introduced into a bacterium like *B. subtilis* for production of the protein encoded by the insert since this host secretes protein into the medium, which allows easy purification (iii) it can be introduced into a eukaryotic host, e.g., yeast, animal cells, etc., either to investigate the function of the insert or (iv) to integrate it into the host genome to achieve one of many diverse objectives.

VARIOUS APPLICATIONS OF rDNA TECHNOLOGY

Recombinant DNA technology essentially involves the industrial application of living organisms for producing useful products or methods for the betterment of humanity. Microorganisms have been exploited to produce many important biomolecules like organic acids, antibiotics, vitamins, hormones, enzymes, steroids and vaccines. Living organisms have also been used as catalyst to carry out many enzymatic biotransformation reactions to produce complex organic molecules which otherwise are difficult to produce using the chemical synthesis route. Apart from this, the metabolic activities of the living organism have been exploited to carry out new biotechnological processes such as extraction of minerals, waste water treatment process, etc. The concept of exploiting microorganisms has been extended to plant and animal cell culture to produce special metabolites. However with the introduction of recombinant DNA technology and better understanding of the genome organization, the application of biotechnology has crossed the species barrier. Efforts have been made to genetically engineer most of the living systems such as bacteria, yeast, fungi, plant and animals to have novel gene products or characteristics. Thousands of genes have been cloned and expressed using recombinant DNA technology. The genetic manipulations using rDNA technology are more precise and one can be more certain about the outcomes compared to other methods, resulting in faster production of organisms with desired traits. Progress in molecular biology and genetic engineering techniques has made an impact in two major areas (a) understanding the biology of the living system by manipulation of genome information and (b) production of useful metabolites or living organisms having desired metabolic characteristics. It has not only helped to improve the product yield of many old biopharmaceutical compounds but has also made possible to produce human proteins in other living organisms like bacteria, fungi and even plants. This has resulted in the production of specific biomolecules in different organisms and also has helped to synthesize genetic material and its related product in the laboratory. In fact, the application of genetic engineering and recombinant DNA technology has led to the generation of new classes of organism called genetically modified organisms (GMO) or live modified organisms (LMO). Moreover, the ability of genetic manipulation of almost all living organisms has led to the genomics evolution with far-reaching applications of the modern biology system. The most notable applications of recombinant technology are development of transgenic animals, plants and microorganisms, production of biomolecules like therapeutic proteins, vaccines, monoclonal antibodies, etc, screening for genetic diseases, gene therapy, bioremediation, forensic science, etc.

Transgenic Microbes

Bacteria were the first organisms to be modified in the laboratory, due to their simple genetics. These organisms are now used particularly in producing large amounts of pure human proteins for use in medicine. Genetically modified bacteria are used to produce the protein insulin to treat diabetes, to produce clotting factors to treat haemophilia, and to produce human growth hormone to treat various forms of dwarfism. These recombinant proteins are much safer than the products they replaced, since the older products were purified from cadavers and could transmit diseases. Transgenic microbes have also been used in recent research to kill or delay tumours, and fight Crohn's disease. In some soils they are used to facilitate crop growth, and to produce chemicals which are toxic to crop pests.

In addition to bacteria, gene therapy uses genetically modified viruses to deliver genes into human cells that can cure diseases. Gene therapy is still relatively new and has been used to treat genetic disorders, and treatments are being developed for a range of incurable diseases such as cystic fibrosis, sickle cell anemia, Parkinson's disease and muscular dystrophy. For example, *Streptococcus mutans* causes tooth decay and cavities by producing acid. Scientists have recently modified *Streptococcus mutans* to produce ethanol could possibly reduce the formation of cavities if properly colonized in the mouth. Current gene therapy technology only targets the non-reproductive cells, meaning that any changes introduced by the treatment cannot be transmitted to the next generation. Gene therapy targeting the reproductive cells—Germ line Gene Therapy—is very controversial and is unlikely to be developed in the near future.

Transgenic Animals

Transgenic animals are just one in a series of developments in the area of biotechnology. A transgenic animal contains in its genome, a gene or genes introduced by one or the other technique of transfection. The gene introduced by transfection is called transgene. Mice have become the main species used in the field of transgenics, for practical reasons, i.e., their small size and low cost of housing in comparison to that for larger vertebrates, their short generation time, and their fairly well-defined genetics. The three principal methods used for the creation of transgenic animals are DNA microinjection, embryonic stem-cell-mediated gene transfer and retrovirus-mediated gene transfer. Transgenics have been produced in a variety of animal species, e.g., mice, rabbits, swine, sheep, goat, cattle, poultry, fish, amphibians, insects and nematodes. However, these activities are so far limited to experimental stages and production technologies, that is, transgenic animals have not yet reached commercial applications. The chief objectives of transgenic production are as follows:

   i. increased milk or meat production,
  ii. model for studying human diseases and therapies and
 iii. molecular farming i.e., transferring several human genes and expressing them as transgenic proteins.

Transgenic Plants

The genetic manipulation of plants has been going on since prehistoric times, when early farmers began carefully selecting and maintaining seeds from their best sow for the next season. Plant breeders have cross fertilized related plants to provide next generation plants with improved characteristics such as higher yield, resistance to diseases and better nutrient content. Recombinant DNA technology can be used for insertion of genes in plants not only from related plant species, but also from unrelated species such as microorganisms. This process of creation of transgenic plants is far more precise and selective than traditional breeding. Application of recombinant technology is primarily for the production of transgenic plants with higher yield and nutritional values, increased resistance to stress and pests. Several commercially important transgenic crops such as maize, soybean, tomato, cotton, potato, mustard, rice, etc. have been genetically modified. During the last few decades, considerable progress has been made to understand the function of genes, isolation of novel genes and promoters as well as the utilization of these genes for the development of transgenic crops with improved and new characters. There are many potential applications of plant genetic engineering. Application of recombinant DNA technology has primarily helped in producing three major types of transgenic plants having improved performances. They are

   1. Development of stress-tolerant plants,
   2. Development of plants having improved yield and
   3. Transgenic plants as a source of biopharmaceuticals.

Transgenic crops with improved nutrition quality have already been produced by introducing genes involved in the metabolism of vitamins, minerals and amino acids.

Screening for Genetic Disorders/Diseases

Recessive mutant phenotypes that follow single-gene inheritance are responsible for more than 500 human genetic diseases. A person homozygous for a trait with parents heterozygous for the same recessive allele will be affected by the disease. Cells derived from the fetus can be screened at an early stage to allow the option of abortion of afflicted fetuses. Many of the enzymes or other proteins that are altered or missing in genetic diseases are now identified. To detect such genetic defects, fetal cells can be taken from the amniotic fluid, separated from other components and cultured to allow the analysis of chromosomes, proteins, enzymatic reactions and other biochemical properties. Testing physiological properties or enzymatic activity in cultured fetal cells limits the screening procedure to those disorders that affect characters or proteins expressed in the cultured cells. However, recombinant DNA technology improves the screening for genetic diseases *in utero*, because the DNA can be analysed directly. In principle, the gene being tested could be cloned and its sequence can be compared with that of a normal gene. The other useful techniques that have been developed are alteration of restriction site by mutation, probing for altered sequences, PCR tests, etc.

Gene Therapy

Gene therapy is a general approach for correcting the defective gene responsible for genetic disorders. In simple words, the function absent in the recipient as a result of a defective gene is introduced into the recipient's chromosomes by vector insert and thereby a genetically cured recipient is generated. This technique is of great potential in humans because it offers the hope of correcting hereditary diseases. Perhaps the controversial application of transgenic technology is in human gene therapy, the treatment and alleviation of human genetic disease by adding exogenous wild-type genes to correct the defective function of mutations. The two basic types of gene therapy applied to humans are germ line and somatic therapy. In the germ-line gene therapy, transgenic cells are introduced into the germ line as well as into the somatic cell population. This therapy not only cures the person treated, but also enables some gametes to carry the corrected genotype. Somatic gene therapy focuses only on the body cells where transgenic cells are introduced into somatic tissue to correct defective function. There are several approaches for inserting normal gene, viz., normal gene may be inserted into a non-specific location within the genome to replace the non-functional gene or an abnormal gene can be substituted by a normal gene through homologous recombination, or abnormal gene can be repaired through selective reverse mutation which returns the gene to normal function. The vectors used for gene therapy are retrovirus, adenovirus, adeno-associated virus and herpes simplex virus. The other nonviral methods for delivering normal gene into the host are direct introduction of therapeutic gene, creation of liposome which carries therapeutic gene which passes through cell membrane, or chemically linking therapeutic DNA which enables it to bind with the cell receptors and pass into the cell through the cell membrane. Currently, gene therapy is experimental due to some experimental failures and controversies.

Bioremediation

Environmental biotechnology uses both naturally occurring and genetically engineered microorganisms for the biodegradation of environmental contaminants.

Bioremediation involves the use of microorganisms to clean up contaminated land, water and air. Recombinant DNA technology helps in improving the efficacy of these processes by genetically engineering the microorganism so that their basic biological processes are more efficient and can degrade more complex chemicals and higher volumes of waste materials. Various techniques employed for improving the microbial activity includes new expression vectors to carry the heterologous genes into the host organism, new mechanisms to control gene expression, containment mechanisms to control persistence of genetically engineered microorganisms (GEM), application of site-directed and random mutagenesis to increase the substrate range and activity of biodegradative enzymes. This technology is also being used in development of methods to track GEMs where bacteria have been genetically modified to emit bioluminescence in response to several chemical pollutants. These are being used to measure the presence of some hazardous chemicals in the environment. Other genetic sensors can also be developed which can be used to detect various chemical contaminants.

Forensic Sciences

The applications of molecular biology in forensics center largely on the ability of DNA analysis to identify an individual from hairs, blood stains and other items recovered from the crime scene. The popular term for these techniques is called genetic fingerprinting or DNA profiling. It is used not only for identification of criminals, but also to study kinship analysis, paternity testing, sex determination, identifying species of endangered animals if hunted/captured, etc. DNA analysis can be done by PCR tests for identifying markers like random fragment length polymorphisms, short tandem repeats, amplified fragment length polymorphisms, Y chromosome analysis and single nucleotide polymorphisms.

SAFETY OF THE RECOMBINANT BIOMOLECULES

Irrespective of the expression system used for the production of recombinant molecules, all biotherapeutics share major safety concern. This is due to the inherent nature of the biological molecules, where the conformation of the molecule is more important along with the chemical nature. Thus, any biomolecule manufactured using a living system needs complete characterization and safety evaluation. The most important criteria during the manufacturing of recombinant molecules using live organisms are the quality, safety and efficacy. These three parameters influence the toxicity and immunogenicity of the final product and thus need critical evaluation. It is also essential that all precautions are taken at each step of biological production taking care of the biosafety aspects. The host organism and waste product coming out of the rDNA process need to be processed according to the Environmental Protection Act of the country so that biodiversity is not affected.

SOME COMMON TECHNIQUES IN MOLECULAR BIOLOGY

I. Isolation of DNA

The use of DNA for analysis or manipulation usually requires that it is isolated and purified to a certain extent. DNA is recovered from cells by the gentlest possible method of cell rupture to prevent the DNA from fragmenting by mechanical shearing. This is usually in the presence of EDTA which chelates the Mg^{2+} ions needed for enzymes that degrade DNA, termed DNase. Ideally, cell walls, if present, should be digested enzymatically (e.g., lysozyme treatment of bacteria), and the cell membrane should be solubilized using detergent. If physical disruption is necessary, it should be kept to a minimum, and should involve cutting or squashing of cells, rather than the use of shear forces. Cell disruption (and most subsequent steps) should be performed at 4°C, using glassware and solutions that have been autoclaved to destroy DNase activity.

After release of nucleic acids from the cells, RNA can be removed by treatment with ribonuclease (RNase) that has been heat-treated to inactivate any DNase contaminants; RNase is relatively stable to heat as a result of its disulphide bonds, which ensure rapid renaturation of the molecule on cooling. The other major contaminant protein is removed by shaking the solution gently with water-saturated phenol or with a phenol/chloroform mixture, either of which will denature proteins but not nucleic acids. Centrifugation of the emulsion formed by this mixing produces a lower, organic phase, separated from the upper, aqueous phase by an interface of denatured protein. The aqueous solution is recovered and deproteinized repeatedly, until no more material is seen at the interface. Finally, the deproteinized DNA preparation is mixed with two volumes of absolute ethanol, and the DNA allowed to precipitate out of solution in a freezer. After centrifugation, the DNA pellet is redissolved in a buffer containing EDTA to inactivate any DNases present. This solution can be stored at 4°C for at least a month. DNA solutions can be stored frozen, although repeated freezing and thawing tend to damage long DNA molecules by shearing. The procedure described above is suitable for total cellular DNA. If the DNA from a specific organelle or viral particle is needed, it is better to isolate the organelle or virus before extracting its DNA, since the recovery of a particular type of DNA from a mixture is usually rather difficult. Where a high degree of purity is required, DNA may be subjected to density gradient ultracentrifugation through caesium chloride which is particularly useful for the preparation of plasmid DNA.

Estimation of DNA It is possible to check the integrity of the DNA by agarose gel electrophoresis and determine the concentration of the DNA by using the fact that 1 absorbance unit equates to 50 μg /ml of DNA and so:

$$50 \times A_{260} = \text{concentration of DNA sample (μg/ ml)}$$

Contaminants may also be identified by scanning UV spectrophotometry from 200 nm to 300 nm. A ratio of 260 nm: 280 nm (approximately 1.8) indicates that the sample is free of protein contamination, and absorbs strongly at 280 nm.

2. Isolation of RNA

The methods used for RNA isolation are very similar to those described for DNA; however, RNA molecules are relatively short, and therefore less easily damaged by shearing, so cell disruption can be rather more vigorous. RNA is, however, very vulnerable to digestion by RNases which are present endogenously in various concentrations in certain cell types and exogenously on fingers. Gloves should therefore be worn, and a strong detergent should be included in the isolation medium to immediately denature any RNases. Subsequent deproteinization should be particularly rigorous, since RNA is often tightly associated with proteins. DNase treatment can be used to remove DNA, and RNA can be precipitated by ethanol. One reagent in particular which is commonly used in RNA extraction is guanidinium thiocyanate which is both a strong inhibitor of RNase and a protein denaturant. It is possible to check the integrity of an RNA extract by analysing it by agrose gel electrophoresis. The most abundant RNA species, the rRNA molecules, 23S and 16S for prokaryotes and 18S and 28S for eukaryotes, appear as discrete bands on the agarose gel and thus indicate that the other RNA components are likely to be intact. This is usually carried out under denaturing conditions to prevent secondary structure formation in the RNA.

Estimation of RNA   The concentration of the RNA may be estimated by using UV spectrophotometry. At 260 nm, 1 absorbance unit equates to 40 µg/ml of RNA and therefore:

$$40 \times A_{260} = \text{concentration of RNA sample (µg/ml)}$$

Contaminants may also be identified in the same way as that for DNA by scanning UV spectrophotometry; however, in the case of RNA, a 260 nm : 280 nm ratio of approximately 2 would be expected for a sample containing no protein. In many cases, it is desirable to isolate eukaryotic mRNA which constitutes only 2–5% of cellular RNA from a mixture of total RNA molecules. This may be carried out by affinity chromatography on oligo(dT)-cellulose columns. Nanodrop spectrophotometer systems have also aided the analysis of nucleic acids in recent years in allowing the full spectrum of information whilst requiring only a very small (microlitre) sample volume.

3. Polymerase Chain Reaction

Polymerase chain reaction (PCR) is an extremely powerful technique, which generates microgram (µg) quantities of DNA copies (up to billion copies) of the desired DNA segment, present even as a single copy in the initial preparation, in a few hours.

PCR is carried out *in vitro*. It utilizes (i) a DNA preparation containing the desired segment to be amplified (ii) two nucleotide primers (about 20 bases long) specific, i.e., complementary, to the two 3′ borders (the sequence present at the 3′ ends of the two strands) of the desired segment (iii) the four deoxynucleoside triphosphates viz., TTP (thymidine triphosphate), dCTP (deoxycytidine triphosphate), dATP (deoxyadnenosine triphosphate) and dGTP (deoxyguanosine triphosphate) and (iv) heat-stable DNA polymerase, e.g., *Taq* (isolated from bacterium *Thermus aquaticus*), *Pfu* (from *Pyrococcus furiosus*) and *Vent* (from *Thermococcus litoralis*) polymerases. *Pfu* and *Vent* polymerases are more efficient than the *Taq* polymerase.

At the start of PCR, the DNA from which a segment is to be amplified, an excess of the two primer molecules, the four deoxyriboside triphosphates and DNA polymerase are mixed together in the reaction mixture. The following operations are now performed sequentially (Figure 7.5).

Step1   The reaction mixture is heated to a temperature (usually 90°–98°C) that assures DNA denaturation. This is the denaturation step.

Step 2   The mixture is now cooled to a temperature (generally 40°–60°C) that permits annealing of the primer to the complementary sequences in the DNA; these sequences are located at the 3′ ends of the two strands of the desired segment. This step is called annealing.

Step 3   The temperature is now so adjusted that the DNA polymerase synthesizes the complementary strands by utilizing 3′−OH of the primers; this reaction is the same as the one that occurs *in vivo* during replication of the leading strand of a DNA duplex. The primers are extended towards each other so that the DNA segment lying between the two primers is copied; this is ensured by employing primers complementary to the 3′-ends of the segment to be amplified. In case of *Taq* polymerase, the optimum temperature for synthesis is between 70° and 75°C; the temperature of reaction mixture is therefore adjusted to this temperature. This situation has the following additional advantages. Between 70 and 75°C, the base-pairing between about 20-bases-long primers and the DNA is much more specific than at 37°C, the optimal temperature for *E.coli* DNA polymerase. This minimizes the changes of annealing of primers to imperfectly matched sequences and thereby amplification of unwanted DNA. The specificity of annealing is further increased by selecting appropriate conditions like ionic strength, primer length, etc. The completion of step 3 completes the first cycle of amplification; each cycle may take few (ordinarily 1–3) minutes.

Step 4   The next cycle of amplification is initiated by denaturation (Step 1), which separates the newly synthesized DNA strands from the old DNA strands.

Step 5   Annealing allows the primers to base-pair with both the new and old strands, the total number of strands being twice their original number.

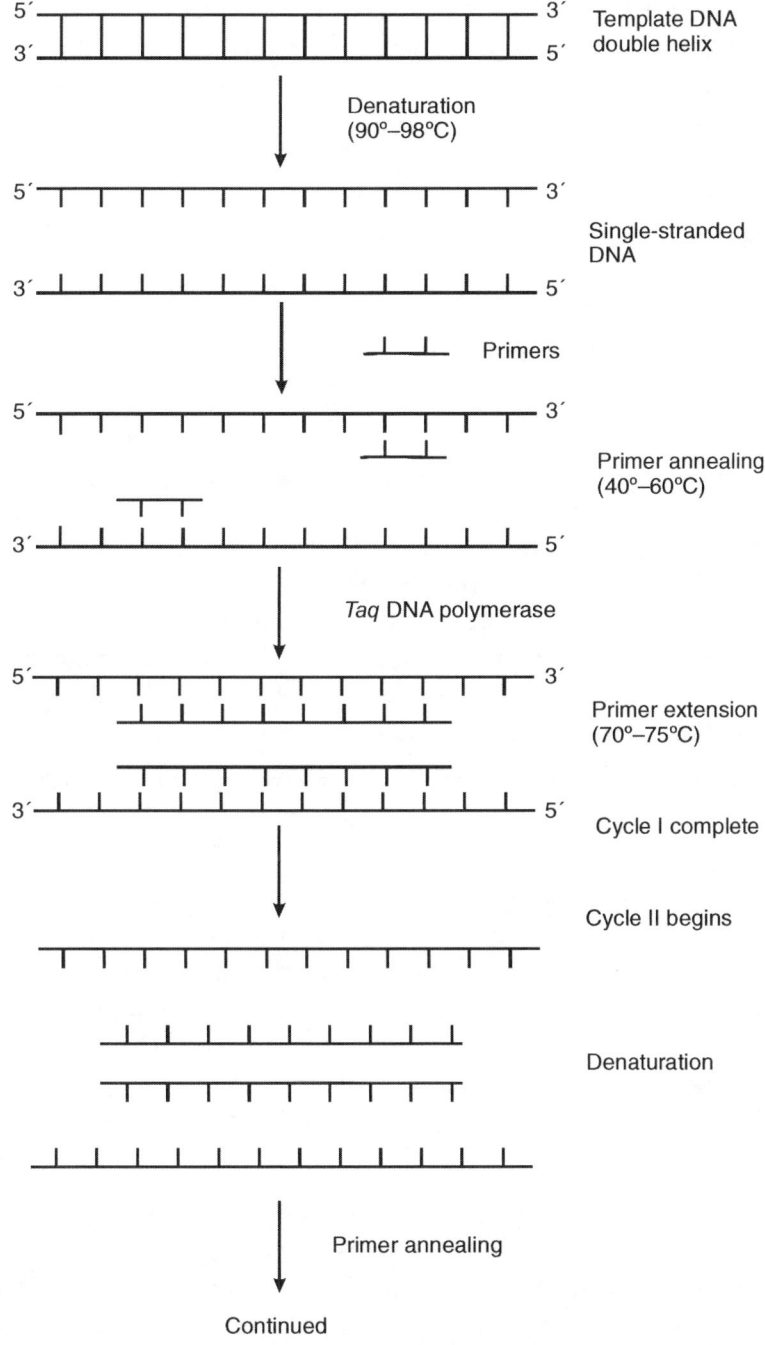

Figure 7.5   Polymerase chain reaction

Step 6   Synthesis of new strands takes place, which doubles the number of copies of the desired DNA segment present at the end of step 1. This completes the second cycle.

Thus, at each cycle, both new and old strands anneal to the primers and serve as templates for DNA synthesis. As a result, at the end of each cycle, the number of copies of the desired segment becomes twice the number present at the end of the previous cycle. At the end of n cycles, 2^n copies of the segment are expected; the real values are quite close to this expectation. The cycle may be repeated up to 60 times, but usually, 20–35 cycles are adequate. After PCR cycles, the amplified DNA segment is purified by gel electrophoresis and can be used for the desired purpose.

BIOETHICAL ISSUES

Bioethics addresses the impact of technology on individuals and societies. In the case of transgenic organisms, a major bioethical issue is freedom of choice. Other broader issues are the ethics of interfering with nature and effects of transgenic organisms on the environment. The changes that are possible with transgenesis surpass the traditional agriculture, although these transgenics interfere with nature. For example, a transgenic tobacco plant emits the glow of a firefly, and a transgenic rabbit given DNA from a human, a sheep, and a salmon secretes a protein hormone that is used to treat bone disorders. If mixing DNA in ways that would not occur in nature is deemed wrong, then transgenesis is unethical. A group opposed to GMOs in New Zealand said, "To interfere with another life-form is disrespectful and another form of cultural arrogance." A more practical protest to transgenic technology is the risk of altering ecosystems.

Bioethical issues related to transgenic animals are of major concern. There is particular concern about the way we exploit animals for food and for development of medical products. For example, the oncomouse is a transgenic mouse to which a oncogene has been added a gene that causes cancer. The mice develop tumours much more frequently than normal and are used in cancer research. But on the other side, we are in need of these animals as they contribute towards our understanding of diseases like cystic fibrosis, heart diseases, AIDS, etc. Hence, the social opinion on transgenic animal research is divided almost in the middle. Use of animals in biotechnological research causes greater suffering to the animals. But most people seem to accept some animal suffering to serve the basic interest and welfare of humankind; this attitude has been termed as interest-sensitive speciesism.

The FDA tests foods to determine their effect on the human digestive system, their biochemical make-up, and their similarity to existing foods. Foods are not judged solely by their origin. The FDA and U.S. Department of Agriculture approved transgenic crops in 1994, and de-regulated the technology two years later, as did the U.S. Environmental Protection Agency. Ironically, as people in wealthier nations object to not having a choice in avoiding genetically modified foods, others complain that the technology is too expensive for farmers in developing nations to use. Another ethical dimension to transgenic organisms is that the methods to create genetically modified seeds, and the seeds themselves, lie in the hands of a few multinational corporations. In the mid-1990s a company sold transgenic plants resistant to the company's herbicides, but that could not produce their own seed, forcing the farmer to buy new seed each year. An international outcry led to the abandonment of this practice, but the use of crops that are resistant to certain herbicides, with a single company owning both seed and herbicide, continues. Some see this as a conflict of interest. More alarming were several incidents in the United States in 2000, when people who object to genetically modified foods vandalized laboratories and destroyed fields of crops, some of which were not even transgenic. So far, foods containing GMOs appear to be safe. They may be easier to cultivate and may permit the development of new variants. However, it will take more time to determine whether or not they have longer-term health and ecological effects.

BIBLIOGRAPHY

Brown, T.A. (1995). *Gene Cloning: An Introduction.* Chapman and Hall, USA. pp. 334.

Primrose, S. B. and Richard, M. (2006). *Principles of Gene Manipulation and Genomics*, Twyman, John Wiley & Sons. 644 pages.

Benjamin Lewin. (2008). *Genes IX.*, Jones and Barlett Publishers, London.

PART II

Microbial Technology—Scale Up and Downstream Processing

- Submerged Culture Fermentation
- Solid-state Fermentation
- Downstream Processing

8

SUBMERGED CULTURE FERMENTATION

INTRODUCTION

Fermentation is defined as any process mediated by or involving microorganisms in which a product of economic value is obtained. Its biochemical meaning relates to the generation of energy by the catabolism of organic compounds. Bioprocesses have been developed from relatively cheap materials for manufacturing enormous range of commercial products such as industrial alcohol and organic solvents as well as expensive speciality chemicals such as antibiotics, vaccines, therapeutic proteins and enzymes. Industrially, the term is used in broader context and refers to any process for the production of a commercially viable product by the mass culture of microorganisms such as bacteria, yeast and fungi. A typical bioprocess is carried out in a reactor or fermenter which provides conducive or controlled environment for chemical or biochemical transformations to occur. The bioprocess or fermentation largely depends upon:

i. The biological properties such as nature of cells (size and morphology), quality of inoculum (age and size), respiration rate, growth, death and product kinetics, formation of byproducts and metabolites, etc.

ii. Chemical properties such as medium constituents, pH, antifoam agents, reaction kinetics, material and energy balances, etc.

iii. Physical properties such as temperature, pressure, power consumption, agitation, heat transfer, mass transfer, gas flow rate, gas hold up, mixing, rheology, etc. and

iv. Engineering properties such as reactor design, control of operation, etc.

Microbial fermentations that are of commercial significance can be classified into the following five major groups:

* Those that produce microbial cells or biomass as the product, e.g., single cell protein, baker's yeast, *Lactobacillus,* etc.

* Those that produce microbial metabolites.

  * Primary metabolites: ethanol, citric acid, glutamic acid, glycerol, acetone, amino acids, vitamins, polysaccharides, etc.

  * Secondary metabolites: all antibiotic fermentations

  * Those that produce industrially useful microbial enzymes, e.g., amylase, protease, pectinase, glucose isomerase, cellulase, hemicellulase, lipase, tannase, lactase, streptokinase, etc.

  * Those that modify a compound which is added to the fermentation, e.g., phenyl acetyl carbinol, steroid biotransformation, S-adenosyl methionine (SAMe), etc.

  * Those that produce recombinant products, e.g., insulin, surface antigen of HBV, interferon, monoclonal antibodies, streptokinase, etc.

ISOLATION AND SCREENING OF MICROORGANISMS FOR INDUSTRIAL FERMENTATION

The success of an industrial fermentation process chiefly depends on the strain used. Since prehistoric times, humans have exploited microorganisms for their betterment. The discovery of penicillin has revolutionized the development of fermentation products, and microbial screening for other useful metabolites such as antibiotics, enzymes, amino acids, organic acids, vitamins, etc. has been intensified. During the past few decades, screening of industrially important microorganisms has evolved steadily. Still it relies on classical techniques such as enrichment, pure culture, mutagenesis, etc. The selection of industrial microorganisms is systematically approached with a strategy. The key elements of the selection strategy are i) defining the activity of interest, ii) surveying known microorganisms with that activity, iii) developing enrichment and screening protocols, iv) identifying sources for new microorganisms and v) developing screening methodology. Primary screening is predominantly a qualitative assay in which a large population of microorganisms is screened either directly or indirectly for a specific type of activity. Secondary screening includes both qualitative and quantitative assays to determine the precise activity of the microorganisms, to verify production or degradation of compounds and to evaluate the production potential of the microorganism identified in the primary screening.

An ideal microorganism should have the following characteristics.

1. It should be pure, and free from phage.
2. It should be genetically stable, but amenable to genetic modification.
3. It should produce both vegetative cells and spores; species producing only mycelium are used mainly in SSF.

4. It should grow vigorously after inoculation.

5. It should produce a single valuable product.

6. It should not produce any toxic by-products.

7. Product should be produced in a short time.

8. It should be amenable for long-term conservation.

9. The risk of contamination should be minimal under the optimum performance conditions.

Isolation of Microorganisms

Isolation of microorganisms from their natural habitats is the first and foremost step in developing a producer strain of commercial interest. Alternatively, microorganisms can also be obtained as pure cultures from culture collection centres. Generally, the microorganisms of industrial importance are bacteria, actinomycetes, fungi and algae. These organisms occur everywhere; but, most common sources of industrial microorganisms are soils from forests, ocean, sea, lake and river mud. The ecological habitat from which a desired microorganism is more likely to be isolated depends on the characteristics of the product desired from it. Some common sites of isolation of desired microorganisms are listed in Table 8.1.

Table 8.1 Sources for isolation of industrial microorganisms

Product	Source
Thermophiles	Hot spring, thermophilic vents, compost, etc.
Alkalophiles	Highly saline soil, wells and seashore, etc.
Acidophiles	Acid soils, fruits, etc.
Osmophiles	Seas and oceans
Yeast	Vineyards and fruit surfaces
Phosphate solubilizers	Agricultural lands, rhizosphere region, etc.
Enzyme producers	Agricultural lands, paper and pulp industry, dairy industry, meat-processing units, fruits and grain-processing units, etc.
Methane producers	Biogas plants, anaerobic waste treatment plants, etc.

A variety of complex isolation procedures have been developed, but no single method can reveal all the microorganisms present in a sample. Different microorganisms can be isolated by using specialized enrichment techniques, e.g., soil treatment (UV irradiation, air drying or heating at 70°–120°C, filtration or continuous percolation, washings from root systems, treatment with detergents or alcohols, preinoculation with toxic agents, selective inhibitors (antibiotics, dyes, salts, etc.), selective nutrients (specific C and N sources), variations in pH, temperature, aeration, etc. The enrichment techniques are designed for selective multiplication of only specific microorganisms of interest present in a sample. These approaches, however, take a long time, and require considerable labour and money. The main methods used routinely for isolation of microorganisms from soil samples are sponging (soil directly), dilution, gradient plate, aerosol dilution, floatation, and differential centrifugation. But these methods are used in conjunction with an enrichment technique.

Screening of Microorganisms for New Products

The next step after isolation of microorganisms is their screening. A set of highly selective procedures, which allow the detection and isolation of microorganisms producing the desired metabolite, constitute primary screening. Ideally, primary screening should be rapid, inexpensive, predictive, specific but effective for a broad range of compounds, and applicable on a large scale.

Primary screening is time-consuming and labour-intensive since a large number of isolates have to be screened to identify a few potential ones. However, this is possibly the most critical step, since it eliminates the large bulk of unwanted isolates that are either nonproducers or producers of unwanted compounds.

Rapid and effective screening techniques have been devised for a variety of microbial products, which utilize either a property of the product or that of its biosynthetic pathway for detection of desirable isolates. Some of the screening techniques are relatively simple. However, for most microbial products of high value, the screening is usually complex and tedious, and often may involve two or more steps. In some cases, it may be desirable to concentrate on a group of organisms expected to yield new products.

Selective isolation of producer microbe and recognition of their metabolite production is the primary step. It involves selection and pre-treatment of the source of interest and enrichment and isolation of producer bacteria on selective media. As an example, the screening of antimicrobials from actinomycetes is described in detail. Actinomycetes are important producers of secondary metabolites including antibiotics, anti-tumour and cytotoxic compounds. Due to their growth requirements and characteristics, various pre-treatments can be used to increase the relative number of actinomycetes on isolation plates by inhibiting the growth of other contaminant organisms found in soil. Drying and dry heating are some of the common methods to reduce the number of undesired gram-negative bacteria which can cause swarming on the isolation plates since the aerial spores of most actinomycetes are relatively resistant to desiccation.

Incubation conditions generally include temperatures of 25°–30°C and a period of 7–14 days during which visible colonies of actinomycetes will be observed. The use of spore-activating and bactericidal agents such as phenol, antibiotics, sodium dodecyl sulphate, yeast extract in isolation media can greatly increase the recovery of actinomycetes by suppressing the counts of bacteria. Some examples of antibiotics used in the isolation of specific actinomycetes include rifampicin, novobiocin and oxytetracycline for *Actinomadura*, *Micromonospora* and *Streptoverticillium* respectively. Antifungal agents such as cycloheximide and nystatin which do not inhibit actinomycetes are also used in isolation media to eliminate fungal contamination.

Following the isolation of producer bacteria, the antimicrobial activities of secondary metabolites may be assessed by overlaying an isolation plate with a single test or indicator organism. A modification of this method known as the agar-spot test has also been used for the detection of antagonistic activity between bacteria. However, the major limitation of both techniques is the possible contamination of selected colonies with the indicator organisms. Alternatively, replica plating of selected colonies may be used to study a range of indicator organisms. But, the limitation in this method is that mobile bacteria can alone be used.

The sensitivity of the isolation medium can be influenced by its nutrient composition, pH, selective agents and incubation temperature. Nonselective or selective enrichments are required to increase the sensitivity of detection of target bacteria. Firstly, the samples are pre-enriched in a nutritive nonselective medium to allow the growth of target bacteria. Pre-enriched samples are then transferred to a secondary selective enrichment medium, where the normal flora is suppressed but the target bacteria are allowed to grow. Finally, the target bacteria is streaked on selective isolation agar and isolated as pure culture. The pure producer culture is then transferred into a broth medium to start the fermentation and optimization process of enhancing the production of active secondary metabolites.

New Screening Technologies

In recent years, more and more microbial genomes are being fully sequenced. With the development of new technologies in functional genomics and proteomics, the discovery of new drug molecules and targets is expected to gain added impetus and momentum. This will be further aided by more powerful bioinformatics and high-throughput screening capabilities. The challenge lies in understanding the structure–function relationship of proteins, receptors and their ligands.

Increasingly, high-throughput target screening and identification make use of sequence-based functional genomic methods. Many proteins contain functional domains such as ATP binding sites and protein–DNA interacting sites that can be identified from sequence similarity with known genes. In the gene chip technology, cDNA or oligonucleotides are deposited on silicon or glass substrates resulting in a dense array of DNA spots on solid supports that allow hybridization to fluorescent or radioactive RNA or DNA samples. The binding between the two sets of nucleic acids can then be captured digitally. Such a high-throughput genomic screen has been used to compare the gene expression patterns between organisms.

Strain Improvement

Microbial strain improvement involves the application of one or a combination of strategies that result in the development of new strains with desired traits. The most commonly sought trait is increased metabolite production. Other desired traits are reduced production of other metabolites and ease of scalability in operation. Much emphasis is placed on improving microbial strains and it has found novel applications in industries. Microorganisms isolated from nature usually produce biologically active molecules in extremely low quantities. The reason is that these molecules are either nonessential for the survival of the microorganism or present in sufficient quantities for satisfying their primary needs. Low productivity of metabolite will result in high manufacturing cost per unit of the product. Therefore, manufacture of commercially important metabolites directly from the microorganisms isolated from nature is often economically not viable. However, there are ways for enhancing the productivity and thus making the process economically viable. This can be achieved by improving the design of fermenters or improving the microbial strains with increased productivity or ability to utilize low-cost raw materials.

The techniques and approaches used to genetically modify strains and to increase the production of the desired product are collectively known as strain improvement or development. Conventionally, strain improvement is achieved through random mutation and screening or selection which is referred to as classical approach. The other methods are targeted approaches for developing improved strains through genetic engineering. Ideally, the classical approach and targeted approaches are integrated to create a synergistic effect for rapid strain development. Strain improvement is based on the following four approaches: (i) mutant selection, (ii) selecting natural variants, (iii) recombination, and (iv) recombinant DNA technology.

Microorganisms undergo slight genetic change with every cell division. They undergo a very large number of cell divisions. After several divisions, the culture media consists of microbes with a wide range of genetic structure. The natural variants which produce maximum product yields can be selected for industrial purposes from this culture media. (For further information about recombination, mutation and rDNA technology see chapters 2, 6 and 7 respectively.)

BIOCHEMICAL REACTIONS AND THEIR KINETICS

Bioprocesses involve many chemical and/or biochemical reactions. Knowledge concerning changes in the composition of reactants and products as well as their rates of utilization and production under given conditions is essential when determining the size of a reactor. It is important, therefore, to have some knowledge of the rates of those enzyme-catalysed biochemical reactions that are involved in the growth of microorganisms, and are utilized for various bioprocesses. In general, bioreactions can occur either in a homogeneous liquid phase or heterogeneous phase, including gas, liquid, and/or solid. Reactions with particles of catalysts, or of immobilized enzymes and aerobic fermentation with oxygen supply represent examples of reactions in heterogeneous phases.

In many bioprocesses, characteristics of the reaction determine to a large extent the economic feasibility of the process. Catalytic reactions in biological systems are driven by catalysts. A catalyst is a substance which affects the rate of reaction without altering the reaction equilibrium or undergoing permanent change itself. Enzymes, enzyme complexes, cell organelles and whole cells perform catalytic roles; the latter may be viable or nonviable, growing or non-growing. Biocatalysts can be of microbial, plant or animal origin. Cell growth is an autocatalytic reaction because the catalyst is a product of the reaction. Reaction parameters such as reaction kinetics and yield of product from substrate should be considered while designing operational strategies in reactors.

Reaction Kinetics

As reactions proceed, the concentrations of reactants decrease. In general, rate of reaction depends on reactant concentration so that the specific rate of conversion decreases simultaneously. Reaction kinetics refers to the relationship between rate of reaction and conditions that affect reaction velocity, such as reactant concentration and temperature. These relationships are described using kinetic expressions. Often, the volumetric rate of a reaction can be expressed as a function of reactant concentrations using the following mathematical form:

$$r_A = kC_A^x C_B^y$$

where,

   k is the rate constant for the reaction,

   C_A and C_B are the concentrations of reactants A and B.

The rate constant is independent of the concentration of reacting species but is dependent on other variables that influence reaction rate, such as temperature. Most reactions speed up considerably as the temperature rises. In the above kinetic expression, the reaction is said to be of order x with respect to component A and order y with respect to component B. The order of the overall reaction is $(x + y)$.

REACTION KINETICS FOR BIOLOGICAL SYSTEMS

The kinetics of many biological reactions are either zero-order, first-order or a combination of these, called Michaelis–Menten kinetics. Kinetic expressions for biological systems are examined in this section. (For more information on Michaelis–Menten kinetics, *see* Chapter 4).

Zero-Order Kinetics

The reaction rate is independent of reactant concentration if a reaction obeys zero-order kinetics. The expression for zero-order kinetics is

$$r_A = k_0$$

where r_A is the volumetric rate of reaction with respect to A and k_0 is the zero-order rate constant. k_0 is a volumetric rate constant with units $kg\ mol\ m^{-3}\ s^{-1}$. Because the volumetric rate of catalytic reaction depends on the amount of catalyst present, when the above expression is used to represent the rate of a cell or enzyme reaction, the value of k_0 includes the effect of catalyst concentration as well as the specific rate of reaction. The equation could be written as

$$k_0 = k'_0 e \text{ or } k_0 = k''_0 x$$

where k'_0 is the specific zero-order rate constant for enzyme reaction and e is the concentration of enzyme. Correspondingly, for cell reaction, k''_0 is the specific zero-order rate constant and x is cell concentration.

First-Order Kinetics

If a reaction obeys first-order kinetics, the relationship between the reaction rate and reactant concentration is as follows:

$$r_A = k_1 C_A$$

where r_A is the volumetric rate of reaction and k_1 is the first-order rate constant with dimensions T^{-1}. Like the zero-order constant, the value of k_1 depends on the catalyst concentration.

In enzyme-catalysed reactions, the kinetics of most enzyme reactions is reasonably well-represented by the Michaelis-Menten equation.

GROWTH KINETICS

There is a little difference between the kinetic equations for enzymes and cells, since cell metabolism depends on the integrated action of a multitude of enzymes. Several phases of cell growth are observed in batch culture and are shown in Figure 8.1.

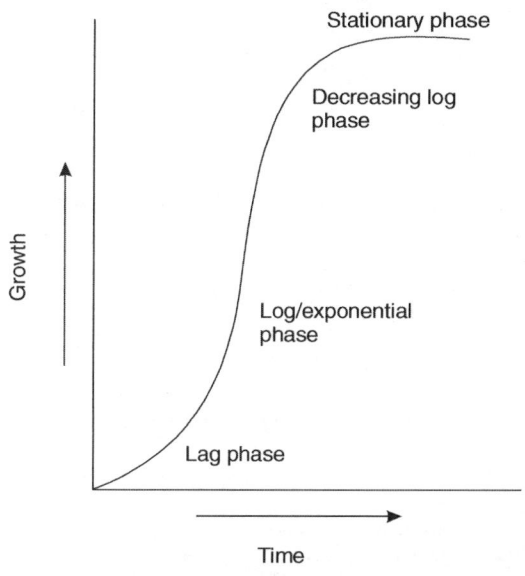

Figure 8.1   Microbial growth phases in batch culture

The different phases of growth are more readily distinguished when the natural logarithm of viable cell concentration is plotted against time. Rate of growth varies depending on the growth phase. During the lag phase immediately after inoculation, rate of growth is essentially zero. Cells use this lag phase to adapt themselves to the new environment; new enzymes or structural components may be synthesized. Following the lag period, growth starts in the acceleration phase and continues through the growth and deceleration phases. If growth is exponential, the growth phase appears as a straight line on a semi-log plot. Growth slows down and cells enter the deceleration phase as nutrients in the culture medium become exhausted or growth inhibitory metabolites accumulate in the medium. After this transition period, the stationary phase is reached, during which no further growth occurs. Subsequently, cells lose viability and undergo lysis, and this phase is termed as the death phase. A summary of the events in a batch culture is shown in Table 8.2.

During the growth and deceleration phases, rate of cell growth is described by the equation

$$r_x = \mu x$$

where,

r_x is the rate of biomass generation measured in kg m^{-3} s^{-1},

x is viable cell concentration and

μ is the specific growth rate measured in kg m^{-3} T^{-1}. μ has the same unit as that of first-order reaction since cell growth is considered as autocatalytic first-order reaction.

Doubling time (t_d) is the term generally used to describe cell growth rates and is expressed as

$$t_d = \frac{\ln 2}{\mu}$$

Determining growth rates in cell culture require measurement of cell concentration or biomass estimation and also can be measured by calculating cell number, dry or wet cell mass, packed cell volume or culture turbidity.

Table 8.2 Summary of batch cell growth

Phase	Description	Specific growth rate
Lag	Cells adapt to the new environment; no or very little growth	$\mu \approx 0$
Acceleration	Growth starts	$\mu < \mu_{max}$
Growth	Growth achieves its maximum rate	$\mu = \mu_{max}$
Decline	Growth slows due to nutrient exhaustion or build-up of inhibitory products	$\mu > \mu_{max}$
Stationary	Growth ceases	$\mu = 0$
Death	Cells lose viability and lyse	$\mu < 0$

EFFECT OF SUBSTRATE CONCENTRATION ON GROWTH

Monod Equation for Growth Kinetics

During the growth and deceleration phases of batch culture, the specific growth rate of cells is influenced by temperature, pH, composition of medium, the rate of air supply and other factors. Often, a single substrate, either carbon or nitrogen source, exerts a dominant influence on rate of growth and is known as the growth-rate-limiting substrate. The Monod equation, similar to the Michaelis–Menton expression, relates the specific growth rate of cells with the concentration of growth-limiting substrate during cell growth:

$$\mu = \frac{\mu_{max}[S]}{K_s + [S]}$$

where,

[S] is the concentration of growth-limiting substrate,

μ_{max} is the maximum specific growth rate having the unit T^{-1},

K_s is the substrate constant, having the same unit as substrate concentration and

μ_{max} and K_s are intrinsic parameters of the cell-substrate system. Values for K_s for several organisms are listed in Table 8.3.

Table 8.3 K_s values for several organisms

Microorganism	Limiting substrate	K_s (mg l^{-1})
Saccharomyces cerevisiae	Glucose	25
Escherichia coli	Glucose	4
	Lactose	20
Aspergillus niger	Glucose	5
Candida sp.	Glycerol	4.5
Pseudomonas aeruginosa	Methanol	0.7
Klebsiella sp.	Carbon dioxide	0.4
	Sulphate	2.7
Bacillus sp.	Glucose	
Cryptococcus sp.	Thiamine	1.4×10^{-7}
Pichia pastoris	Glycerol	
	Methanol	

Often the values of K_s are very small, of the order of mg/litre for carbohydrate substrates and µg/litre for other compounds such as amino acids. The level of growth-limiting substrate in culture media is normally much greater than K_s. As a result, growth can be approximated using zero-order kinetics with growth rate independent of substrate concentration until [S] reaches very low values.

$$\mu = \frac{\mu_{max}}{K_s}$$

When s is greater than K_s, $\mu = \mu_{max}$ and K_s is usually very small compared to the starting substrate concentration, [S] remains $>> K_s$ during most of the culture period. This explains why μ remains constant and equal to μ_{max} in batch culture until the medium is virtually exhausted of substrate. When s finally falls below K_s, transition from growth to stationary phase can be very abrupt as the low level of residual substrate is rapidly consumed by the large number of cells present.

Though Monod equation is valid for relating the specific growth rate of cells with the concentration of growth-limiting substrate during cell growth, it has limited applicability when rapid changes in growth conditions occur at extremely low substrate levels.

BIOREACTOR

Bioreactor is an apparatus in which practical biochemical reactions or fermentations are performed, often using enzymes and/or living cells. Bioreactors which use living cells are usually called fermenters. The apparatus used for waste water treatment using biochemical reactions is another example of a bioreactor. Since most biochemical reactions occur in the liquid phase, bioreactors usually handle liquids. Processes in bioreactors often also involve a gas phase, as in the case of aerobic fermenter. Some bioreactors handle particles such as immobilized enzymes or cells either suspended or fixed in a liquid phase. Bioreactors range from simple stirred or non-stirred open containers to complex integrated systems involving varying levels of advanced computer control. The fermenter is custom-designed for the intended processes and is equipped with top-quality components.

Fermentation reactions are multiphase, involving a gas phase (containing nitrogen, oxygen and carbon dioxide), one or more liquid phases (aqueous medium and liquid substrate) and solid microphase (the microorganisms and possibly solid substrates). All phases must be kept in close contact to achieve rapid mass and heat transfer. In a perfectly mixed bioreactor, all reactants entering the system must be immediately mixed and uniformly distributed to ensure homogeneity inside the reactor.

FUNCTIONS OF FERMENTER

The main function of a fermenter is to provide a controlled environment for the growth of microorganisms or animal cells, to obtain the desired product. In designing and constructing a fermenter, a number of crucial points must be considered:

* The vessel should be capable of being operated aseptically for a number of days, should be reliable for long-term operation, and should meet the requirements of containment regulations.

* The material of which the bioreactor is made is continuously in contact with the solutions entering the bioreactor or the actual culture, and must be corrosion-resistant to prevent trace metal contamination.

* The material must be non-toxic so that slight dissolution of the material or components does not inhibit culture growth.

* The material must withstand repeated sterilization with high-pressure steam.

* The bioreactor stirring system, entry ports and plates must be easily machinable and sufficiently rigid and not be deformed or broken under mechanical stress.

* Adequate aeration and agitation should be provided to meet the metabolic requirements of the microorganisms. However, the mixing should not cause damage to the organisms.

* Power consumption should be as low as possible.

* A system of temperature and pH control should be provided.

* Sampling facilities should be provided.

* Evaporation losses from the fermenter should not be excessive.

* The vessel should be designed to require the minimal use of labour in operation, harvesting, cleaning and maintenance.

* Ideally, the vessel should be suitable for a range of processes, but this may be restricted because of containment regulations.

* The vessel should be constructed to ensure smooth internal surfaces, using welds instead of flanges wherever possible.

* The vessel should be of similar geometry to both smaller and larger vessels in the pilot plant or plant to facilitate scale-up.

* There should be adequate service provisions for individual plants.

FERMENTER AND ITS COMPONENTS

Although many different types of fermenters have been described in the literature, only a few have proved to be satisfactory for industrial aerobic fermentations. Pilot-scale and industrial-scale vessels are normally constructed of stainless steel or at least have a stainless steel cladding to limit corrosion. The most commonly used ones are based on a stirred upright cylinder with sparger aeration. Thus, fermenter vessel can be produced in a range of sizes from one dm³ to thousands of dm³. A detailed view of the fermenter and its components is shown in Figure 8.2.

Vessel Shapes

Vessel shapes are determined based on the nature of operation. In principle, vessels are subdivided into pressurized and non-pressurized vessels. Non-pressurized vessels are used for the storage of media and do not have to meet stringent design criteria and in most cases, they are thin-walled and wide. In the production fermenter employing microbial cultures, narrow and tall vessel shapes are used. Depending on diameter and size, they are designed with either flat or domed bottoms. Narrow vessel shapes allow a high-power input resulting in good mixing, efficient aeration, and high heat transmission. For cell culture requiring gentle treatment, wider shapes are employed. The shape of the vessel bottom depends on a number of criteria. The most important are the function of the bottom in the process, e.g., mixing parameters in the vessel, idling properties and the operating conditions, e.g., pressure to which the bottom is exposed. For operations employing high pressure, the better suited are round-shaped bottoms. Most fermenters have torispherical bottoms. A fermenter requires some kind of supporting structure either in the form of legs designed as pipe constructions whose ends are permanently bolted to the floor or the legs can be supported to extended metal platforms with or without wheels.

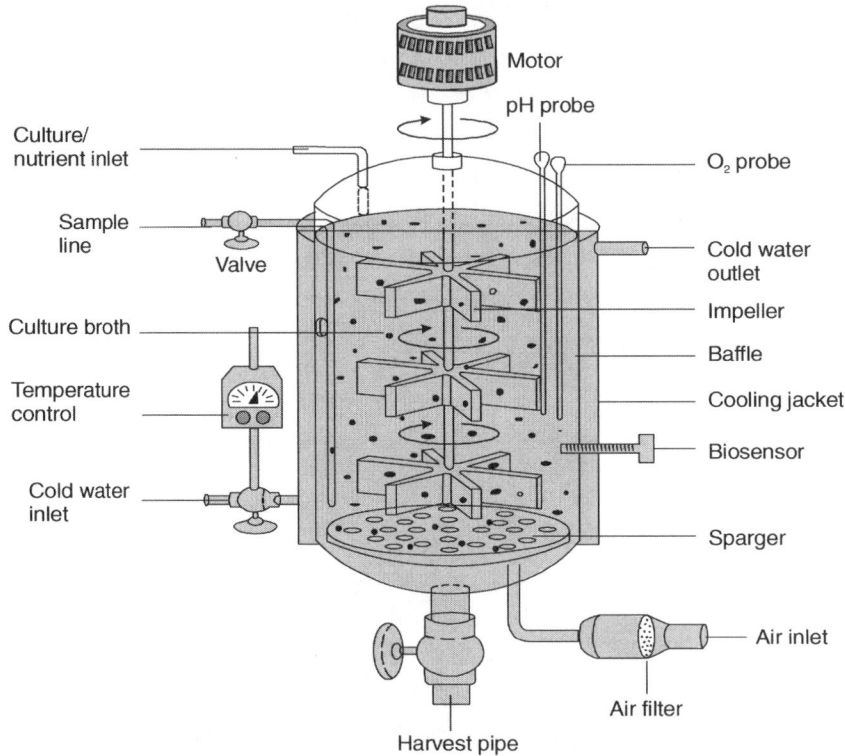

Figure 8.2   Fermenter and its components

Temperature Control

Heat will be produced by microbial activity and mechanical agitation, and if the heat generated by these two processes is not ideal for the particular manufacturing process, it may have to be removed from the system. A basic distinction is made between heating surfaces on the outer wall of the reactor and interior exchange surfaces. On a laboratory scale, little heat is normally generated and excess heat has to be provided by placing the fermenter in a thermostatically controlled bath, or by the use of internal heating coils or a heating jacket through which water is circulated. Fermenters are often designed with double jacket as heat exchange surface and it serves the function of insulation. The insulation is primarily used for temperature control. This also protects operators against the risk of burns. In addition, insulation reduces heat radiation and thus prevents excessive heating of the service rooms.

Heat exchangers are useful in providing adequate temperature control, and for this purpose, plate heat exchanging elements in heating circuits are used. The heating circuit medium and the steam are routed through two separate adjacent chambers using the counterflow principle. The very thin-walled and specially shaped exchangers generate turbulent flow and thus expedite the transfer of heat. After a certain operating time, the efficiency of a heat exchanger may continuously deteriorate as a result of deposits on the surface due to hard water. To remove depositions, it is necessary to rinse the heat exchangers periodically with rinsing agents containing acid solutions.

Aeration and Agitation

Most industrial microbial processes are aerobic, and are mostly carried out in aqueous medium containing salts and organic substances; usually these broths are viscous, showing a non-Newtonian behaviour. In these processes, oxygen is an important nutrient used by microbes for growth, maintenance and metabolite production, and scarcity of oxygen affects the process performance. Therefore, it is important to ensure an adequate delivery of oxygen from a gas stream to the culture broth. During aerobic processes, the oxygen is transferred from a rising gas bubble into a liquid phase and ultimately to the site of oxidative phosphorylation inside the cell. Efficient oxygen transfer during the fermentation process is enabled by aeration and agitation systems of the fermenter.

The primary purpose of aeration is to provide microorganisms in submerged culture with sufficient oxygen for metabolic requirements, while agitation should ensure that a uniform suspension of microbial cells is achieved in a homogeneous nutrient medium. The type of aeration–agitation system used in a particular fermenter depends on the characteristics of the fermentation process under consideration. Although fine bubble aerators without mechanical agitation have the advantage of lower equipment and power costs, agitation may be dispensed with sufficient aeration, e.g., in processes where broth of low viscosity and low total solids are used. Thus, mechanical agitation is usually required in fungal and bacterial fermentations. Non-agitated fermentations are normally carried out in vessels of a height to diameter ratio of 5:1. In such vessels, aeration is sufficient to produce high turbulence; but a tall column of liquid does require greater energy input for the production of compressed air.

The structural components of the fermenter involved in aeration and agitation are:

* Agitator or impeller

* Stirrer glands and bearings
* Baffles
* Sparger

The agitator (Impeller)   The agitator is required to achieve a number of mixing objectives, e.g., bulk fluid and gas-phase mixing, air dispersion, oxygen transfer, heat transfer, suspension of solid particles and maintaining a uniform environment throughout the vessel contents. To design a fermenter for achieving these conditions, knowledge of the most appropriate agitator, air sparger, baffles, the best positions for nutrient feeds, acid or alkali for pH control and antifoam addition is essential. Processes vary greatly in their stirring requirements: some need very intensive stirring movements with high-power input (e.g., microbial cultures), while others depend on gentle stirring (e.g., cell cultures). A vast array of different impeller types is available for the key stirring tasks such as homogenizing, dispersing, suspending, etc.

Disc turbines, vane discs, open turbines of variable pitch and propellers are widely used in fermentation processes. The stirring task depends primarily on the flow pattern. Every impeller is characterized by the typical flow pattern it generates. It is used wherever the specific flow pattern achieves the optimal mixing effect for the corresponding process. The disc turbine with a radial primary flow direction is used predominantly where turbulent flow is needed. It is especially suitable for dispersing liquid/liquid substances and to achieve a high phase-related exchange surface which determines the rate of substance transport. The propeller with axial primary flow direction involves only very small shear forces. This leads to very gentle but nevertheless thorough mixing. For this reason, the propeller is recommended especially for the cultivation of animal cells. Pitched blade impellers are particularly suitable for homogenizing, suspending, and dispersing liquid/liquid and solid/liquid substances, as well as the gentle stirring media. In case of very intensive stirring tasks, it must be ensured that all interior components are mounted safely.

Stirrer and bearings   Every stirrer function requires an appropriate stirrer system which basically comprises the following four components: drive (motor), shaft bearing, stirrer shaft and impeller. The core element of every drive is the shaft bearing. The satisfactory sealing of the stirrer shaft assembly has been one of the most difficult problems to overcome in the construction of fermentation equipment which can be operated aseptically for long periods. A number of different designs has been developed to obtain aseptic seals. The stirrer shaft can enter the vessel from the top, side or bottom of the vessel. Top entry is most commonly used; but, bottom entry may be advantageous if more space is needed on the top plate for entry ports, and the shorter shaft permits higher stirrer speeds to be used by eliminating the problem of the shaft whipping at high speeds. Originally, bottom-entry stirrers have been considered undesirable as the bearing would be submerged. Mechanical seals that seal dynamically can be used for bottom entry provided that they are routinely maintained and replaced at recommended intervals.

Baffles   Four baffles are usually incorporated into agitated vessels of all sizes to prevent a vortex or rotary flow and the formation of thrombi caused by rotary drives so as to improve aeration efficiency. In vessels over 3 dm^3 diameter, six or eight baffles may be used. Baffles are metal strips roughly one-tenth of the vessel diameter and attached radially to the wall. The agitation effect is only slightly increased with wider baffles, but drops sharply with narrower baffles. Baffles are often attached to the interior walls to promote mixing and thus increase heat transfer and possibly reaction rates. Baffles should be installed so that space exists between them and the vessel wall to ensure scouring action around and behind the baffles to minimize microbial growth on the baffles and the fermenter walls. Extra cooling coils may be attached to baffles to improve the cooling capacity of a fermenter without unduly affecting the geometry. Baffles may be permanently installed or removable.

Sparger   Many processes require aeration of the medium. This is brought about by ring spargers with aeration holes to generate many small, homogeneously distributed bubbles, sintered pipes to generate minuscule bubbles and to protect sensitive media and bubble-free aeration systems for oxygenation by diffusion. In the case of shear-sensitive cell cultures, gentle stirrer movements and aeration are particularly important to avoid mechanical destruction of the cells. Bubble-free aeration requires a very large exchange surface whereas ring spargers generate air bubbles with small exchange surface with effective gas transfer. A sparger is a device for introducing air into the liquid in a fermenter and the choice of the system depends on the aeration rate, rheological properties of the medium and the type of culture. There are three basic types of spargers used: the porous sparger, the orifice sparger (a perforated pipe) and the nozzle sparger (an open or partially closed pipe). A combined sparger-agitator may be used in laboratory fermenters. The bubble size produced from spargers is always 10 to 100 times larger than the pore size of the aerator block. The throughput of air is low because of the pressure drop across the sparger, there is also the problem of the fine holes getting blocked by growth of the microbial culture. To ensure adequate gas transfer, aeration can be accompanied with agitation that creates stirring movement.

Air Filter System

Fermentation processes require a supply of air into the medium. In principle, air is preconditioned inside the air-feed line so that the required quantity reaches the medium in sterilized form. Sterile air will be required in very large volumes in many aerobic fermentation processes. Although there are a number of ways of sterilizing air, only two methods such as heat and filtration are generally accepted. Sterilization of the exhaust gas can be achieved by 0.2 µm filters on the outlet pipe. Filter housings and air filter cartridges are available in a multitude of designs and dimensions. The type of air filter selected for a particular process depends on the most diverse factors: type of filtration (particle, micro or ultrafiltration), medium (gas or liquid) and required filter surface (aeration rate). Air supply lines are designed as simple air feeds or involve multiple gas mixing units with manual or automatic control. Like many other product lines, the air supply unit must be designed to allow steam sterilization. Since filters are sensitive to humidity, care should be taken that they are not clogged by condensate formation.

Sampling and Feed Ports

To prevent contamination when operating a fermenter, it is essential that both the addition vessel and the fermenter are maintained at a positive pressure and the addition port is equipped with a steam supply. Several ports must be available in the fermentor lid for applications such as addition of acid/alkali solutions, feed solutions, antifoam agents, media, supply of air or oxygen, etc. A sterile barrier must be maintained between the fermenter contents and the exterior when the sample port is not being used, and it must be sterilizable after use. Addition of nutrients and acid/alkali to small fermenters is normally made via silicone tubes which are autoclaved separately and a peristaltic pump is used after aseptic connection.

Sensor Probes

Sensors probes are used to monitor processes in the reactor. Controls for all medium-related parameters such as temperature, oxygen, turbidity, pH, conductance etc. are located in the lower part of the reactor. Additional physical parameters such as pressure, broth level, foam, etc. can normally be read on the upper portion of the reactor. For the measurement of these parameters, many components based on various measuring principles are available and the appropriate sensor is selected on the basis of operating conditions, accuracy and design. Double "O" ring seals have been used for many years to provide an aseptic seal for glass electrodes in stainless steel housings in fermenters.

Foam Control

Most cell cultures produce a variety of foam-producing and foam-stabilizing agents such as proteins, polysaccharides and fatty acids. Foam build-up in fermenters is very common, particularly in aerobic systems. Foaming causes a range of reactor-operating problems and foam control is an important parameter in fermentation process design. Excessive foam overflowing from the top of the fermenter provides a route for entry of contaminating organisms and causes blockage of outlet gas lines. Liquid and cells trapped in the foam represent a loss of bioreactor volume; conditions in the foam may not be favourable for metabolic activity. In addition, fragile cells can be damaged by collapsing foam. Addition of special antifoam compounds to the medium is the most common method of reducing foam build-up in fermenters since they are surface tension-lowering substances. Commonly used antifoam agents are silicone oil, soybean oil, PPG 2000, etc. Since antifoams reduce the rate of oxygen transfer, mechanical foam dispersal is generally preferred. Mechanical foam breakers, such as high-speed discs rotating at the top of the vessel, break down foam by an impact mechanism inside the fermenter.

Valves

Valves attached to fermenters and ancillary equipment are used for controlling the flow of liquids and gases in a variety of ways. The valves may be of the following types:

* Simple ON/OFF valves which are either fully open or fully closed.
* Valves which provide coarse control of flow rates.
* Valves which may be adjusted very precisely so that flow rates may be accurately controlled.
* Safety valves which are constructed in such a way that liquids or gases will flow in only one direction.

Besides, non-return valves (NRV) are used wherever a reflux of the flowing medium has to be avoided. NRVs are fittings that prevent reflux by a rebound mechanism. The path is opened by the flow. In case of congestion, the valve is closed by the weight of the mechanism, by a spring return, or by the reflex of the medium itself.

The bottom of the fermenter is equipped with bottom drain valves for draining the reactor. It is important that the dimensions of the drain be large enough for fast drainage. Usually bellow type bottom drain valves are well-suited and widely employed designs.

Every pressure device should be protected by a safety element to ensure that the permissible service pressure is never exceeded. In most cases, this is achieved with a safety valve that is accurately set and sealed by a calibration authority. These valves are fitted with sight glass/rupture discs that break when tolerance limit of pressure exceeds due to blockage of the exhaust line and the exhaust air escapes into an exit line.

Steam Traps

Steam traps are used wherever steam condenses. In all steam lines, it is essential to remove any steam condensate which accumulates in the piping to ensure optimum process conditions. This may be achieved by incorporating steam-trap devices, which will collect and remove automatically any condensate, but not the steam, at appropriate points in steam lines.

Reflux Cooler and Air Exhaust Unit

The air escaping from the fermenter has been heated and is therefore saturated with humidity. This leads to a loss of liquid, which is undesirable in many processes. This loss can be greatly reduced by using a reflux cooler. In a cooler, the escaping exhaust air is cooled down until the humidity condenses and flows back into the fermenter. It is essential that the design of the unit as well as the dimensions of the piping and the layout of the return pipes allow a trouble-free reflux of the medium. The flow rate of the exhaust air must be controlled accordingly.

Exhaust air from the fermenter is discharged through the exhaust air line and the unit ensures that only filtered air can escape. To minimize the loss of liquid, cooling elements (reflux coolers) may be used which condense the exhaust air and route the liquid back into the medium. The air exhaust unit should be sterilizable and controllable in the manual or automatic modes. This unit should be able to prevent vacuum formation in the fermenter as a result of cooling processes, for instance, after sterilization.

Seals

Every detachable connection requires a seal. Greater the number of seals, greater the risk of contamination through leaks. Sealed connections should not only be perfectly tight but also optimally positioned and easy to disassemble and service. Non-detachable connections (e.g., welded metring lines) have the disadvantage of being permanently installed, which may cause problems in case of damage due to mechanical or chemical impact. Detachable, sealed connections can always be retrofitted or replaced when damaged and thus are more flexible. The tightness of sealed connections can be checked with procedures such as pressure drop or leak-tightness tests.

Good hygienic design means that gaps and dead spaces are minimized, media can drain, and surfaces can be easily cleaned. This must always be weighed against other factors such as safety, manufacturing capacity, handling and costs. The design of a sealed connection should therefore consider not only optimal positioning, but other factors as well. Stainless steel tends to deform under the influence of heat, and this may have a negative effect on the surfaces of sealed connections.

COMPONENTS OF A FERMENTATION PROCESS

Regardless of the type of fermentation, an established process may be divided into the following six basic components:

* The formulation of media to be used in culturing the process organism during the development of the inoculum and in the production fermenter.

* The sterilization of the medium, fermenter and ancillary equipment.

* The production of an active, pure culture in sufficient quantity for inoculation in the production vessel.

* The growth of the organism in the production fermenter under optimum conditions for product formation.

  ❋ The extraction of the product and its purification.

  ❋ The disposal of effluents produced by the process.

We will discuss about each of these components in detail in the following sections.

MEDIUM FORMULATION

Medium formulation is an essential stage in the design of successful laboratory experiments, pilot-scale development and manufacturing processes. The constituents of a medium must satisfy the elemental requirements for cell biomass and metabolite production and there must be an adequate supply of energy for biosynthesis and cell maintenance. The first step to consider is an equation based on the stoichiometry for growth and product formation. Thus, for an aerobic fermentation:

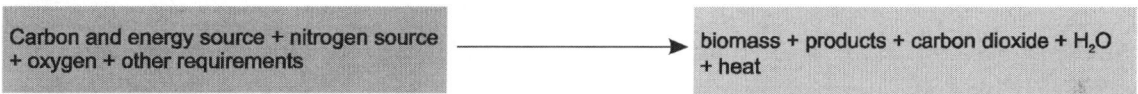

Carbon and energy source + nitrogen source + oxygen + other requirements   ⟶   biomass + products + carbon dioxide + H_2O + heat

This equation should be expressed in quantitative terms, which is important in the economical design of media if component wastage is to be minimal. Thus, it should be possible to calculate the minimal quantities of nutrients which will be needed to produce a specific amount of biomass.

Knowledge of the elemental composition of a process microorganism is required for the solution of the elemental balance equation. Absolute minimum quantities of nitrogen (N), sulphur (S), phosphorus (P), magnesium (Mg) and potassium (K) must be included in the initial medium recipe. Trace elements such as iron (Fe), zinc (Zn), copper (Cu), manganese (Mn), cobalt (Co), molybdenum (Mo), boron (B), selenium (Se) and vanadium (Va) may also be needed in smaller quantities. Some nutrients are frequently added in substantial excess of that required, e.g., P, K; however, others are often near limiting values, e.g., Zn, Cu. The concentration of P is deliberately raised in many media to increase the buffering capacity. Some microorganisms cannot synthesize specific nutrients, e.g., amino acids, vitamins or nucleotides. Once a specific growth factor has been identified, it can be incorporated into a medium in adequate amounts as a pure compound or as a component of a complex mixture. The carbon substrate has a dual role in biosynthesis and energy generation. The other major nutrient which will be required is oxygen which is provided by aerating the culture.

Water

Water is the major component of all fermentation media and is needed in many of the supplementary services such as heating, cooling, cleaning and rinsing. Clean water of consistent composition is, therefore, required in large quantities from reliable permanent sources.

When assessing the suitability of a water supply, it is important to consider pH, dissolved salts and effluent contamination.

Energy Sources

Energy for growth comes from either the oxidation of medium components or from light. Most industrial microorganisms are chemo-organotrophs and therefore, the commonest source of energy will be the carbon source such as carbohydrates, lipids and proteins. Some microorganisms can also use hydrocarbons or methanol as carbon and energy sources.

Carbon Sources

The main product of a fermentation process will often determine the choice of carbon source, particularly if the product results from its direct dissimilation. The purity of the carbon source may also affect the choice of substrate. The method of media preparation, particularly sterilization, may affect the suitability of carbohydrates for individual fermentation processes. It is better to sterilize sugars separately because they may react with ammonium ions and amino acids to form black nitrogen-containing compounds which will partially inhibit the growth of many microorganisms. Starch suffers from the handicap that when heated in the sterilization process, it gelatinizes, giving rise to very viscous liquids, and only concentrations of up to 2% can be used without modification.

Some examples of carbon sources are as follows:

Carbohydrates It is common practice to use carbohydrates as the carbon source in microbial fermentation processes. The most widely available carbohydrate is starch obtained from maize grain. It is also obtained from other cereals, potatoes and cassava. Maize and other cereals may also be used directly in a partially ground state. Barley malt is the main substrate for brewing ale and lager beer in many countries. Sucrose is obtained from sugar cane and sugar beet. It is commonly used in fermentation media in a very impure form as beet or cane molasses. The use of lactose and crude lactose in media formulations is now extremely limited. Corn steep liquor is a by-product after starch extraction from maize. Although primarily used as a nitrogen source, it does contain lactic acid, small amounts of reducing sugars and complex polysaccharides.

Oils and fats Typical oil contains approximately 2.4 times the energy of glucose on a per weight basis. Examples of oil used in fermentation media include glycerol trioleate, soyabean oil or rapeseed oil, methyl oleate, etc.

Hydrocarbons and their derivatives There has been considerable interest in hydrocarbons. Initial work has been done using n-alkanes for production of organic acids, amino acids, vitamins and co-factors, nucleic acids, antibiotics, enzymes and proteins. Methane, methanol and n-alkanes have all been used as substrates for biomass production.

Nitrogen Sources

Most industrially used microorganisms can utilize inorganic or organic sources of nitrogen. Inorganic nitrogen may be supplied as ammonium salts or nitrates. Ammonium salts such as ammonium sulphate will usually produce acid conditions as the ammonium ion is utilized and free acid will be liberated. On the other hand, nitrates will normally cause an alkaline drift as they are metabolized.

Organic nitrogen may be supplied as amino acid, protein or urea. In many instances, growth will be faster with a supply of organic nitrogen and a few microorganisms have an absolute requirement for amino acids. However, amino acids are more commonly added as complex organic nitrogen sources which are non-homogeneous, cheaper and readily available.

Other proteinaceous nitrogen compounds serving as sources of amino acids include corn steep liquor, soya meal, peanut meal, cotton-seed meal, and yeast extract. Chemically defined amino acid media devoid of protein are necessary for the production of certain vaccines when they are intended for human use.

Minerals

All microorganisms require certain mineral elements for growth and metabolism. In many media, magnesium, phosphorus, potassium, sulphur, calcium and chloride are essential components and because of the concentrations required, they must be added as distinct components. Others such as cobalt, copper, iron, manganese, molybdenum and zinc are also essential but are usually present as impurities in other major ingredients. There is obviously a need for batch analysis of media composition to ensure that this assumption can be justified, otherwise there may be deficiencies or excesses in different batches of media.

Chelators

Many media cannot be prepared or autoclaved without the formation of a visible precipitate of insoluble metal phosphorus. The problem of insoluble metal phosphate(s) may be eliminated by incorporating low concentrations of chelating agent such as ethylene diamine tetraacetic acid (EDTA), citric acid, polyphosphates, etc. into the medium. These chelating agents preferentially form complexes with the metal ions in a medium. The metal ions then may be gradually utilized by the microorganisms.

Growth Factors

Some microorganisms cannot synthesize a full complement of cell components and therefore, they require preformed compounds called growth factors. The growth factors most commonly required are vitamins, but there may also be a need for specific amino acids, fatty acids, or sterols. Many of the natural carbon and nitrogen sources used in media formulations contain all or some of the required growth factors.

Buffers

The control of pH may be extremely important if optimal productivity is to be achieved. A compound may be added to the medium to serve specifically as a buffer, or may also be used as a nutrient source. Many media are buffered at about pH 7.0 by the incorporation of calcium carbonate. If the pH decreases, the carbonate is decomposed. Phosphates play an important role in buffering. However, high phosphate concentrations are critical in the production of many secondary metabolites.

The balanced use of the carbon and nitrogen sources will also form a basis for pH control as buffering capacity can be provided by the proteins, peptides and amino acids, such as in corn-steep liquor. The pH may also be controlled externally by addition of ammonia or sodium hydroxide and sulphuric acid.

Precursors

Some chemicals, when added to certain fermentations, are directly incorporated into the desired product, e.g., phenylacetic acid is still the most widely used precursor in penicillin production.

Inhibitors

When certain inhibitors are added to fermentations, more of a specific product may be produced, or a metabolic intermediate which is normally metabolized is accumulated. In most cases, the inhibitor is effective in increasing the yield of the desired product and reducing the yield of undesirable related products. Inhibitors have also been used to affect cell wall structure and increase the permeability for release of metabolites. The best example is the use of penicillin and surfactants in glutamic acid production.

Inducers

The majority of enzymes which are of industrial interest are inducible. Induced enzymes are synthesized only in response to the presence of an inducer in the environment. Inducers are often substrates such as starch or dextrins for amylases, maltose for pullulanase and pectin for pectinases.

Oxygen Requirements

It is sometimes forgotten that oxygen, although not added to an initial medium as such, is nevertheless a very important component of the medium in many processes, and its availability can be extremely important in controlling growth rate and metabolite production.

The medium may influence the oxygen availability in a number of ways including the following:

1. Fast metabolism: The culture may become oxygen-limited because sufficient oxygen cannot be made available in the fermenter if certain substrates, such as rapidly metabolized sugars which lead to a high oxygen demand, are available in high concentrations.

2. Rheology: The individual components of the medium can influence the viscosity of the final medium and its subsequent behaviour with respect to aeration and agitation.

3. Antifoams: Many of the antifoams in use will act as surface-active agents and reduce the oxygen transfer rate.

Trace Elements

The role of trace elements in medium formulation can be significant. Cultured cells normally require Fe, Zn, Cu, Se, Mn, Mo and Va. These are often present as impurities in other media components.

Osmolality

The optimum range of osmotic pressure for growth is often quite narrow and varies with the type of cell and the species. It may be necessary to adjust the concentration of NaCl when major additions are made to a medium.

Media Optimization

In most cases, a fermentation medium that supports the production of the metabolite of interest is first developed based on the previous literature, the researchers' past experience and trial and error. In industrial settings, this medium becomes the starting point for further optimization. The primary aim is to increase the product titre and to improve the economy of a fermentation process. Secondary goal includes the development of a simplified metabolite-purification process.

The basic nutrients for the growth of microorganisms include carbon, nitrogen and minerals. Depending on the organism and the fermenter conditions used for growth and metabolite production, additional nutrients such as aminoacids, vitamins, nucleotides or even special chemicals may be required. There is an array of carbon and nitrogen sources available; some are in relatively pure form, whereas others occur in complex forms such as the byproducts of the food and agricultural industries. Pure or high quality raw materials are more consistent from batch to batch. The byproducts of the food and agricultural industries are often less consistent in quality. Pure or higher quality carbon and nitrogen sources are generally used for high-value products such as therapeutic proteins, whereas the less expensive raw materials are used for the production of low-cost and high-volume commodity items such as organic acids or bulk chemicals. When developing fermentation media, cost, availability and consistency of lot have to be taken into consideration. In addition, fermentation media should facilitate better downstream processing.

Detailed investigation is needed to establish the most suitable medium for an individual fermentation process. All microorganisms require water, sources of energy, carbon, nitrogen, minerals, vitamins and oxygen if aerobic. On a small scale, it is relatively simple to devise a medium containing pure compounds, but the resulting medium, although supporting satisfactory growth may be unsuitable for use in a large scale process.

On a large scale, one must normally use sources of nutrients to create a medium which will meet as many as possible of the following criteria:

1. It will produce the maximum yield of product or biomass per gram of substrate used.
2. It will produce the maximum concentration of product or biomass.
3. It will permit the maximum rate of product formation.
4. There will be minimum yield of undesired products.
5. It will be of a consistent quality and be readily available throughout the year.
6. It will cause minimal problems during media making and sterilization.
7. It will cause minimal problems in other aspects of the production process particularly aeration and agitation, extraction, purification and waste treatment.

The use of cane molasses, beet molasses, cereal grains, starch, glucose, sucrose and lactose as carbon sources and ammonium salts, urea, nitrates, corn steep liquor, soya bean meal, slaughterhouse waste and fermentation residues as nitrogen sources, have tended to meet most of the above criteria for production media because they are cheap substrates. However, other more expensive pure substrates may be chosen if the overall cost of the complete process can be reduced; because it is possible to use simpler procedures.

It must be remembered that the medium selected will affect the design of the fermenter to be used. The problem of developing a process from the laboratory to the pilot scale and subsequently to the industrial scale must also be considered. A laboratory medium may not be ideal in a large fermenter with a low gas-transfer pattern. A medium with a high viscosity will also need a higher power input for effective stirring. Besides meeting requirements for growth and product formation, the medium may also influence pH variation, foam formation, oxidation–reduction potential and the morphological form of the organism. It may also be necessary to provide precursors or metabolic inhibitors. The medium will also affect product recovery and effluent treatment.

Different combinations and sequences of process conditions need to be investigated to determine the growth conditions which produce the biomass with the physiological state best constituted for product formation. There may be a sequence of phases each with a specific set of optimal conditions. Medium optimization by the classical method of changing one independent variable (nutrient, antifoam, pH, temperature, etc) while fixing all the others at a certain level can be extremely time consuming and expensive for a large number of variables. When more than five-independent variables are to be investigated, the Plackett-Burman design may be used to find the most important variables in a system, which is then optimized in further studies. The next stage in medium optimization would be to determine the optimum level of each key independent variable which has already been identified by the Plackett-Burman design. Recently, new techniques such as statistical experimental design, genetic algorithms (GA) and particle swam optimization (PSO) have become available to assist in the media optimization for fermentation processes.

Plackett-Burman designs are experimental designs developed by Robin L.Plackett and J.P. Burman for investigating the dependence of some measured quantity on a number of independent variables/factors and this is to minimize the variance of the estimates of these dependencies using a limited number of experiments. The Plackett-Burman design provides an efficient way to identify the most important ones from a large number of variables. Response surface methodology (RSM) explores the relationships between several explanatory variables and one or more response variables. The method was developed by G.E.P. Box and K.B. Wilson. The main idea of RSM is to use a sequence of designed experiments to obtain an optimal response. This model is easy to estimate and apply, even when little information is known about the process by using fractional factorial designs for first degree polynomial model. More complicated design, such as central composite design can be implemented to estimate a second degree model to optimize a response.

STERILIZATION OF THE FERMENTER

Once the design problems of aeration and agitation have been solved, it is essential that the design meets the requirements of the degree of asepsis and containment demanded by the particular process. It will be necessary to be able to sterilize and keep sterile a fermenter and its contents throughout a complete growth cycle. The fermenter should be so designed that it may be steam sterilized under pressure. The medium may be sterilized in the vessel or separately and subsequently added aseptically. If the medium is sterilized *in situ*, its temperature should be raised prior to the injection of live steam to prevent the formation of large amounts of condensate. This may be achieved by steam being introduced into the fermenter coils or jacket. Heating loops or circuits are used to heat or cool a medium, and they provide a way to control cultivation temperatures or execute sterilization processes. Controlled heating circuit has both components of hot-water/steam and cold water heat exchangers. The main elements of a heating circuit include: safety device to prevent exceeding the permissible service pressure, recirculation pump for recirculating the cooling/heating agent, and at least one power component for heating. Power components can be electrical or hot water/steam heat exchangers. The selection of these components depends on the energy sources available on site as well as on the size of the unit to be heated or cooled. Cooling is generally achieved by perfusing the circuit with cold fresh water while simultaneously letting off hot circuit water until the desired temperature is attained. As every point of entry to and exit from the fermenter is a potential source of contamination, steam should be introduced through all the entry and exit points except the air outlet from which steam should be allowed to leave.

INOCULUM DEVELOPMENT

Inoculum production is a critical stage in an industrial fermentation process. The preparation of a population of microorganisms from a dormant stock culture to an active state of growth that is suitable for inoculation in the final production stage is called inoculum development. The inoculum is prepared in a stepwise process employing increasing volumes of media. As the first step, inoculum is taken from a working stock culture to initiate growth in a suitable liquid medium, which mainly favours rapid cell growth but not product formation. The master culture is reconstituted and plated on to solid medium; approximately, ten colonies of typical morphology of high producers are selected and inoculated on to slopes as the sub-master cultures, each sub-master culture being used for a new production run. At this stage, shake flasks may be inoculated to check the productivity of the culture and the results of such tests should be known before the developing inoculum eventually reaches the production plant. A sub-master culture is used to inoculate a shake flask (250 or 500 cm^3 containing 50 or 100 cm^3 medium) which, in turn, is used as inoculum for a larger flask, or a laboratory fermenter, which may then be used to inoculate a pilot-scale fermenter. Culture purity checks are carried out at each stage to detect contamination as early as possible. Although the results of these tests may not be available before the culture has reached the production plant, at least it will be known at which stage in the procedure the contamination has occurred.

Bacterial vegetative cells and spores are suspended, usually, in sterile water, which is then added to the broth. In case of nonsporulating fungi and actinomycetes, the hyphae are fragmented and then transferred to the broth. Inoculum development is generally done using flask cultures; flasks of 50 ml to 12 litres may be used and their number can be increased as per need. Wherever needed, small fermenters may be used. Inoculum development is usually done in a stepwise sequence to increase the volume to the desired level. At each step, inoculum is used at 0.5–5.0% of the medium volume; this allows a 20–200-fold increase in inoculum volume at each step. Typically, the inoculum used in the production stage is about 5% of the medium volume.

TYPES OF FERMENTATION PROCESS

A fermentation process is a biological process, in which the growth of specific cells is attained in a controlled sterile environment. Industrial fermentation processes may be divided into three main types. These are batch, fed-batch and continuous fermentations (Figure 8.3). In batch reactors, all components, except gaseous substrates such as oxygen, pH-controlling substances and antifoaming agents, are placed in the reactor at the beginning of the fermentation. During the process, there are neither input nor output flows. In fed-batch process, nothing is removed from the reactor during the process, but one substrate component is added in order to control the reaction rate by its concentration. In a continuous process, there are both input and output flows, but the reaction volume is kept constant.

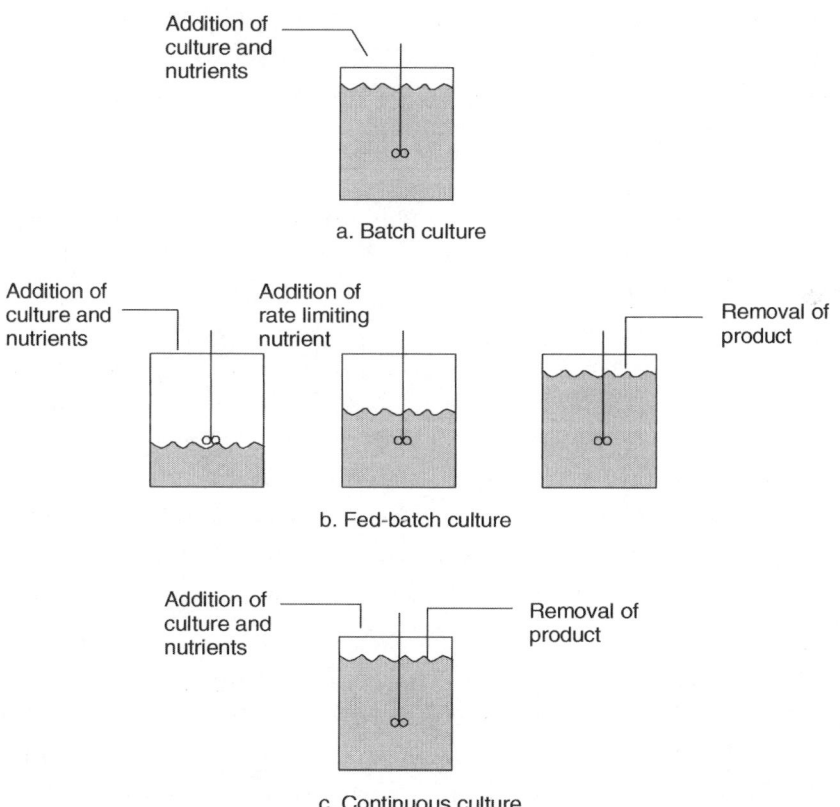

Figure 8.3   Batch, fed-batch and continuous fermentation

Batch Fermentation Process

A typical batch culture is a closed system in which a limited amount of nutrients is present initially in the medium. The medium is inoculated with the desired microorganism which is allowed to grow under defined conditions. The culture passes through all the four phases of growth mentioned earlier, i.e., lag phase, log phase, stationary phase and death phase. During the growth of the microorganism in this medium, no nutrients are added except acid or alkali for maintaining the pH and air for growth of aerobic microorganisms. Normally, the carbon substrate present in the medium serves as the limiting nutrient for growth.

The procedure for a batch fermentation is first to inoculate a small flask of nutrient broth with a pure culture and allow it to grow. A small amount of the culture in the original flask is pipetted out during the exponential growth phase, or log phase, and it is used to inoculate the next flask. This process is repeated a few times to ensure that the culture is acclimatized before it is employed for fermentation process. To reduce the shock resulting from a drastic change in the growth environment, the composition of the media used in preparing the inoculum should optimally be identical to that used in the main process. The quantity of inoculum used generally varies between 1 to 5% with different microbes and it is preferable to use minimal quantity of inoculum, especially if the product is growth-associated.

The fermenter is filled with the medium containing raw materials that are to be fermented. The temperature, pH and pressure for microbial fermentation are properly adjusted, and occasionally nutrient supplements are added to the medium. The medium is steam-sterilized in a pure culture process. After ensuring both quality and quantity, the inoculum of a pure-culture is added to the fermenter from a separate pure culture vessel in a sterile manner. The physiological condition of the inoculum when it is transferred to the next culture stage can have a major effect on the performance of the fermentation process. The optimum time of transfer must be determined experimentally and then procedures established so that inoculation with an ideal culture may be achieved routinely. These procedures include the standardization of culture conditions and monitoring the state of an inoculum so that it is transferred at the optimum time, i.e., in the correct physiological state.

The most widely used criteria for the transfer of inoculum is biomass, and parameters such as packed cell volume, dry weight, wet weight, turbidity, respiration, residual nutrient concentration and morphological form have also been used. Fermentation proceeds, and after the proper time, the contents of the fermenter are taken out for further processing. The fermenter is cleaned and the process is repeated. Thus, each fermentation is a discontinuous process divided into batches.

Fed-Batch Fermentation Process

Fed-batch reactors are widely used in industrial applications because they combine the advantages from both batch and continuous processes. In fed-batch culture, nutrients are continuously or semi-continuously added to a system, while effluent is removed discontinuously. It is usually used to overcome substrate inhibition or catabolite repression so as to increase the product formation and also to produce high cell density. Before starting a fed-batch process, batch fermentation should be performed to get details of the microbial fermentation such as specific growth rate, nutrient requirements, catabolite repression by sugars, consumption rate of nutrients especially limiting nutrient, induction of expression of the product, etc. From batch fermentation, the operator should have knowledge of the following:

* Physico-chemical parameters such as temperature, agitation, pH, medium composition, etc.
* Specific needs of precursors, inducers or other enrichment factors.
* The different growth phases, rates of substrate consumption and product formation, by-product accumulation, etc.
* The relationship between the biomass and product formation (growth and non-growth-associated product) and the oxygen uptake rates.
* Limiting substrate for growth and relationship between the specific growth rate and the limiting substrate concentration.
* Eventual inhibitions from the substrate and/or product.

Two basic approaches to the fed-batch fermentation can be used: the fixed-volume fed-batch culture and the variable-volume fed-batch culture.

Fixed-volume fed-batch   In this type of fed-batch process, the limiting substrate is fed without diluting the culture. The culture volume can also be maintained practically constant by feeding the growth-limiting substrate in undiluted concentrated form accompanied by appropriate oxygen supply. Once the fermentation reaches a certain stage, i.e., when aerobic conditions cannot be maintained anymore, the culture is removed and the biomass is diluted to the original volume with fresh medium containing the feed substrate. The dilution decreases the biomass concentration and results in an increase in the specific growth rate due to consumption of feeding of nutrients. As feeding continues, the growth rate declines gradually due to increase in biomass. Finally, the culture approaches the maximum sustainable level in the vessel with respect to growth and volume of the medium, at which point the culture may be diluted again to start another cycle.

Variable-volume fed-batch   In this mode of process, the volume changes with the duration of fermentation due to continuous feeding of substrate. The feed can be provided according to one of the following options:

i. The same medium used in the batch mode is added;

ii. A solution of the limiting substrate at the same concentration as that in the initial medium is added; and

iii. A very concentrated solution of the limiting substrate is added at a rate less than (i) and (ii).

Advantages of fed-batch fermentation

❋ Production of high cell densities due to extended duration, which are particularly important in the production of growth-associated products.

❋ Controlled conditions in terms of feeding of substrate and oxygen during the fermentation.

❋ Control over the production of by-products or catabolite repression effects due to limited provision of substrates solely required for product formation.

❋ The mode of operation can overcome and control deviations in the organism's growth pattern as found in batch fermentation.

❋ Allows the replacement of water loss by evaporation.

❋ Alternative mode of operation for fermentations leading to accumulation of toxic substances because cells can only metabolize a certain quantity at a time. These substances are metabolic by-products of fermentation of substrates that further interfere with the growth and product formation.

❋ Increase of antibiotic-marked plasmid stability by providing the correspondent antibiotic during the timespan of the fermentation.

❋ No additional special piece of equipment is required as compared with the batch fermentation mode of operation.

Disadvantages of fed-batch fermentation

❋ It requires previous analysis of the microorganism, its requirements and the understanding of its physiology with the productivity.

❋ It requires a substantial amount of operator skill for the set-up, definition and development of the process.

❋ In a cyclic fed-batch culture, care should be taken in the design of the process to ensure that toxins do not accumulate to inhibitory levels and that nutrients other than those incorporated into the feed medium can be limiting. Also, if many cycles are run, the accumulation of non-producing or low-producing variants may result.

❋ The quantities of the components to control must be above the detection limits of the available measuring equipment.

Continuous Fermentation Process

In batch cultures, during the early stages of logarithmic growth phase, the conditions remain relatively constant but during the later stages when the cell growth is quite large, drastic changes take place in the chemical composition of the culture medium. For many studies such as those on physiological processes involving synthesis of an enzyme, exponentially growing cells are needed. It is, therefore, necessary to maintain the cultures in constant environment for long periods. Such systems are possible with a continuous culture which is essentially an open system in which the culture volume is maintained constant by adding fresh medium continuously and removing the spent culture medium continuously at the same rate. When such a system is in equilibrium, various parameters of the system such as culture volume, cell number and concentration of nutrients remain constant. Under such a situation, the system is said to be in steady state.

Exponential growth in batch culture may be prolonged by the addition of fresh medium to the vessel, provided that the medium has been designed such that growth is substrate-limited (i.e., by some component of the medium) and not toxin-limited. Exponential growth will proceed until the additional substrate is exhausted. This exercise may be repeated until the vessel is full. However, if an overflow device is fitted to the fermenter such that the added medium displaces an equal volume of culture from the vessel, then continuous production of cells could be achieved. If medium is fed continuously to such a culture at a suitable rate, a steady state is achieved eventually, and formation of new biomass by the culture is balanced by the loss of cells from the vessel. The flow of medium into the vessel is related to the volume of the vessel and the dilution rate, D, is defined as:

$$D = F/V$$

where F is the flow rate ($dm^3\ h^{-1}$) and V is the volume (dm^3).

Thus, D is expressed in units h^{-1}.

The net change in cell concentration over a time period may be expressed as:

$$dx/dt = \text{growth} - \text{output}$$

or

$$dx/dt = \mu x - DX$$

Under steady-state condtions, the cell concentration remains constant.

Thus,

$$dx/dt = 0$$
$$\mu x = Dx$$
$$\mu = D$$

Thus, under steady-state conditions, the specific growth rate is controlled by the dilution rate, which is an experimental variable. It will be recalled that under batch culture conditions, an organism will grow at its maximum specific growth rate and therefore, it is obvious that a continuous culture may be operated only at dilution rates below the maximum specific growth rate. Thus, within certain limits, the dilution rate may be used to control the growth rate of the culture.

The growth of the cells in a continuous culture of this type is controlled by the availability of the growth-limiting chemical component of the medium and thus the system is described as a chemostat. An alternative type of continuous culture to the chemostat is the turbidostat, where the concentration of the cells in the culture is kept constant by controlling the flow of medium such that the turbidity of the culture is kept within certain narrow limits. This may be achieved by monitoring the biomass with a photoelectric cell and feeding the signal to a pump supplying medium to the culture such that the pump is switched on if the biomass exceeds the set point and it is switched off if the biomass falls below the set point. Systems other than turbidity may be used to monitor the biomass concentration, such as CO_2 concentration or pH in which case it would be more correct to term the culture a biostat. The chemostat is the more commonly used system because it has the advantage over the biostat of not requiring complex control systems to maintain a steady state. However, the biostat may be advantageous in continuous enrichment culture in avoiding the total washout of the culture in its early stages.

Continuous culture can outperform batch culture by eliminating the inherent down time for cleaning and sterilization and the long lags before the organisms enter a brief period of high productivity. It is possible to maintain very high rates of product formation for long times with continuous cultivation. Continuous culture is superior to batch culture in several ways. Interpretation of results is difficult for batch culture because of changing concentrations of products and reactants, varying pH and redox potential, and a complicated mix of growing, dying, and dead cells. Data from continuous cultures have much less complexity because there are dynamic equilibria from the steady state. Cause and effect relationships tend to be obvious. Although continuous culture gets much more productivity from the bioreactor, there is not so great an improvement over batch culture in terms of the total amount of tanks and resources because there must be equipment for make-up and sterilization to support the continuously operated vessel. Problems associated with this mode of processing are possibility of contamination while adding or recovering broth, strain reversion and reduction in the productivity especially if the product is a growth-associated one.

Applications of continuous cultivation　Commercial continuous fermentation has its best prospects with strains that are used for developing either cell mass or to secrete an enzyme or metabolite. Processes with finely tuned mutants for special products are carried out in batches to minimize reversion to less productive strains. Another barrier to progress for continuous process is the cost factor and is mainly due to establishment of each research station. Advancement in research and development requires multiple vessels for screening many variables, but there are usually very few continuous fermenters because the cost of pumps, reservoirs, sterilizers and controls is relatively high. Furthermore, much labour is needed to start a run, to keep the instruments working, and to monitor the process and results. Continuous industrial microbial processes are much less common than batch processes, but most biological waste treatment steps are operated continuously because the waste stream keeps flowing in.

RHEOLOGICAL PROPERTIES AFFECTING FERMENTATION PROCESSES

Rheology is the study of the flow of matter, particularly the liquid state. Growth and product formation is largely influenced by oxygen transfer that in turn depends on medium rheology. Changes in rheology of fermentation broths are caused by variation of one or more of the following properties:

    i. cell concentration
    ii. cell morphology, including size, shape and mass
    iii. flexibility and deformability of cells
    iv. osmotic pressure of the suspending fluids
    v. concentration of polymeric substrate
    vi. concentration of poromeric product and
    vii. rate of shear

The rheological behaviour of a fermentation broth is of paramount importance in describing the transport phenomena in the fermenter. In order to achieve effective mass transfer and mixing of medium, either the reduction in medium viscosity or higher turbulent flow in the culture should be attained in the medium. Reduction in the medium viscosity is not feasible during the process. However, effective mass transfer and mixing of medium could be ensured by higher turbulent flow by providing optimal stirrer speed (N), and the factors that are to be considered are vessel geometrics, optimal agitation, more surface area of air bubbles, higher oxygen diffusivity, effective mass transfer and mixing in the medium. Problems associated with poor bulk mixing and gas–liquid mass transfer would be less in smaller fermenter but they are aggravated in large scales.

FERMENTATION PROCESS SCALE-UP

The performance of improved strains selected from a screening process using small-scale fermentation vessels must be validated in pilot-scale fermenters. Once validated, the improved strains will be used for commercial production. The ultimate commercial success is dependent on the ability to scale up the process, first from laboratory to pilot plant level and then to full commercial scale. The process of transferring a laboratory fermentation process to an industrial operation is referred to as scale-up. The goal of scale-up is to re-create in large production fermenters the optimal conditions in which the improved strains are cultivated in the laboratory. This is accomplished by controlling environmental conditions, including power input, mixing ability, oxygen transfer, shear stress, heat transfer, media sterilization and seed culture preparation. The achievement of successful process scale-up must fit within a range of physical and economic constraints.

Historically, scale-up of improved strain candidates has centred on maintaining the same physical environment for the growing cells as in small fermentation vessels. This goal, however, is difficult to achieve because of the geometry of large-scale fermenters having low surface-to-volume ratios compared to model fermentation vessels. The scale-up might be successful by maintaining constant impeller tip speed to achieve equal shear stress, constant agitation power per unit volume of broth, constant mixing time and constant oxygen mass transfer. Optimization of fermentation processes continues to play an essential role in maximizing the potential of each improved strain.

PRODUCT RECOVERY OR DOWNSTREAM PROCESSING

The choice of product recovery process is based on the following criteria:

* The intracellular or extracellular location of the product.
* The concentration of the product in the fermentation broth.
* The physical and chemical properties of the desired product (as an aid to selecting separation procedures).
* The intended use of the product.
* The minimal acceptable standard of purity.
* The magnitude of bio-hazard of the product or broth.
* The impurities in the fermenter broth.
* The marketable price for the product.

The above-mentioned aspects will be discussed in Chapter 10 on downstream processing.

BIBLIOGRAPHY

Casida, L.E., Jr. (2009). *Industrial Microbiology.* New Age International (P) Ltd. New Delhi.

Mcneil, B. and Harvey, L.M. (2008). *Practical Fermentation Technology.* Wiley, USA.

Smith, J.E. (2009) "Bioprocess/fermentation technology." *Biotechnology.* Cambridge University Press, USA. pp. 49–71.

Stanbury, P.F., Hall, S. and Whitaker, A. (1999). *Principles of Fermentation Technology.* Butterworth-Heinemann, USA.

9

SOLID-STATE FERMENTATION

INTRODUCTION

Solid-state fermentation (SSF) process can be defined as the growth of microorganisms on moist solid materials in the absence or near absence of free-flowing water. However, the substrate must possess enough moisture to support growth and metabolism of the microorganisms. The lower moisture level at which SSF can occur is approximately 12%, because below this level, all biological activities cease. The upper limit is a function of absorbency and hence moisture content, which varies depending upon the nature of the substrate. A majority of the SSF processes involve filamentous fungi, although some involve bacteria and yeast also. SSF processes can be classified into two major categories based on the type of microorganisms involved, viz., natural or indigenous SSF and pure-culture SSF (individual strains or a mixed culture). Pure cultures are generally used in industrial SSF processes as they help in optimum substrate utilization for the targeted product, whereas mixed cultures are used for bioconversion of agro-industrial residues. A majority of the SSF processes involve aerobic organisms. These processes have been used for the production of food, animal feed, pharmaceutical and agricultural products.

SSF comprises two different types of substrates. The most commonly used system involves cultivation on a natural material that acts as both support and energy source and the less frequently used one involves cultivation on an inert support impregnated with a nutrient solution. Since natural substrates create problems in the fermentation kinetic studies due to their heterogeneous nature, such studies could be performed better using synthetic inert solid substrates. The use of a defined liquid medium and an inert support with a homogeneous physical structure improves controlling and monitoring the process and the reproducibility of fermentations. The substrates most commonly used in SSF processes are often products or by-products of agriculture, forestry or food processing such as cereal grains, legume seeds, corn, soybeans, wheat bran, soy grits,

Table 9.1 Historical events of Solid State Fermentation (Pandey, 1992)

Year	Developments
2000 BC	Bread making by Egyptians Cheese making using *Penicillium roquefortii*
3000 BC	Koji preparation by using steamed rice as solid substrate and *Aspergillus oryzae*, as source of inoculum
17th century	Preparation of Miso and Tempeh in South-East Asian countries using steamed and cracked legumes as solid substrates by inoculating with *Rhizopus* sp.
18th century	Production of vinegar from apple pomace Production of gallic acid used in printing by mould fermentation.
19th century	Development of composting and solid waste treatment using SSF. Trickling filter for sewage disposal was invented. A digestive enzyme, takadiastase, produced by *Aspergillus oryzae* using wheat bran by SSF.
20th century	Production of primary metabolites using microorganisms by SSF introduced
1900–1920	Production of fungal enzymes mainly amylases and kojic acid. Kojic acid used as an insecticide, antibiotic and analytical reagent.
1920–1940	Production of fungal enzymes, gluconic acid and citric acid. Rotary drum fermenter developed
1940–1950	Considered as Golden Era of Fermentation Industry. Discovery and development of wonder drug, penicillin, by SSF and SmF
1950–1960	Steroid transformations in SSF using fungal cultures
1970-1980	Production of Single Cell Protein (SCP) and production of mycotoxins for use in cancer research
1980-present	Development of column type fermenter, development of bioprocesses such as bioremediation, and biodegradation of hazardous materials, biodetoxication of agro-industrial wastes, biotransformation of crops and crop residues for nutritional enrichment, large scale production of bioactive secondary metabolites such as antibiotics, alkaloids, enzymes, organic acids, biopesticides, biosurfactants, biopharmaceuticals, biofuels and aromatic compounds etc.

lignocellulosic materials such as straws, saw dust, etc. Most of these substrate materials are polymeric molecules, insoluble or sparingly insoluble in water, but are mostly cheap, easily available and represent a concentrated source of nutrients for microbial growth. Many of these fermentations have great antiquity and in many instances, there are records dating back to hundreds of years (Table 9.1).

CHARACTERISTICS OF SSF AND SMF

SSF processes are distinct from submerged fermentation (SmF) culturing, since microbial growth and product formation occurs at or near the surface of the solid substrate particle having low moisture content. Thus, it is crucial to provide optimized water content and control the water activity of the fermenting substrate which in turn affects the microbial activity adversely. Moreover, water has profound impact on the physico-chemical properties of the solids and this, in turn, affects the overall process productivity. The distinguished characteristics of SSF and SmF are tabulated in Table 9.2.

Table 9.2   Characteristics of SSF and SmF

Characteristics	Solid-state	Submerged
Substrate	Static	Agitated
Water usage	Limited	Unlimited
Oxygen supply by	Diffusion	Aeration
Inoculum size	Large	Small
Nutrients	Complex	Defined, complex, semisynthetic
Trace elements	Usually not required	Mostly essential
Specific activity	High	Moderate
Products	Highly concentrated	Lower concentration
Catabolite repression	Negligible	High
Volume of fermentation mash	Smaller	Larger
Liquid waste produced	Negligible	Significant volume
Physical energy requirement	Low	High
Human energy requirement	High	Low
Capital investment	Low	High

FACTORS INFLUENCING SSF

There are several important aspects, which should be considered for the development of any bioprocess by SSF. The factors governing can be divided into biological and physico-chemical factors. Biological factors are related to the biology, metabolic process and reproduction of microorganisms. Physico-chemical factors include temperature, pH, moisture content, aeration, etc. The various factors include:

    i.   Choice, type and size of suitable microorganisms

    ii.   Choice of suitable substrate and its pre-treatment

   iii. Particle size of the substrate

   iv. Moisture content and water activity of the substrate

    v.   pH of the substrate

   vi. Temperature of fermenting matter

  vii. Nutritional factors

 viii. Aeration and agitation

i. Choice, type and size of suitable microorganisms

The ability of the microorganisms for growing on a solid substrate is determined by diverse parameters such as water activity requirement, adherence capacity and penetration into the substrate, and ability to assimilate mixtures of different polysaccharides due to the complex nature of the substrates used. Among the different groups of microorganisms used, the filamentous fungi are shown to be particularly suitable and are the best adapted microorganisms for SSF owing to their physiological, enzymological and biochemical properties.

The hyphal mode of fungal growth gives the filamentous fungi the power to penetrate into the solid substrates. This also gives them a major advantage over unicellular microorganisms for the colonization of the substrate and the utilization of the available nutrients. In addition, their ability to grow at low water activity and high nutrient concentration makes fungi efficient and competitive in natural microflora for bioconversion of solid substrates.

Fungi are very efficient enzyme producers, and can produce aromatic compounds and health-promoting substances of interest to the food industry, and biopesticides. From a practical point of view, vegetative growth is preferred over sporulation. However, bacteria and yeasts have also been used in traditional cultivation in SSF processes. Bacteria have been used for enzyme production, composting, ensiling and some food processes such as natto, soybean paste, vinegar, sausages, etc.

Yeasts have been mainly used for ethanol production and protein enrichment of agricultural residues. The uses of various microflora are listed in Table 9.3.

Based on the type of product being produced, different types of inoculums can be used. There are several advantages in the use of spores rather than vegetative cells for inoculum. They can serve as a biocatalysts in bioconversion reactions, convenient to inoculate, greater flexibility in the coordination of inoculum preparation, prolonged storability for subsequent use, and higher resistance to mishandling during transfers. However, it has some disadvantages such as longer lag time, different optimal conditions for spore germination and vegetative growth, and larger inoculum size requirement. The spores are metabolically dormant, and hence the metabolic activities must be induced and the appropriate enzyme systems must be synthesized before the fungus begins to utilize the substrate and grow. However, some organisms require vegetative inocula.

Table 9.3 Commonly used microbes in SSF

Microflora	SSF Process
Bacteria	
Bacillus sp.	Composting, natto, amylase
Pseudomonas sp.	Composting
Serratia sp.	Composting
Streptococcus sp.	Composting
Lactobacillus sp.	Ensiling, food
Clostridium sp.	Ensiling, food
Yeast	
Endomycopsis burtonii	Cassava, rice
Saccharomyces cerevisiae	Food, ethanol
Schwanniomyces castelli	Ethanol, amylase
Fungi	
Alternaria sp.	Composting
Aspergillus sp.	Composting, industrial, food
Fusarium sp.	Composting, gibberellins
Monilia sp.	Composting
Mucor sp.	Composting, food, enzyme
Rhizopus sp.	Composting, food, enzymes, organic acids

(Contd.)

Table 9.3 (Continued)

Microflora	SSF Process
Fungi	
Phanerochaete chrysosporium	Composting, lignin degradation
Trichoderma sp.	Composting, biological control, bioinsecticide
Beauveria sp., *Metarrhizium* sp.	Biological control, bioinsecticide
Amylomyces rouxii	Cassava, rice
Aspergillus oryzae	Koji, food, citric acid
Rhizopus oligosporus	Tempeh, soybean, amylase, lipase
Aspergillus niger	Feed, proteins, amylase, citric acid
Pleurotus oestreatus, P. sajor-caju	Mushroom
Lentinus edodes	Shiitake mushroom
Penicillium notatum, P. roquefortii	Penicillin, cheese

Based on the type of product being produced, different types of inoculums can be used. There are several advantages in the use of spores rather than vegetative cells for inoculum. They can serve as a biocatalysts in bioconversion reactions, convenient to inoculate, greater flexibility in the coordination of inoculum preparation, prolonged storability for subsequent use, and higher resistance to mishandling during transfers. However, it has some disadvantages such as longer lag time, different optimal conditions for spore germination and vegetative growth, and larger inoculum size requirement. The spores are metabolically dormant, and hence the metabolic activities must be induced and the appropriate enzyme systems must be synthesized before the fungus begins to utilize the substrate and grow. However, some organisms require vegetative inocula.

ii. Choice of Substrate and its Pre-treatment

The selection of a suitable substrate for SSF depends on several factors mainly related to cost and availability. All solid substrates have a common basic macromolecular structure, viz., starch, cellulose, lignocellulose, pectin and other polysaccharides. The most commonly used system involves cultivation on a natural material, and the less frequently used one involves cultivation on an inert support impregnated with a liquid medium. The characteristics of natural and inert substrates are described in detail (Table 9.4).

Table 9.4 Characteristics of natural and inert substrates used in SSF

Natural substrate	Inert substrate
They are by-products from agriculture or agro-industries.	The inert solid support can be of artificial or natural origin.
They are heterogeneous water-insoluble materials such as grains and grain by-products, cassava, potato, beans and sugar beet pulp, which are rich in amylose or lignocelluloses.	They are homogeneous materials which include sugar cane bagasse, coir pith, oil cakes, hemp, inert fibres, resins, polyurethane foam and vermiculite which are impregnated with a liquid medium, which contains all the nutrients such as sugars, lipids, organic acids, etc.
Cheap and easily available.	Expensive.
It serves both as a support and a nutrient source.	It serves only as support.
Pretreatment is necessary for its increased utilization by the microorganism.	No pretreatment is required.
The heterogeneous physical structure and undefined nutrient content creates problems in fermentation kinetic studies.	The use of a defined liquid medium and an inert support with a homogeneous physical structure improves controlling and monitoring of the process and the reproducibility of fermentation.
During microbial growth, the solid medium is disintegrated causing changes in the geometric and physical characteristics of the medium, and consequently reducing heat and mass transfer.	Use of an inert support, with a more or less constant physical structure throughout the process, enables improved control of heat and mass transfer.
Product recovery is difficult.	Product recovery is less complicated.

Several pretreatment methods used to convert the raw substrate into a suitable form include:

  i.   Size reduction by sifting, grinding, rasping or chopping.

  ii. Damage to outer substrate layers by grinding, pearling or cracking.

  iii. Chemical or enzymatic hydrolysis of polymers to increase substrate availability.

iv. Supplementation with nutrients (phosphorus, nitrogen, salts) and setting the pH and moisture content.

v. Cooking or vapour treatment for macromolecular structure pre-degradation and elimination of major contaminants.

In both substrate types, the success of the process is directly related to the physical characteristics of the support, which favour both gases and nutrients diffusion and the anchorage of the microorganisms. The physical characteristics of the solid matrix such as particle size and shape, porosity and consistency of the material must be taken into account because of their influence on the development of SSF.

iii. Particle Size

Particle size of the substrate is important as it is related to substrate characterization and system capacity to interchange with microbial growth and heat and mass transfer during SSF process. Moreover, it affects the surface area to volume ratio of the particle, which determines the fraction of the substrate, which is initially accessible to the microorganism and the packing density within the surface mass. The surface area to volume ratio increases as the particle size decreases. The size of the substrate determines the void space, which is occupied by air. Since the rate of oxygen transfer into the void space affects growth, the substrate should contain particles of suitable size to enhance mass transfer. Generally, smaller substrate particles would provide larger surface area for microbial action but too small particles may result in substrate agglomeration, which may interfere with microbial respiration/aeration and thus result in poor growth. Smaller particle size is also advantageous for heat transfer and exchange of oxygen and carbon dioxide between the air and the solid surface. At the same time, larger particles also provide better respiration/aeration efficiency but provide limited surface for microbial action. In relation to particle size, it must be remembered that in SSF process, it does not remain constant and tends to diminish.

iv. Moisture and Water Activity (a_w)

Fungi are known to favour a moist environment for their growth. An optimum moisture level has to be maintained, as lower moisture tends to reduce nutrient diffusion, microbial growth, enzyme stability and substrate swelling. Higher moisture levels lead to particle agglomeration, gas transfer limitation and competition from bacteria. The existence of an optimum moisture content of the medium even for bacterial cultures has been stressed as it has profound effects on growth kinetics, and on the physicochemical properties of solids, which, in turn, affect productivity. In general, the moisture levels in SSF processes vary between 30 and 85%. For bacteria, the moisture of the solid matrix must be higher than 70%, and in the case of filamentous fungi, it could be as wide as 20–70%. This signifies the necessity for the development of specific SSF processes. The water requirements of microorganisms should be defined in terms of the water activity (a_w)

rather than the water content of the solid substrate. Water activity is defined as the vapour pressure of a liquid divided by that of pure water at the same temperature. a_w represents the availability of water for reaction in the solid substrate. The water activity is highly dependent upon the water-binding properties of the substrate. The water activity of solid substrate can decrease during SSF as a result of dehydration of the solid substrate and accumulation of solutes in the substrate. Reduced a_w has a marked effect on microbial growth as it extends the lag phase, decreases the specific growth rate, and results in low amount of biomass production. Bacteria require higher values of a_w for growth than fungi, thereby enabling fungi to compete more successfully at the a_w values encountered in SSF processes. The water activity is measured using different methods and equipment, which basically imply to measure the respective pressure between the mixture and the gaseous phase in equilibrium. It is important that the water activity remains close to "one" for optimal spore production. The importance of water in the system is due to the fact that the great majority of viable cells are characterized by moisture content of 70–80%. The optimum water activity may be different for growth and product formation; therefore, this condition offers the possibility of manipulating the water activity during fermentation. As the moisture content of the substrate greatly depends on the water activity, any small change in a_w could lead to great effect on the moisture. Fermentation also leads to increase in a_w of the substrate due to co-production of water; still, it may not help maintain the moisture levels in the substrate due to evaporative losses. The importance of moisture and water activity implies that it is necessary to consider the exact quantities to be added to the substrate, while preparing a substrate.

v. pH

Another important factor in any fermentation process is pH, and it may change in response to metabolic activities. The reason is the secretion of organic acids that will cause the pH to drop. On the other hand, the assimilation of organic acids, which may be present in some media, will lead to an increase in pH. Each microorganism possesses a pH range for its growth and activity with an optimum value within the range. Filamentous fungi have reasonably good growth over a broad range of pH, 2.0–9.0, with an optimal range of 3.8–6.0. On the other hand, yeasts have a pH optimum between 4.0 and 5.0 and can grow in a large pH range of 2.5–8.5. This typical pH versatility of fungi can be beneficially exploited to prevent or minimize bacterial contamination, especially when choosing a lower pH. A very noticeable example of the influence of pH in microbial development is offered by the use of *Lactobacillus* sp. as a preserving agent of storage fibres. In fungal SSF, unlike the homogeneous three-phase system of submerged fermentation, *in situ* pH control is practically impossible due to the heterogeneous three-phase system, and lack of proper equipment and electrodes to determine the pH in solid materials. The pH change will also occur due to the nitrogen source selected as well as the growth characteristics. An attempt to overcome the problem of pH variability during SSF process, however, is obtained by substrate formulation considering the buffering capacity of the different components employed or by the use of buffer formulation with components that have no deleterious influence on the biological activity. Use of urea as a nitrogen source rather than ammonium salts is one way of controlling the pH.

vi. Temperature

The most important of all the physical variables affecting SSF performance is temperature, because growth and production of enzymes or metabolites are usually sensitive to temperature. As in the case of pH, fungi can grow over a wide range of temperatures between 20°C and 55°C, and the optimum temperature for growth could be different from that for product formation. Normally, the fermentation processes develop with mesophilic (temperature tolerance approximately up to 50°C) microbial strains. This critical criteria in SSF, viz., temperature control, is quite difficult to effect because the conventional convection or conductive cooling devices are inadequate for dissipating metabolic heat due to poor thermal conductivity of most solid substrates, as well as the predominantly static nature of SSF. Moreover, the rate of heat generation is directly proportional to the level of metabolic activity in the system. In combination with local moisture content and availability of void spaces, this could lead to thermal gradients. Therefore, the key issue in SSF is heat removal and hence, most studies on reactor designs are focussed on maximizing heat removal. The problem is aggravated in large-scale systems where heat generation leads to serious moisture losses, low yields or loss of fungal activity. So far, the best method developed to deal with heat build–up is by means of forced aeration, which plays multiple roles in SSF. The critical step would be the flow rate and to humidify or dehumidify the air since low or high moisture content brings about adverse effects. Evaporative cooling has been employed in many cases as it has greater efficiency than convection and conduction, and has the ability to remove up to 80% of the heat generated. It, however, results in large moisture loss and consequent drying of the solids and must therefore be combined with water addition to ensure moisture–content control. The significance of temperature in the development of a biological process determines some important effects such as protein denaturation, enzyme inhibition, acceleration or inhibition on the production of a particular metabolite and cell death.

vii. Nutritional Factors

Nutrients can regulate sporulation through metabolic effects and these include carbon and nitrogen sources, minerals, and vitamins or cofactors. Carbon represents the energy source that will be available for the growth of the microorganism. Nitrogen sources stimulate fungal conidiation while minerals and growth factors improve fungal sporulation. Besides, some compounds synthesized by the microorganism itself are also found to induce its own sporulation. While dealing with media formulations, it is necessary to take into account the biomass composition. Cellular biomass presents an average of 40–50% carbon, 30–50% oxygen, 6–8% hydrogen and 3–12% nitrogen. Other elements such as phosphorus, sulphur and metals are also important, although in small quantities. It is believed that the ratio between the carbon and nitrogen (C:N) is the most crucial for a particular process to obtain a specified product. The ratios of C/N in the biomass and in the medium are related to the yield as well as biomass composition. Normally, C/N ratio of 10–20 is frequently found suitable for fermentation processes. When dealing with the production partially or product not associated with the growth, C/N ratio can vary. Various natural and chemical nutrient sources are used in fermentation processes (Table 9.5).

Table 9.5 List of commonly used nutrient sources

Source	Nutrients	
	Natural	**Chemical**
Carbon	Corn sugar, starch, cellulose, pectin, sugar cane, sugar beet molasses	Sucrose, lactose, raffinose, maltose, cellobiose, malt extract, glycerol, ethanol, potassium acetate, glucose, galactose, polygalacturonic acid, starch, maltodextrin, hydrocarbons
Nitrogen	Peptones, amino acids, soybean meal, corn steep liquor, yeast extract, tryptone	Ammonium tartrate, ammonium oxalate, ammonium sulphate, ammonium nitrate, ammonium chloride, sodium nitrate, urea
Phosphorus and sulphur	Corn steep liquor	Phosphate and ammonium sulphate salts
Vitamins and growth factors	Yeast, yeast extract, wheat germ meal, cotton seed meal, beef extract, corn steep liquor	Thiamine, biotin, calcium pantothenate, folic acid
Trace elements		Boron, sodium, nickel, calcium, potassium, iron, zinc, copper, magnesium, manganese, molybdenum, etc.

viii. Aeration and Agitation

Aeration and agitation have significant influence in SSF due to oxygen demand in the aerobic process, and heat and mass transport phenomena in a heterogeneous system. Aeration fulfils four main functions in SSF: maintaining aerobic conditions, desorbing carbon dioxide and other volatile metabolites, regulating the substrate temperature as well as the moisture level. The rate of aeration is therefore determined by factors such as the growth requirements of the microorganism, the production of gaseous and volatile metabolites, and heat evolution. Many operating parameters and medium characteristics can affect O_2 transfer rates, including the air pressure and flow rate, the porosity of moist solids, the bed depth of the moist fermenting solids, perforations in the culture vessel, the moisture content of the medium, the reactor geometry and impeller rotational speed and geometry. Considering the aeration levels, it is very important that they can be expressed through intensive values to make independent levels from the scale in which the process is developing. A criterion to express the aeration in a particular system is obtained by the airflow intensity. The other point to be taken into consideration is the air quality at the entrance to the fermenter. Airflow is the main tool to remove the heat evolved during the process. Water evaporation and heat transfer with the surrounding environment are the processes to keep the process temperature in proper limits. A variable flow of air and its quality can improve temperature control.

SSF allows free access of oxygen to the substrate, and no oxygen limitation is observed. Because of the rapid rate of oxygen diffusion into the water film, and also because of the very high surface of contact between gas phase, substrate and the aerial mycelium, aeration may be easier in SSF. The control of the gas phase and airflow is a simple and practical approach to regulate gas transfer. The theoretical respiratory quotient (RQ) of aerobic microorganisms is 1.0, and RQ below 1.0 is an indication of the insufficient oxygen transfer, which will reflect in the poor growth of the microorganisms. Oxygen uptake rate (OUR) and/or carbon dioxide production rate (CDPR) have been extensively used in SSF for estimation of growth. These rate measurements provide the advantage of a fast response time and are directly linked to the metabolism of the microorganism. From these data, microbial biomass for each microorganism is estimated by evaluating its yield coefficients and these coefficients, may change as a function of growth rate.

Agitation is one of the most important parameters in aerobic fermentation since it ensures homogencity with respect to temperature and gaseous environment and provides a gas–liquid interfacial area for gas-to-liquid as well as liquid-to-gas transfers. Agitation also enhances mass and heat transfer and enables the possibility of homogeneous water addition, allowing the compensation of water loss caused by evaporation. Consequently, solid supports with less absorption capacities can be used in mixed reactors and the other solid support characteristics should be taken into account when selecting the substrate. It should not form agglomerates since aeration and mixing will become more difficult when the solid support particles stick together. The solid support should retain its structure and be able to withstand the shear force caused by the agitation. Further, the microorganism should also be able to withstand the shear forces. On some solid supports, the microbial activity is adversely affected by mixing whereas in others, no adverse effect is noticed.

It must be emphasized, however, that agitation is not used in many aerobic SSF processes carried out in static reactors such as tray fermenters. In contrast, agitation is usually an essential part of periodically or continuously agitated SSF bioreactors. The type of process, reactor design and the product concerned govern the requirement of agitation in SSF systems. Although a number of products are produced in greatly enhanced yields in agitated systems, poor growth has also been reported. Agitation is also known to have adverse effects on substrate porosity due to the compacting of the substrate particles, disruption of fungal attachment to the solids, and damage to fungal mycelia due to shear forces in SSF systems. Moreover, agitation may promote or prevent aggregate formation of the fermenting mass depending on the nature of the solids. Application of intermittent or very slow rather than continuous agitation is found to be more appropriate to prevent damage to the mycelia and disruption of mycelial attachment to solids in certain cases.

BIOMASS AND ITS MEASUREMENT

Biomass is a fundamental parameter in characterizing microbial growth and biomass measurement is essential to follow the growth kinetics in SSF. This is required to analyse and diagnose the status of fermentation and to develop mathematical models of the bioreaction to calculate specific productivities. Direct determination of biomass in SSF is very difficult due to the problem of separating the microbial biomass from the substrate. This is especially true for SSF processes involving fungi, because the fungal hyphae penetrate into and bind tightly to the substrate and hence it has been estimated indirectly from physical measurements such as temperature, effluent gas composition, light reflectance, composition changes including estimation of cell components, as analysed by infrared spectrophotometry and by variation of the dielectric properties. Other indirect methods include determination of biological activities such as ATP, enzymatic activities, DNA assay, respiration rate (O_2 consumption and CO_2 release), immunological activity, nutrient consumption, and analyses of cell constituents such as chitin, glucosamine, nucleic acids, ergosterol and protein. But, some of these approaches have drawbacks as enzymatic activity and ergosterol content vary during fungal development, and nutrient consumption can be applied only in sterile conditions.

Although direct estimation of biomass is a difficult process, complete recovery of fungal biomass can be achieved under artificial circumstances in membrane filter culture. The membrane filter prevents penetration of fungal hyphae into the substrate and the entire fungal mycelium can be recovered by peeling it off the membrane and weighing it directly or after drying. Microscopic observation using a scanning electron microscope is another way of estimating fungal growth. Image analysis by computer software, confocal microscopy, based on specific reaction of fungal biomass with specific fluorochrome probes and reflectance infrared spectrophotometry are the other methods of estimating fungal growth in SSF.

In the indirect method, measurement of respiratory metabolism is used for estimation of biomass, as this metabolic activity is growth-linked. Carbon compounds in the substrate are metabolized and converted into biomass and carbon dioxide. Production of carbon dioxide causes a decrease in the weight of the fermenting substrate, and the amount of weight loss can be correlated to the amount of growth that has occurred. Growth estimation based on carbon dioxide release or oxygen consumption assumes that the amount of biomass produced per unit of gas metabolized must be constant. The measurement of carbon dioxide released or oxygen consumed is more powerful when coupled with the use of a correlation model that correlates biomass with a measurable parameter. But, application of this model requires the use of numerical techniques to solve the differential equations. The metabolic activity in SSF is very important for studying all theoretical and practical aspects of respirometric measurement of fungal biomass cultivated in SSF. Another metabolic activity of equal importance for estimation of biomass is production of extracellular enzymes. A good correlation between growth and hydrolytic enzymes such as amylase, cellulase and pectinases, and organic acids is also reported. The biomass can also be determined by measuring specific components such as protein, nucleic acids and ergosterol.

Glucosamine, a monomer of chitin and a cell wall component, is yet another reliable compound for fungal biomass estimation. Glucosamine content is constant irrespective of the age of the cultivation and culture conditions but appears to depend on the medium composition, cultivation method used, and the age of the mycelium, and is affected by the nature of the carbon source. These drawbacks are circumvented by using the membrane model system so that glucosamine and biomass dry weight could be measured simultaneously during cultivation on a substrate with the same composition as that used in bioreactor cultivations. A comparison of the estimation of glucosamine and ergosterol, however, has revealed that determination of the latter is more convenient. Moreover, the extraction procedures and processing time for glucosamine exceeds 24 h. Other methods of biomass estimation include evaluation of mycelial growth based on the difference in electric conductivity between biomass and the substrate and monitoring the pressure drop during SSF. Biomass estimation is more suited to processes where the main product is the biomass itself, as for secondary metabolite production; these are the best indicators of growth and expression in laboratory or plant activities. In general, oxygen uptake and carbon dioxide release methods seem to be the most promising methods for biomass estimation in aerobic SSF. No method is ideally suited to all situations and hence the most appropriate method to a particular SSF application must be chosen in each case based on the simplicity of the procedure, cost and accuracy.

Bioreactors for SSF

In the SSF process, the fermenter provides the environment for the proper growth and activity of microorganisms, which carry out the biological reactions. SSF processes could be operated in batch, fed-batch or continuous modes, although batch processes are the most common. An important aspect to be considered during the construction of a bioreactor is the sensitivity of the substrate and/or the microorganism to the shear forces generated by mixing. In the design and fabrication of solid-state fermenters, factors to be considered are

 i. materials and methods of construction of the fermentation vessel,
 ii. ability to withstand pressure,
 iii. ability to withstand agitation, rotation, or shaking,
 iv. substrate type,
 v. process variables, and
 vi. extent of control required.

Besides, the other factors to be considered are

 i. techniques of inoculation,
 ii. sampling and transfer systems,
 iii. sterilization of equipment,
 iv. fermenting medium,
 v. aeration and monitoring, and
 vi. measurement and control of various parameters.

It could lead to the development of reliable commercial fermentation equipment. Several types of fermenters have been designed, constructed and used in many small, pilot and large-scale applications of SSF. Some are used for producing substrates for edible mushrooms, commercial enzymes, animal feeds, and for soil bioremediation. Laboratory studies have generally been carried out in flasks, beakers, Petri dishes, glass jars and columns. Wide-mouthed Erlenmeyer flasks, Roux bottles, and roller bottles have also been used regularly. Generally, the only means for heat removal from the substrate bed in these fermenters is forced aeration, but for small-scale uses, simple trays with no forced aeration are sometimes used. Unlike submerged fermentation, in SSF, the choice of a suitable reactor system is difficult, given the heterogeneity of the substrate matrix. Therefore, selection will depend on factors such as substrate type, process variables, extent of control required, etc.

There are several types of fermenters designed and used for various purposes but the basic concept almost always revolves around tray type or drum type bioreactor. On the basis of mixing and aeration, SSF bioreactors can be divided into 4 major groups:

Group I

This group includes bioreactors in which the bed is static, or mixed only very infrequently and air is circulated around the bed, but not blown forcefully through it. These are often referred to as tray bioreactors. Tray reactors consist of flat trays (Figure 9.1).

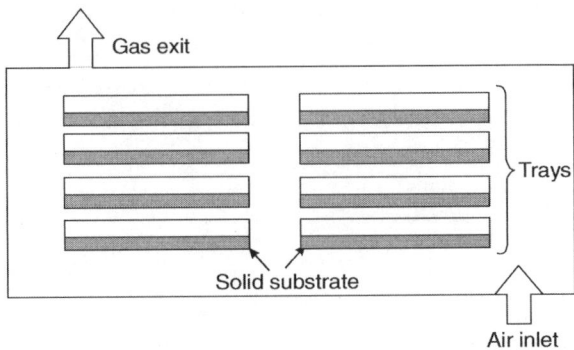

Figure 9.1   Tray bioreactor

The substrate is spread onto each tray forming a thin layer, only a few centimetres deep. The reactor is kept in a chamber at constant temperature through which humidified air is circulated. The main disadvantage of this configuration is that numerous trays and large volume are required, making it an unattractive design for large-scale production.

Group II

This includes bioreactors in which the bed is static or mixed only very infrequently and air is blown forcefully though the bed. These are typically referred to as packed-bed bioreactors (Figure 9.2).

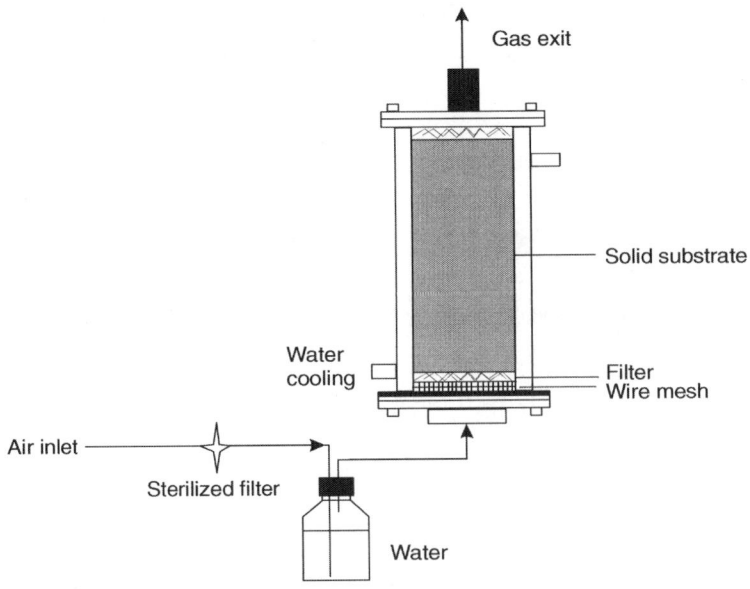

Figure 9.2   Packed bed bioreactor

It is usually composed of a column of glass or plastic with the solid substrate retained on a perforated base. Through the bed of substrate, humidified air is continuously forced. It may be fitted with a jacket for circulation of water to control the temperature during fermentation. This is the configuration usually employed in commercial Koji production. The main drawbacks associated with this configuration include difficulties in obtaining the product, non-uniform growth, poor heat removal and scale-up problems.

Group III

In this group, the bed is continuously mixed or mixed intermittently with a frequency of minutes to hours, and air is circulated around the bed, but not blown forcefully through it (Figure 9.3).

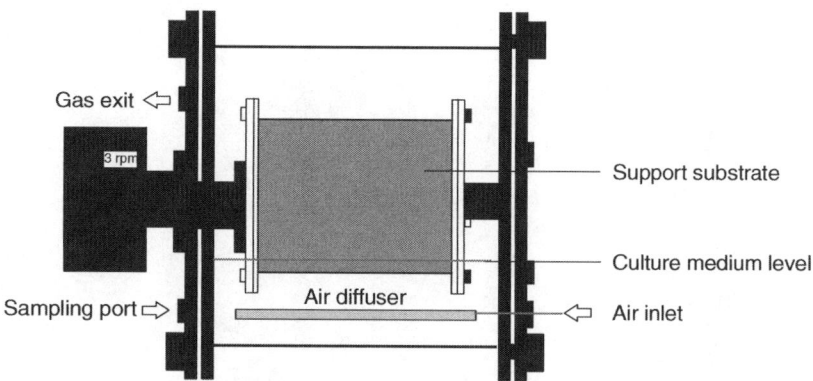

Figure 9.3   Horizontal drum bioreactor

This type of bioreactor design allows adequate aeration and mixing of the substrate, whilst limiting the damage to the inoculum or product. Mixing is performed by rotating the entire vessel or by various agitation devices such as paddles and baffles. Its main disadvantage is that the drum is filled to only 30% capacity, otherwise mixing is inefficient.

Group IV

This group includes bioreactors in which the bed is agitated and air is blown forcefully through the bed as shown in Figure 9.4. This type of bioreactor can typically be operated in two mixing modes, continuous and intermittent.

Various designs, such as gas-solid fluidized beds, the rocking drum and various stirred-aerated bioreactors fulfil these criteria. Fluidized bed reactor is designed in such a way as to avoid the adhesion and aggregation of substrate particles and to supply a continuous agitation with forced air. Although the mass heat transfer, aeration and mixing of the substrate is increased, damage to inoculum and heat build-up through sheer forces may affect the final product yield.

The tunnel and rotating drum reactors are employed for batch or continuous SSF processes. There are some problems associated with solid beds such as poor mixing, heat transfer, characteristics and material handling; hence, SSF bioreactor systems are yet to reach a high degree of development. There are four types of reactors to perform SSF processes and each in its own design tries to make conditions more favourable for fermentation under solid-state conditions.

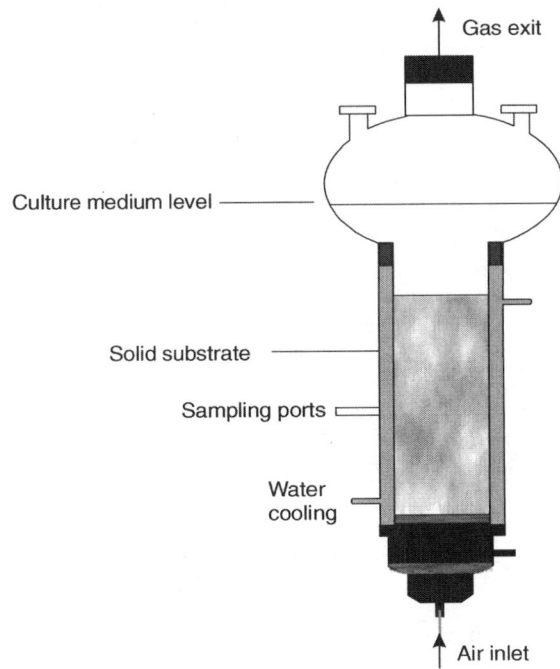

Figure 9.4   Fluidized bed bioreactor

In SSF, a wide variety of solid substrates are employed, which possess differences in composition, size, mechanical resistance, porosity and water holding capacity. The SSF bioreactors must be constructed with a strong material which must be anticorrosive and nontoxic to the process organism. It should have a low cost. The entry of contaminants into the process as well as the uncontrolled release of organisms into the process must be avoided by a careful design of seals and filtration of the inlet air stream. Other important aspects to be considered during the construction of a bioreactor are effective regulation of aeration, mixing and heat removal. This could avoid the problems related to ineffective heat removal, evaporative loss of water from the substrate bed and thermal gradients which affect the yield and quality of the desired product. The control of operational parameters such as temperature, water activity and O_2 concentration and the maintenance of uniformity within the substrate should be as effective as possible. Therefore, a bioreactor system should be designed to facilitate the substrate preparation, its sterilization, inoculation, loading and unloading of the bioreactor, product and biomass recovery.

Modelling in SSF system is another important aspect. The concept of modelling is the search for mathematical expressions to establish the relationship between two different variables that characterize the system. Modelling of bioreactors employed in SSF processes can play a crucial role in the analysis, design and development of bioprocesses. The growth measurement parameters, analysis of cellular growth, determination of substrate consumption, etc. are difficult to measure due to the heterogeneous nature of the substrate. Hence, to know more detail about fermentation kinetics, a synthetic model substrate can be used. It is well known that the fermentation kinetics is extremely sensitive to the variation in ambient and internal gas compositions. So the cellular growth of the microorganisms can be determined by measuring the change in gaseous compositions inside the bioreactor. This can also be determined by substrate digestion, heating and centrifuging the substrate using light reflectance, DNA measurement, glucosamine level for biomass, protein content, O_2 uptake rate and CO_2 evolution rate.

EXTRACTION OF FERMENTED SOLIDS

For the success of commercial production of enzymes by SSF, there is a need for efficient extraction and downstream processing techniques, which have a significant influence on the cost of production as well as the overall economics of the process. The technique of extraction of the fermented solids is an important step for achieving highly concentrated extracts. The efficiency of leaching depends on a number of factors such as pretreatment of the fermented solids, solvent efficiency, diffusivity of solute and solvent, retention of solvents by solids, mixing of solids and solvents, solid : solvent ratio, solid–solvent contact time, contact temperature and pH of the system. The leaching step can largely determine the economics of SSF process and it is therefore essential to choose the correct leaching technique. The extraction of the fermented solids for recovery of the product is generally carried out by using simple percolation method, multiple-contact counter-current leaching, repeated extraction, supercritical fluid extraction, plug flow column extractor and aqueous two-phase partitioning. Percolation and counter-current leaching techniques are used most often.

Simple Percolation

The substrate with mouldy growth is transferred to a glass column measuring 24 × 4 cm, the bottom end of which is lightly plugged with glass wool. Water is added to different batches of the mouldy substrate in different ratios 1:4, 1:8, 1:10, 1:12 and 1:15 (w/v) in two aliquots. The first aliquot is added and allowed to be in contact for 30 min and the extract is then collected. The second aliquot is then added and allowed to be in contact for another 30 min. The two extracts obtained from the different solid : solvent ratios are mixed and filtered through Whatman filter paper No.4 and the clear filtrate is used for enzyme assay and the percent recovery is calculated.

Multiple-Contact Counter-Current (MCCC) Method

The mouldy substrate is weighed separately into four 500 ml beakers designated as A1, B1, C1 and D1. About 200 ml of distilled water is added to A1, stirred, mixed intermittently for 30 min., and pressed, and the resulting enzyme extract is designated as I stage enzyme extract. The total residue is transferred from A1 to A2 and 200 ml of fresh distilled water is added and allowed to be in contact for 30 min. The solids are then pressed and the resulting extract is added to fresh mouldy substrate in beaker B1 and kept in contact for 30 min with intermittent stirring. The total residue in beaker A2 is again transferred to A3 and 200 ml of fresh distilled water is added and allowed to be in contact for 30 min. The solids in beaker B1 are pressed and the resulting extract is designated as II stage enzyme extract. The total residue in B1 is transferred to beaker B2 and allowed to be in contact for 30 min. with the enzyme extract obtained by pressing the solids in beaker A3. The total extract obtained from B2 is transferred to beaker C1 containing fresh mouldy substrate and is allowed to be in contact for 30 min. The residue in B2 is then transferred to beaker B3. This stepwise process is repeated till four stages of extracts are obtained to get maximum recovery of the enzyme (Figure 9.5).

Extraction by Plugflow Method

A glass column of 24×4 cm provided with a stopcock at the outlet is loaded with 40 g of the mouldy substrate over a glass wool layer to prevent the escape of solid particles in the extract. The mouldy substrate is wetted with 4.4 volumes of its weight of distilled water before being loaded onto the column. The solvent is fed at a constant flow rate of 3 ml/min into the column using a peristaltic pump. The extract in different fractions of 50 ml is collected from the bottom outlet of the column. The extracts are filtered through Whatman filter paper No. 4 and the enzyme activity in the filtrate is calculated. The extraction efficiency is calculated by extracting 40 g of mouldy substrate in flask with 400 ml of the solvent under static conditions for 5 h (Figure 9.6).

Extraction Under Static and Agitated Condition

Batches of 5 g of the mouldy substrate are taken in a series of 250 ml Erlenmeyer flasks, and water is added in the ratio of 1:10 (w/v) and kept under static as well as agitated conditions @150 rpm at 30°C up to 6 h. The flasks are taken out at 1 h intervals and the contents are filtered through Whatman filter paper No.4 to obtain a clear extract and the enzyme activity is estimated. The percent recovery is then calculated.

Supercritical Fluid Extraction (SCFE)

SCFE is a technology suitable for extraction and purification of a variety of compounds, particularly those that have low volatility and/or are susceptible to thermal degradation.

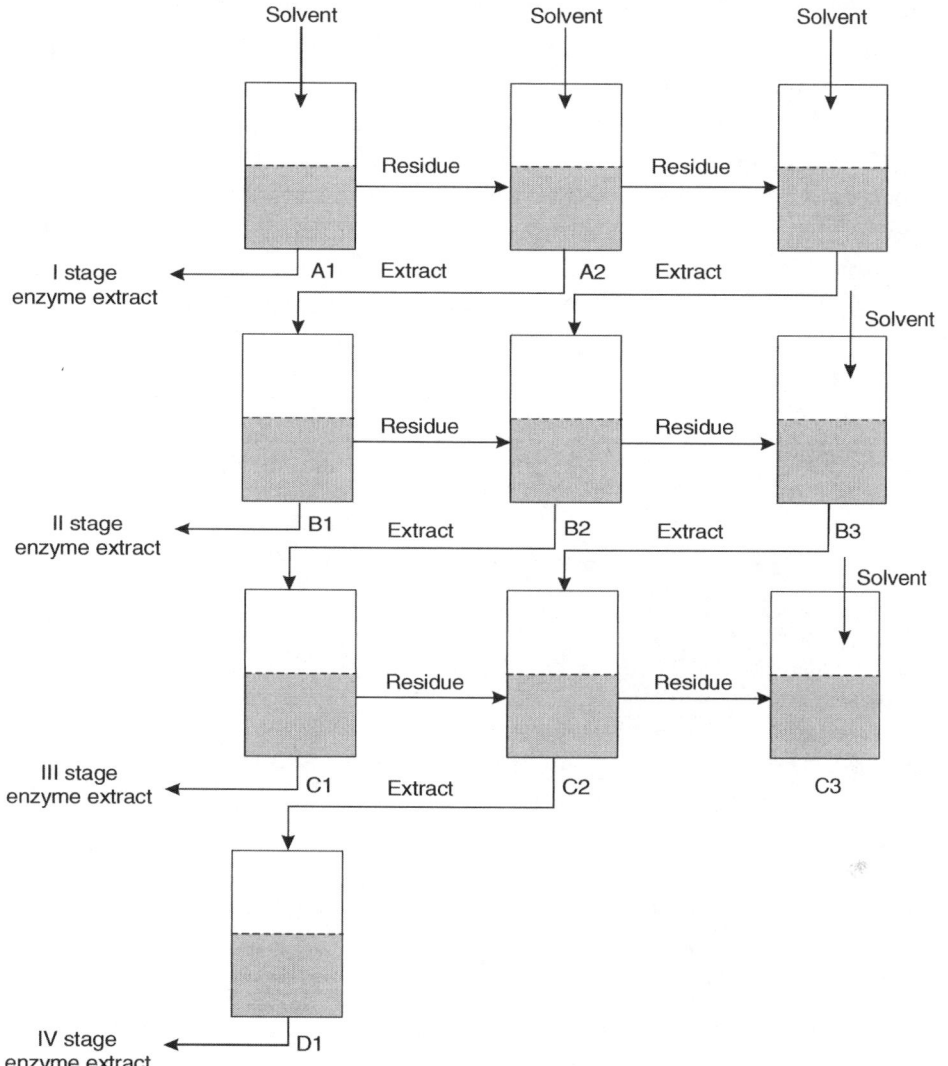

Figure 9.5 Multiple contact counter-current system (Mala, 1999)

It is an operation with a fluid at temperature and pressure near the critical point. Supercritical fluids have a higher diffusivity and lower density, viscosity, and surface tension. Liquid CO_2 is the most frequently used solvent for SCFE, because of its practical advantages such as nontoxic and nonflammable character, environmental safety, availability, low cost at high purity, and suitability for extracting heat-labile, natural compounds with low volatility and polarity. When the extract is recovered in the separators, CO_2 is easily separated because of its high volatility. SCFE allows the extraction of active ingredients from plants and other natural substrates with a better reproduction than conventional operations. Thermal degradation and decomposition of labile compounds are avoided, due to the operation at reduced temperature, whereas the absence of light and oxygen prevents oxidation reactions. Extracts from supercritical treatments with liquid CO_2 can be regarded as all natural, and the products earmarked necessary for food applications have the GRAS (Generally Regarded As Safe) status. Supercritical fluid-processed materials do not require separate sterilization stages, since gram-positive and gram-negative bacteria can be inactivated at mild temperatures. The high-pressure gradient during pressure release can yield extracts free of living microorganisms and their spores, with a longer shelf life than standard solvent extracts. The major disadvantages of SCFE are the high critical pressure, the expensive equipment, and the low dielectric constant, which suggests poor solvent power.

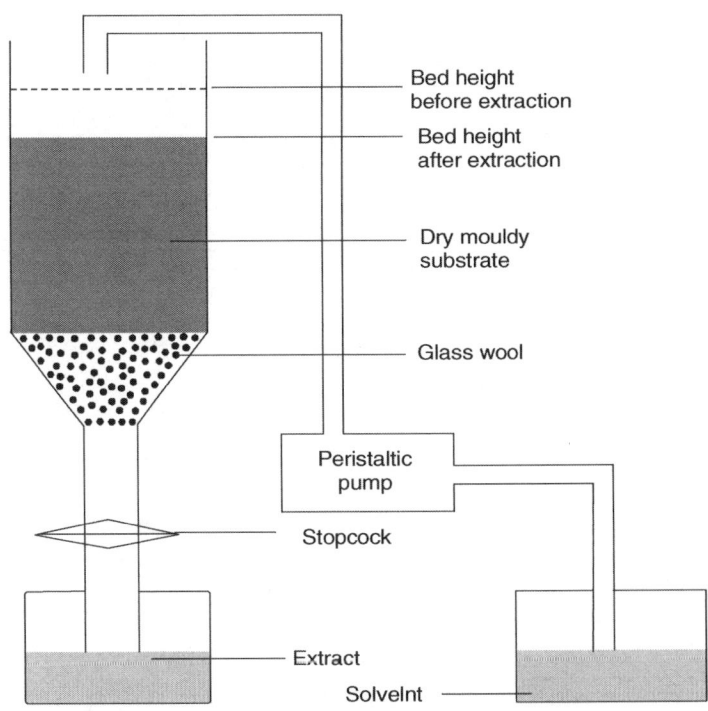

Figure 9.6 Plug flow extraction system (Mala, 1999)

Aqueous Two-Phase System (ATPS)

ATPS provides a rapid, easily scalable method for separation of soluble proteins from insoluble materials and other undesired proteins. The method can be operated in continuous mode. ATPS aims at high throughput and seeks to circumvent problems associated with diffusion limitations experienced with most chromatographic methods. It can handle large volumes, is easily scalable and can hold high biomass load in comparison with other separation techniques. Phase formation in aqueous two-phase systems is based on the incompatibility of certain polymer mixtures in forming homogeneous solutions. Rather, under specific concentrations of the mixing polymers, two immiscible phases are formed in equilibrium with each other. Phase systems are prepared from stock solutions of polyethylene glycol (PEG) 50% (w/w) and potassium phosphate 40% (w/w). The phosphate stock solution is prepared by mixing appropriate amounts of K_2HPO_4 and NaH_2PO_4 until the desired pH is achieved. 1 g samples are prepared, mixed thoroughly and centrifuged at room temperature (28°–32°C) for 5 min. at 2000×g to speed up phase separation. The centrifuged samples are allowed to stand for 30 min. according to standard protocols. Total protein is added to a final concentration of 0.1 g/l. Control systems without protein are used for reference. The volume of the top and bottom phases is measured and the protein concentration is determined in each phase. Samples from the top and bottom phase are removed and used for SDS-PAGE and chromatographic purification.

Downstream processing of the SSF products will be discussed in Chapter 10.

APPLICATIONS OF SSF

In industrial applications, SSF can be utilized in a controlled way to produce the desired product. Current trends have focussed on the application of SSF for the development of bioprocesses such as bioremediation and biodegradation of hazardous compounds, biological detoxification of agro-industrial residues, bioconversion of biomass, biotransformation of crop residues for nutritional enrichment, biopulping, and production of value added products such as biologically active secondary metabolites including antibiotics, alkaloids, plant growth factors, enzymes, organic acids, biopesticides, biosurfactants, biofuel, aroma compounds, etc. SSF system, termed as a low-technology system, appears to be a promising one for the production of value-added, low-volume, high cost products such as biopharmaceuticals.

Table 9.6 Industrial production of various compounds by SSF

Sector	Product	Substrate used	Microorganism used
	Mushroom production	Straw	*Agaricus bisporus* *Lentinus edodes* *Volvariella volvaceae*
Agro-food industry	Sauerkraut	Cabbage	Lactic acid bacteria
	Soy sauce	Soybeans and wheat	*Aspergillus oryzae*
	Tempeh	Soybeans	*Rhizopus oligosporus*

(Contd.)

Table 9. 6 (Continued)

Sector	Product	Substrate used	Microorganism used
	Ontjom	Peanut press cake	*Neurospora sitophila*
	Cheeses	Milk curd	*Penicillium roquefortii*
	Single Cell Protein	Cellulosic waste such as straw, bagasse, saw dust, starch hydrolysate	*Chaetomium cellulolyticum, Fusarium graminearum*, mushrooms
	Sake (rice wine)	Rice	*Aspergillus oryzae*
	Ang-kak (Red rice)	Rice	*Monascus purpureus*
	Leaching of metals	Low-grade ores	*Thiobacillus* sp.
	Organic acids such as citric acid, fumaric acid, lactic acid, itaconic acid	Cane sugar, molasses, wheat bran, potato starch, cassava	*Aspergillus niger, Rhizopus* sp., *Lactobacillus* sp.
Industrial uses	Enzymes such as α-amylases, proteases, pectinases, lipases, xylanases, cellulases, glucoamylases	Wheat bran, wheat straw, oilcakes, rice bran, sugar cane baggase	*Aspergillus* sp., *Bacillus licheniformis, Bacillus subtilis, Rhizopus oligosporus, Pseudomonas* sp.
	Biofuel – Ethanol	Cereal flour, fruit pomace, sweet sorghum, cassava, potato starch	*Saccharomyces cerevisiae*
	Bioactive products such as antibiotics, mycotoxins, bacterial endotoxins, gibberellins, steroid intermediates	Cereal grains, baggase, wheat bran, rice husk, coconut waste, etc.	*Penicillium chrysogenum, Streptomyces* sp., *Aspergillus niger, Bacillus thuringiensis, Gibberella fujikuroi, Mycobacterium* sp.
	Aroma compounds	Sugar cane bagasse, cassava bagasse, coffee husk	*Rhizopus oryzae, Ceratocystis fimbriata, Bacillus subtilis*
Environmental applications	Bioremediation and biodegradation of hazardous wastes	Caffeinated residues, pesticides, polychlorinated biphenyls	*Bacillus* sp., *Pseudomonas* sp., *Phanerochaete chrysosporium, Trametes versicolor*
	Biological detoxification of agro-industrial wastes	Canola meal, cassava peels, coffee husk and pulp	*Rhizopus oligosporus, Rhizopus* sp., *Phanerochaete chrysosporium*

SSF is the better process for spore production than SMF and it provides spores with better yield, morphology and high stability. Spores are the reservoirs of metabolites and can be used as biocatalysts for bioconversion reactions and also as biocontrol agents. Biorefineries have added more value to SSF, as biomass is the only foreseeable source of energy to meet the needs of the future generation, which adds to the importance of agro-residual waste. Table 9.6 shows various products produced by SSF using different microorganisms using various bioreactors.

ADVANTAGES AND DISADVANTAGES OF SSF

The practical advantages of SSF over SmF include the following:

 i. Non-aseptic conditions.
 ii. Use of raw materials as substrates.
 iii. Resembles the natural habitat for microorganisms.
 iv. Use of a wide variety of matrices (which vary in composition, size, mechanical resistance, porosity and water-holding capacity).
 v. Low capital cost.
 vi. Low energy expenditure.
 vii. Less expensive downstream processing (in case, extraction of the product is necessary).
 viii. Requires less solvent and lower recovery cost than SmF.
 ix. Less water usage and lower wastewater output.
 x. Potential higher volumetric productivity.
 xi. Higher concentration of the products.
 xii. High reproducibility.
 xiii. Lesser fermentation space (the volumetric loading of the substrate is much higher in SSF than in SmF because the moisture level of the SSF is lower, resulting in compact fermenter or fermentation facility).
 xiv. Easier control of contamination.
 xv. Generally simpler fermentation media.

The disadvantages of SSF are the following:

 i. Difficulty in agitation of the substrate bed, resulting in heterogeneously distributed physiological, physical and chemical environment in the substrate bed.
 ii. Difficulties in fermentation control, mainly in heat build-up.

iii. The solid nature of the substrate causes problems in monitoring of the process parameters such as moisture, aeration, pH, biomass, etc.

iv. Difficulty in rapid determination of microbial growth and other fermentative parameters.

v. Limited types of microorganisms that can grow at low moisture levels and

vi. Fermentation of bacteria using SSF is difficult since chances of contamination are more.

BIBLIOGRAPHY

Couto, S.R. and Sanroman, M.A. (2006). "Application of solid-state fermentation to food industry—A review". *J. Food Engg.* 76, 291–302.

Díaz-Reinoso, B., Moure, A., Domínguez, H and Parajó, J.C. (2006). "Review: Supercritical CO_2 Extraction and Purification of Compounds with Antioxidant Activity". *J. Agric. Food Chem.* 54, 2441–2469.

Krishna, C. (2005). "Solid-State Fermentation Systems - An Overview". *Crit. Rev. Biotechnol.* 25, 1–30.

Mala, J.G.S. (1999). "Studies on lipase from *Aspergillus niger*: Isolation, Production, Characterization and Application". Ph.D thesis submitted to the University of Madras, pp. 79–101.

Pandey, A. (1992). "Recent process developments in solid-state fermentation." *Process Biochem.* 27, 109–117.

Pandey, A., Soccol, C.R. and Larroche, C. (2008). *Current developments in solid state fermentation*, Asiatech Publishers Inc, New Delhi.

Smith, J.E. (2009) "Bioprocess/fermentation technology." *Biotechnology*. Cambridge University Press, US. pp. 49–71.

Mitchell, D.A., Krieger, N. and Berovic, M. (Eds.). (2006). *Solid state fermentation bioreactors: Fundamentals of design and operation.* Springer-Verlag, Germany.

10
DOWNSTREAM PROCESSING

INTRODUCTION

The production of therapeutic proteins, which is made possible by discoveries in biotechnology, generated sales exceeding $100 billion in 2010. In addition, biotechnology has led to marked improvement and expansion in the biochemical process industry for production of enzymes, diagnostics, chemicals, pharmaceuticals and foods. Continued introduction of new technology necessitates innovation in process development, scale-up and downstream processing. An integral and cost-intensive part of these processes is associated with downstream processing for product isolation and purification.

Downstream processing refers to the recovery and purification of valuable bioproducts such as proteins, enzymes and antibiotics from animal, plant or microbial sources. It is an essential step in the manufacture of pharmaceuticals such as antibiotics, therapeutic proteins, hormones (e.g., insulin and human growth hormone), antibodies, vaccines and industrial enzymes. Downstream processing is usually considered a specialized field in biochemical engineering, though many of the key technologies are being developed by chemists and biologists to upscale the downstream processes of biological products. Downstream processing and analytical separation both refer to the separation or purification of biological products, but at different scales of operation and for different purposes. Downstream processing implies manufacture of a purified product fit for a specific use, generally in marketable quantities, while analytical separation refers to purification for the sole purpose of measuring a component or components of a mixture, and may deal with small sample quantity.

For certain applications, biological products can be used as crude extracts with a little or no purification. However, biopharmaceuticals typically require exceptional purity, making downstream processing a critical component of the overall process. From the regulatory viewpoint, biopharmaceutical firms are more concerned about the impurities though they occur in low concentration as they can possibly interfere with the health of human beings. Currently, proteins and antibiotics are the most important biopharmaceuticals. Blood plasma fractionation was the first full-scale biopharmaceutical industry with an annual production in the 100 ton scale. Precipitation with organic solvents has been and continues to be the principal purification tool in plasma fractionation, although, recently, chromatographic separation processes have also been integrated into this industry. Anti-venom antibodies and other anti-toxins extracted from animal sources are additional examples of early biopharmaceuticals, also purified by a combination of precipitation, filtration and chromatography. In contrast, current biopharmaceuticals are almost exclusively produced by recombinant DNA technology. Chromatography and membrane filtration serve as the main tools for purification of these products.

Bioproducts widely vary in their physical, chemical and biological properties and several strategies of downstream processing are available. The choice of the recovery process is based on the following criteria:

* Prior to adopting a strategy, it is essential to have knowledge on the nature of the bioproduct, whether nucleic acids, proteins, antibiotics, organelles, microbial cells, etc., that has to be recovered and its physical and chemical properties such as molecular weight, density, ionic status, function, etc.

* Is the microbial derived product, an intracellular, membrane-bound protein or is it extracellular?

* If it is an intracellular protein, it has to be confirmed whether the protein is active and in soluble form or is in inclusion bodies that are usually misfolded, insoluble and inactive. For proteins present in inclusion bodies, pretreatment with enzymes or chemicals is necessary to impart correct folding of the protein.

* The concentration of the product in the fermentation broth. Is the product an inducible one? If so, the duration to achieve maximal yield of product from the microbe during the course of fermentation has to be known.

* It is to be decided whether the product has to be recovered in partially pure or pure form, keeping in view the end use or application of the product. Pharmaceutical products must be made pure and devoid of impurities. However, other industries may use partially pure preparations.

* The strategy to be adopted must not affect the function of the product and the product should pass the minimal acceptable standard of purity.

* Quantitative and qualitative aspects of the downstream process have to be considered. Though complete recovery of product (quantitative) with its function (qualitative) is the most important objective of downstream process, this is not usually possible. This is mainly due to inclusion of multiple and sequential steps/ strategies not only to recover products in pure form but also to avoid impurities

and undesirable components. Each step contributes to loss of product and its function. While adopting a strategy, it is ensured that the product must retain its function, or the damage caused to the product should be minimum and acceptable.

* It is preferable to adopt techno-economically feasible strategies.

* The optimized process finds its use in scale-up.

STAGES IN DOWNSTREAM PROCESSING

Downstream processing operations are divided into the following main categories to obtain a product from its natural state as a component of a tissue, cell or fermentation broth through progressive improvements in purity and concentration.

Removal of insolubles is the first stage and involves the capture of the product as a solute in a particulate-free liquid, e.g., separation of cells, cell debris or other particulate matter from fermentation broth containing an antibiotic. Typical unit operations to achieve this are filtration, centrifugation, sedimentation, flocculation, electro-precipitation and gravity settling. Besides, additional operations such as grinding and homogenization or cell disintegration are required to recover products from plant and animal tissues, and intracellular components of microbial cells are included in this category.

Product isolation is the removal of unwanted components or impurities, whose properties vary markedly from that of the desired product. For most products, water is the chief impurity and isolation steps are designed to remove most of it, reducing the volume of material to be handled and concentrating the product. Solvent extraction, adsorption, ultrafiltration, aqueous two-phase extraction and fractional precipitation are some of the unit operations involved.

Product purification is done to separate those contaminants that resemble the product very closely in physical and chemical properties. Consequently, steps adopted at this stage are expensive to carry out and they require sensitive and sophisticated equipment. This stage contributes a significant fraction of the entire downstream processing expenditure. Examples of operations include affinity chromatography, size exclusion chromatography, ion exchange chromatography, reverse phase chromatography, electrophoresis, etc.

Product finishing describes the final processing steps dealing with packaging of the product in a form that is concentrated, stable, easily transportable and convenient. Crystallization, lyophilization and spray-drying are typical unit operations. Depending on the product and its intended use, this stage may also include operations to sterilize the product and remove or deactivate trace contaminants, which might compromise product safety.

A few product recovery methods combining two or more stages may be considered, e.g., adsorption and affinity chromatography accomplish the removal of insoluble and non-specific components as well as product isolation in a single step. In the following sections, some main aspects of downstream operations will be discussed.

REMOVAL OF MICROBIAL BIOMASS AND INSOLUBLES

Microbial cells and other insoluble materials are normally separated from the harvested broth by filtration or centrifugation. Because of the small size of microbes, it will be necessary to consider the use of filter aids to improve filtration rates, while heat and flocculation treatments are employed for increasing sedimentation rates in centrifugation.

FOAM SEPARATION

Foam separation depends on using methods which exploit differences in surface activity of materials. The material may be whole cells or macromolecules such as a protein or colloid and it is selectively adsorbed or attached to the surface of gas bubbles rising through a liquid to be concentrated or separated and finally removed by skimming. It may be possible to make some materials surface-active by the application of surfactants such as long-chain fatty acids, amines and quarternary ammonium compounds. Materials made surface-active and collected are termed colligends whereas the surfactants are termed collectors. When developing this method of separation, the important variables which may need experimental investigation are pH, air-flow rates, surfactants and colligend–collector ratios.

PRECIPITATION

Precipitation is conducted at various stages of the product recovery process. It is particularly a useful process in that it allows enrichment and concentration in one step, thereby reducing the volume of the material for further processing. It is possible to obtain some products (or to remove certain impurities) directly from the broth by precipitation, or to use the technique after a crude cell lysate has been obtained.

Typical agents used in precipitation render the compound of interest insoluble, and these include:

* Acid and bases to change the pH of a solution until the isoelectric point of the compound is reached and pH equals pI, when there is no overall charge on the molecule and its solubility is decreased.

* Addition of salt at low ionic strength can increase solubility of a protein by neutralizing charges on the surface of the protein, reducing the ordered water around the protein and thereby increasing entropy of the system. Salting out is a method of separating proteins based on the principle that proteins are less soluble at high salt concentrations. The salt concentration needed for the protein to precipitate out

of the solution differs from protein to protein. This process is also used to concentrate dilute solutions of proteins.

There are both hydrophobic and hydrophilic amino acids in protein molecules. After protein folding in aqueous solution, hydrophobic amino acids usually form protected hydrophobic areas while hydrophilic amino acids interact with the molecules of solvation and allow proteins to form hydrogen bonds with the surrounding water molecules. If enough of the protein surface is hydrophilic, the protein can be dissolved in water. When the salt concentration is increased, some of the water molecules are attracted by the salt ions, which decrease the number of water molecules available to interact with the charged part of the protein. As a result of the increased demand for solvent molecules, the protein-protein interactions get stronger than the solvent–solute interactions; the protein molecules coagulate by forming hydrophobic interactions with each other. This process is known as salting out.

Salts such as ammonium and sodium sulphate are used for the recovery and fractionation of proteins. The salt removes water from the surface of the protein revealing hydrophobic patches, which come together, causing the protein to precipitate. Most hydrophobic proteins will precipitate first, thus allowing fractionation to take place. Unwanted proteins can be removed from a protein solution mixture by salting out, as long as the solubility of the protein in various concentrations of salt solution is known. After removing the precipitate by filtration or centrifugation, the desired protein can be precipitated by altering the salt concentration to the level at which the desired protein becomes insoluble. The effect of salt on different proteins may differ. Certain proteins precipitate from solution under conditions in which others remain quite soluble. Once the protein is precipitated, it can be pelleted out by centrifugation and the pellet can be redissolved in buffer for further purification steps.

* Addition of organic solvents like methanol acid in precipitation of dextrans, mucilaginous substances and polysaccharides out of the broth. Chilled ethanol and acetone can be used in the precipitation of proteins mainly due to changes in the dielectric properties of the solution.

* Non-ionic polymers such as polyethylene glycol (PEG) can be used in the precipitation of proteins and they are similar in behaviour to organic solvents.

* Polyelectrolytes can be used in the precipitation of a range of compounds, in addition to their use in cell aggregation.

* Protein-binding dyes (triazine dyes) bind to and precipitate certain classes of proteins.

* Affinity precipitants are an area of current interest in that they are able to bind and precipitate compounds selectively.

FILTRATION

Filtration is one of the most common processes used at all scales of operation to separate suspended particles from a liquid or gas, using porous medium which retains the particles from a liquid or gas to pass through. It is possible to carry out filtration under a variety of conditions, but a number of factors will obviously influence the choice of the most suitable type of equipment to meet the specified requirements at minimum overall cost, including:

* The properties of the filtrate, particularly its viscosity and density
* The nature of the solid particles, particularly their size and shape, the size distribution and packing characteristics
* Solids–liquids ratio
* The scale of operation
* The need for batch or continuous operation
* The need for pressure or vacuum suction to ensure an adequate flow rate of the broth/liquid

Membrane Filtration

Membrane filtration is a technique, which is used to separate particles from a liquid or fermentation broth for the purpose of purifying it. This filtration method has a number of applications in several industries especially, chemical and pharmaceutical, ranging from treating wastewater to recovery of enzymes and proteins. In membrane filtration, a solvent is passed through a semipermeable membrane. The membrane's permeability is determined by the size of the pores in the membrane, and it acts as a barrier to particles which are larger than the pores, while the rest of the solvent can pass freely through the membrane. The result is a cleaned and filtered fluid on one side of the membrane, with the removed solute on the other side. There are several different approaches to membrane filtration such as ultrafiltration, microfiltration, and reverse osmosis and in all cases, the goal is to obtain a filtered solvent. Microfiltration that employs molecular weight cut-off (MWCO) of above 0.44 µm is useful in separating the biomass and other insolubles, whereas ultrafiltration is useful in recovering macromolecules and their complexes of 1–300 kDa molecular weight.

The size of the pores has to be carefully calculated to exclude undesirable particles, and the size of the membrane has to be designed for optimal operating efficiency. Membranes are also prone to clogging as the pores get slowly filled up with trapped particles. The system must provide access for easy cleaning and maintenance so that it can be kept in good working order. Filter aids such as diatomaceous earth and charcoal are useful in improving filtration rates.

Many membrane filtration systems are designed for industrial uses. One of the major advantages of such a system is that it does not require the use of chemicals or additives, which cuts down on operating costs. Additionally, membrane filtration requires minimal energy and it can in fact be designed to run on almost no energy, with a pressurized system which takes advantage of gravity and forces the solvent through the membrane at a steady rate.

Successive membrane filtration is also gaining interest, in which the solvent passes through a series of membranes. In this approach, the pores get progressively smaller, removing more and more impurities from the fluid. This technique reduces clogging of the system as the solvent is slowly filtered. It has an added advantage of fitting into a compact space, because the membranes can all be very small and still work efficiently.

Ultrafiltration (UF)

Industries such as chemical, pharmaceutical, food and beverage processing and waste water treatment employ ultrafiltration in order to recycle flow or add value to products by eliminating undesirable constituents on the basis of molecular weight. UF is a type of membrane filtration, in which hydrostatic pressure forces a liquid or fermentative broth against a semipermeable membrane with defined pore size. Suspended solids and solutes of high molecular weight are retained, while water and low molecular weight solutes pass through the membrane. This separation process is used in industry and research for purifying and concentrating macromolecular (10^3–10^6 Da) solutions, especially protein solutions. UF is not fundamentally different from microfiltration except in terms of the size of the molecules it retains. UF is applied in cross-flow or dead-end mode and separation in UF undergoes concentration polarization. UF's main advantage is its ability to purify, separate and concentrate target macromolecules in continuous systems. UF does this by pressurizing the solution flow. The solvent and other dissolved components that pass through the membrane are known as permeate. The components that do not pass through are known as retentate. Depending on the molecular weight cut-off of the membrane used, macromolecules may be purified, separated or concentrated in either fraction.

Currently, the study of UF processing is done mainly in laboratory set-ups because it is very prone to membrane fouling caused by increased solute concentration at the membrane surface [(either by macromolecular adsorption to internal pore structure of membrane or aggregation of protein deposit on surface of membrane), which leads to concentration polarization (CP)]. Concentration polarization is the major problem in UF that decreases permeate flux during operation.

Tangential Flow Filtration (TFF) System

A cascade TFF system is defined as a two-stage TFF system (Figure 10.1). The first stage clarifies the feedstock using a microfiltration membrane (MF) while the second stage concentrates the permeate from the first stage using an UF membrane. The permeate from the second stage is recycled for use as the diafiltration medium for the first stage. The product of interest is in the retentate from the second stage. Cascading TFF systems are extremely effective for clarification and recovery of recombinant proteins from microbial fermentation (lysates/whole cell suspensions) and cell cultures. Recovery of recombinant proteins from solutions with high concentration of particulate matter (up to 30% solids) is ideal using a cascade TFF system. Normal flow filtration (NFF) methods are impractical for processing fluids with high concentration of suspended solids. Larger surface areas are required to remove the high quantity of solids and these filters may also plug or the solids may even break through. While centrifuges can be employed for primary clarification, additional secondary clarification may be required prior to concentration of protein using UF.

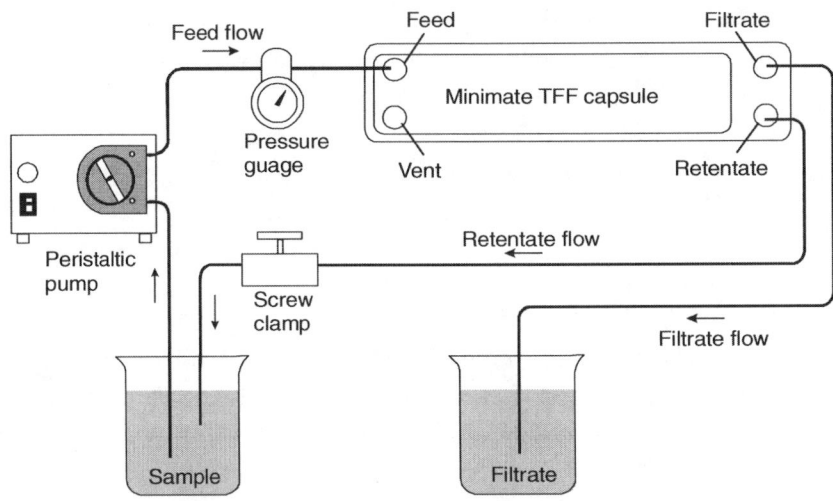

Figure 10.1 Tangential Flow Filtration system

REVERSE OSMOSIS (RO)

RO is a separation process where the solvent molecules are forced by an applied pressure to flow through a semipermeable membrane in the direction opposite to that dictated by osmotic forces and hence it is termed reverse osmosis. It is used for the concentration of smaller molecules when ultrafiltration is not found suitable. Concentration polarization is again a problem and must be controlled by increased turbulence at the membrane surface.

DIALYSIS AND FACTORS AFFECTING DIALYSIS RATE

Dialysis is the movement of molecules by diffusion from high concentration to low concentration through a semipermeable membrane. Only those molecules such as salts and sugars that are small enough are able move through the membrane pores and reach equilibrium with the entire volume of solution in the system. Once equilibrium is reached, there is no further net movement of the substance because molecules will be moving through the pores into and out of the dialysis unit at the same rate. By contrast, large molecules that cannot pass through the membrane pores will remain on the same side of the membrane as they are when dialysis is initiated. To remove additional unwanted substances, it is necessary to replace the dialysis buffer so that a new concentration gradient can be established. Once the buffer is changed, movement of particles from high (inside the membrane) to low (outside the membrane) concentration will resume until equilibrium is once again reached. With each change of dialysis buffer, substances inside the membrane are further purified by a factor equal to the volume difference of the two compartments.

Factors that affect the completeness of dialysis include (1) dialysis buffer volume, (2) buffer composition, (3) the number of buffer changes, (4) time, (5) temperature and (6) particle size vs. pore size. Substances that are very much smaller than the pore size will reach equilibrium faster than substances that are only slightly smaller than the pores. The molecular weight cut-off (MWCO) describes the molecular weight at which a compound will be 90% retained following overnight dialysis. The MWCO is determined by testing many different proteins of known molecular weight. In general, these MWCO apply to globular proteins. More linear proteins may be able to pass through the pores, even though their molecular weight exceeds the stated MWCO. To compensate for this, a dialysis device with a smaller MWCO is chosen. For DNA or RNA, a MWCO not greater than one-third the molecular weight should be used to prevent excessive sample loss.

In downstream operations of fermentation, dialysis is usually performed after salting out of the broth. The broth subjected to salting out results in concentrated protein solution rich in salt content. Subsequently, dialysis is used to remove excess salt and impurities from protein solution. In some cases, removal of salts may enhance activity of the proteins or enzymes and this justifies the requirement of the dialysis step.

CENTRIFUGATION

Centrifugation is a process that uses centrifugal force to separate and purify mixtures of biological particles in a medium. It is the principal and essential step employing centrifuge for isolation of cells, subcellular organelles, molecular complexes and macromolecules such as proteins, nucleic acids, etc. There are two main types of centrifugation depending on sample volume, viz. analytical and preparative centrifugation. Analytical centrifugation is concerned with the study of purified macromolecules, and preparative centrifugation is applied to the separation of cells or biomass, organelles, tissues and other insoluble particles.

Centrifugal Force and Sedimentation

A centrifuge uses centrifugal force (g) to isolate suspended particles from their surrounding medium on either a batch or a continuous-flow basis. It is to be noted that the effect of sedimentation is due to the influence of the Earth's gravitational field ($g = 981$ cm s^{-2}) versus the increased rate of sedimentation in a centrifugal field ($g > 981$ cm s^{-2}). The centrifugal field is expressed in multiples of the gravitational field, g (981 cm s^{-2}). Particles or cells in a liquid suspension generally settle or sediment due to gravity ($1 \times g$) with respect to function of time. However, the length of time required for such separations is impractical because particles that are extremely small in size or molecular weight will not separate at all in solution, unless subjected to high centrifugal force. When a suspension is rotated at a certain speed or revolutions per minute (RPM), centrifugal force causes the particles to move radially away from the axis of rotation. The force on the particles (compared to gravity) is called Relative Centrifugal Force (RCF). For example, an RCF of $1000 \times g$ indicates that the centrifugal force applied is 1000 times greater than the Earth's gravitational force.

The rate of sedimentation depends upon the applied centrifugal field (G; in cm s^{-2}) that is determined by the radial distance (r; in cm), of the particle from the axis of rotation and the square of the angular velocity (ω), of the rotor (in radians per second):

$$G = \omega^2 r$$

$$\omega = \frac{2\pi s}{60}$$

where, s is the rotor speed.

RCF is the ratio of applied centrifugal field (G) and the speed to the standard acceleration of gravity (g).

$$RCF = \frac{4\pi^2 (\mathrm{Re}v/\min)^2 r}{(3600 \times 981)} = \frac{G}{g}$$

RCF units are dimensionless and denote multiples of g.

The following points are worthy of consideration prior to designing a centrifugation process.

* The more dense a substance is, the faster it sediments in a centrifugal field.
* The more massive a substance is, the faster it moves in a centrifugal field.
* The greater the centrifugal force is, the faster the substance sediments.
* The greater the frictional coefficient is, the slower the substance moves.

✽ The sedimentation rate of a given substance is zero when the density of the substance and the surrounding medium are equal.

Biological substances exhibit a drastic increase in sedimentation when they undergo acceleration in a centrifugal field. The rate of sedimentation is also influenced by the nature of the substance or particle, i.e., its density and radius, and the viscosity of the surrounding medium. Stoke's Law describes these relationships for the sedimentation of a particle. According to the law, the rate of sedimentation of spherical particles suspended in a fluid of Newtonian viscosity characteristics is proportional to the square of the diameter of the particles; thus, the rate of sedimentation of a particle under gravitational force is:

$$V = \frac{2r^2}{9} \frac{(\rho_p - \rho_m)}{\eta} xg$$

Biological substances moving through a viscous medium experience a frictional drag, whereby the frictional force acts in the direction opposite to sedimentation and equals the velocity of the particle multiplied by the frictional coefficient. This frictional coefficient depends on the size and shape of the particle. While the sample moves towards the bottom of a centrifuge tube in a fixed-angle rotor, its velocity increases due to the increase in radial distance. At the same time, the particles also encounter a frictional drag that is proportional to their viscosity.

Types of Centrifugal Separations

1. *Differential centrifugation*   Separation is achieved primarily based on the size of the particles in differential centrifugation (Figure 10.2). This type of separation is commonly used in simple pelleting and in obtaining a partially pure preparation of subcellular organelles and macromolecules. For the study of subcellular organelles, tissues or cells are first disrupted to release their internal contents. This crude lysate is referred to as a homogenate. During centrifugation of a cell homogenate, larger particles sediment faster than smaller ones and this provides the basis for obtaining crude organelle fractions by differential centrifugation. A cell homogenate can be centrifuged at a series of progressively higher g-forces and times to generate pellets of partially purified organelles.

When a cell homogenate is centrifuged at 1000 × g for 10 min. unbroken cells and heavy nuclei pellet to the bottom of the tube. The supernatant can be further centrifuged at 10,000 × g for 20 min to pellet subcellular organelles of intermediate velocities such as mitochondria, lysosomes and microbodies. Some of these sedimenting organelles can be obtained in partial purity and are typically contaminated with other particles. Repeated washing of the pellets by resuspending in isotonic solvents and re-pelleting may result in removal of contaminants that are smaller in size. Obtaining partially purified organelles by differential centrifugation serves as the preliminary step for further purification using other types of centrifugal separation (density gradient separation).

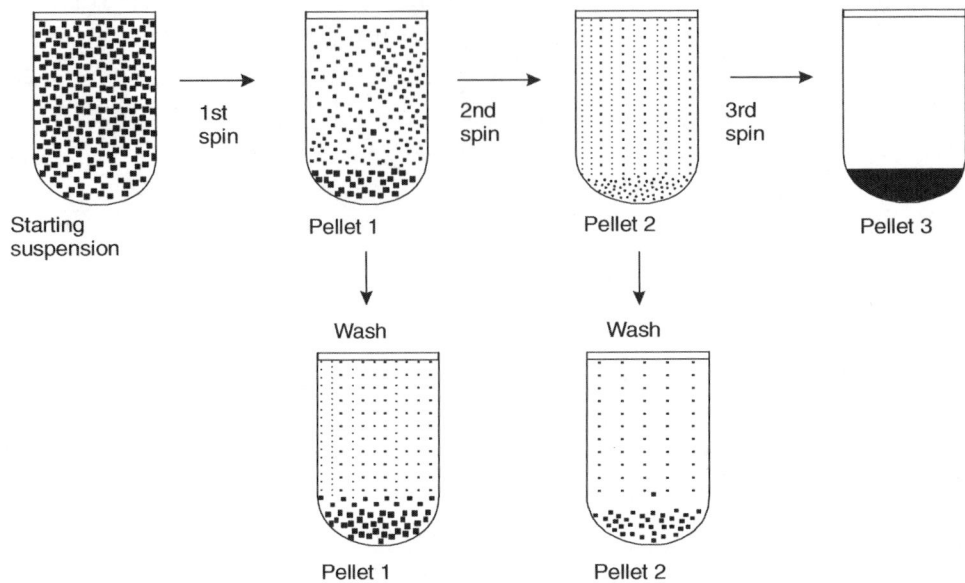

Figure 10.2 Differential centrifugation

2. *Density gradient centrifugation* Density gradient centrifugation is the preferred method to purify subcellular organelles and macromolecules. Density gradients can be generated by placing layer after layer of gradient media such as sucrose in a tube with the heaviest layer at the bottom and the lightest at the top in either a discontinuous or continuous mode. The cell fraction to be separated is placed on top of the layer and centrifuged. Density gradient separation can be classified into two categories and they are i) Rate-zonal (size) separation and ii) Isopycnic (density) separation.

i. Rate-zonal (size) separation Rate-zonal separation takes advantage of particle size and mass instead of particle density for sedimentation. Figure 10.3 illustrates a rate-zonal separation process and the criteria for successful rate-zonal separation. Examples of common applications include separation of cellular organelles such as endosomes or separation of proteins, such as antibodies. For instance, antibody classes have very similar densities, but different masses. Thus, separation based on mass will separate the different classes, whereas separation based on density will not be able to resolve these antibody classes.

Criteria for successful rate-zonal centrifugation

✳ Density of the sample solution must be less than that of the lowest density portion of the gradient.

* Density of the sample particle must be greater than that of the highest density portion of the gradient.
* The path length of the gradient must be sufficient for the separation to occur.
* If a too long run is performed, particles may all pellet at the bottom of the tube.

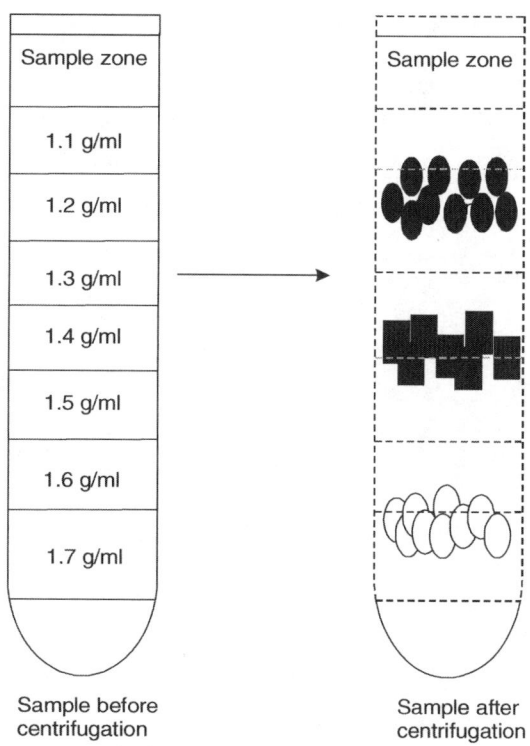

Figure 10.3   Rate-zonal separation

ii. Isopycnic separation   In this type of separation, a particle of a particular density will sink during centrifugation until a position is reached where the density of the surrounding solution is exactly the same as the density of the particle. Once this quasi-equilibrium is reached, the length of centrifugation does not have any influence on the migration of the particle. A common example for this method is separation of nucleic acids in a cesium chloride (CsCl) gradient. Figure 10.4 illustrates the isopycnic separation and criteria for successful separation. A variety of gradient media can be used for isopycnic separations.

Criteria for successful isopycnic separation

* Density of the sample particle must fall within the limits of the gradient densities.
* Any gradient length is acceptable.

❊   The run time must be sufficient for the particles to band at their isopycnic point. Excessive run times have no adverse effect.

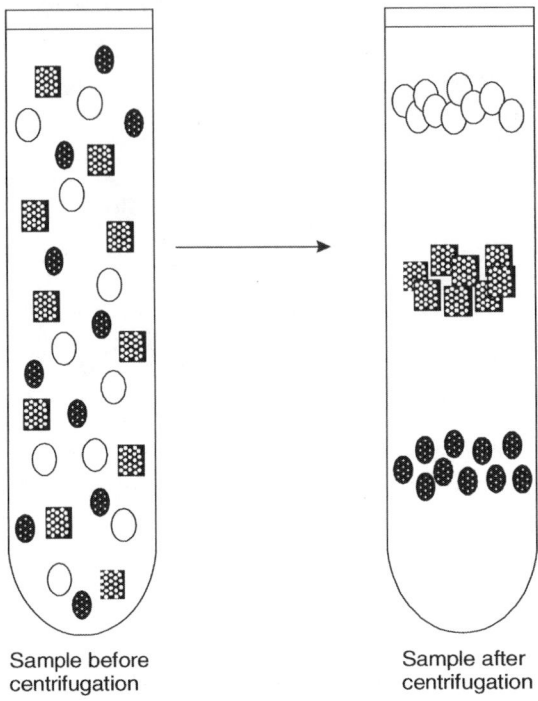

Sample before
centrifugation

Sample after
centrifugation

Figure 10.4   Isopycnic separation

Rotor Categories

Rotors can be broadly classified into three common categories namely 1) swinging-bucket rotors, 2) fixed-angle rotors and 3) vertical rotors (Figure 10.5). Each type of rotor has strengths and limitations depending on the type of separation.

1. In swinging-bucket rotors, the sample tubes are loaded into individual buckets that hang vertically while the rotor is at rest. When the rotor begins to rotate, the buckets swing out to a horizontal position. This rotor is particularly useful when samples are to be resolved in density gradients. The longer path length permits better separation of individual particle types from a mixture. However, this rotor is relatively inefficient for pelleting.

2. In fixed-angle rotors, the sample tubes are held fixed at the angle of the rotor cavity. When the rotor begins to rotate, the solution in the tubes reorients. This rotor type is most commonly used for pelleting applications. Examples include pelleting bacteria, yeast, and other mammalian cells. It is also useful for isopycnic separation of macromolecules such as nucleic acids.

3. In vertical rotors, sample tubes are held in vertical position during rotation. This type of rotor is not suitable for pelleting applications but it is the most efficient for isopycnic (density) separations due to the short path length. Applications include plasmid DNA, RNA, and lipoprotein isolations.

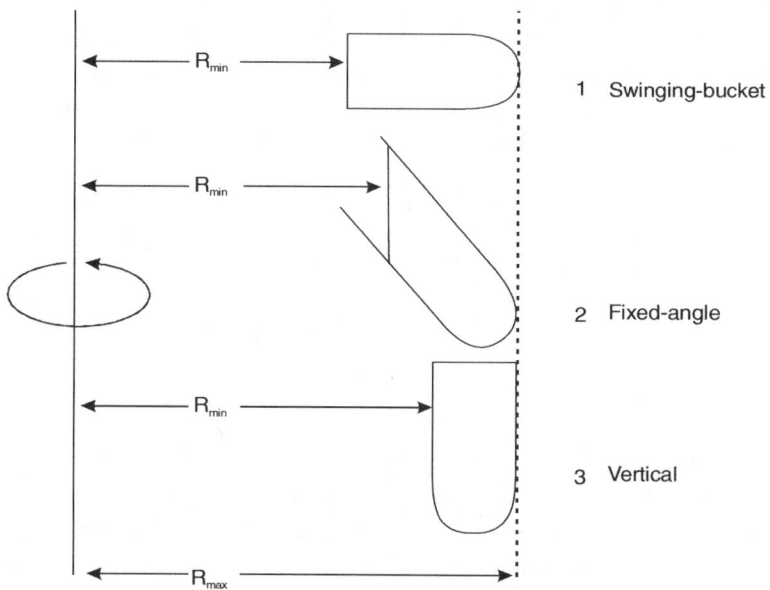

Figure 10.5 Types of rotors

Continuous Centrifuge for Industrial Processes

1. *Tubular-bowl centrifuge* The tubular-bowl centrifuge is widely used in the enzyme manufacture and pharmaceutical industries. Feed enters under pressure through a nozzle at the bottom, is accelerated to rotor speed, and moves upwards through the cylindrical bowl. As the bowl rotates, particles travelling upward are spun out and collide with the walls of the bowl. Solids are removed from the liquid if they move with sufficient velocity to reach the wall of the bowl within the residence time of the liquid in the machine. As the feed rate is increased, the liquid layer moving up the wall of the centrifuge becomes thicker; this reduces performance of the centrifuge by increasing the distance a particle must travel to reach the wall. Liquid from the feed spills over a weir at the top of the bowl; solids which have collided with the walls are collected separately. When the thickness of sediment collecting in the bowl reaches the position of the liquid-overflow weir, separation efficiency declines rapidly. This limits the capacity of the centrifuge. Tubular centrifuges are applied mainly for difficult separations requiring high centrifugal forces. Solids in tubular centrifuges are accelerated by forces between 13,000 and 16,000 times the force of gravity.

2. *Disc-stack bowl centrifuge*   An alternative to the tubular centrifuge is the disc-stack bowl centrifuge, which is commonly employed in bioprocessing. These centrifuges contain conical sheets of metal called discs, which are stacked one on top of the other with clearances as small as 0.3 mm. The discs rotate with the bowl and their function is to split the liquid into thin layers. Feed is released near the bottom of the centrifuge and travels upwards through matching holes in the discs. Between the discs, heavy components of the feed are thrown outward under the influence of centrifugal forces as lighter liquid is displaced towards the centre of the bowl. As they are flung out, the solids strike the undersides of the discs and slide down to the bottom edge of the bowl. At the same time, the lighter liquid flows in and over the upper surfaces of the discs to be discharged from the top of the bowl. Heavier liquid containing solids can be discharged either at the top of the centrifuge or through nozzles around the periphery of the bowl. These centrifuges develop forces of 5000–15000 times gravity. In practical operations at appropriate flow rates, the minimum particle diameter separated is about 0.5 μm.

CELL DISRUPTION

Downstream processing of fermentation broths usually begins with separation of cells by filtration or centrifugation. The next step is considered based on location of the desired product. The location of many of the bioproducts such as recombinant proteins and enzymes are intracellular, and to release them from biomass, effective cell disruption modes are essential. A variety of methods are available to disrupt cells. Mechanical options include grinding with abrasives such as glass beads, high-speed agitation, high-pressure pumping and ultrasound. Non-mechanical methods such as osmotic shock, freezing and thawing, enzymatic digestion of cell walls, and treatment with solvents and detergents can be adopted. The yield will vary depending upon the method of cell disruption and properties of physicochemical products. A widely used method for cell disruption is high-pressure homogenization that employs pressure energy. Shear forces generated in this treatment are sufficient to completely disrupt many types of cells. This method employs a high-pressure pump that incorporates an adjustable valve with restricted orifice through which cells are forced at pressures up to 600 atm. Cells pass through a high-energy density field that downsize dispersed particles to the order of magnitude of micrometres and nanometres. Homogenizing valves are designed and dimensioned to obtain the required degree of micronization and dispersion to the lowest possible pressure, based on the range of applications. The strong dependence of protein release on pressure suggests that high-pressure operation is beneficial; complete disruption in a single pass may be possible if the pressure is sufficiently high. Reduction in the number of passes through the homogenizer is generally preferable because multiple passes produce fine cell debris which can cause problems in subsequent clarification steps. Release of protein is markedly dependent on temperature; protein recovery increases at elevated temperatures up to 50°C. However, cooling to 0-4°C is recommended during operation of homogenizers to minimize protein denaturation.

CHROMATOGRAPHY

Chromatography is a separation procedure for resolving mixtures and isolating components. The basis of chromatography is differential migration, i.e., the selective retardation of solute molecules during passage through a bed of resin particles. As solvent flows through the column, the solutes travel at different speeds depending on their relative affinities for the resin particles. As a result, they will be separated and appear for collection at the end of the column at different times. The pattern of solute peaks emerging from a chromatography column is called a chromatogram. The fluid carrying the solutes through the column or used for elution is known as the mobile phase; the material which stays inside the column and effects the separation is called the stationary phase.

Principle

The basis of all forms of chromatography is the distribution or partition coefficient (K_d), which describes the way in which a compound (the analyte) is distributed between two immiscible phases. For two such phases, stationary (A) and mobile (B), the value for this coefficient is a constant at a given temperature and is given by the expression:

$$\frac{\text{Concentration in phase A}}{\text{Concentration in phase B}} = K_d$$

The term effective distribution coefficient is defined as the total amount, as distinct from the concentration, of analyte present in one phase divided by the total amount present in the other phase. It is in fact the distribution coefficient multiplied by the ratio of the volumes of the two phases present. If the distribution coefficient of an analyte between two phases A and B is 1, and if this analyte is distributed between 10 cm^3 of A and 1 cm^3 of B, the concentration in the two phases will be the same, but the total amount of the analyte in phase A will be 10 times the amount in phase B.

All chromatographic systems consist of the stationary phase, which may be a solid, gel, liquid or a solid/liquid mixture that is immobilized, and the mobile phase, which may be liquid or gaseous, and which is passed over or through the stationary phase after the mixture of analytes to be separated has been applied to the stationary phase. During the chromatographic separation, the analytes continuously pass back and forth between the two phases so that differences in their distribution coefficients result in their separation.

Chromatography is a high-resolution technique and therefore it is suitable for recovery of high-purity therapeutics and pharmaceuticals. Liquid chromatography is of great relevance to bioprocessing and it can take a variety of forms. It finds application both as a laboratory method for sample analysis and as a preparative technique for large-scale purification of biomolecules. There have been rapid developments in the technology of liquid chromatography aimed at isolation of recombinant products from genetically engineered organisms. Chromatographic methods available for purification of proteins, enzymes, peptides, amino acids, vitamins, nucleic acids, alkaloids, steroids, pigments and many other biological materials include adsorption chromatography, partition chromatography, ion exchange chromatography, gel filtration chromatography and affinity chromatography. These methods differ in the principal mechanism by which molecules are retarded in the column.

Modes of Chromatography

Chromatographic separations may be carried out in one of the two modes: planar and column chromatography.

Planar chromatography   In this mode, the stationary phase is present as or on a plane. The plane can be a layer of solid particles spread on a support such as a glass plate (thin layer chromatography) or a paper, serving as such or impregnated by a substance as the stationary bed (paper chromatography). Different compounds in the sample mixture travel at different distances based on their degree of interaction with the stationary phase as compared to that of the mobile phase. The specific retention factor (R_f) of each chemical can be used to aid in the identification of an unknown substance. Planar chromatography can be of two types namely, thin layer chromatography and paper chromatography.

Thin layer chromatography (TLC)   The mobile phase is a solvent and the stationary phase is a thin layer of adsorbent like silica gel, alumina, or cellulose, which is coated onto a glass, plastic or metal foil plate. The mixture of analytes is applied as a spot or band near the edge of the coated plate and the mobile liquid phase passes across the plate, held either horizontally or vertically, by capillary action, causing the analytes to migrate at characteristic rates to the opposite end. This mode of chromatography has the practical advantage over column chromatography in that a number of samples can be studied simultaneously. TLC is simple to carry out but has been largely superseded by high performance liquid chromatography (HPLC) for many applications. However, it continues to find extensive use in the fields of peptide mapping and natural products research.

Paper chromatography   Separations in paper chromatography involve the same principles as those in TLC, but the difference is that the paper is used as a stationary phase. It is used for separating and identifying mixtures that are or can be coloured, especially pigments. In paper chromatography, like TLC, substances are distributed between a stationary phase and a mobile phase. Two-way paper chromatography, also called two-dimensional chromatography, involves using two solvents and rotating the paper 90° in between. This is useful for separating complex mixtures of similar compounds, for example, amino acids. Paper chromatography is a technique that involves placing a small dot or line of sample solution onto a strip of chromatography paper. The paper is placed in a jar containing a shallow layer of solvent and sealed. As the solvent rises through the paper, it meets the sample mixture which starts to travel up the paper with the solvent. This paper is made of cellulose, a polar substance, and the compounds within the mixture travel farther if they are non-polar. More polar substances bond with the cellulose paper more quickly, and therefore do not travel as far.

Column chromatography   In this type, the stationary phase is packed into a glass or metal column. The mixture of analytes is then applied and the mobile phase, commonly referred to as the eluent, passes through the column either by gravity feed or by use of a pumping system or applied gas pressure. This is the most commonly used mode of chromatography from an analytical biochemical point of view. The stationary phase is either coated onto discrete small particles (the matrix) and packed into the column or applied as a thin film to the inside wall of the column. As the eluent flows through the column, the analytes separate on the basis of their distribution coefficients and emerge individually in the elute as it leaves the column.

Column chromatography can be classified into gas and liquid chromatography. A typical column chromatographic system using a gas or liquid mobile phase consists of the following components.

* **Stationary phase**   Chosen to be appropriate for the analytes to be separated, either solid or liquid.
* **Column**   This may be either of the conventional type, filled with the matrix stationary phase or of the micropore type in which the stationary phase is coated directly on the inside wall of the column.
* **Mobile phase and delivery system**   It is chosen to complement the stationary phase and hence to discriminate between the sample analytes and to deliver a constant rate of flow into the column (gas or liquid).
* **Injector system**   To deliver test samples to the top of the column in a reproducible manner.
* **Detector and Recorder**   To give a continuous record of the presence of the analytes in the elute as it emerges from the column. Detection is usually based on the measurement of a physical parameter such as visible or ultraviolet absorption or fluorescence. A peak on the chart recorder represents each separated analyte.
* **Fraction collector**   For collecting the separated analytes for further biochemical studies.

Gas chromatography (GC)   GC is also known as gas-liquid chromatography (GLC). It is mainly used for separating and analysing compounds that can be vaporized without decomposition. In GC, the mobile phase is a carrier gas, usually an inert gas such as helium or an unreactive gas such as nitrogen. The stationary phase is a layer of liquid or polymer on an inert solid support, inside a small-diameter glass tube (a capillary column) or a solid matrix inside a larger metal tube (a packed column). A known volume of gaseous or liquid analyte is injected into the head of the column using a microsyringe. As the carrier gas carries the analyte molecules through the column, this motion is inhibited by the adsorption of the analyte molecules either onto the column walls or onto packing materials in the column. The rate at which the molecules progress along the column depends on the strength of adsorption, which in turn depends on the type of molecule and the stationary phase materials. This causes each compound to elute at a different time, known as the retention time of the compound. A detector is used to monitor the outlet stream from the column; thus, the time at which each component reaches the outlet and the amount of that component can be determined (Figure 10.6).

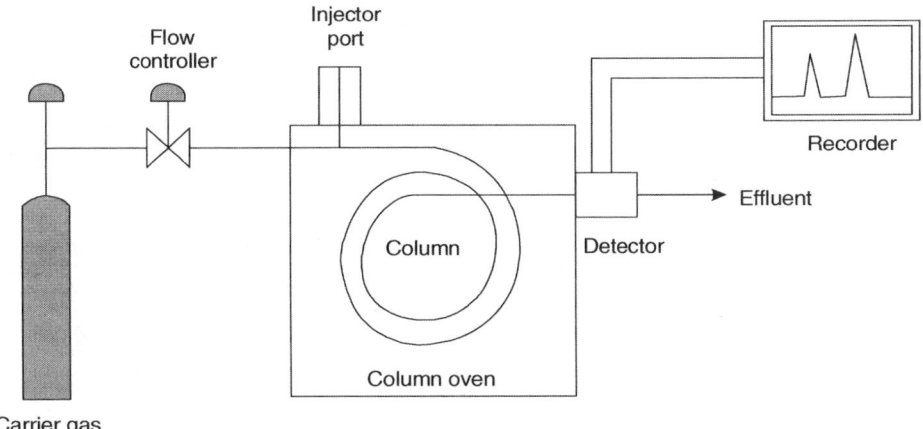

Figure 10.6   Gas chromatography

Liquid chromatography   It is a chromatographic technique that can identify, quantify and purify the individual components from a mixture of compounds. It utilizes smaller column size, smaller media inside the column and higher mobile phase pressures. It typically employs different types of stationary phases, a pump that moves the mobile phase(s) and analyte through the column, and a detector to provide a characteristic retention time for the analyte. Analyte retention time varies depending on the strength of its interactions with the stationary phase, the ratio/composition of solvents used and the flow rate of the mobile phase. The sample to be analysed is introduced in small volumes into the stream of mobile phase. The solution moved through the column is slowed by specific chemical or physical interactions with the stationary phase present within the column. The velocity of the solution depends on the nature of the sample and on the composition of the stationary (column) phase. The time at which a specific sample elutes, i.e., comes out of the end of the column is called the retention time. The use of smaller particle size column packing which creates higher backpressure increases the linear velocity giving the components less time to diffuse within the column, improving the chromatogram resolution. Common solvents used include any miscible combination of water or various organic liquids, the commonly used ones being methanol and acetonitrile.

The various types of liquid chromatography are described in detail below:

i. *Adsorption chromatography* Biological molecules have various tendencies to adsorb onto polar adsorbents such as silica gel, alumina, diatomaceous earth and charcoal. Performance of the adsorbent relies strongly on the chemical composition of the surface, i.e., the types and concentrations of exposed atoms and groups. The order of elution of sample components depends primarily on molecular polarity. Because the mobile phase is in competition with solute for adsorption sites, solvent properties are also important. Polarity scales for solvents are available to aid mobile phase selection.

ii. *Partition chromatography* Partition chromatography relies on the unequal distribution of solute between two immiscible solvents. This is achieved by fixing one solvent (the stationary phase) to a support and passing the other solvent containing solute over it. The solvents make intimate contact allowing multiple extractions of solute to occur. Several methods are available to chemically bond the stationary solvent to supports such as silica. When the stationary phase is more polar than the mobile phase, the technique is called normal phase chromatography. When non-polar compounds are being separated, it is usual to use a stationary phase which is less polar than the mobile phase; this is called everse phase chromatography (RPC). A common stationary phase for RPC is hydrocarbon with 8 or 18 carbons bonded to silica gel; these materials are called C_8 and C_{18} packings, respectively. The stationary phase is essentially inert and only non-polar hydrophobic interactions are possible with analytes. Solvent systems most frequently used are water–acetonitrile, water–methanol and mixture of water, methanol and tetrahydrofuran; aqueous buffers are also employed to suppress ionization of sample components. Elution is generally in the order of increasing solute hydrophobicity. This technique is highly selective but requires the use of organic solvents. Some proteins are permanently denatured by solvents and will lose functionality during RPC. Therefore, this method is not recommended for all applications, particularly if it is necessary for the target protein to retain activity.

iii. *Ion exchange chromatography* The basis of separation in this procedure is electrostatic attraction between the solute and dense clusters of charged groups on the column packing. Ion exchange chromatography can give high resolution of macromolecules and is used commercially for fractionation of antibodies and proteins. Column packings for low molecular weight compounds include silica, glass and polystyrene; carboxymethyl and diethylaminoethyl groups attached to cellulose, agarose or dextran provide suitable resins for protein chromatography and they serve cation and anion exchanger respectively. Choosing an ion exchanger depends on the pI value of the target protein and the pH of buffer used. Solutes are eluted by changing the pH or ionic strength of the liquid phase; salt gradients are the most common way of eluting proteins from ion exchangers. Proteins move through the column at rates determined by their net charge at the pH being used. With cation exchangers, proteins with a more negative net charge move faster and elute earlier, e.g., an acidic protein having a pI of 5.0 will carry a net negative charge at pH 7.0 using phosphate or Tris buffer. The protein will bind to an anion exchanger (DEAE) and will be repelled from a cation exchanger (CM). When the protein mixture containing the target protein is subjected to ion exchange chromatography, neutral and basic proteins are removed in flow-through and the target protein is recovered with increasing salt concentration based on the ionic strength (Figure 10.7).

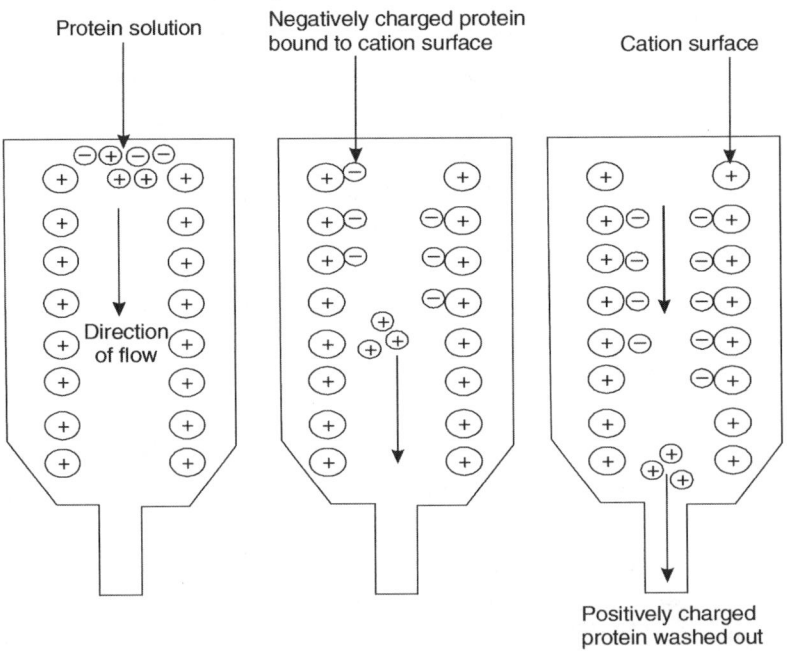

Figure 10.7   Schematic representation of ion exchange chromatography (Anion exchange)

iv. Gel filtration or size exclusion chromatography (SEC)   SEC is a chromatographic method in which molecules in solution are separated by their size and shape but not by molecular weight (Figure 10.8). It is usually applied for the separation and analysis of large molecules or macromolecular complexes such as proteins, polysaccharides, lipophilic components and industrial polymers. Typically, when an aqueous solution is used to transport the sample through the column, the technique is known as gel filtration chromatography. The name gel filtration chromatography is used when an organic solvent is employed as a mobile phase. Molecules in solution are separated in a column packed with gel particles of defined porosity. This method typically uses a gel medium, usually, cross linked polyacrylamide, dextran or agarose and filter under low pressure. One requirement for SEC is that the analyte does not interact with the surface of the stationary phases. The speed with which the components travel through the column depends on their effective molecular size. Large molecules are completely excluded from the gel matrix and move rapidly through the column to appear first in the chromatogram. Small molecules are able to penetrate the pores of the packing, traverse the column very slowly, and appear last in the chromatogram. Molecules of intermediate size enter the pores but spend less time there than the small solutes.

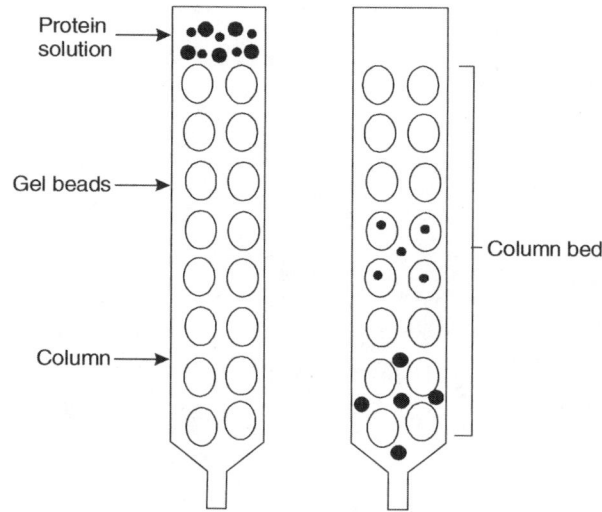

Figure 10.8   Gel filtration or size exclusion chromatography

v. *Affinity chromatography*   This separation technique exploits the binding specificity of biomolecules. Enzymes, hormones, receptors, antibodies, antigens, binding proteins, lectins, nucleic acids, vitamins, whole cells and other components capable of specific and reversible binding are amenable to highly selective affinity purification. Column packing is prepared by linking a binding molecule called a ligand to an insoluble support; when sample is passed through the column, only solutes with appreciable affinity for the ligand are retained. The ligand must be attached to the support in such a way that its binding properties are not seriously affected; molecules called spacer arms are often used to set the ligand away from the support and make it more accessible to the solute. Many ready- made support–ligand preparations are available commercially and are suitable for a wide range of proteins. Conditions for elution depend on the specific binding complex formed; elution usually involves a change in pH, ionic strength or buffer composition. Enzyme proteins can be desorbed using a compound with higher affinity for the enzyme than the ligand, e.g., a substrate or substrate analogue. Affinity chromatography using antibody ligands is called immuno-affinity chromatography (Figure 10. 9).

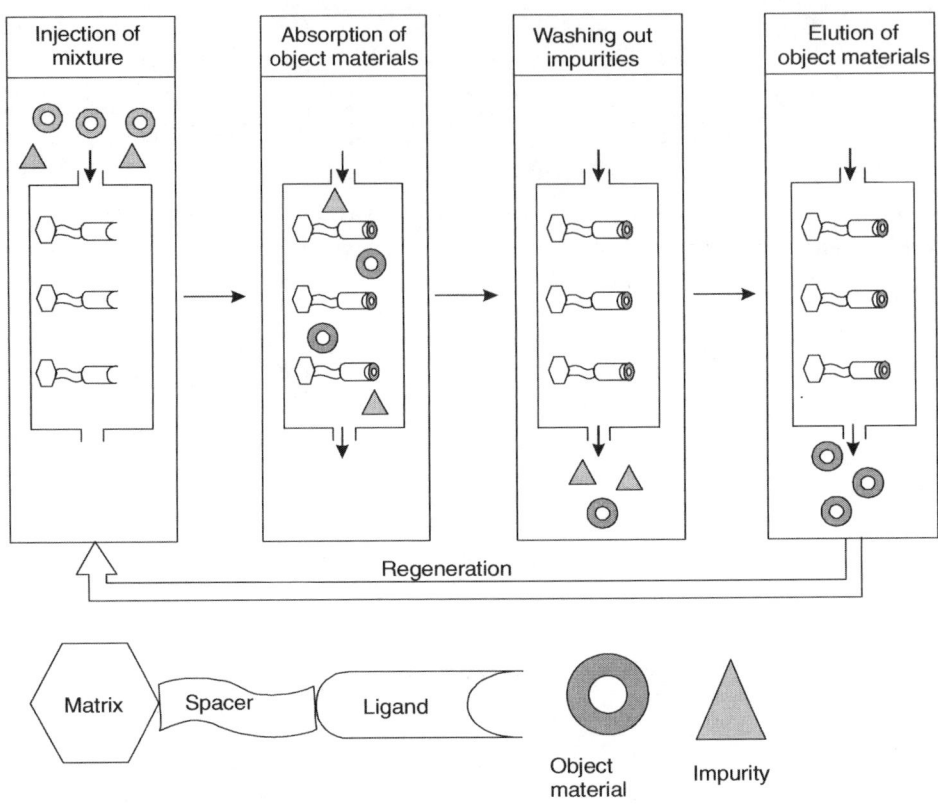

Figure 10.9 Affinity chromatography

vi. *Hydrophobic Interaction chromatography* The principle of hydrophobic interaction chromatography (HIC) is similar to reverse-phase chromatography (RPC) and both are related liquid chromatography (LC) techniques based on the interactions between solvent-accessible non-polar groups (hydrophobic moiety) on the surface of biomolecules and the hydrophobic ligands (alkyl or aryl groups) covalently attached to the gel matrix. In practice, however, they are different. Adsorbents for RPC are more highly substituted with hydrophobic ligands than HIC adsorbents. The degree of substitution of HIC adsorbents is usually in the range of 10–50 mmoles/ml gel of C_2–C_8 alkyl or simple aryl ligands, compared to several hundred mmoles/ml gel of C_4–C_{18} alkyl ligands for RPC adsorbents. Consequently, protein binding to RPC adsorbents is usually very strong, which requires the use of non-polar solvents for their elution. RPC has found extensive applications in analytical and preparative separations of mainly peptides and low molecular weight proteins that are stable in aqueous organic solvents. Hydrophobic ligands coupled to cross linked agarose matrices (sepharose) are butyl, octyl, alkyl and phenyl groups. HIC is an alternative way of exploiting the hydrophobic properties of proteins, working in a more polar and less denaturing environment. Compared to RPC, the polarity of the complete system of HIC is increased by decreased ligand density on the stationary phase and by adding a salt such as ammonium sulphate to the mobile phase.

Considerations of liquid chromatography for separation of biomolecules

* ✳ Choice of stationary phase will depend to a large extent on the type of chromatography employed.
* ✳ For high capacity, the solid support must be porous with high internal surface area.
* ✳ Solid support used as matrix should be insoluble and chemically stable during operation and cleaning.
* ✳ The particles should exhibit high mechanical strength and show little or no non-specific binding.

ELECTROPHORESIS OF PROTEINS

Electrophoresis is defined as the migration of a charged particle under the influence of an electric field. Many important biomolecules such as amino acids, peptides, proteins, nucleic acids possess ionizable groups and, therefore, at any given pH, they exist in solution as electrically charged species (cations or anions). Under the influence of an electric field, these charged particles will migrate either to the cathode or anode, depending on the nature of their net charge. Electrophoresis is carried out in an appropriate buffer, which is essential to maintain a constant state of ionization of the molecules being separated. Any variation in pH would alter the overall charge and hence the rate of migration of the molecules in the applied field. Proteins can be separated by electrophoresis on the basis of (a) size using SDS-PAGE-denatured electrophoresis, (b) shape using Native PAGE, a non-denatured gel electrophoresis and (c) charge using isoelectric focusing.

Sodium Dodecyl Sulphate-polyacrylamide Gel Electrophoresis (SDS-PAGE)

SDS-PAGE is the most widely used method for analysing protein mixtures qualitatively. This method is based on the separation of proteins according to size. It is particularly useful for monitoring protein purification and determining the relative molecular mass of proteins. When heated under denaturing conditions, proteins become unfolded and coated with the anionic detergent sodium dodecyl sulphate (SDS), acquiring net negative charges irrespective of their intrinsic electrical charges (Figure 10.10).

Samples to be run on SDS-PAGE are boiled for 5 min. in sample buffer containing β-mercaptoethanol and SDS. β-mercaptoethanol reduces disulphide bridges that are holding together the protein tertiary structure, and the SDS binds strongly to the denatured protein. Each protein in the mixture is therefore fully denatured by this treatment and opens up into a rod-shaped structure with a series of negatively charged SDS molecules along the polypeptide chain. On average, one SDS molecule binds for every two amino acid residues. The original native charge on the molecule is therefore completely flooded by the negatively charged SDS molecules.

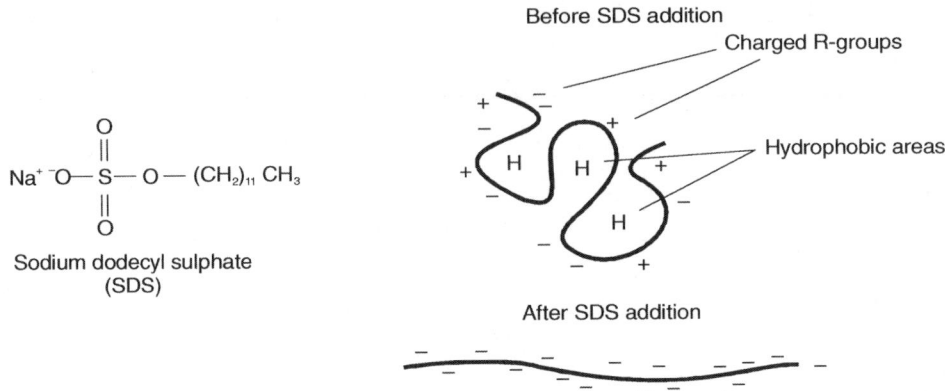

Figure 10.10 Effect of the addition of SDS

Once the samples are loaded, a current is passed through the gel. The negatively charged protein molecules migrate from cathode to the positively charged electrode (anode) in an electric field. After gel electrophoresis, the size of the polypeptide can be estimated by comparing its migration distance with that of the standard molecular weight proteins. The sample buffer also contains an ionizable tracking dye, bromophenol blue, that allows the electrophoretic run to be monitored. The smaller the protein, the more easily it can pass through the pores of the gel, whereas large proteins are successively retarded by frictional resistance due to the sieving effect of the gels. Being a small molecule, the bromophenol blue dye is totally unretarded and therefore indicates the electrophoresis front.

The acrylamide gels of 6–14%, based on the size of the protein, can be prepared by varying the concentrations of polymerizing acrylamide with a cross-linker (bis-acrylamide) in the presence of a catalyst, tetramethylenediamine (TEMED), and an initiator (ammonium persulphate) with a suitable gel buffer (Tris). Solutions are normally degassed prior to polymerization. Oxygen molecules inhibit polymerization. Additionally, heat will be produced during polymerization. The rate at which the gels polymerize can be controlled by varying the concentrations of TEMED and ammonium persulphate.

The most widely used system of gel electrophoresis is the discontinuous gel electrophoresis. In this system, three-fourths of the gel is separating gel and one-fourth is stacking gel, and there are two buffer systems for making the gel. If the samples are loaded directly on the top of the gel, the sharpness is lost and the protein band in the gel will be as broad as possible. This problem is overcome by polymerizing a stacking gel on the top of the resolving gel. When electrophoresis starts in such a system, the proteins and ions migrate into the stacking gel. The proteins concentrate in a very thin zone called the stack between the fast moving chloride ion (Cl^- present in Tris) and the slow-moving glycine ion. As a result, the bands produced are very sharp and clear. When the dye reaches the bottom of the gel, the current is turned off, and the gel is removed from between the glass plates and shaken in an appropriate staining solution (usually Coomassie Brilliant Blue; CBB) and then washed in destaining solution. The destaining solution removes unbound background dye from the gel, leaving stained proteins visible as blue bands on a clear background. Alternatively, the bands can be visualized under an illuminator. Visualization of the protein bands is carried out by incubating the gel either with CBB R-250 stain or with silver staining. Though CBB R-250 is suitable for detecting protein in a band containing 100 ng, the highly sensitive silver staining method is suitable for detecting protein in the range of 2–5 ng.

Native (buffer) Gels

SDS-PAGE is the most frequently used gel system for studying proteins. But this method is not applicable to detect a protein of interest (often an enzyme) on the basis of its biological activity, since the protein is denatured by the SDS-PAGE procedure. Hence, in this case, it is necessary to use non-denaturing conditions. This can be accomplished by using native polyacrylamide gels without SDS and the proteins are not denatured prior to loading. Since all the proteins in the sample being analysed carried their native charge at the pH of the gel (normally pH 8.7), proteins separate according to their different electrophoretic mobilities and the sieving effects of the gel. It is, therefore, not possible to predict the behaviour of a given protein in a buffer gel but, because of the range of different charges and sizes of proteins in a given protein mixture, good resolution is achieved.

Isoelectric Focusing (IEF) Gels

This method is ideal for the separation of amphoteric substances such as proteins because it is based on the separation of molecules according to their different isoelectric points or in other terms, on the basis of charges on the surface as a function of pH. This method has high resolution, being able to separate proteins that differ in their isoelectric points by as little as 0.01 of a pH unit.

The most widely used system for IEF utilizes horizontal gels of 0.15 to 1 mm thickness using glass plates or plastic sheets. Separation is carried out in a polyacrylamide gel in the presence of carrier ampholytes, which establish a pH gradient increasing from the anode to the cathode. Ampholytes are complex mixtures of synthetic polyaminopolycarboxylic acids. Ampholytes are available in different pH ranges covering either a wide band (e.g., pH 3–10) or various narrow bands (e.g., pH 7–8), and a pH range is chosen such that the samples being separated will have their isoelectric points (pI values) within this range. As protein contains both positive (amines) and negative (carboxyl) charge-bearing groups, the net charge of the protein will vary as a function of pH. A pH gradient is established in the field of protein separation. As the protein migrates into an acidic region of the gel, it will gain positive charge through protonation of the carboxylic and amino groups. At some point, the overall positive charge will cause the protein to migrate away from the anode (+) to a more basic region of the gel. As the protein enters a more basic environment, it will lose positive charge and gain negative charge, via amino and carboxylic acid group de-protonation, and consequently, will migrate away from the cathode. In the end, the protein reaches a position in the pH gradient where its net charge is zero (defined as its pI or isoelectric point) or zwitterion form. At that point, the electrophoretic mobility is zero. Migration will cease, and concentration equilibrium of the focused protein is established. Following electrophoresis, the gel must be stained to detect the proteins. However, this cannot be done directly, because the ampholytes will stain too, giving a totally blue gel. The gel is therefore first washed with fixing gel that allows the much smaller ampholytes to be washed out. The gel is stained with CBB and then destained. The pI of a particular protein may be determined conveniently by running a mixture of proteins of known isoelectric point on the same gel. A number of mixtures of proteins with differing pI values are commercially available, covering the pH range 3.5–10. After staining, the distance of each band from one electrode is measured and a graph of distance for each protein against its pI (effectively the pH at that point) plotted. By means of this calibration line, the pI of an unknown protein can be determined from its position on the gel.

Two-dimensional Polyacrylamide Gel Electrophoresis (2-D PAGE)

This technique combines the principles of IEF (first dimension), which separates proteins in a mixture according to charge (pI) and the size separation principle of SDS-PAGE (second dimension). The combination of these two techniques to give two-dimensional (2-D) PAGE provides a highly sophisticated analytical method for analyzing protein mixtures. Using this method, a range of 1000 and 10000 proteins from a cell or tissue extract are resolved. The isoforms of a protein can be easily isolated with 2D PAGE. 2D PAGE is finding huge applications in proteomics.

Isotachophoresis

Isotachophoresis is a technique used in analytical chemistry to separate charged particles which migrate at equal speeds (Greek: *iso* = equal, *tachos* = speed, *phoresis* = migration). It is a further development of electrophoresis. It is a powerful separation technique using a discontinuous electrical field to create sharp boundaries between the sample constituents (Figure 10.11). In isotachophoresis, the sample is introduced between a fast leading electrolyte and a slow terminating electrolyte. After application of an electric potential, a low electrical field is created in the leading electrolyte and a high electrical field in the terminating electrolyte. The pH at sample level is determined by the counter-ion of the leading electrolyte that migrates in the opposite direction. In the first stage the sample constituents migrate at different speeds and start to separate from each other. The faster constituents will create a lower electrical field in the leading part of the sample zone and vice versa. Finally, the constituents will completely separate from each other and concentrate at an equilibrium concentration, surrounded by sharp electrical field differences. Specific spacer or marker molecules are added to the sample to separate physically the sample constituents.

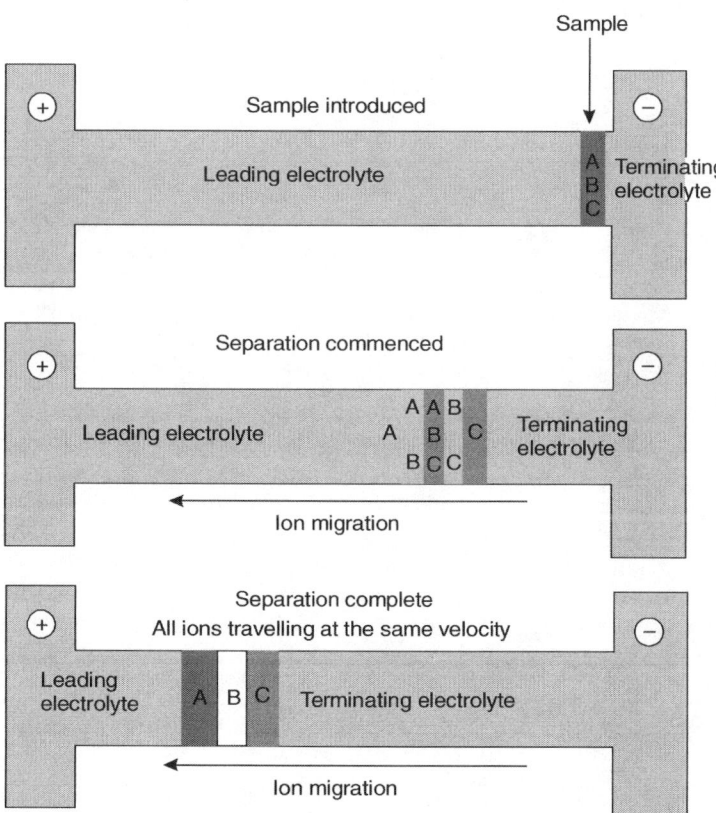

Figure 10.11 Isotachophoresis

BIBLIOGRAPHY

Desai, M.A. (2000). *Downstream Processing of Proteins: Methods and Protocols* (*Methods in Biotechnology*), Humana Press, USA. p. 240.

Scopes, R.K. (2010). *Protein Purification: Principles and Practice*. Springer-Verlag, USA. p. 399.

Wilson, K. and Walker, J. (2010). *Principles and Techniques of Biochemistry and Molecular Biology*. Cambridge University Press, USA. p. 744.

PART III

MICROBIAL TECHNOLOGY—APPLICATIONS

- Enzyme Technology—Medical Applications
- Enzyme Technology—Industrial Applications
- Understanding of the Constituents of Skin for Application of Microbial Technology in Leather Industry
- Microbial Control in Curing Process
- Enzymes in Soaking
- Dehairing—Conventional and Enzymatic Methods
- Bating—State of Art
- Degreasing—Analysis of Different Systems
- Recent Trends in Waste Management
- Protocols for Enzyme Evaluation
- What is Ahead

11

ENZYME TECHNOLOGY–
MEDICAL APPLICATIONS

The global market for medical enzymes is estimated at $6 billion in 2010 and is expected to grow at a compound annual growth rate of 3.9%, to reach $7.2 billion in 2015 (BCC Research report, June 2011). Development of medical applications for enzymes has been at least as extensive as those for industrial applications. The variety of enzymes and their potential therapeutic applications are considerable. In medicine, they are useful because of their unique specificity, thus avoiding side effects when used on a patient.

Medical applications of enzymes are shown as four categories: i) As diagnostic aids in myocardial infarction (MI), arthritis, jaundice, cancer, etc. ii) As enzyme markers in genetic disorders, iii) As diagnostic markers in dental medicine and iv) As therapeutic agents.

DIAGNOSTIC AIDS IN MI, ARTHRITIS, JAUNDICE, CANCER, PANCREATIC DISORDERS AND SKIN DISEASES

Creatine Kinase-MB (CK-MB)

Chest pain can result from indigestion or from a serious heart problem. The determination of enzyme levels in the serum is often helpful in pinpointing the body tissue or organ which has been damaged or is malfunctioning. When a tissue is injured or diseased, some cells of that tissue are destroyed and enzymes are released into the bloodstream. An increase in the enzyme level in the blood indicates that the particular tissue has been damaged.

Compared with other tests for MI diagnosis, creatine kinase-MB (CK-MB) has the advantage of early increase, high sensitivity as well as specificity. It invariably shows a rapid increase in serum in the early hours after admission with chest pain. Use of CK-MB as a marker helps to diagnose MI within 6 h of the onset of symptoms and >60% of infarctions are detectable within the first hour of admission to the cardiac intensive care unit. When the heart muscle dies during MI, it releases many molecules into the bloodstream, one being CK. CK is shown to exist in three molecular forms MM, MB and BB. While total CK is recognized as nonspecific, CK-MB is the most specific, accurate and cost-effective means of detecting MI (Rosalki *et al.*, 2004).

During recent years, CK-MB activity assays have been replaced by CK-MB mass assays which measure the protein concentration of CK-MB, rather than its catalytic activity. The advantage of CK-MB mass assay is that analytical interferences which lead to false positive results are minimal (Mair, 1997).

Lactate Dehydrogenase (LDH)

An increase in total serum LDH activity could result from the damage to the heart muscle, skeletal muscle, pancreas or liver. To differentiate the tissue damaged, the levels of individual isozymes of LDH are determined. A large increase in the serum level of LDH-1 indicates that heart muscle has been damaged as in myocardial infarction. Liver disease is indicated when elevated levels of LDH-5 are observed. (*For more information, see* Chapter 4).

Acid Phosphatase

Acid phosphatase hydrolyses phosphate groups from biomolecules. It is a lysosomal enzyme, with an optimum at acid pH. Acid phosphatase is found throughout the body. Male prostate gland has 100 times more acid phosphatase than any other body tissue. Acid phosphatase assay is carried out to check whether prostate cancer has spread to other parts of the body and also to monitor the effectiveness of treatment. Acid phosphatase level is supplemented by the prostate specific antigen (PSA) test.

Tartrate-Resistant Acid Phosphatase (TRAP)

TRAP is a glycosylated monomeric metalloenzyme expressed in mammals with a molecular weight of approximately 35 kDa. It is differentiated from other mammalian acid phosphatases by its resistance to inhibition by tartrate, by its molecular weight and by its characteristic purple colour. TRAP is found in osteoclasts and released into circulation during bone resorption. Naphthol phosphate is specifically hydrolysed by TRAP and not by other phosphatases. TRAP level is increased in rheumatoid arthritis, osteoporosis and metabolic bone diseases.

Alkaline Phosphatase (ALP)

ALP is an enzyme classified under hydrolases and it is shown to remove phosphate groups from nucleotides, proteins and alkaloids. The process of removing the phosphate group is known as dephosphorylation and ALP functions under alkaline pH. Higher ALP levels in serum are observed (i) when bile ducts are blocked as in the case of obstructive jaundice and (ii) when there is active bone formation as in the case of PAGET's bone disease or rheumatoid arthritis. Lower levels of ALP are less common than elevated levels.

Aspartate Aminotransferase (AST)

AST, also known as serum glutamate oxaloacetate transaminase (SGOT), is a pyridoxal phosphate (PLP) dependent enzyme. It catalyses the following reaction:

$$\text{Aspartate} + \alpha\text{-ketoglutarate} \rightleftharpoons \text{Oxaloacetate} + \text{Glutamate}$$

Significant increase in the serum level (10–100 times normal) of AST indicates severe damage to heart cells (MI) or liver (viral hepatitis or toxic liver necrosis or jaundice).

Alanine Aminotransferase (ALT)

ALT, also known as serum glutamate pyruvate transaminase (SGPT) catalyses the following reaction:

$$\text{Alanine} + \alpha\text{-ketoglutarate} \rightleftharpoons \text{Pyruvate} + \text{Glutamate}$$

Increased serum level of ALT indicates a severe liver disease, usually viral hepatitis, toxic liver necrosis or jaundice.

Lipase

i. *Pancreatic disorders*   Level of lipases in serum can be used as a diagnostic tool for detecting conditions such as acute pancreatitis and pancreatic injury (Lott and Lu, 1991). Acute pancreatitis usually occurs as a result of alcohol abuse or bile duct obstruction. Although serum trypsin level, ultrasonography, computed tomography and endoscopic retrograde cholangio-pancreatography are more accurate laboratory indicators for pancreatitis, serum lipase and amylase levels are still used to confirm the diagnosis of acute pancreatitis (Munoz and Katerndahl, 2000).

ii. *Skin disorders*   Higaki and Morahashi (2003) have examined *Propionibacterium acnes* lipase in skin diseases. Butyric acid production is higher in axillary seborrheic dermatitis (ASD) than in other dermatitis and that in acne vulgaris (AV) is significantly higher than in controls. *P. acnes* lipase is the pathogenic factor in AV and fatty acids produced by lipase might be the pathogenic factor in ASD.

ENZYME MARKERS IN GENETIC DISORDERS

Phosphoribosyl-α-Pyrophosphate (PRPP) Synthetase

Patient with gout disease is shown to excrete three times the normal amount of uric acid per day and has markedly increased levels of PRPP in the red blood cells. PRPP is an intermediate in the biosynthesis of adenosine monophosphate (AMP) and guanosine triphosphate (GTP). Uric acid arises directly from degradation of AMP and GMP. *In vitro* assays reveal that the patients' red blood cell PRPP synthetase activity is increased threefold (Becker *et al.*, 1973).

Orotate Phosphoribosyl Transferase and Orotidine-5'-Phosphate Decarboxylase

In hereditary orotic aciduria, there is a double enzyme deficiency in the pyrimidine biosynthetic pathway leading to accumulation of orotic acid. Both orotate phophoribosyl transferase and orotidine-5'-phosphate decarboxylase are deficient, causing decreased *in vivo* levels of cytidine triphosphate (CTP) and thymidine triphosphate (TTP). In these enzyme deficiency diseases, the patients are pale, weak and fail to thrive. Administration of the missing pyrimidines as uridine or cytidine promotes growth and general well-being and decreases orotic acid excretion (Webster *et al.*, 1995).

Cystathionase

Cystathionase is either deficient or inactive in the genetic disease cystathioninuria. Cystathionase catalyses the following reaction:

$$\text{Cystathionine} \longleftrightarrow \text{Cysteine} + \alpha\text{-ketobutyrate}$$

Deficiency of cystathionase leads to accumulation of cystathionine in the plasma. Since cystathionase is a pyridoxal phosphate-dependent enzyme, vitamin B_6 is administered to patients. Many respond to vitamin B_6 therapy with a fall in the plasma level of cystathionine (Pascal *et al.*, 1975).

Pyruvate Dehydrogenase

Children diagnosed with pyruvate dehydrogenase deficiency usually exhibit elevated serum levels of lactate, pyruvate and alanine which produce a chronic lactic acidosis. Such patients frequently exhibit severe neurological defects, resulting in death. The diagnosis of pyruvate dehydrogenase deficiency is usually made by assaying the enzyme complex in cultures of skin fibroblasts taken from the patient. This situation is corrected by administering dichloroacetate, an inhibitor of pyruvate dehydrogenase (Patel and Harris, 1995).

Glucose-6-phosphate Dehydrogenase (G6PD)

G6PD catalyses the oxidation of glucose-6-phosphate to 6-phosphoglucanolactone wherein $NADP^+$ acts as a coenzyme. It is an important enzyme for the maintenance of intergrity of red blood cells. A deficiency or inactivity of this enzyme leads to hemolytic anemia.

When certain drugs such as antimalarials, antipyretics or sulphur antibiotics are administered to susceptible patients, an acute hemolytic anemia might result in 48–96 h. Susceptibility to drug-induced hemolytic disease may be due to a deficiency in G6PD activity in erythrocytes (Lazzatio and Meta, 1995).

DIAGNOSTIC MARKERS IN DENTAL MEDICINE

Salivary Enzymes and Periodontal Disease

There are important enzymes associated with cell injury and cell death such as AST, ALT, LDH, CK, ALP, ACP and gamma glutamyl transferase (GGT). Changes in enzymatic activity reflect metabolic changes in the gingival crevicular fluid in inflammation. When the activity levels of AST, ALT, LDH, CK, ALP, ACP and GGT have been assayed in patients with periodontal disease, the results show a significant increase in the activity of the above enzymes (Todorovic *et al.*, 2006). After conventional periodontal therapy, the activities of all salivary enzymes are significantly decreased as shown in Table 11.1.

Table 11.1 Differences between CK, LDH, AST, ALT, GGT, ALP, ACP activity (U/L ± SD) in saliva of healthy patients and patients with periodontal disease, before and after periodontal treatment (Todorovic *et al.*, 2006)

Enzyme	Healthy patients	Patients with periodontal disease (before treatment)	Patients with periodontal disease (after treatment)
CK (U/L)	3.60 ± 1.95	44.25 ± 12.16*	23.12 ± 5.13*
LDH (U/L)	99.50 ± 12.02	1015 ± 114.40*	215.25 ± 28.12*
AST (U/L)	21.20 ± 6.76	184.30 ± 78.14*	50.25 ± 14.18*
ALT (U/L)	7.30 ± 1.76	98.15 ± 20.72*	67.21 ± 13.16*
GGT (U/L)	4.50 ± 1.95	12.19 ± 4.55*	8.03 ± 1.21*
ALP (U/L)	7.30 ± 2.05	38.40 ± 9.89*	25.12 ± 7.34*
ACP (U/L)	20.53 ± 4.01	81.76 ± 15.40*	42.25 ± 10.11*

*-Statistically siginificant p<0.001

Aspartate Aminotransferase (AST) Levels in Saliva and Gingival Crevicular Fluid with Periodontal Disease Progression

Periodontal disease is one of the most common inflammatory diseases of the oral cavity, characterized by the progressive destruction of the alveolar bone and soft tissues surrounding the teeth. The relationship between AST levels in saliva and gingival crevicular fluid (GCF) with periodontal disease progression has been studied in a large number of patients. The results show that when compared between the AST levels in saliva and GCF, GCF is shown to have a higher level of AST. Hence, AST level in GCF could serve as a potential biochemical marker for periodontal disease progression (Dewan and Bhatia, 2011).

Glucosyl Transferase (GTF) Inactivation Reduces Dental Caries

Dental caries has been an intractable disease in spite of intense research. The metabolic acids produced by *Streptococcus mutans* demineralise the tooth surface and lead to dental caries. The enzyme GTF produced by *S. mutans* is the key factor in this process, the level of which gets elevated. A simple method of inactivating GTF that actively participates in dental caries is by taking advantage of a Fenton reaction which requires metal ions such as iron or copper and peroxide as shown below:

i. Ferrous ions react with molecular oxygen (by oxidation) and yield superoxide radical ions.

$$Fe^{2+} + O_2 \rightarrow Fe^{3+} + O_2^-$$

ii. By dismutation reaction, superoxide radical ions produce hydrogen peroxide.

$$2O_2^- + 2H^+ \rightarrow H_2O_2 + O_2$$

iii. Ferrous ions react with hydrogen peroxide and produce hydroxyl radical ions via the Fenton reaction

$$Fe^{2+} + H_2O_2 \rightarrow Fe^{3+} + OH^- + OH^-$$

The hydroxyl radical ions produced via the Fenton reaction inactivate GTF, an enzyme responsible for the production of dental caries (Devulapalle and Mooser, 2001).

THERAPEUTIC AGENTS

Theoretically, it should be possible to give active enzymes to individuals who cannot make them and to use enzyme therapy in cases where a person produces toxic levels of a chemical that can be enzymatically destroyed. But, this is difficult since there is the problem of getting the enzyme to the tissue where it is needed unless properly encapsulated. Orally administered enzymes are denatured by gastric secretion. Even if they pass through the stomach, the intact proteins cannot cross the intestinal cell membranes. There are some special cases in which enzyme treatment is appropriate.

Digestive Enzymes from Pancreas

A mixture of digestive enzymes (amylase, trypsin and lipase), in specially coated capsules, is given as a supplement to patients with digestive disorders. Purified preparations of trypsin and chymotrypsin are used to clean the wound after severe burn and these enzymes degrade the proteins in the dead cells, clotted blood and pus at the burn site. When trypsin and a DNA-degrading enzyme are administered by inhalation as an aerosol, they liquefy the thick sputum that collects in the respiratory tract of patients with asthma and cystic fibrosis.

Hyaluronidase

Hyaluronidases are a family of enzymes that degrade hyaluronic acid. By catalysing the hydrolysis of hyaluronan, hyaluronidase lowers the viscosity of hyaluronan and increases tissue permeability. It is used in medicine in combination with other drugs to speed up their dispersion and delivery. Hyaluronidase is used (i) in ophthalmic surgery in combination with local anaesthetics (ii) to increase the absorption rate of parenteral fluids given by hypodermoclysis (iii) as an adjunct in subcutaneous urography for improving resorption of radio-opaque agents and (iv) for extravasation of hyperosmolar solutions.

Asparaginase

Asparaginase catalyses the hydrolysis of asparagine to aspartic acid. It functions as a drug for the treatment of acute lymphoblastic leukemia (ALL). All leukemic cells are unable to synthesize the nonessential amino acid asparagine, whereas normal cells make their own asparagine. Thus, leukemic cells require high amount of asparagine. These leukemic cells depend on circulating asparagine. Asparaginase catalyses the conversion of L-asparagine to aspartic acid and ammonia, thus depriving the leukemic cell of circulating asparagine.

Streptokinase (SK)

Immediately after the onset of a heart attack in patients, streptokinase is recommended for intravenous administration, thus reducing the amount of damage to the heart muscle. SK stimulates a cascade system responsible for the production of active plasmin, a proteolytic enzyme which digests fibrin, the main structural component of blood clots. As streptokinase is a bacterial enzyme, the body has the ability to build up immunity to it. Hence, this medication should not be given after four days from the first administration, as it may not be very effective and in addition, it can cause an allergic reaction. For this reason, it is usually administered only when the first heart attack occurs. When more than one heart attack occurs, the drug tissue-plasminogen activator (TPA) is recommended.

Peptide Synthesis Using Protease

i. Synthesis of aspartame using thermolysin

An economically relevant peptide synthesis is that of aspartame, a low-calorie sweetener used to replace sugar and reduce the caloric content in soft drinks.

Aspartame is a dipeptide containing L-aspartic acid and methyl ester of L-phenylalanine and is 180 times sweeter than sucrose. Aspartame synthesis is carried out by thermolysin, a protease (Oyama, 1992).

ii. Porcine insulin is converted to human insulin by replacing the c-terminal alanine (b30) residue by a threonine in a single step catalysed by trypsin.

iii. Another example of a kinetically controlled peptide synthesis is the α-chymotrypsin-catalysed production of kyotorphin (Tyr-Arg), an analgesic dipeptide (Fischer *et al.*, 1994).

BIBLIOGRAPHY

Becker, M.A., Kostel, P.J., Meyer, I.J. and Seegmiller, J.E. (1973). "Human phosphoribosylpyrophosphate synthetase: increased enzyme-specific activity in a family with gout and excessive purine synthesis." *Proc. Natl. Acad. Sci. U.S.A.* 70, 2749–2752.

Devulapalle, K.S. and Mooser, G. (2001). "Glucosyltransferase inactivation reduces dental caries." *J. Dent. Res.* 80, 466–469.

Dewan, A. and Bhatia, P. (2011). "Evaluation of aspartate aminotransferase enzyme levels in saliva and gingival crevicular fluid with periodontal disease progression—A pilot study." *J. Int. Oral Health.* 3, 19–24.

Fischer, A., Bommarius, A.S., Drauz, K. and Wandrey, C. (1994). "A novel approach to enzymatic peptide synthesis using highly solubilizing Na-protecting groups of amino acids." *Biocatalysis.* 8, 289–307.

Higaki, S. and Morahashi, M. (2003). "*Propionibacterium acnes* lipase in seborrheic dermatitis and other skin diseases and Unsei-in." *Drus Exm. Clin.Res.* 29, 157–159.

Lazzatio, L. and Meta, A. (1995). *The Metabolic and Molecular Basis of Inherited Disease,* 7th edn. C.R.Scriver, A.L.Beaudet, W.S.Sly and D.Vale. (Eds.). McGraw-Hill, New York.

Lott, J.A. and Lu, C.J. (1991). "Lipase isoforms and amylase isoenzymes—assays and application in the diagnosis of acute pancreatitis." *Clin. Chim.* 37, 361–368.

Mair, J. (1997). "Cardiac troponin I and troponin T: are enzymes still relevant as cardiac markers?" *Clin. Chim. Acta.* 257, 99–115.

Munoz, A. and Katerndahl, D.A. (2000). "Diagnosis and management of acute pancreatitis." *AM. Fam. Physician.* 62, 164–174.

Oyama, K. (1992). *Chirality in Industry.* Collins, A.N., Sheldrake, D.N. and Crosby, J. (Eds.). Wiley, Chichester. pp. 237–247.

Pascal, T.A., Gaull, G.E., Beratis, N.G., Gillam, B.M., Tallan, H.H. and Hirschhom, K. (1975). "Vitamin B6-responsive and unresponsive cystathioninuria: two variant molecular forms." *Science.* 190, 1209–1211.

Patel, M.S. and Harris, R.A. (1995). "Mammalian alpha-keto acid dehydrogenase complexes: gene regulation and genetic defects." *FASEB J.* 9, 1164–1172.

Rosalki, S.B., Roberts, R., Katus, H.A., Giannitsis, E and Ladenson, J.H. (2004). "Cardiac biomarkers for detection of myocardial infarction: perspectives from past to present." *Clin. Chem.* 50, 2205–2213.

Todorovic, T., Dozic, I., Barrero, M.V., Ljuskovic, B., Pejovic, J., Marjanovis, M. and Knezeic, M. (2006). "Salivary enzymes and periodontal disease." *Med. Oral Pathol. Oral Cir. Bucal.* 11, E115–E119.

Webster, D.R., Becroft, D.M.O. and Suttie, D.P. (1995). *The Metabolic and Molecular Basis of Inherited Disease,* 7th edn. C.R.Scriver, A.L.Beaudet, W.S.Sly and D.Vale. (Eds.). McGraw-Hill, New York.

12

ENZYME TECHNOLOGY—
INDUSTRIAL APPLICATIONS

INTRODUCTION

Enzyme technology has multivarious industrial applications. The early use of enzymes in various processes relied on plant and animal sources. There are 3000 different enzymes described to date and of these, only about 5% are exploited commercially. Microbial enzymes, due to their versatile applications ranging from household detergents to fine chemical industry, have gained greater focus than enzymes obtained from plant and animal sources and they account for more than 75% of the total sale of enzymes worldwide. The market for microbial enzymes is growing spectacularly because of improved understanding of biochemistry and bioprocessing of potent enzymes. Enzymes from microbial sources are more preferred over plant and animal sources since production from the latter sources is not readily inducible and scalable besides difficulties related to recovery, consistency, susceptibility to genetic manipulation, seasonal fluctuations, land availability and animal ethical issues. In addition, the microorganisms in diversified ecosystems yield a variety of enzymes that differ in their substrate specificity as well as optimum functional requirements.

Rapid development in microbial enzyme technology occurred in the mid-1950s. The major advantages of microbial enzymes are as follows:

* Basic knowledge of the characteristics of enzyme has been rapidly expanding, leading to the potential for using microbial enzymes as industrial catalysts.
* Most of the industrially important enzymes could be produced from some microorganism.
* Advent of genetic engineering and recombinant microorganisms has resulted in a vast expansion of the enzyme industry.

Soluble enzymes occupy a special place in industrial applications, notably in the areas of (i) detergent formulations (proteases, lipases and cellulases) (ii) food manufacturing (starch hydrolysis, bread, cheese and fruit juice manufacture), (iii) paper and pulp processing (iv) textile manufacture and (v) degumming of oil. When we consider industrial applications of enzymes, detergents (35–40%), food (40–45%) and starch (10–15%) have a major share.

The enzyme industry is the result of rapid developments in biotechnology seen over the past four decades (Kirk *et al.*, 2002). The enzyme industry is a highly diversified industry that is still growing both in terms of size and complexity and the details are shown in Table 12.1.

Table 12.1   Enzymes used in various industrial segments, and their applications (Kirk *et al.*, 2002)

Industry	Enzyme class	Application
Detergent (laundry and dish wash)	Protease	Protein stain removal
	Amylase	Starch stain removal
	Lipase	Lipid stain removal
	Cellulase	Cleaning, colour clarification, anti-redeposition (cotton)
	Mannanase	Mannanan stain removal (reappearing stains)
Starch and fuel	Amylase	Starch liquefaction and saccharification
	Amyloglucosidase	Saccharification
	Pullulanase	Saccharification
	Glucose isomerase	Glucose to fructose conversion
	Cyclodextrin-glycosyltransferase	Cyclodextrin production
	Xylanase	Viscosity reduction (fuel and starch)
Food (including dairy)	Protease	Milk clotting, infant formulas (low allergenic), flavour
	Lipase	Cheese flavour
	Lactase	Lactose removal (milk)
	Tannase	Tea cream solubilization
	Pectin methyl esterase	Firming fruit-based products
	Pectinase	Fruit-based products
	Transglutaminase	Modify visco-elastic properties

(Contd.)

Table 12.1   (Continued)

Industry	Enzyme class	Application
Baking	Amylase	Bread softness and volume, flour adjustment
	Xylanase	Dough conditioning
	Lipase	Dough stability and conditioning (*in situ* emulsifier)
	Phospholipase	Dough stability and conditioning (*in situ* emulsifier)
	Glucose oxidase	Dough strengthening
	Lipoxygenase	Dough strengthening, bread whitening
	Protease	Biscuits, cookies
	Transglutaminase	Laminated dough strengths
Animal feed	Phytase	Phytate digestibility—phosphorus release
	Xylanase	Digestibility
	β-Glucanase	Digestibility
Beverage	Pectinase	Depectinization, mashing
	Amylase	Juice treatment, low-calorie beer
	β-Glucanase	Mashing
	Acetolactate decarboxylase	Maturation (beer)
	Laccase	Clarification (juice), flavour (beer), cork stopper treatment
Textile	Cellulase	Denim finishing, cotton softening
	Amylase	Desizing
	Pectate lyase	Scouring
	Catalase	Bleach termination
	Laccase	Bleaching
	Peroxidase	Excess dye removal
Pulp and paper	Lipase	Pitch control, contaminant control
	Protease	Biofilm removal
	Amylase	Starch-coating, de-inking, drainage improvement

(Contd.)

Table 12.1   (Continued)

Industry	Enzyme class	Application
	Xylanase	Bleach boosting
	Cellulase	De-inking, drainage improvement, fibre modification
Fats and oils	Lipase	Transesterification
	Phospholipase	Degumming, lyso-lecithin production
Organic synthesis	Lipase	Resolution of chiral alcohols and amides
	Acylase	Synthesis of semisynthetic penicillin
	Nitrilase	Synthesis of enantiopure carboxylic acids
Personal care	Amyloglucosidase	Antimicrobial (combined with glucose oxidase)
	Glucose oxidase	Bleaching, antimicrobial
	Peroxidase	Antimicrobial

DETERGENT INDUSTRY

One of the most important and profitable applications of enzymes is in detergents, where the total global market has been around 50,000 million rupees. Commercial detergent products are used in laundering and dishwashing. To provide desirable benefits, enzymes must be stable and must function well in the presence of anionic and nonionic surfactants, chelators and oxidizing agents. Hence, enzymes have been engineered to meet these challenges.

Essentially, all the enzymes found in today's detergents are hydrolytic in nature and they catalyse the hydrolysis of chemical bonds present within a polymeric substrate. Commonly used hydrolases are proteases, amylases, lipases, cellulases and mannanases.

Proteases are hydrolases that catalyse the hydrolysis of peptide bonds. Stains such as blood, grass, spinach and keratin from collar and cuff soil are relevant for laundry applications while baked-on egg soils are of interest for dishwashing applications. Nearly all commercial proteases are serine proteases from the *Bacillus* family and they contain the catalytic triad of amino acids (i.e., aspartic acid, histidine and serine) in their active sites. Commercial detergent proteases differ in their pH and temperature optima, bleach sensitivity and dependence on Ca^{2+} and Mg^{2+} ion concentration for stability. Autoproteolysis and compatibility with other detergent enzymes are of concern during the formulation and storage of detergents, particularly those in liquid form. This problem has been solved by adding reversible protease inhibitors such as boric acid and propylene glycol to the detergent. Upon dilution of the detergent in the wash, the inhibitors are released from the active site, allowing the enzyme to perform its function.

Amylases are also hydrolases that catalyse the hydrolysis of glucosidic linkages in gelatinized starch polymers. Starch polymers are found in foods such as pasta, fruit, chocolate, baby food, barbeque sauce and gravy. The most common detergent amylase is the α-amylase which hydrolyses the α–1–4–glucosidic bonds in starch. Most commercial amylases are derived from *Bacillus* or *Aspergillus* genera.

Lipases are a fairly new addition to the commercial detergent market. Lipases break down triglycerides into glycerol and fatty acid, thereby increasing their water solubility. Oil stains are among the most difficult to remove via the conventional surfactant technology commonly found in detergents. Stain removal benefits are observed on greasy stains such as butter and lipstick and on body soils such as sebum.

The efficacy of lipase from *A. niger* MTCC 2594 as an additive in laundry detergent formulations has been assessed. A washing protocol with four critical factors, viz., detergent concentration, lipase concentration, buffer pH and washing temperature is chosen. The optimal conditions for the removal of olive oil from cotton fabric are: 1% detergent, 75 U of lipase, buffer pH of 9.5 and washing temperature of 25°C. Under optimal conditions, the removal of olive oil from cotton fabric is 33% and 17.1% at 25°C and 49°C, respectively, in the presence of lipase over treatment with detergent alone. Hence, lipase from *A.niger* could be effectively used as an additive in detergent formulation for the removal of triglyceride soil both in cold and warm wash conditions (Saisubramanian *et al.*, 2006).

The use of cellulases in detergents started in the late 1980s. Cellulases can be either endocellulases or exocellulases and they function by catalysing the cleavage of β-1,4–glycosidic bonds in cellulose. Commercial cellulases come from both bacterial and fungal sources. New cotton consists of smooth fibres; but with prolonged use and washing, microfibrils or broken strands of fibre create a "fuzz" or roughness on the fabric surface. Cellulases remove this and improve the appearance and smoothness of the fabric. Cellulases also restore colour of cotton that has been washed several times and to give jeans the so-called "stone wash" look.

The recent introduction of a new enzyme class into a detergent has been the addition of mannanase. This enzyme helps to remove various food stains containing guar gum, a commonly used stabilizer and thickening agent in food products (Kirk *et al.*, 2002).

These properties of bacterial hydrolases make them suitable for use in the detergent industry.

STARCH HYDROLYSIS AND FRUCTOSE PRODUCTION

Two enzymes carry out conversion of starch to glucose, bacterial α-amylase first cuts the large alpha-1, 4-linked glucose polymers into shorter oligomers at high temperature and this phase is called liquefaction. Saccharification is the next phase and glucoamylase hydrolyses the oligomers into glucose. This is done by fungal enzymes, which operate at lower pH and temperature than α-amylase. Sometimes, additional debranching enzymes like pullulanase are also added to improve the glucose yield.

Large volumes of glucose syrups are converted by glucose isomerase after Ca^{2+} removal to fructose-containing syrup. This is done by bacterial enzymes, which need Mg^{2+} ions for activity. Large-scale chromatographic separation is employed to separate fructose from glucose. Alternatively, fructose is concentrated to 55% and used as a high-fructose corn syrup in soft drink industry.

FOOD PROCESSING AID

Acrylamide is a chemical compound that is formed in starchy food when it is baked or fried. During heating, asparagine present in starchy foods is converted to acrylamide in a process known as Maillard reaction. By adding asparaginase before baking or frying the food, asparagine is converted to aspartic acid and ammonia and the, formation of acyrlamide is significantly reduced. As a food processing aid, asparaginase can effectively reduce the level of acrylamide up to 90% in a range of starchy foods without changing the colour or taste.

BREAD MAKING

The process involves the mixing of wheat flour (mainly starch and protein) with yeast and water. Starch consists of D-glucose units linked by α-1,4 glycosidic bonds, with α-1,6 bonds at branching points. α-amylase and β-amylase present in the flour cleave some of the α-1, 4 bonds and the products are glucose, maltose and some oligosaccharides. Glucose and maltose are then metabolized by yeast, and carbon dioxide is formed which distends the protein framework of the dough, and makes it ready for baking. Supplementation is then done by fungal α-amylase from *A. oryzae*. Proteases from *A.oryzae* may be introduced at this stage to cause limited breakdown of the wheat protein (gluten), then shortening mixing time and enabling more uniform dough to be obtained.

INSTANT TEA MANUFACTURE

Tannase has a promising application in instant tea manufacture. Tannins, accounting for 25% of the water-soluble ingredients in tea leaves, interact with caffeine, resulting in the development of turbidity upon cooling the tea extract. Precipitate formation in the extracts is common as creaming and under certain conditions, the precipitate may flocculate. The principal constituents in cream are theaflavins, thearubigins and caffeine. Cold soluble tea should be clear and bright when reconstituted but cream causes an undesirable haze. Cream is often removed or solubilized for producing tea extract concentrates or soluble tea powders which are widely used for iced beverages. Cream is hydrolysed by tannase to low molecular weight components, reducing turbidity and increasing cold water solubility (Boadi and Neufeld, 2001).

In the conventional chemical process of tea cream solubilization, the tea decoction is chilled to 5°C and centrifuged. The pellet (cream fraction) obtained is subjected to alkali treatment by adding 20% aqueous sodium hydroxide and the temperature is increased to 90°C with the simultaneous aeration for 60 min. Neutralization is carried out using mild acid (Sheetal et al., 2001). The process necessarily needs high inputs of temperature and alkali. In addition, the chemical oxygen demand (COD) of the effluent is bound to be high due to the use of alkali and acid. Comparatively, the enzymatic process is safe.

DAIRY INDUSTRY

Lipases are extensively used in the dairy industry for the hydrolysis of milk fat. Current applications include the flavour enhancement of cheeses, acceleration of cheese ripening, manufacturing of cheeselike products and the lipolysis of butterfat and cream. The addition of lipases that primarily release short chain (mainly C_4 and C_6) fatty acids leads to the development of a sharp, tangy flavour, while the release of medium-chain (C_{12} and C_{14}) fatty acids tends to impart a soapy taste to the product. The flavour ingredients synthesized by lipases are acetoacetate, beta-keto acids, methyl ketones, flavour esters and lactones (Hasan *et al.*, 2006).

BREWING

Barley seed is allowed to germinate under moist conditions to produce malt. The reserve starch is broken by amylase present to give glucose and maltose. The grains are then roasted to prevent further growth and the soluble material is extracted by water to produce the wort. Yeast is then added to produce ethanol by alcoholic fermentation of glucose and maltose.

Neutrase, a neutral protease, is insensitive to the natural plant proteinase inhibitors and is therefore useful in the brewing industry. The bacterial neutral proteases are characterized by their high affinity for hydrophobic amino acid pairs. Their low thermotolerance is advantageous for controlling their reactivity during the production of food hydrolysates with a low degree of hydrolysis.

Fungal acid proteases have an optimal pH between 4.0 and 4.5 and they are stable between pH 2.5 and 6.0. They are particularly useful in the cheese making industry due to their narrow pH and temperature specificities. In view of the accompanying peptidase activity and their specific function in hydrolysing hydrophobic amino acid bonds, fungal neutral proteases supplement the action of plant, animal, and bacterial proteases in reducing the bitterness of food protein hydrolysates. Fungal alkaline proteases are also used in food protein modification.

WINE MAKING

The main tannins present in wines are catechins and epicatechins, which can create a complex with galactocatechins and other galoyl derivatives. Fifty percent of the colour of the wine is due to the presence of tannins. However, if these compounds are oxidized to quinones by contact with air, they could form an undesirable turbidity and this causes problems in the quality. Hence, use of tannase is ideal for removing the tannins (Aguiler and Gutierrez-Sanchez, 2001).

CHILLPROOFING OF BEER

In the chillproofing of beer, tannase is used to remove tannins, which are present as anthocyanidins. An undesirable turbidity is presented due to the presence of tannins and this problem could be resolved by the use of tannase.

TEXTILE INDUSTRY

Desizing   Cotton yarns are starch-coated prior to weaving to prevent their breakage on mechanical looms and this is known as "sizing". Before dyeing, the woven material is treated with amylases to remove this protective coating. This operation, known as "desizing", can also be accomplished with alkalis or acids. But these chemicals attack the cellulose fibres. Enzymatic action is more effective and easily controlled than hydrolysis by acids and alkalis. Bacterial enzymes are preferred since they withstand higher temperatures.

Degumming of silk fibres   Although degumming of silk fibres is usually carried out by using hot concentrated soap solution, proteolytic enzymes can also be used to obtain similar results. Silk fibres are bound by a gummy proteinous substance known as "sericin" which can be removed by the action of proteolytic enzymes in the degumming process. Enzymatic degumming of silk is comparatively milder than continued action of hot concentrated soap solution.

Preparation of denim fabric   Denim is a special soft cotton-based fibre where the dye is partially faded away. Originally, volcanic lava stones were used in the preparation of Denim from an indigo-dyed cotton fibre to achieve a high degree of dye fading. The stones caused considerable damage to fibres and machines. When cellulase enzymes are used, the same effect is achieved as a result of alternating cycles of desizing and bleaching enzymes and chemicals in washing machines. This process is known as "biostorming".

Use of oxidases in bleaching   Chlorine-based chemicals are replaced by hydrogen peroxides as bleaching agents. To degrade excess peroxide, catalase, which degrades hydrogen peroxide, is used. Another approach is to use laccase, a polyphenol oxidase from fungi, for bleaching textiles.

Modification of polyester In textile industry, polyester has certain key advantages including high strength, soft hand, stretch resistance, stain resistance, machine washability, wrinkle resistance and abrasion resistance. Synthetic fibres have been modified enzymatically for use in the production of yarns, fabrics, rugs and other consumer items. This means that, for example, the characteristics of a polyester fibre are so modified that such polyesters are more susceptible to post-modification treatments. The use of polyesterase to improve the ability of a polyester fabric to uptake chemical compounds such as cationic compounds, fabric finishing compositions, dyes, antistaining compounds, antimicrobial compounds and antiperspirant compounds has been demonstrated (Hasan *et al.*, 2006).

ANIMAL FEED

Addition of enzymes to animal feed started in the 1980s. Barley contains β-glucan which causes high viscosity in the chicken gut. Initially, barley-based feed diets were chosen for the addition of β-glucanase. A major observation, after the addition of β-glucanase, is the increase in animal weight gain and also an increased feed conversion ratio.

Later, when wheat-based diets were taken up for testing, xylanase enzymes were found to be effective. Available metabolizable energy is increased to 7–10% when xylanase is added to wheat-based broiler feed. Feed-enzyme preparation contains a mixture of glucanases, xylanases, proteinases and amylases. Advantages of enzyme addition are:

  i. Reduction of viscosity
  ii. Increase in the absorption of nutrients
  iii. Reduction in the amount of faeces

Another development in the feed industry is the addition of phytase. Phytase, a phosphodiesterase, liberates phosphate from phytic acid, a common compound in plant-based feed materials. The advantage of adding phytase is that there is reduction of phosphorus in faeces resulting in reduced environmental pollution. Another advantage is that phosphorus need not be added to the feed diet. In addition to broiler feed industry, enzymes are also used in pig feeds and turkey feeds.

FAT AND OLEOCHEMICAL INDUSTRY

Lipases, apart from their natural function of hydrolysing carboxylic ester bonds, can catalyse esterification, interesterification and transesterification reactions in nonaqueous media.

The world production of butter fat is around 5.5 million tonnes per annum. Cocoa butter fat is a high-value product because triacylglycerols (TAG) are high in stearates which give a melting point of 37°C. Hence, cocoa butter fat melts in the mouth to give a smooth 'mouth appeal' effect, e.g., chocolate. Palm oil TAGs are high in palmitates and they give a melting point of 23°C. So, it is a low-value product. Conversion of palm oil into cocoa butter fat substitute can be achieved by interesterification using lipase and it is a commercial process (Hasan *et al.*, 2006).

Use of lipases to carry out industrial hydrolysis of tallow has a number of advantages. The heat required is around 50°C and the consumption of fossil fuels to make steam is much less. Because of the low temperature, there is less degradation of unsaturated fatty acids. Depending on the specificity of the lipase and the raw material, partial hydrolysis could yield a concentrated or purified mixture of fatty acids and/or partial glycerides with unique properties not found in bulk fatty acids obtained from total hydrolysis of tallow (Hasan *et al.*, 2006).

Lipases have one of the most important applications in organic syntheses. 1-Butyl oleate is produced by direct esterification of butanol and oleic acid to decrease the viscosity of biodiesel in winter use. Trimethylol-propane esters are synthesized for use as lubricants. Transesterification reactions in organic solvent systems using lipases have opened up the possibility of enzyme-catalysed production of biodegradable polyesters (Linko *et al.*, 1998).

PHARMACEUTICAL INDUSTRY

(i) *Use of lipases*   Lipases can be used to resolve racemic mixtures and to synthesize the chiral building blocks for pharmaceuticals, agrochemicals and pesticides. Some lipases retain their activity in nonpolar organic solvents. This property can be used in the hydrolysis of water-insoluble esters, as in the resolution of racemic mixtures through stereospecific hydrolysis (Kirchner *et al.*, 1985).

Chirality is a key factor in the efficacy of many drugs. Production of single enantiomers of drug intermediates has become increasingly important in pharmaceutical industry. Lipase from *C. antarctica* has been used for the kinetic resolution of racemic flurbiprofen by the method of enantioselective esterification with alcohols (Zhang *et al.*, 2005).

Baclofen, (RS)-beta-(aminomethyl)4-chloro benzene propanoic acid, is used in pain therapy and also as a muscle relaxant. It produces two isomers and lipase from *C. cylindracea* has been used as a catalyst for resolving racemic mixtures (Muralidhar *et al.*, 2001).

(ii) *Use of tannases (tannin acyl hydrolases)*   One of the major commercial applications of tannases is the hydrolysis of tannins to gallic acid, a key intermediate in the synthesis of an antibiotic, trimethoprim (Mahendran *et al.*, 2006). Trimethoprim is an antibacterial agent and when given along with sulphamethoxazole, it has a broad spectrum of action.

POULTRY INDUSTRY—HYDROLYSIS OF FEATHER MEAL

Poultry feathers are available in plenty and a major portion of this is left unutilized. They are notably deficient in lysine and methionine. Tryptophan appears to be marginally deficient in all keratins. The commercial utilization of feathers has been mainly confined to cooked (autoclaved) feather meal which is used as a partial replacement for fish meal or meat meal in practical rations for poultry feeding. A multiple proteinases concentrate from *Streptomyces moderatus* has been used for the hydrolysis of cooked feather meal and the properties of the cooked feather meal have been much improved by enzyme hydrolysis. When the hydrolysed feather meal, supplemented with lysine and methionine, is used as the protein source in chick diets, increased protein efficiency ratio and net protein utilization values are observed.

EFFLUENT TREATMENT

Lipases are utilized in activated sludge and aerobic waste processes, where thin layers of fat must be continuously removed from the surface of aerated tanks to permit oxygen transport. This skimmed fat-rich liquid is digested with lipases from *C. rugosa* (Bailey and Ollis, 1986). Effluent treatment is also necessary in industrial processing units such as abattoirs, food processing industry, leather industry and poultry waste processing (Godfrey and Reichelt, 1983). Lipase is shown to be used for the degradation of lipid-rich waste water (Dharmsthiti and Kuhesuntisuk, 1998).

BIBLIOGRAPHY

Aguiler, C.N. and Gutierrez-Sanchez, G. (2001). "Review: Sources, properties, applications and potential uses of tannin acyl hydrolase." *Food Sci. Tech. Int.* 7: 373–382.

Bailey, J., E. and Ollis, D.F. (1986). *Biochemical Engineering fundamentals. Applied enzyme catalysis*, 2nd edn. McGraw-Hill, New York. pp. 157–227.

Boadi, D.K. and Neufeld, R.J. (2001). "Encapsulation of tannase for the hydrolysis of tea tannins." *Enz. Microb. Technol.* 28: 590–595.

Dharmsthiti, S. and Kuhasuntisuk, B. (1998). "Lipase from *Pseudomonas aeruginosa* LP602: biochemical properties and application for wastewater treatment." *J. Industr. Microbiol. Biotechnol.* 21, 75–80.

Godfrey, T. and Reichelt, J. (1983). *Industrial enzymology: the application of enzymes in industry*. The Nature Press, London. pp. 170–465.

Hasan, F., Shah, A.A. and Hameed, A. (2006). "Industrial applications of microbial lipases." *Enz. Microb. Technol.* 39, 235–251.

Kirchner, G., Scollar, M.P. and Klibanov, A.M. (1985). "Resolution of racemic mixtures via lipase catalysis in organic solvents." *Biosci. Biotechnol. Biochem.* 107, 7072–251.

Kirk, O., Borchert, T.V. and Fuglsang, C.C. (2002). "Industrial enzyme applications." *Current Opinion in Biotechnol.* 13, 345–351.

Linko, Y.Y., Lamsa, M., Wu, X., Uosukaine, E., Seppala, J. and Linko, P. (1998). "Biodegradable products by lipase biocatalysis." *J. Biotechnol.* 66, 41–50.

Mahendran, B., Raman, N and Kim, D.J., (2006). "Purification and characterization of tannase from *Paecilomyces variotii*: hydrolysis of tannic acid using immobilized tannase." *Appl. Microbiol. Biotechnol.* 70: 444–450.

Muralidhar, R.V., Chirumamilla, R.R., Ramachandran, V.N., Marchant, R. and Nigam, P. (2001). "Racemic resolution of RS-baclofen using lipase from *Candida cylindracea.*" *Meded Rijksuniv Gent Fak Landbouwkd Toegep Biol Wet.* 66, 227–232.

Saisubramanian, N., Edwinoliver, N.G., Nandakumar, N., Kamini, N.R. and Puvanakrishnan, R. (2006). "Efficacy of lipase from *Aspergillus niger* as an additive in detergent formulations: a statistical approach." *J. Ind. Microbiol. Biotechnol.* 33, 669–676.

Sheetal, S.M., Vijay, S., and Prakash, D.V. (2001). "Process for producing tea concentrate." US Patent No. 6296887.

Zhang, H.Y., Wang, X., Ching, C.B. and Wu, J.C. (2005). "Experimental optimization of enzymic kinetic resolution of racemic flurbiprofen." *Biotechnol. Appl. Biochem.* 42, 67–71.

13

CONSTITUENTS OF SKINS—THEIR ROLE IN LEATHER PROCESSING

INTRODUCTION

Leather industry is one of the major foreign-exchange earners with the export target of around ₹1,80,000 million for 2011. It is well known that the leather industry is one of the major polluting industries in India, since all the process steps in leather manufacture are chemical-oriented. Breakthrough technologies are necessary to cause a paradigm shift from chemical processing to bioprocessing, and one of the best options for this transition is microbial technology.

Animals such as goat, sheep, cow and buffalo are slaughtered in the slaughterhouse or abattoir. Immediately after slaughter, 'flaying' is carried out and the meat goes for human consumption while the skins and hides are taken to tanneries for processing into leather. It is not always possible to transport the raw skins and hides to tanneries immediately after "flaying" since the source of collection and the tanneries are not generally located in the same area. The time delay between the flaying operations and the start of pretanning processes varies from weeks to months. If the raw skins and hides are not preserved or cured immediately after flaying, they undergo putrefaction. Usually, bacteria, rather than fungi, are the major cause of putrefaction. The purpose of 'preservation' or 'curing' of the raw hides and skins is to render them resistant to putrefaction during storage and transport till they are processed into leather. The quality of the leather is dependent on proper preservation of hides and skins. Hence, microbial control is an important step in preserving the hides and skins before they are taken up for leather processing.

"Dehairing" or removal of hair is the primary operation in the pretanning process of leather manufacture. Conventionally, skins and hides are chemically treated with a mixture of lime and sodium sulphide and the hair is removed as a pulp. By this chemical method, valuable hair is lost. In addition, dissolved hair contributes to high BOD load while COD is increased to a very high level due to the use of lime and sodium sulphide. These problems are obviated by the use of enviro-friendly microbial proteases in the dehairing process. The reader has to understand the mechanism of dehairing by both chemical and enzymatic methods.

There are more than 3000 tanneries located in India with a total processing capacity of 7,00,000 tons of hides and skins per year.

It should be understood that the skins and hides, immediately after 'flaying', undergo many chemical as well as enzymatic treatments in the pretanning processes. The reader has to understand that, by these processes, whether the structure and chemical composition of hides and skins are really altered. Hence, structure and chemical composition of collagen, reticulin, elastin, keratin and proteoglycans are discussed in this chapter.

The term beamhouse process refers to the processes the skins/hides undergo in the tannery as a preparative step before tanning. This includes soaking, trimming, fleshing, dehairing, liming, deliming, bating, degreasing and pickling. The term dates back to the time when the hair was removed from the skins placed on a sloping curved table or on a large log wood using a two handed blunt knife. The working of the skins on the beam is still in use. Beamhouse operations are very important and as the tanners put it, "Leather is made in the beamhouse". Complex principles of both biochemistry and inorganic chemistry are underlined in the beamhouse operation and the mechanisms of some of the operations are yet to be fully understood.

It is well known that raw hides and skins as received in the tannery for processing are composed of both soluble and insoluble proteins. They are also called globular and fibrous proteins. The fibrous proteins are collagen, elastin, keratin and reticulin, while the globular proteins are albumin, globulin and mucoids. Collagen, being the main protein which makes leather, should be retained fully in order to make the finished leather structurally very strong. The other fibrous proteins and all the globular proteins are to be removed as far as possible to get the desired end product with proper physical properties. For example, in order to make sole leather, retention of soluble proteins is desirable, whereas, to make soft leathers such as glove or garment leathers, soluble proteins should be removed from skins completely. Removal of these proteins can be achieved by chemical treatments. But, besides chemical treatments, certain enzymatic treatments are also necessary to get optimum results. One such treatment is technically called bating. The material used for this treatment is known as Enzyme bate, which contains a potent proteolytic enzyme, an inert carrier of the enzyme and deliming salts such as ammonium chloride or ammonium sulphate or both.

Dehairing of hides and skins is traditionally done by the chemical process using lime and sodium sulphide. The process is technically called Liming process. Lime, due to its sludge-forming character, gives rise to pollution problems. Sulphide is toxic, and high alkalinity of lime and sulphide reduces the strength of wool or hair thereby making the by-product useless. It has been shown that any crude potent proteolytic enzymes can be successfully used for dehairing of hides and skins. Apart from solving the pollution problem, the enzyme dehairing system has many other advantages.

A fat-splitting enzyme, lipase, has been shown to remove the excess natural fat from greasy skins by hydrolysing the grease present in the skin. This process replaces the solvent degreasing method since the latter is expensive and hazardous.

It is, therefore, evident that hydrolytic enzymes play a significant role in the pretanning processes of leather manufacture. For commercial use in leather manufacture, these enzymes or enzymatic formulations need not be pure but must be cheap as compared to that of commercial chemicals used in leather industry. In order to use enzymes in leather processing, it is imperative to know the requirement of a particular enzyme, its nature and specificity, the speciality of substrates on which the enzyme will act and finally the end results which are to be achieved.

LAYERS OF SKIN

Leather is made from the raw hides and skins of animals such as cow, goat, sheep, buffalo, etc. Structurally, all the hides and skins are more or less the same except for minor differences. Skin and hide are anisotropic, i.e., their structure and properties widely vary throughout their dimension. The cross section of the skin comprises three layers (Figure 13.1 a and b).

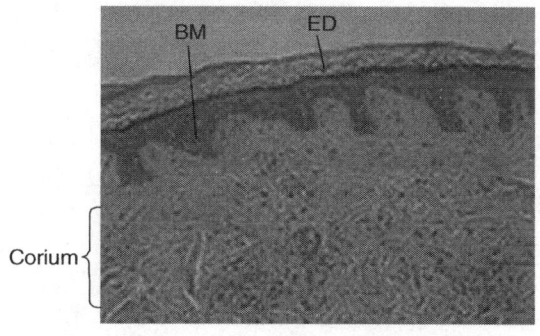 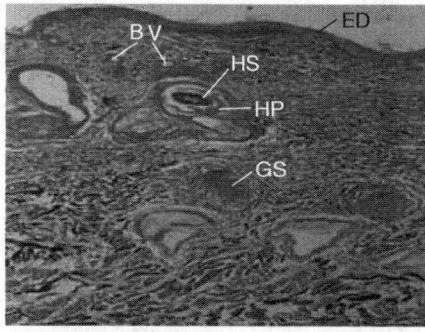

Figure 13.1   (a) Cross section of skin—distinctive layers of epidermis, basement membrane and corium across the cross section of skin [Hematoxylin and eosin staining (×175)]. (b) collagen fibre bundles (purple) and epidermis and non-collagenous proteins (pink) [Masson's trichrome staining (×175)] (ED: Epidermis; BM: Basement membrane BV; Blood vessels; HS: Hair shaft; HF: Hair follicle; GS: Glandular structures of fat and sebum) (*See* Plate 1.1)

On the surface of the intact skin lies the keratinous epidermal layer comprising the epidermis and its appendages such as hairs, hair root, sheaths, etc. Immediately below it, the basement membrane is to be found, attaching the epidermal layer to the underlying dermis or corium. In the tanning process, the corium is transformed into leather. During the dehairing process, the epidermal layer must be removed from the dermis. A further component of raw skin or hide is the subcutaneous or hypodermis layer comprising connective tissue and fat which is removed by fleshing. Hides and skins are usually presented to the tanner with adhering flesh and fat. This must be removed at the earlier stage of pretanning processes as it is a barrier to the even penetration of chemicals and enzymes, which would cause non-uniformity of leather properties.

Keratin is stabilized through disulphide(—S—S—) bonds. The fully developed keratin in hair and the upper part of the epidermal layer is highly resistant to chemical or biological attack, except sulphide treatment, which breaks down by the disulphide bonds. The basement membrane is a thin layer of 50–100 nm between the epidermal layer (including the hair roots) and the dermis. It is attached to the surface layer or grain of the dermis. It consists of a special type of collagen as well as various glycoproteins and proteoglycans which are interlinked in a close network.

The structure of the basement membrane and consequently that of the corium–epidermis junction is mainly based on protein–protein links, although protein–carbohydrate interaction and a few sulphur bridges also play a predominant role. The basement membrane thus becomes a target for selective proteolytic enzymes, and an attack on the basement membrane is an essential feature in the enzymatic dehairing process.

Breaking the corium–epidermis junction is achieved by destroying or modifying the basal cell layer of the epidermis, the hair bulb and root sheaths and/or the basement membrane network. Fine hair is more deeply anchored in the dermis than the fully developed hair.

The dermis or corium is the thickest layer and is tightly connected to the epidermis by a basement membrane (Figure 13.2). This is the main layer of the skin or hide constituting about 98% of the total thickness of the hide, and only 2% is constituted by the epidermis. It harbours many hair follicles, hair roots, sweat glands, sebaceous glands, apocrine glands, lymphatic vessels and blood vessels. Also, collagenous, elastic and reticular fibres are densely concentrated in the dermis that weave and spread throughout the layer (Figure 13.2 and 13.3). Corium is principally composed of collagen fibres, which are present in the form of bundles.

Figure 13.2 and Figure 13.3 show the presence of reticular (corium; lower dermis) and elastic fibres (upper dermis) in black respectively. Collagen fibre bundles are shown in pink colour. Grey colour represents epidermis and non-collagenous proteins.

Figure 13.2   Cross section of skin showing reticulin (RT) [Gomori staining (×175)] (*See* Plate 1.3)

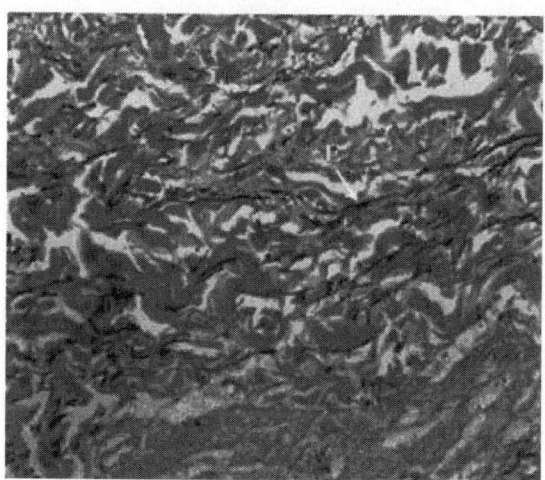

Figure 13.3   Cross section of skin showing elastin (E) [Verhoeff's staining (×520)] (*See* Plate 1.3)

The fibres of the bundles are encased in sheaths of another kind of fibre which consists of fine threads of a protein called reticulin (Figure 13.2). At various points along the length of the fibre bundles, the reticular threads penetrate the sheaths, encircle the fibre bundles and bind them together in the form of garters. The fibre bundles are woven with one another at an angle to the horizontal surface of the corium. The fibre bundles diminish in size as they approach the upper surface of the corium and at the level of the hair roots, they turn upwards and branch into smaller bundles and individual fibres. At the uppermost surface just below the top, called the grain layer, the bundles are grouped into small fibres only. The fibres are composed of extremely fine fibrils of about 0.0005 mm diameter.

The fibrils are composed of micelles about 0.000,002 mm diameter. The collagen fibrils in this layer are cemented together by proteoglycans and hyaluronic acid substances. In the corium, there are principally two layers, viz., grain layer and the reticular layer. The grain layer which is called thermostat layer or corium minor keeps the body temperature constant through the action of sweat glands and fat glands present in it. The grain layer occupies about 1/5th of the total thickness of the corium. Reticular layer, also called corium major, is below the grain layer and it is the main portion of the corium, constituting about 4/5th of its thickness. It is called reticular layer on account of its netlike woven structure. The fibres of this layer are longer than those of the corium minor.

CHEMICAL CONSTITUENTS OF SKINS AND HIDES

The fundamental structural element of animal skins and hides is an elongated fibril about 100 nm thick composed mostly of collagen. These collagen fibrils are embedded in a softer interfibrillar matrix consisting mainly of proteoglycans bound to a collagen fibril surface. Most of the leather making processes attempt to pack these fibrils and fill the spaces among them without damaging the structure. In the dehairing process, the hair along with epidermis, non-collagenous proteins and other cementing substances is removed from the skin.

Collagen

Collagen is a predominant group of animal proteins abundantly found in fibrous tissues such as tendon, ligament and skin, cornea, cartilage, bone, placenta, blood vessels, etc. It is the main component of connective tissue, making up about 20% to 30% of the whole-body protein content. Collagen is a generic name for a family of about 28 distinct collagen types, each serving different functions in different connective tissues. The major collagen present in skin is type I collagen, the term collagen refers to type I collagen unless and otherwise specified. Other collagens also have a role in leather processing, and they will be dealt with wherever necessary. Collagen is approximately 300 nm long and 1.5 nm in diameter, made up of three polypeptide strands (called alpha chains) of about 1050 amino acids long, each possessing the conformation of a left-handed helix (its name is not to be confused with the commonly occurring alpha helix, a right-handed structure). These three left-handed helices are twisted together into a right-handed coiled coil, a triple helix or "super helix", which is a cooperative quaternary structure stabilized by numerous hydrogen bonds (Figure 13.4).

All the common amino acids are found in skin but hydroxyproline is almost uniquely present in collagen and elastin when compared to other proteins. Tryptophan, an aromatic amino acid, is absent and because of this, the collagen is deficient as foodstuff; however, it is present in keratin, a major protein of hair and epidermis. The content of hydroxyproline provides the basis of quantitating the collagen content in skin. A distinctive feature of collagen is the regular arrangement of amino acids in each of the three chains. The sequence often follows the pattern Glycine (Gly) – Proline (Pro) – X or Gly – X – Hydroxyproline (Hyp) where X may be any other amino acid residue other than Gly, Pro and Hyp. The formation of right-handed super helix is possible only because of the high glycine content, which has the smallest α-carbon side chain, a hydrogen atom, so that glycine is always situated in the centre of the triple helix. Pro or Hyp constitute about one-sixth of the total sequence and glycine accounts for the one-third of the sequence. The structure of the collagen is primarily attributed to the content of glycine, proline and hydroxyproline.

Shrinkage temperature of raw skin depends on the contents of Pro and Hyp amino acids and this is due to steric conformation, restriction by the ring structure of these amino acids. Hydrothermal stability is imparted by high content of hydroxyproline and proline; most importantly, the content of hydroxyproline rather than proline is important in stabilizing the collagen structure by hydration as they play an effective role in binding water. If water is removed from the triple helices by drying, collagen becomes thermally fragile, thus forming brittled fibres. Consequently, the fibrous structure is severely impaired resulting in the leather becoming stiff and this reduces the value of the material. Hydrothermal stability of collagen is defined as the influence of wet heat on the integrity of the macromolecules in terms of the denaturation transition. Collagen responds to wet heat by shrinking; in the case of native collagen, this is at 65°–70°C. Degradation of collagen by wet heat causes the transition from helix to random coil, which causes gelatinization of collagen followed by solubilization. This is usually an irreversible effect which severely damages the quality and integrity of the collagen materials.

Regions of about 20 amino acids that are not helical can be found at the C-terminal end of collagen triple helices and they are called telopeptide regions. They play an important role in holding collagen macromolecules together. They can be removed if collagen is treated with proteases. During the production of collagen synthesis for extracellular matrix assembly in connective tissues, the collagen is synthesized as monomer procollagen and then the monomers self-assemble into a fibrous form.

Skin is primarily composed of the protein collagen and the potential for chemical modification of this protein offers the tanner the opportunity to convert an unappealing starting material to a desirable product.

Fibrils, Fibril Bundles and Fibres

The triple helices are bound together in bundles called fibrils and they are the smallest units of collagen structure (Figure 13.4).

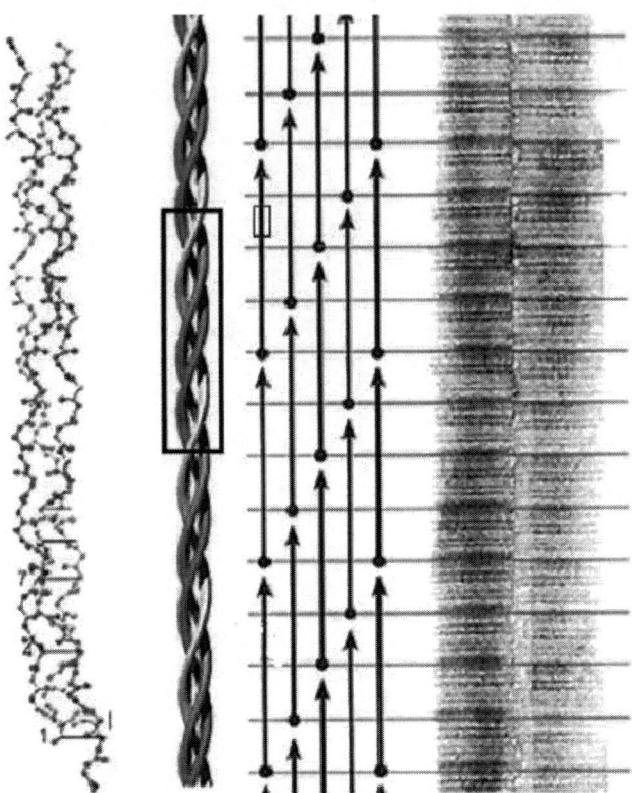

Figure 13.4 Schematic arrangement of collagen triple helices to form fibrils

Fibrils have a diameter of 50–200 nm, depending on the collagen type. The ends of adjacent collagen molecules are displaced from one another by a distance 67 nm, which produces the striated appearance seen in electron micrographs, with bands spaced about 67 nm apart. Fibrils are arranged together to form fibril bundles which are the constructing units of fibres. Splitting and opening up of fibre structure at this level, at the junctions between fibril bundles, is essential for producing softness and strength in the final leather, particularly in deposition of lubricating agents and to facilitate penetration of stabilizing chemicals and dyeing agents. Fibril bundles come together to form fibres that can be seen under light microscopy (Figure 13.5). The fibres characteristically divide and rejoin with other fibres throughout the corium structure.

Reticulin

Reticulins are also called collagen type III macromolecules that can be found as fibres intertwined along with type I collagen fibres to provide flexibility to the fibre structure. It is observed that foetal skin has a high proportion of reticulin i.e., 30–40% compared to about 15% in mature skin and this suggests that the flexibility of fibre structure in mature skin is reduced. It differs from collagen type I as the triple helix of reticulin is made up of three αI(III) chains and hence, they form finer fibres.

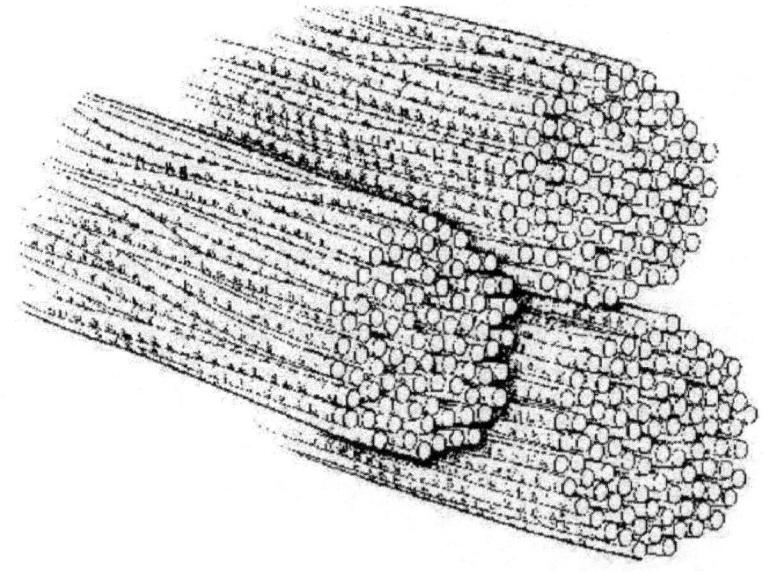

Figure 13.5 Arrangement of collagen fibrils into fibers

Elastin

This helical protein having structural resemblance with collagen is abundantly found in grain layer and it controls elasticity as well as contributes to the physical properties of the skin and leather. It contains more non-polar amino acids than collagen and hence, is more hydrophobic. However, it contains less hydroxyproline content compared to collagen suggesting that its structure depends less on hydrogen bonding. Elastin strands twist together to form double helix and these helices, after excluding water due to high hydrophobicity, form crosslinked network via desmosine residues and adopt a random and disordered conformation having high degree of flexibility and contractability. Removal of elastin allows the grain to relax by reducing the contractability of the material resulting in flattening and increase in the area of the skin. Alkaline liming as well as several dehairing enzyme preparations do not remove elastin network. This objective of producing soft leather by degradation of elastin is achieved during bating, wherein elastolytic enzyme preparations are employed. Please refer to the Chapter on bating enzymes.

Keratin

Keratin differs from other proteins in skin and it is a principal protein component of hair and epidermis. It is a helical protein containing less glycine and proline when compared to collagen and has no hydroxyproline. However, the contents of cystine and tyrosine are higher than that present in elastin and collagen. The total sulphur content in keratin, collagen and elastin are 3.9%, 0.2% and 0.3% respectively. During the growth of hair, cysteine can be converted to cystine by oxidation. It forms fibrils created from protein chains linked via disulphide bonds, which are the target of many of the dehairing techniques. Typical hair contains three structures that can be distinguished by structure and function. They are (i) hair bulb, (ii) prekeratinized hair and (iii) cuticle. The hair bulb is the site where the hair grows and is attached to hair root or papilla. It contains proteins but not the matured keratins, and the hair bulb can be the target of dehairing agents. Situated above and attached to the bulb is the prekeratinized hair, in which disulphide links are established in the growing hair and the hair is made of soft keratin. It may be the target in hair-save dehairing methods. The cuticle is the outer surface of the hair staple and it is made up of hard and mature keratin that contains a high degree of disulphide links. During hair-burn dehairing, it is broken and present in suspended solids in effluent. Though conventional hair-burn dehairing methods attack all these three structures, protease-mediated dehairing actions are confined to the structures of hair bulb and prekeratinizied hair.

Proteoglycans

Proteoglycans (PG) are interfibrillary cementing substances located on the surfaces of the collagen fibrils of skin in a highly ordered manner and tie the adjoining collagen fibrils together through strong electrostatic interactions. Decorin, a leucine-rich dermatan sulphate PG, is the most abundant interfibrillary substance present in the skin, and it is having a protein core of 36 kDa and an anionic sulphated glycosaminoglycan (GAG) chain of about 15 kDa covalently linked to N-terminus via O-serine glycosidic linkage. Besides, several small oligosaccharide chains are attached to protein core through N-glycosidic linkage. The protein core of decorin non-covalently interacts with hyaluronic acid, a high molecular weight non-sulphated GAG, which leads to the formation of PG aggregates.

OTHER CHEMICAL CONSTITUENTS OF SKINS AND HIDES

The other chemical constituents of raw hides and skins are water, fats, mineral matter and proteins. By far, the largest proportion is water which constitutes about 70% of the weight of a fresh skin or hide. Water is present in two forms, viz., bound and free. Only the free water and not the bound water can be pressed out of the skin by considerable pressure. The fat content of the hides and skins is usually small but it can be as high as 30% on the dry skin weight in some kind of woolly sheep skins. Raw hides and skins contain a small percentage of mineral matter, about 0.35 to 0.5% on the raw weight. This consists mainly of phosphates, carbonates, sulphates and chlorides of sodium, potassium, magnesium and calcium.

Collagen is the most important protein for a leather chemist. Except collagen, all the fibrous as well as globular proteins are eliminated during pretanning operations. The albumins are soluble in water and can be removed by washing the hides and skins in water. The globulins are soluble in dilute salt solution and they are removed automatically when salted hides and skins are washed in water. The glycoproteins which constitute the cementing interfibrillary substance of hides and skins are soluble in dilute alkali and are, therefore, to a large extent removed during the liming process. What is left behind is removed during bating by the action of enzymes. The keratins are of two kinds, the softer and the harder. Both can easily be removed by dissolving in a solution of Sodium sulphide and calcium hydroxide. Elastin and reticulin are either partially or wholly removed by treatment with dehairing/bating enzyme. Reticulin also can be removed by the action of sodium sulphide.

BIBLIOGRAPHY

Covington, A.D. (2009). *Tanning Chemistry: The Science of Leather.* RSC Publishing, Cambridge, U.K.

Dutta, S.S. (1985). *An Introduction to the Principles of Leather Manufacture.* Indian Leather Technologists Association, Calcutta.

O´ Flaherty, F., Roddy, W.T. and Lollar, R.M. (1978). *The chemistry and Technology of Leather.* Vol. I Krieger Publishing Co., Malabar, Florida.

Sarkar, K.T. (1981). *Theory and Practice of Leather Manufacture.* Ajoy Sorkar, Chennai.

14

MICROBIAL CONTROL IN CURING PROCESS

INTRODUCTION

'Preservation' and 'curing' are commonly used terms to imply some form of chemical treatment of raw hides and skins to prevent microbial contamination as well as microbial activities. It is not always possible to transport the raw stock to tanneries immediately after "flaying" because the sources of collection and the tanneries are not generally located in the same area. The time-delay between the flaying operation and the start of pretanning processes varies from weeks to months. If the raw stock is not cured immediately after flaying, it undergoes complete 'putrefaction' within two to three days. Usually, bacteria, rather than fungi, are the major cause of putrefaction since they have a doubling time of 2–3 h at 30°C and a high multiplication rate, compared to fungi, as the latter may take days to start the degradation process (Cooper and Galloway, 1973). The purpose of preservation or curing the raw stock is to render them resistant to putrefaction during storage and transport till they are processed into leather. The quality of the leather is dependent on proper preservation of hides and skins. Without affecting the leather quality, tanners can opt for suitable changes in preservation methods not only to make the curing economical but also to protect the environment (Vijayalakshmi *et al.*, 2009).

One of the major sources of pollution from tannery effluent is the use of common salt or sodium chloride in curing. Conventional curing of hides and skins employs large quantity of salt, i.e., 30–50% of the weight of flared skin. This process has a major environmental constraint as the salt is discharged into the effluent after the subsequent 'soaking' operation. This is a major source of pollution from tanneries in terms of high total dissolved solids (TDS) and chloride content in the effluent discharge. As the dissolved sodium chloride is difficult to be treated and removed from waste water, such discharge of tannery waste water onto land leads to significant increase in the salinity of the soil

(Daniels, 1997). The TDS and chloride content of the effluent vary between 150–300 g/kg and 90–150 g/kg of hides processed respectively (Rajamani, 1998).

PUTREFACTION AND DEGRADATION OF HIDES AND SKINS

Hides and skins are rich in proteins, fats, carbohydrates, pigments, etc. and these can be good nutrient sources for microbes especially bacteria. Once the hides and skins are flayed from the dead animals, putrefaction starts by the action of various hydrolytic enzymes such as proteases, lipases and carbohydrases that are secreted by bacteria. Opportunistic microbes derived from the environment, viz., air, water, soil, manure and extraneous filth and the microbial flora inhabiting in the flayed skins form the 'microbial consortium' which start the process of decomposition of skins (Birbir and Ilgaz, 1995). The microbes that are described in the context of curing and putrefaction principally represent bacteria. Bacteria secreting one or more of the above enzymes have the potential to degrade the skin effectively. The bacteriological conversion of protein and other hide substances to different organic compounds is called putrefaction and the substances that are produced as a result of putrefaction include indole, amines, amino acids, fatty acids, fermented products such as acetate, sugars and odorous ammonia and sulphides. The bacterial mediated putrefaction continues until the above-said products are finally converted to gas components such as CO_2, H_2O, N_2, H_2, O_2, H_2S, NH_3 etc.

AUTOLYSIS

Skin plays a vital role in temperature regulation and excretion of metabolites when the animal is alive. It is also a barrier for many microbes preventing their entry into the tissues and circulatory system of animals. Once the skin is removed or flayed from the animal, the function of temperature regulation stops immediately. Freshly flayed raw stock suffers temperature abuse, causing dramatic change in the metabolic process. Vital functions such as transport as well as exchange of oxygen and nutrients, and removal of metabolites from the dermal tissue are stopped and this results in accumulation of toxic materials. Enzyme-controlled processes are stopped due to inactivation of coenzymes. These conditions readily lead to activation of intracellular proteolytic enzymes such as cathepsins and collagenases, resulting in self-digestion of cells in hides and skins through a process called 'autolysis'. The quantity of autolytic enzymes does not increase with time as do bacterial enzymes (Bailey, 1999a). If the material is not subjected to cold storage, the process of autolysis increases as the ambient temperature favours enzymatic degradation. The secondary process accompanying autolysis is the action of putrefactive bacteria for which autolysis products are used as nutrients.

DEGRADATIVE ENZYMES

Among the proteolytic enzymes, collagenases play a prime role in effecting deterioration of skin and these enzymes are activated during autolysis or secreted by microflora present in the skin. Collagenases degrade collagen in native collagen fibrils, which are the fundamental elements of extracellular matrix (ECM) of connective tissues. (Refer to collagen structure in Chapter 13). True collagenases may cleave simultaneously all three chains or attack a single strand under physiological conditions of pH and temperature (Harper, 1980). Mammalian or tissue collagenases split collagen in its native triple helical conformation at a specific site yielding fragments of TCA (tropocollagen A) and TCB (tropocollagen B) representing ¾ and ¼ lengths of the tropocollagen molecule respectively. Tissue collagenases produce a scission per α-chain whereas bacterial collagenases, e.g., collagenases from *Clostridium,* produce multiple scissions per chain. An increasing number of microorganisms, many of which are putative human pathogens, have been reported to produce enzymes which can degrade collagen and they include *Bacillus cereus, Clostridium, Serratia marcescens, Proteus mirabilis, Streptococcus* sp., *Staphylococcus* sp. etc. (Harrington, 1996). *S. marcescens is* shown to produce collagenase capable of degrading type I collagen but not reconstituted collagen fibrils (Molla *et al.,* 1986). It has also been suggested that bacterial proteinases may be capable of activating latent mammalian collagenases, thus contributing to the degradation of collagen indirectly. Collagenases are useful in identifying the collagen types. Gelatinases play a degradative role by acting specifically on type IV collagen.

The stromelysins are connective tissue metalloproteases that display broad ability to cleave extracellular matrix proteins but are unable to cleave the triple helical fibrillar collagens. They are activated during autolysis of skin and effect deterioration. Cathepsin B is an intracellular cysteine protease, having similar action as chymotrypsin in cleaving the amino terminal nonhelical peptide of collagen (*Refer* to the mechanism of action of chymotrypsin in Chapter 17). The pH optimum of the enzyme against collagen in solution is found to be around 4.5.

Elastases are proteolytic enzymes exhibiting specificity on elastin and its fibers. The elastic network is essential for imparting elasticity and tensile strength to the skin, and degradation of the elastin accounts for loss in physical and chemical properties. Elastases are metalloenzymes and they function efficiently in the presence of metal elements. They are activated during autolysis of skin connective tissue or secreted by skin, inhabiting bacteria such as *Pseudomonas* sp. and *Bacillus* sp. Elastase from *Pseudomonas aeruginosa* has collagenolytic activity against type I and III (Heck *et al.,* 1986).

Several proteases degrade proteoglycans, which are interfibrillar cementing substances of collagen fibrils. Collagen fibrils are more vulnerable to attack by collagenases once they are exposed after the degradation of proteoglycans.

Lipases secreted by microflora act on the fat bodies present in the flesh side of the skin. They modify the surface of flesh side by degradation, resulting in penetration of other microbial degradative enzymes such as proteases and carbohydrases. Carbohydrases, especially amylase, degrade glycoproteins and mucoid substances of skin.

LONG- AND SHORT-TERM PRESERVATION METHODS

The aim of preservation techniques is to denature the enzymes having degradative action or to render them non-functional so that no deterioration of raw stock takes place. To retain the raw stock quality over a long period, i.e., several months, long- term preservation methods such as salting and drying are adopted depending on the storage conditions. Preservation for a period of one week to a month without deterioration is usually done for the purpose of transporting or exporting the raw stock. The conditions favouring putrefaction are moisture content, pH, temperature, nature of microflora, type of enzymes secreted by microbes, microbes and enzymes favouring autolysis, etc. There are several options available for effective preservation depending on the need, duration of preservation, raw stock quality, etc. (Covington, 2009).

CURING OF HIDES AND SKINS

Hides and skins that are obtained after flaying are called green or fresh hides and they contain 70–80% water content, which is ideal for promoting bacterial and fungal growth as well as their proliferation. Even low moisture content, i.e., 25% favours the growth of spore-forming bacteria and fungi that are adapted to dry conditions. The spores produced by them are capable of surviving for months to years and germinate once the optimal conditions return. The reduction of moisture is an important feature of preservation methods and it is generally done by salting and drying of skins. Besides, temperature of above 25°C is optimum for microbial growth though growth still occurs at 4°C at a reduced rate.

Microbes need optimal conditions for growth in terms of pH, temperature and moisture content and in the absence of chemical preservatives such as microbicides, bacterial/fungal static agents, enzyme inhibitors, etc. Optimal conditions for growth vary among microbes. Once growth is established on the skins, bacteria enter into exponential growth phase and multiply using the degraded constituents of skins by secreting extracellular enzymes. This process of bacterial multiplication and enzyme mediated putrefaction of skins continue until there is a depletion of nutrients. It is expected that during the course of putrefaction, the pH of the pelt could be reduced due to accumulation of fermented products and this could pave way for establishment of moulds and fungi to start another round of degradation. Then, the microbes enter into stationary phase and by this time, the skin is nearing complete degradation. The degradation end products are characteristic of the nature and type of microflora performing the putrefaction.

Bacteria and moulds that operate on a fresh skin include pathogenic, halophilic and non-pathogenic spore producers (Table 14.1).

Table 14.1 Bacteria and moulds that operate on a fresh or unpreserved skin

Pathogenic bacteria	*Staphylococcus aureus*
	Staphylococcus pyogenes
	Serratia marcescens
	Serratia salinaria
	Serratia cutirubra
	Alcaligenes marshalli
	Alcaligenes odorans
	Pseudomonas aeruginosa
	Moraxella sp.
Halophilic and halotolerant bacteria	*Sarcina lutea*
	Sarcina litoralis
Bacteria producing 'Red Heat' effect	*Serratia marcescens*
	Myxococcus rubescens
Gram-positive, spore-forming bacteria (Pathogenic and non-pathogenic *Bacillus* sp.)	*Bacillus cereus*
	B.megaterium
	B.anthracis
	B.subtilis
	B.sphaericus
	B.laterosporus
	B.pantothenticus
	B.licheniformis
	B.coagulans
	B.pumilis
Other bacterial species	*Micrococcus lutes*
	Micrococcus roseus
	Rhodococcus sp.
	Trichophyton purpureum
	Trichophyton violaceum
	Trichophyton sulfureum
Moulds and yeasts	*Penicillium* sp.
	Aspergillus sp.
	Mucor sp.
	Candida sp.
	Cryptococcus sp.

The above-said microbes, except fungi, secrete enzymes that are functionally active in the pH range of 7.0-8.5 and these enzymes diffuse through both the grain and flesh sides of the skins. Subsequently, they attack the collagen-rich reticular network, the interfibrillar protein and collagen fibrils. This leads to a greater or lesser degree of degradation of pelts. It is also important to note that many of the bacterial and fungal organisms are highly pathogenic to humans, and can initiate a potential infection in humans. Hence, choosing a right curing procedure is essential for not only keeping the skins microbe-free but also to ensure safety in the working environment.

Altering the pH of the pelt may be a viable option to reduce the rate of degradation of skins considerably. However, acidification can result in acid swelling in the absence of electrolytes. Besides, pH reduction may also favour the growth of fungal organisms and moulds (Table 14.1). Moulds can grow on unpreserved hides or skins and on leather, especially if the pH value falls below 7.0.

The observation of bacteria within the skin is the clear sign of skin status and quality (Birbir and Ilgaz, 1996). Usually, it is considered normal if there are microbes on the grain and flesh sides because these microbes are chiefly obtained from the environment and their growth can be controlled by choosing a right curing procedure. However, it is not a good sign to observe bacteria within the corium. If they reach the grain–corium junction, grain damage as well as loss of hide substance in the final leather is inevitable since the constituents are subjected to attack by proteolytic enzymes secreted by bacteria (Covington, 2009). The quality and state of pelts or the extent of degradation can be determined by quantifying the nitrogen content, uronic acids, free fatty acid content, collagen content, etc. Qualitative assessment can be performed by viewing microscopically the surface and cross sections of the skins. Special staining techniques will be useful for examining the individual contents such as collagen, elastin, reticulin, hyaluronic acid and proteoglycans of pelts (Sivasubramanian *et al.*, 2008).

pH CONTROL

As the flayed skins are rich in collagen and other structural proteins, they are readily subjected to putrefaction by proteases secreted by bacteria. Different proteolytic enzymes hydrolyse protein at different pH values within the range of 3.0 to 10.0. Raw skins can be temporarily preserved at low pH value in the pickled condition for transportation and storage purposes. After removal of hair and flesh, the pelts are brought to a very low pH of 1.5–2.5 by treating the stock with 1–2% H_2SO_4 and 10–14% NaCl. This operation is known as "pickling". As a means of bacterial control, the method is very effective and the predominant commercial use of this is in dewoolled sheep skins and not for cattle hides. This mode of preservation tends to impart good softness to the final leather and this could only be considered for specific types of leather such as garment, suede, etc.

Pickled pelts are susceptible to fungal attack due to decrease in pH, and this can be prevented by using fungicides. Besides, very-long-term storage will significantly weaken the leather through acid hydrolysis, which starts in the pickled pelt if the temperature is raised beyond 25°C. The concentration of salt is relatively low compared to wet salting, but it is sufficient enough to resist acid swelling.

A salt-free pickling preserving system using acetic acid and sodium sulphite has been reported for short-term preservation. This method relies on acidity (pH < 4.0) to reduce bacterial activity as well as the biocidal action of sulphur dioxide to sterilize the hide (Bailey and Hopkins, 1977).

Alkali treatment is not widely practised, because at pH 10.0 and above, there is a risk of hair immunization and it adversely affects hair removal. Some alkaline additives such as sodium carbonate, sodium hydroxide, etc. for salt are known, and it is common for soaking to be conducted at pH 9–10, not only to aid partial rehydration, but also to reduce bacterial and enzyme activity. Moreover, controlling the operation is difficult since the method increases the rate of alkali-catalysed hydrolysis of collagen and hence it is not popular.

NON-SALT CURING METHODS

Kanagaraj and Chandra Babu (2002) have extensively reviewed the different alternatives to salt-curing techniques.

Drying

Drying is the conventional way of preserving skins and hides and it has become less common because the method is generally uncontrollable. Completely dried skins have very less moisture and will not facilitate microbial growth and enzyme activity. However, spore-forming microbes are retained in dormant stage and they germinate once the pelt is wet back initiating the process of deterioration. The completely dried skins are also vulnerable to putrefaction once the moisture is restored. Practically, it is easier to dry skins rather than hides as the former is less in thickness when compared to the latter.

Methods of drying include sun-, air- and chemical drying and the most primitive one is sun-drying, which is practised more commonly in some tropical countries where the climate is conducive to drying rather than in temperate regions. It is substituted by salting, which is reliable and convenient to perform. By sun-drying, the moisture content of the pelts is reduced to about 10%, depending on the relative humidity of the atmospheric air. The drying cannot be done under strong sunlight because in such case, an impermeable layer is formed on the flesh side that prevents further drying. So, the raw stock is always dried under mild sun. If drying is carried out by spreading the skins on the ground with flesh side up, the dried material crumbles and becomes difficult to rehydrate resulting in inferior leather (Sarkar, 1981). Instead, the skins are dried by big and medium curers under stretched condition in frames. Sun-dried skins or flints are relatively slow in absorbing water. The difficulty which is usually met with such class of skins is the non-uniformity in soaking.

Though this practice is simple and economical, it is difficult to control. The success of the practice is limited by various parameters such as the rate of evaporation of moisture, surface characteristics, the air temperature, the surface temperature, internal structural complexity, moisture gradient or variance throughout the skin, the rate of air flow, etc. Besides, it is difficult to assess the quality of the curing method. Another demerit of this method is that the skins are susceptible to insect infestation and damage by rats. To prevent damage by insects, the dry cured hides are dipped in 0.25% of sodium arsenate for 1 min and the hides are re-dried by spreading (Sarkar, 1981).

Curing of skins by chemical drying is carried out in which dehydration is achieved by the use of ethyl alcohol followed by ethyl ether. Reducing the moisture content of the skins using acetone and preserving them using low concentrations of boric acid is helpful in preserving skins for more than a month (Suguna *et al.*, 2009). The use of laboratory desiccant, silica gel, for drying is also reported wherein silica gel along with 5% salt on skin weight, preferably with a biocidal agent facilitates reduction in moisture content from 70% to 35% in goat skins at an ambient temperature of 31°C. This method can be useful for preserving skins for two weeks with the reduction of total TDS and chlorides (Kanagaraj *et al.*, 2001). Also, use of acrylate polymer at a concentration of 3% on skin weight has been reported for drying and it is found to be effective for short-term storing of skins up to two weeks (Quadery, 1999). It is reported that a stepwise drying of raw goat skins under vacuum in the temperature range of 30-45°C results in acceptable moisture level of 20% (Rai *et al.*, 2009). The dried skins are found to have a shelf life of more than five weeks. But, the above-said methods are applicable only for short-term preservation since the effect of dehydration is partial and not complete.

The drying operation should be rapid, consistent and even throughout the skin and should not leave moist areas in the interior which will decay during storage. The hydrodynamic properties of pelts are primarily determined by interfibrillar cementing substances such as proteoglycans, glycoproteins and mucoids and during the process of drying, these water-holding components protect the collapsed fibre structure from sticking, acting as a rehydratable barrier between the collagen elements. If the process is not even, moisture can be retained in these substances resulting in deterioration.

Freeze-drying

It is a modern way of preserving most precious fur skins wherein skins are dried by freezing. Evaporation occurs in high vacuum and water is removed to obtain materials in liquid-free state. The main difference between freeze-drying and air-drying is that during air-drying, the surface of the skin is the surface of evaporation whereas in freeze-drying, evaporation occurs on the fibre surface and among fibres where the ice crystals are present. Another difference is lack of liquid motion. Goods dried in this way are very stable and after moistening, they take the shape equal to that of a fresh one. Almost no change in physical and chemical properties is observed. A main disadvantage of this method is that it is time-consuming and limited by space. It is not applicable for huge quantity of raw stock, especially hides, and is suitable only for preserving fur skins.

Microwave Drying

This method prevents denaturation of collagen and makes the skin free from bacteria as drying of skin takes place in vacuum much faster than air-drying. Moreover, no chemicals are used and hides are preserved for longer period than conventional method. It also permits drying to a point when all the parts of the hide are uniformly moist and still flexible. It also permits splitting without consuming much time and moisture equilibration. The disadvantage in this method is high cost.

Temperature Control

Skins can be preserved and stored at a temperature that lowers bacterial and enzyme activities. Many of the bacteria deteriorate the skin and they require 35°–37°C for their growth. Lowering the temperature below 20°C reduces drastically the growth and proteolytic action of bacteria.

Chilling

Hides can be chilled to 2°–4°C by blowing moist air to prevent drying of raw stock. This method is suitable for transport, and skins remain fresh up to a month (Haines, 1981). Alternatively, a transport heavy vehicle can be equipped with refrigeration facility to accommodate huge quantity of skins, and this serves dual purpose of preservation and transport within the country (Bailey, 1999a). Here, the skins are iced down immediately after flaying and placed in insulated bins for transport. This permits delays up to 4 days and extends considerably the distance these hides can be moved. Use of coolants, e.g., carbon dioxide snow (–80°C) and liquid nitrogen (–196°C) can be used to chill the pelts. But, use of liquid nitrogen is less effective as the rapid evaporation of liquid creates a gas layer between coolant and the pelt surface, reducing heat transfer out of the pelt. This is convenient for both short-term preservation for transport but involves significant capital cost for establishing the cooling system and cold-storage facility. This method of chilling and freezing has been in practice in the colder regions of the world, viz., USA and Russia (Covington, 2009).

Super Chilling with Dry Ice

Raw stock is placed on a conveyer belt, fed into the spraying unit and sprayed with dry ice. Application is through specially designed spray nozzles, and at a temperature of –35°C, the dry ice rapidly reduces the temperature throughout the skin. This avoids the wetting problems of conventional ice treatment and provides uniformity of chilling, with the objective of safe short-term preservation for a minimum of 72 h. Cold gas evolves as the dry ice evaporates and is blown on both sides of the skin for pre-chilling. This reduces the carbon dioxide spray needed for the final stage of super-chilling. The technical advantage of this method could prove to be remarkable, removing at a stroke the problem of excessive wetting, ice indentation and water run-off.

Freezing

Preserving at a temperature of –15° to –20°C completely stops the proteolytic activity. Hence, an effective method of curing is to deep-freeze the washed and cleaned skins at –20°C (Stephens, 1987). Cleaning is necessary prior to freezing the skins because dirt, dung, blood and adhered flesh could be sources of bacteria and enzymes if not removed. The skins can be stored up to one year by this method. This method has many limitations. It is too expensive and is more suitable for skins than hides. Special equipments as well as transport units and trained labour are required for transporting and handling the chilled materials respectively. Skins must be frozen quickly, to avoid the formation of large ice crystals that can disrupt the fibre structure. Storage duration is limited due to formation of ice crystals, and at –20°C, the raw materials become brittle resulting in spoilage of materials if not carefully handled. Thawing of frozen skins takes more than 48 h if the quantity is huge. Moreover, if huge quantity needs to be thawed in the float, freezing of float happens due to absorption of latent heat. The mass of skin and float then needs to be thawed evenly without inducing an adverse temperature gradient which is typically a major practical problem.

Icing

Ice crystal can be applied to raw stock for short preservation up to a week. If the crystals are small, contact area with the pelt is more and hence effective chilling is achieved. However, rapid melting of ice and consequent wetting of the raw stock cannot be prevented; this is a serious disadvantage of the method. Ice provides insulating effect for the centre of the woolskin pile but the edges are soon unprotected. Though it is relatively cheap, it is only suitable for very short-term preservation. Besides, expenditure for installing an ice plant, electricity and storage facility do not make the method attractive among tanners.

To avoid the problem of rehydration by melting ice, British tanners have developed a preservative formulation in which they add a suitable water-soluble biocide during ice formation. Melting releases the biocide, conferring protection against bacteria. The biocide used in the method is 1,3-dihydroxy-2-bromo-2-nitropropane which forms emulsion in water and gets separated during melting.

Biocides

Biocides are toxic materials or inhibitors that can prevent the bacterial, fungal and their associated enzymatic activities on skin constituents. Biocides either derived naturally or chemically, are used for short-term preservation of skins up to 1–2 weeks. The effectiveness of biocide depends on the mode of action, its concentration and storage temperature. Skins of both sides can be sprayed with a solution containing the biocide. There are two effective types of chemical treatments for bacteria: bacteriostats which limit the growth of organisms at any stage and bactericides which kill them.

Several chemicals are known to have microbicidal action and they either belong to the phenolic group or are heavy metal compounds. 2-(cyanomethylthio)benzothiazole (TCMTB), methylene bis(thiocyanate) (MBT), triazine, sodium tri- or penta-chlorophenate, sodium orthophenylphenate, naphthalene, trichlorobenzene and 1,2-benzisothiazolin-3-one (BIT) are effective microbicidal agents useful in curing the skins. Though mercury compounds and chlorinated phenols are found to be effective, they are not only toxic to humans but also detrimental to the environment and are no longer in use for this purpose in many countries. Safer but less effective substitute materials are available in the market and this includes boric acid, sodium chlorite, zinc chloride, biguanide, dithiocarbamate and 1,3-dihydroxy-2-bromo-2-nitropropane (bronopol) (Hughes, 1974; Knight and Cooke, 2003). Recently, it has been shown that cetyltrimethylammonium bromide (CTAB) at a concentration of 5% is used as a short-term preservative agent for stripped goat skins (Ganesh Babu *et al.*, 2009).

A powerful crosslinking agent, formaldehyde, that kills almost all microbes at a concentration of 0.25% can be useful as an effective preservative. The leather obtained by this preservation mode is lightly firmer. Excess formaldehyde may cause difficulty in dehairing (Sharphouse and Kinweri, 1978).

Besides toxicity, some of the biocides, viz., aldehyde and triazine, have mild tanning properties and cause difficulties in dehairing, swelling and splitting of fibres during liming. Since curing by biocides offers only a short-term preservation option, skins that are to be transported to longer distance need long-term preservation methods. In such case, the use of biocide along with salt may be a viable option. Central leather Research Institute (CLRI), Chennai, has developed a neem-oil-based method of curing of skins and this is useful for preserving skins for a considerable period of time without any detrimental effect on leather quality.

A salt-free pickling preserving system using acetic acid with sodium sulphite has shown an improvement in preservation (Bailey and Hopkins, 1977). The sulphite-acetic acid treatment rapidly reduces the microbial population on the fresh hides and maintains a low level of viable organisms after application. Sulphur dioxide is the active material in the elimination of microbial activity. In aqueous solution, sulphur dioxide is present as dissolved gas, bisulphite ion and/or sulphite ion. The proportion of each is dependent on the pH of the solution. At a high pH, sulphite ions are formed and this has no preservative effect. At a low pH, sulphite exists in dissolved gas form, which rapidly escapes from the solution. Use of acetic acid in addition to sodium sulphite results in a buffered solution of pH 4.5 and at this pH, sulphur dioxide is present in equilibrium with bisulphite ion, and sulphur dioxide is slowly released into the atmosphere. Using a closed system to store hides provides maximum effectiveness in this method of preservation.

Plant-based Antimicrobial Formulations

An eco-friendly and cost-effective curing method based on plant formulation from the leaves of *Acalypha indica* is reported as an alternative to salt curing to preserve raw stock for short-term preservation up to two weeks (Vijayalakshmi *et al.*, 2009). Crude methanol and ethanol solvent extracts of the leaves are prepared in powder form and they are applied (15% w/w based on wet weight) on the skins. Plant product can be scrapped off from skin before leather processing and it can be composted and used as manure. This study reports monitoring at different temperatures, *viz.*, 25°, 30°, 37° and 42°C and duration up to 14 days of raw stock subjected to bacterial degradation. This study interestingly reports that several bacterial strains are observed during the course of skin degradation but not the fungal strains.

Curing by Electron Beam Irradiation

Electron beam irradiation preservation produces a hide with fresh hide characteristics combined with an extended shelf life. The procedure is to rinse hides with biocide, seal them in polythene bags and irradiate with 1.4 MRad at 10 MeV, which provides protection for an indefinite period of time if done properly (Ross and Herer, 1992; Bailey *et al.*, 2001). The high-energy electrons kill all the microbes on the hide preventing them from damaging the hide. This method has been approved in USA for eliminating gram-negative bacteria. However, tanners object to this method for the following reasons: 1. Each hide has to be individually enclosed in a polythene bag to eliminate recontamination after irradiation. Once the pack is opened or damaged, the skin is subjected to decay by recontamination by microbes. 2. Irradiation treatment reduces the tensile strength up to 10% in the final leather. 3. Cost of the preservation, expensive electron irradiation equipment and its associated infrastructure and solid waste disposal problems are the other demerits of this method.

Effects of Moisture and Chilling on Irradiation

Refrigeration significantly increases the shelf life of irradiated samples compared to room temperature storage. A combination of refrigeration and irradiation provides remarkable preservation for six months even at the lowest level of irradiation tested (0.6 MRad) whereas non-irradiated samples begin to deteriorate within two weeks. Also, reducing the moisture content of hides lowers the required level of irradiation needed for effective preservation.

Gamma Irradiation

This is useful for sterilizing hides in bulk quantity while electron beam irradiation is suitable for hides individually packed in polythene bags. Gamma irradiation provides complete sterilization at a dose of about 20 kGy (kilo Gray), providing protection for 2–3 months. The radiation is capable of damaging the skin by reducing the tear strength, but apparent effect can be observed above 30 kGy (Bailey, 1999b). This method also shares the problems of electron beam irradiation method discussed earlier in this chapter.

Ozone for Short-term Preservation

Ozone is known to be a strong oxidant as well as a disinfecting agent. Not much information is available on the effect of ozone for preservation of raw skins and hides. In this study, samples of fresh goat skins are treated with ozone for 0.5, 1.0, 1.5 and 2.0 hours and the microbial activity is monitored for a period of two weeks. Experimental skins are exposed to ozone gas at 2 g/h in a specially designed glass column reactor. Two samples are exposed to ozone; while one sample is kept in open atmosphere, the other is kept in a sealed bag at room temperature. In addition, a conventional salting method has also been followed. The results show that ozone treatment for 30 min. could effectively eliminate the microorganisms present in the raw skins. The ozonized samples kept in the open environment are well-preserved while those kept in the sealed bag are putrified within a duration of two days. Ozone-treated skins kept in the open environment are also shown to be resistant to microbial action for more than two weeks. Ozone technology has more potential for short-term preservation of skins (Sivakumar *et al.*, 2010).

SALTING

Globally, salting is the most common and conventional method of preserving hides and skins and the final cured materials are known as wet salted hides and skins. Bacteria and their enzymes require more than 25% of moisture content to start putrefaction. Since raw stock contains 60–70% moisture, these materials undergo putrefaction readily unless cured properly. Salt curing preserves hides by a combination of removal of water from the hide and lowering the water activity of the remaining moisture. Water activity (a_w) is a measure of availability of water to microorganisms in a particular environment for their growth and metabolic activities. Generally, bacteria and fungi require an optimum a_w of above 0.8 and 0.6 respectively (Frazier and Westhoff, 2008). Except halophiles, salt curing effectively inactivates microflora by removing water from microbes present on hide. It is important to note that salt is a bacteristat rather than bactericide because salt does not kill bacteria but reduces their activities. The bacteria can be reactivated when the pelt is rehydrated.

Salt influences the fibre structure by its dehydration effect by improving the leather quality. Salted and fresh raw stock have similar physical properties and area yield of the leather except for the coarser break in fresh raw stock. This coarser break in fresh pelts is due to finely split fibre structure resulting in more opening up and makes the area loose. The mechanism is unknown (Haines, 1981).

Wet Salting

Wet and dry salting are the two methods of salt curing widely practised in the trade. The term "wet salting" refers to the application of dry sodium chloride to the freshly flayed skin or hide, wet with the natural moisture content. Satisfactory curing depends on the amount of salt used, salt quality, grain size and even distribution of salt throughout the skin. Rock or sea salt can be used for curing and the salt grains should neither be too fine nor too coarse. If the salt grains are big, the curing is not proper due to slow solubility of salt. On the other hand, fine salt dissolves rapidly and the resultant brine flows out without curing the hide. The International Council of Tanners has recommended that the ideal size of salt grain is 2 to 3 mm; but for cattle hides, a mixture containing not more than 25% of fine grain salt (< 0.5 mm) and about 50% of medium grain (1–2 mm) has been suggested.

It is important to remember that the flayed hides are required to be cooled down to atmospheric temperature since, if curing is done before dissipation of body heat, bacterial action on the hide will be triggered. Also, premature salting of the skins before sufficient cooling will result in bacterial attack, particularly in salted and stacked skins, because more heat is generated in stacking.

The presence of the adhering flesh layer prevents penetration of curing agents into the skins making the curing less effective. Hence, it is ideal that the flesh is removed before curing. Compared to grain side, the well fleshed skins contain more amount of collagen but they are less compact and less dense on the flesh side. So, flesh side is given preference for salt diffusion during salt curing processes.

Dung, dirt and blood adhere to the flayed skins promoting bacterial growth. These skins should be washed clean with salt water and the excess water is allowed to drain off before proper curing. Dehydration can also be induced by wet salting which is performed by means of a homogeneous mixture of salt and chemicals (100 parts) to which antiseptic products (1 part) such as naphthalene, sodium carbonate and boric acid are added to prevent microbial damage to the skin. The quantity of the mixture is about 30% on the basis of green weight of hide. This mixture is evenly distributed over the flesh side, and often folded into quarters, with additional salt in between each layer and on top of the pile. Generally, hides are treated with 30% salt (on their weight), while skins are treated with about 40–50% of salt (on skin weight), because skins contain high moisture content. No area should be left exposed without salt. Another piece of hide is then spread on it with flesh side up and salting and rubbing continued as before. The process is continued till the height of the pile reaches 5' ft. The piling helps to squeeze brine out of the pelts, leaving 40–50% moisture. The equilibrium of salt distribution throughout the entire cross section is reached between 24 to 48 h (Cooper, 1973). If the hides are removed from the salt before curing is complete, those areas of the hide will receive a poor cure. Over time, these areas begin to deteriorate, while properly cured areas are well-preserved. Tanneries handling huge quantity of materials adopt automated approaches in salt curing in which the skin is presented by conveyor to a curtain of fluidized solid salt. By this wet salted "stacking" method, skins can be preserved up to 1 month and care must be taken to ensure that storage conditions should not favour rapid moisture absorption of skins.

Part of the applied salt dissolves in the water present on the surface of the flesh side and draws further quantity of moisture from the hide due to osmotic effect of strong salt solution and salt gradually penetrates into the hide due to diffusion. The removal of water reduces the viability of bacteria and the high concentration of salt within the skin creates an osmotic effect at the cell wall of bacteria, reducing their viability further. It is necessary to achieve at least 90% saturation (i.e., about 30% w/v in the moisture content in the skin, about 13% on hide weight), within the skin, to ensure there is no microbial activity. Since 20 to 25% moisture content of hide is drawn by the salt and gets drained out from the pile in the form of saturated salt solution, excess amount is used always in the practice of salt curing. The curing floor is never made flat; instead, slight inclination is provided for satisfactory drainage of the brine solution.

Kanagaraj *et al.* (2001) have developed a short-term preservation technique using silica gel (15%) in place of salt as a curing agent alone and in combination with a biocide such as *p*-chlorometacresol (PCMC). The proportion of silica gel (10%) and PCMC (0.1%) has been studied at an ambient temperature of 31°C for two weeks. The result of this study show that this technique is as efficient as salt curing for the preservation of skins, and it does not pose any problem either in soaking or in the leather manufacturing process.

Another attempt has been made to evolve a less–salt (5% salt with sodium metabisulphite [SMBS]) and salt-less (SMBS only) curing system for pollution abatement

(Kanagaraj *et al.*, 2005). SMBS is shown to effectively preserve the raw goat skin without affecting the collagen matrix. In addition, SMBS (1.0–1.5% w/w) could be used even at temperatures in the range of 35°–45°C without affecting the quality of skins.

Drum Curing

Like the concrete mixer, the specially designed drum with spiral steel shelf inside can accommodate 50 to 200 fleshed green hides. The skins or hides are tumbled effectively in this closed drum with 25–50% salt and an additive (SMBS or bactericides) for 4–6 h duration—4 h for skins and 6 h for hides. Rotation speed must be kept low to avoid heat generation (Covington, 2009). A drum of 11' ft height and 8' ft width can cure 15000 lbs of hides at a time with 30 H.P energy. The operation is monitored by measuring the brine solution as the curing proceeds until 90% saturation is achieved. Though less common and limited to big tanneries, this wet salting drum curing method has several advantages: More saturation of pelt moisture in short duration, i.e., 4 h is achieved in this method than in stacking, which takes almost 48 h. Either fine or coarse salt can be used and the amount of salt solution left in the drum after curing is less. The brine can be reused if it is boiled, filtered and sun-dried.

Dry Salting

Though dry salting is not commonly practised as a method of raw stock curing, in tropical countries where salt is cheap and plentiful, this method is still used. Dry salting is a combination of any type of wet salting and sun-drying, wherein the moisture content is typically reduced to about 25%. In India, dry salting of hides and skins is carried out mostly with Khari salt which is not hygroscopic and does not absorb moisture even during monsoon season as it contains sodium sulphate as the main constituent. Skins are dry salted with Khari salt and preserved normally for a period of 6–12 months. This method has the advantages of reduced weight for transporting and the preserved raw stock is easier to rehydrate than rawstock that has been subjected to only drying. However, it is reported that common use of impure salt could produce 'red heat', a red pigmentation on the flesh side due to the presence of halophilic bacteria that produce rhodopsin pigment (Bailey and Birbir, 1996) and drying may lead to rancid fat and alkaline hydrolysis. Several halophilic and halotolerant bacteria degrade skins by secreting extracellular hydrolytic enzymes such as proteases, lipases and amylases (Sanchez *et al.*, 2003).

Types of Salt

Salt is so inexpensive to produce; but the major cost is the transportation which often exceeds the manufacturing cost of the salt itself. As mentioned already, the success of salt curing primarily depends on the characteristics of the salt used. The quality and the constituents vary widely among different salts that will markedly influence the preservation and properties of the final leather.

Regardless of the type of salt used, there are limits to the impurity content of the salt, viz., calcium (0.1%), magnesium (0.1%) and iron (0.01%). Both Ca^{2+} and Mg^{2+} can react with phosphate and carbonate in the pelt, fusing the fibre structure and creating a fault known as "hard spot". Common salt is not so common and there are about 120 different grades of salt available. All of these are derived from three types of salt, viz., marine salt, rock salt and vacuum salt.

Marine salt   This grade of salt is prepared by evaporating sea water, a common source of salt for many purposes; consequently, it is highly contaminated with potassium, calcium and magnesium salts that serve as nutrients for certain halophilic bacteria, particularly the red and violet hide-staining varieties (Bailey and Birbir, 1996). It also contains chlorides and sulphates of calcium and magnesium that make the salt hygroscopic and consequently interfere with the dehydration, resulting in a poor curing of hide. Sea salt prepared for the food industry is sterilized by the boiling and evaporating processes. But, the salt destined for industrial applications is evaporated by solar action and it is likely to be contaminated with halophilic bacteria, which require saturated salt solution of >4 M or 20% for their growth (Birbir *et al.*, 1996). Hence, marine salt is strictly unsuitable for preserving raw stock.

Rock salt   This is the ideal salt suitable for curing hides. This grade of salt is mined and crushed to the required crystal dimensions. There may be appreciable concentrations of calcium and magnesium salts. Rock salt is available in the form of coarse, medium and fine crystals.

Vacuum salt   This is of high purity (99.9% NaCl) and therefore is a relatively high-cost product, most commonly used for making brine solution. It is characterized by its small crystal size, < 1 mm, and free-flowing nature. The crystal size makes the vacuum salt unsuitable for preservation because in the presence of moisture, the crystals fuse and grow. The effect tends to reduce the surface free energy. As a result of this mechanism the salt can create a fused mass, thus reducing the area of contact between the salt and the pelt surface, resulting in the slow-down of the preserving process and the risk of bacterial growth.

Granular salt   Granular salt also has high purity, but the crystal size is larger, 2-4 mm, and it is cheaper than vacuum salt. The crystal size means that the rate of crystal growth is slower and contact with the surface is maintained. The main use is for raw stock preservation.

Khari salt   The salt is formed by soil salts dissolving in the rain water and migrating to the surface of the earth, where it is formed as a crust after the water has evaporated. This material is very impure and cannot be considered as a technical grade of salt.

It contains variable amounts of sodium, magnesium, chloride, sulphate, bicarbonate and carbonate ions, typically a mixture of sodium and magnesium sulphates. It is also contaminated with halophilic bacteria and is generally unfit for curing the hides. Khari-salt-cured hides always give a dirty look due to the presence of earthly matter in it. (*See* 'Dry salting' discussed earlier in this chapter for more information.)

Demerits of Salt Curing

Wet or dry salt curing has several disadvantages that are listed below:

1. High percentage of salt is required for curing and this excess salt increases the pollution problems when they are discharged into the environment. Dissolved solids are the most difficult-to-remove pollutants in the effluent.

2. The excess salt used increases the transport cost. The major cost of purchasing salt is the transportation which often exceeds the cost of the salt itself.

3. Salt is dirty and reuse is counterproductive.

4. Even reuse is possible by boiling, filtration and reverse osmosis; but, the methods are very expensive and scale-up of these methods is not economically feasible.

5. If landfill approaches are adopted for solid or sludge disposal, land contamination occurs. Ground water level and quality of ground water will come down due to contamination. Contaminated land becomes infertile for plant growth since the sodium ions will strip out the mineral micronutrients.

6. If unused excess salt is discharged into sea, marine life will be disrupted.

Therefore, waste salt is a material that has truly no value. Regardless of all the other perceived environmental issues, this is the biggest problem facing the global leather industry. Hence, it is necessary to look for alternative curing methods instead of using salt.

USE OF ALTERNATIVE OSMOLYTES IN CURING

Hides can also be cured with potassium chloride (KCl) instead of sodium chloride since both the salts share physical and chemical properties among themselves (Bailey, 1995; Bailey and Gosselin, 1996). KCl exhibits osmotic pressure as that of NaCl, and both can be used as effective agents for curing (Bailey and Joseph, 1996). The leathers produced from both curing methods have similar qualitative, physical and chemical properties. However, the main disadvantage of using NaCl is that the disposal of sodium ions present in NaCl into the environment has negative effects on plant growth. In contrast, potassium ions can serve as nutrients for plant growth, and plants readily take them up from the soil. Hence, the use of KCl for curing will not only solve the environmental problem but also allow the effluent waste stream to be disposed of to agricultural or horticultural land. The main disadvantages of using KCl against NaCl as a curing agent are: KCl has low solubility in ambient or low temperature as compared to NaCl. Also, its solubility is temperature-dependent, whereas solubility of the latter is not dependent on temperature. The use of KCl at 4 M does not affect the growth of halotolerant bacteria and growth effects are almost similar to that of NaCl (Birbir and Bailey, 2000; Vreeland *et al.*, 1998). High cost of KCl compared to NaCl makes the curing option economically not feasible. Recently, a curing method using sodium silicate or waterglass is reported as a substitute for salt, wherein sodium silicate solution is neutralized with sulphuric and formic acids and dried. The resultant soak liquor can be used for plant growth without any adverse effects (Münz, 2007).

EVALUATION AND GRADING OF RAW OR CURED HIDES AND SKINS

Hides and skins, being natural biomaterials, vary in quality from animal to animal and even from one location to another of the same skin or hide. Attempts to correlate finished leather quality in terms of extent of microbial degradation, histological characteristics, chemical composition and physical properties of raw or cured skins and hides are not entirely successful. The other quantitative approaches taken into consideration to determine the quality of raw stock are hydroxyproline and collagen content, proteolytic activity and estimation of microbial contamination.

The traditional method of assessing hides and skins is by visual inspection and sometimes with the help of a grading box. There are four grades viz. primes, seconds, rejections and double rejections. The factors taken into consideration for grading are quality of hair, skin substance and defects such as warbles, sores, ulcers, scabs and hair slips, etc. These are well-recognized specifications for different grades, which the exporters follow. Standards of grading vary among different exporters. Skins free from all defects are classified as primes or first quality. Those having one or two minor defects are classified as seconds and rest as rejections and double rejection.

BIBLIOGRAPHY

Bailey, D.G. and Hopkins, W.J. (1977). "Cattle hide preservation with sodium sulfite and acetic acid." *J. Am. Leather Chem. Assoc.* 72, 334.

Bailey, D.G. (1995). "Preservation of cattle hides with potassium chloride." *J. Am. Leather Chem. Assoc.* 90 (1), 13–21.

Bailey, D.G. and Joseph, A.G. (1996). "The preservation of animal hides and skins with potassium chloride." *J. Am. Leather Chem. Assoc.* 91, 317–333.

Bailey, D.G and Birbir, M. (1996). "The impact of halophilic organisms on the grain quality of brine-cured cattle hides." *J. Am. Leather Chem. Assoc.* 91, 47–51.

Birbir, M., Kallenberger, W., Ilgaz, A. and Bailey, D.G. (1996). "Halophilic bacteria isolated from brine cured cattle hides." *J. Soc. Leather Technol. Chem.* 80, 87–90.

Bailey, D.G. and Gosselin, J. (1996). "The preservation of animal hides with potassium chloride." *J. Am. Leather Chem. Assoc.*, 91, 317–333.

Bailey, D.G. (1999a). "Preservation of hides and skins." In: Leafe, M.K. (ed.). *Leather Technologists Pocket Book.* The Society of Leather Technologist and Chemists, East Yorkshire, England, pp. 5–21.

Bailey, D.G. (1999b). "Gamma radiation preservation of cattle hides. A new twist on an old story." *J. Am. Leather Chem. Assoc.*, 94, 259–267.

Birbir, M. and Bailey, D.G. (2000). "Controlling the growth of extremely halophilic bacteria on brine-cured cattle hides." *J. Soc. Leather Technol. Chem.* 84, 201–204.

Bailey, D.G., DiMaio, G.L., Gehring, A.G and Ross, G.D. (2001). "Electron beam irradiation preservation of cattle hides in a commercial-scale demonstration." *J. Am. Leather Chem. Assoc.* 96, 382–392.

Birbir, M. and Ilgaz, A. (1995). "Isolation and identification of bacteria adversely affecting hide and leather quality." *J. Soc. Leather Technol. Chem.*, 80. 147–153.

Birbir, M.and Ilgaz, A. (1996). "Isolation and identification of bacteria adversely affecting hide and leather quality." *J. Soc. Leather Technol. Chem.*, 80. 147–153.

Cooper, D.R. (1973). "A New Look at Curing." *J. Soc. Leather Technol. Chem.* 57, 19–25.

Cooper, D.R. and Galloway, A.C. (1973). "The quality and yield of leather from machine fleshed hides." *J. Soc. Leather Technol. Chem.* 57, 147–151.

Galloway, A.C. and D.R. Cooper. (1974). "Fungicides for treating wet blue hides". *J. Soc. Leather Tech. Chem.* 58–69.

Covington, A.D. (2009). Curing and preservation of hides and skins. In: *Tanning chemistry: The science of leather.* RSC Publishing, Cambridge, UK, pp. 72–93.

Daniels, R. (1997). "Overview: Avoiding salts." *World Leather* 10, 41.

Frazier, W.C and Westhoff, D.C. (2008). *Food Microbiology*, 4th edn. Tata McGraw Hill Companies Inc., New York.

Ganesh Babu, T., Nithyanand, P., Chandra Babu, N.K. and Karutha Pandian, S. (2009). "Evaluation of Cetyltrimethylammonium Bromide as a potential short term preservative agent for stripped goat skin". *World J. Microbiol. Biotechnol.*, 25, 901–907.

Haines, B.M. (1981). "Conservation of cattle hides by freezing". *J. Soc. Leather Technol. Chem.* 65, 41–54.

Harper, E. (1980). "Collagenases". *Annu. Rev. Biochem.* 49: 1063–1078.

Harrington, D.J. (1996). "Bacterial collagenases and collagen-degrading enzymes and their potential role in human disease". *Infect. Immun.* 64, 1885–1891.

Heck, L.W., Morihara, K., McRae, W.B. and Miller, E.J. (1986). "Specific cleavage of human type III and IV collagens by *Pseudomonas aeruginosa* elastase." *Infect. Immun.* 51:115–118.

Hughes, R. (1974). "Temporary preservation of hides using boric acid". *J. Soc. Leather Technol. Chem.* 58, 100–103.

Kanagaraj, J., Chandrababu, N.K., Sadulla, S., Rajkumar, G.S., Visalakshi, V. and Chandrakumar, N. (2001). "Cleaner techniques for the preservation of raw goat skins." *J. Cleaner Production*, 9, 261–268.

Kanagaraj, J. and Chandrababu, N.K. (2002). "Alternatives to salt curing techniques – a review." *Journal of Scientific and Industrial Research.* 61, 339–348.

Kanagaraj, J., Sastry, T.P., and Rose, C. (2005). "Effective preservation of raw goat skins for the reduction of total dissolved solids." *Journal of Cleaner Production.* 13, 959–964.

Knight, D.J and Cooke, M. (2003). *The Biocides Business.* Wiley-VCH.

Molla, A., Matsumoto, K., Oyamada, I., Katsuki, T. and Maeda, H. (1986). "Degradation of protease inhibitors, immunoglobulins, and other serum proteins by Serratia protease and its toxicity to fibroblasts in culture." *Infect. Immun.* 53: 522–529.

Münz, K.H. (2007). "Silicates for raw hide curing." *J. Am. Leather Chem. Assoc.* 102(1), 16–21.

Quadery, A.H. (1999). M.Sc. Thesis, University College of Northampton.

Rai, C.L., Surianarayanan, M., Sivasamy, A. and Chandrasekaran, F. (2009). "Pollution prevention by salt free preservation of raw goatskins through stepwise drying." *Int. J. Env. Eng.* 1: 295–305.

Rajamani, S. (1998). Cleaner tanning technologies, in UNIDO Report, 2: 9–18.

Ross, G.D and Herer, A.S. (1992). "Treatment of raw animal hides and skins." United States Patent No. 5096553.

Sanchez, P.C., Martin, S., Mellado, E. and Ventosa, A. (2003). "Diversity of moderately halophilic bacteria producing extracellular hydrolytic enzymes." *J. Appl. Microbiol.*, 94: 295–300.

Sarkar, K.T. (1981). "Hides and skins." In: *Theory and Practice of Leather Manufacture.* Macmillan India Press, Chennai.

Sharphouse, J.H. and Kinweri, G. (1978). "Formaldehyde - preservation of raw hides and skins". *J. Soc. Leather Technol. Chem.* 62, 119–123.

Sivakumar, V., Balakrishnan, P.A., Muralidharan, C. and Swaminathan, G. (2010). "Use of ozone as a disinfectant for raw animal skins – application as short term preservation in leather making." *Ozone Sci. Eng.* 32, 449–455.

Sivasubramanian, S., Manohar, B.M. and Puvanakrishnan, R. (2008). "Mechanism of enzymatic dehairing of skins using a bacterial alkaline protease." *Chemosphere.* 70, 1025–1034.

Stephens, L.J. (1987). "The effect of preservation by freezing on the strength of kangaroo leathers". *J. Am. Leather Chem. Assoc.* 82, 45–49.

Suguna, L., Rathinasamy, V., Iyappan, K., Chandrababu, N.K. and Mandal, A.B. (2009). "Alternatives to Salt Preservation: Preservation by Reducing Moisture Content". *Asian J Water, Environment and Pollution.* 6: 119–122.

Vijayalakshmi, K., Judith, R. and Rajakumar, S. (2009). "Novel plant based formulations for short term preservation of animal skins". *J. Sci. Ind. Res.* 68, 699–707.

Vreeland, R.H., Angelini, S. and Bailey, D.G. (1998). "Anatomy of halophile induced damage to brine cured cattle hides." *J. Am. Leather Chem. Assoc.* 93, 121–131.

15

ENZYMES IN SOAKING

INTRODUCTION

Soaking is the first operation in the tannery in which the fresh or preserved raw hides and skins are treated with water for making them clean and soft. Raw stock of hides and skins are received in the tannery in one of the following four conditions. (i) fresh immediately after flaying (ii) wet salted (iii) dry salted and (iv) dried. The restoration of moisture lost due to curing and preservation and the removal of extraneous matter are of prime importance in soaking, where water rehydrates the dried inter-fibrillary protein and loosens the cementing substance of the fibres. The collagen fibres of the dermal connective tissue and keratinous layer of epidermis also take up water to become more flaccid and flexible. The duration of the soaking period, the number of changes of water and the use of antiseptics are all dependent on the thickness of the hides and skins, regional temperature, packing of collagen fibres, viz., loose and dense, curing methods and often greasiness of the skins (Puvanakrishnan and Dhar, 1988).

Dry hides absorb water very slowly and consequently, much time is required to make them absorb the required amount of water to be soft. If, however, some amount of sodium hydroxide or sodium sulphide is added to the soak water, the hides and can absorb water more quickly. Usually, 1–2 parts of NaOH and 1–1.5 parts of Na_2S are used per 1000 parts of water to make soak liquor for dry and dry salted hides and skins. Usually, skins and hides are soaked for about 12 to 24 h and for thick hides, it can be extended to 36 h using preservatives to prevent deterioration of skins and hides by microbes. After the 36 h period of soaking, if they are not soft enough, they can be made soft by "breaking over the beam" or "dry drumming' methods. In soaking dry hides, it should be borne in mind that they should be made almost as soft as green/fresh hides. When hides are dried, the fibres become hard and horny. By soaking, sufficient water–should be introduced into the fibres and into the interfibrillary cementing substance to make them soft and loosen the collagen fibrous structure by removing water soluble proteins and predominant cementing substances such as hyaluronic acid (HA), a non-sulphated glycosaminoglycan in the skin (Cantera, 2001b; Sivasubramanian *et al.*, 2008a). Besides, this process eliminates salt that is used for curing skins and hides. The replenished water serves as a vehicle for penetration of the chemicals employed subsequently to effect hair loosening, plumping and alkaline action.

In addition, soaking removes blood, manure and urine which are ideal nutrients for bacterial growth and which, along with certain salts in soaks, may cause dark stain to the finished leathers. Sand, stones and parasites are also removed, thereby avoiding the possibility of damage to machines used in subsequent operations. It has been observed that curing salts, if not removed in soaking but transferred to the lime liquors, have been found to reduce the alkaline plumping action and also the dehairing rate. Flesh and fatty adipose tissue can be easily removed mechanically from the skins only after proper soaking and this is called green fleshing. Presence of stains and damage to skin cause uneven penetration of chemicals used in the subsequent operations and this results in reduction in the quality and value of leather product.

No single standard procedure for washing and soaking is followed, and tanneries in different parts of the world adopt different procedures. The factors that require considerations are the type and nature of raw material such as source of animal, water content of the raw stock, substances that are to be dissolved in the process, etc., the method of curing or preservation, soak water characteristics such as concentration of salt in soak water, pH, rate of diffusion of substances, etc., float-to-pelt ratio, effect of antiseptics, duration, temperature prevailing in the region and labour associated with the immediate and subsequent operations.

SOAKING OF RAW HIDES AND SKINS

Freshly flayed hides and skins are called green which are soft enough and therefore, they do not require regular soaking. Since green hides and skins are usually contaminated with dung, dirt, blood and urine, they need thorough washing. If the hides and skins appear to be in partially putrefied condition as judged from the hair slip or putrid smell, the time of soaking should be short. The best method is to wash the stock in a drum to clean as quickly as possible.

SOAKING OF WET SALTED HIDES AND SKINS

Wet salted hides and skins lose nearly 30–35% of natural water by curing. But, these hides retain sufficient quantity of water and hence, there is no problem in soaking back. A few hours treatment in plain water will rehydrate wet salted hides and skins quite well.

In winter, it is often desirable to put wet salted hides in a pit of water in the evening and leave them overnight. Next morning, the stock is washed in a drum with two changes of water. During summer, wet salted hides and skins dry up considerably. In such cases, it is advisable to use wetting agents to facilitate soaking and also to emulsify fats in hides and skins.

SOAKING OF DRY SALTED AND DRIED HIDES AND SKINS

The effectiveness of the soak primarily depends on the amount of water used, since salt removal is associated with the quantity of water used. During soaking of salted skins and hides, salt is effectively partitioned between the two phases between the soak liquor and the rehydrated pelt and the solubilization rate depends on the difference in concentration between them.

Dry hides and skins from the tropics require considerably longer period of soaking for complete rehydration. Since these hides and skins are dried without any salt, there may be a partial gelatinization of the fibres. Mechanical action should be avoided in the early stages of soaking since the fibres are too stiff and mechanical flexing will result in breaking of the fibres. Long soaking time (36–48 h) is necessary to wet back dry hides and skins, the water being changed after initial soak of 24 h. A disinfectant is added to the initial soak in order to allow rehydration and prolong soaking of hides and skins with less danger of bacterial attack. For proper soaking of dried hides and skins, some chemical, physical and enzymatic actions are necessary to not only rupture the cross linkages between collagen fibres but also solubilize and remove the interfibrillary cementing substances.

EFFECT OF DIFFERENT FACTORS ON SOAKING

Temperature

The temperature of the water used for washing and soaking should range between 10° and 30°C. Raising the temperature of the water aids in the dispersion of the globular proteins and accelerate the soaking process. However, soaking at higher temperatures will result in looseness of the leather, gelatinization and solubilization of collagen material and general coarseness of the fibre (Thorstensen, 1985). The ease with which dry hides and skins can be soaked depends upon the temperature at which they have been dried. Globular protein like albumin is coagulated and rendered insoluble during drying at high temperature. But, the gelatinous fibres do not coagulate and hence they are soaked back with difficulty if dried at high temperature. Hides dried at lower temperature can be softened in a shorter duration compared to the hides dried at higher temperature (Sarkar, 1981).

Formerly, flint hides were soaked by the enzymatic action of the putrefactive bacteria of old soak liquor that imparts softness and renders the hardened tissues soluble. This process is regarded as dangerous, as the action varies with the temperature and is difficult to control. Some portion of the hide may be too softened, while others may be insufficiently soaked (Sarkar, 1981).

Water

The water used in soaking should be free from organic matter and iron salts, the latter being responsible for the iron stain in leather. The organic matter promotes bacterial growth. Soft to medium hard water at 18°–20°C ensures the best working condition. The amount of water used for washing and soaking varies from 4 to 20 L per kg of stock washed. Soaking float is typically 200–300% on salted pelt weight, depending on the raw material and its state of cleanliness. Because of this variation, tanners determine the extent of washing by the clarity of the effluent rather than by an exact time control. Use of sea water having more than 25000 ppm of salinity and high total dissolved solids (TDS) for the soaking process is reported (Vedaraman and Iyappan, 2006) wherein the salt of alkali earth metals such as calcium hydroxide and magnesium hydroxide are added along with sea water. However, the major limitation associated with sea water is that it contains significant quantities of salts and minerals that affect leather processing adversely. Also, the use of untreated sea water results in poor opening up of fibre structure as it represses swelling and non-loosening of flesh adhered to the raw material, resulting in inadequate defleshing.

Generally, wet and dry salted hides consume more water in the soaking process than dried and fresh stock as they absorb more water and consequently, the process is slower in salted hides than fresh stock. Multiple soaking may be required for dry salted hides so as to periodically eliminate salt liquor as well as to impart uniform hydration of material. Multiple soaking gradually reduces the osmotic pressure inside the soaked material as the process diffuses salt out of the material resulting in plumpness. The rate of salt diffusion depends on the difference in the concentration between the bulk solution and the solution within the pelt and the difference is usually smaller in liquor containing salt than a process employing fresh water. The purpose of rehydration is to fill up the fibre structure with water, ensuring that all elements of skin are wetted evenly in order to facilitate uniform penetration of dehairing as well as other chemicals through the cross section. Improper rehydration during the soaking process does not produce leather of uniform properties thereby resulting in reduction in the quality and value of leather.

pH

For proper soaking, alkaline soak baths are recommended. Since acid soak causes difficulties in subsequent liming process, alkaline soak is widely accepted. If the pH value of the alkaline soak liquor is kept above 12.0 in the absence of any dehairing agent, part of the keratin undergoes an internal structure change and becomes resistant to alkali and to dehairing agents. This makes dehairing process more difficult. This immunization of keratin is avoided if the pH of the alkaline soak liquor is kept below 12.0. Swelling of dermis in the presence of NaOH is higher than lime since sodium collagenate is more dissociated when compared to calcium collagenate, which causes ionic imbalance resulting in more water uptake (Bienkiewicz, 1990; Thanikaivelan, 2001). However, the layers of skin swell unevenly when the rate of swelling is high, resulting in loose grain (Valeika *et al.*, 1996). The increased swelling at alkaline pH in the presence of sodium hydroxide is also attributed to the presence of a high concentration of hydroxyl ions rather than any specific effect due to the sodium ion (Bowes and Kenten, 1950).

In acid soaking, the common acids, viz., hydrochloric acid, sulphuric acid, sulphurous acid (saturated) and formic acid are generally used. Alkaline soaking agents which are commonly used are sodium hydroxide, potassium hydroxide, sodium sulphide, sodium arsenite, sodium carbonate, sodium bisulphite and borax (Dutta, 1985).

Disinfectants

Unless necessary precautions are taken during the soaking process, as a result of the microbial activities, the hides and skins may undergo various kinds of damage. Minor damages can be understood with perceptible effects, such as putrid smell and matt, and lustreless or blind sections in the grain; serious damages include hair-slip, loose grain and reduced firmness, and heavy putrefaction such as pitting holes, putrefaction marks on the grain and loosening of the grain. Further, numerous defects such as looseness, weak fibre, weak grain, pitted grain, loss of grain layer, stains, uneven dyeing, uneven buffing and spueing may occur in the salted hides because of fungal effect (Ozyaral and Birbir, 2005).

In soaking, removal of the curing salt and rehydration of the skin favour the possibility of bacterial growth. The cured hide carries with it a wide variety of bacteria, ranging from halophilic to pathogenic (causing infectious animal diseases) and may damage the hides and skins. Hides cured with clean salt do not require the addition of disinfectant in soaking. Skins preserved with salt and the soak liquor containing salt arrest microbial growth (bacteriostatic effect) due to dehydration. The liquor generated after soaking exhibits a very high salinity to the extent of 15–25% due to the presence of excess salt in the hides/skins that are to be processed. In case of fresh hides or skins, there can be no bacteriostatic effect and hence they are at greater risk of bacterial damage. Chlorinated aromatic compounds are generally used as disinfectants. Phenol is not advisable since any tanning action at this stage will give hard leather and make hair removal difficult besides causing dyeing problems. They also inhibit enzymes. The disinfectants normally used in soaking are sodium hypochlorite (oxidizing agent), sodium trichlorophenate, sodium pentachlorophenate and formic acid. In the case of multiple soak, one may consider inclusion of microbicide in every float change (Covington, 2009). The presence of fungal flora especially in the form of spores cannot be ruled out; however, they do not create any damage to the pelt as their growth rate is very low and may take days to impart damage, if any, to produce. A number of fungicides, including 2-(thiocyanomethylthio) benzothiazole, in the wide range of biocidal products now available for preventing defects of biological origin in hides and leathers, have been compared in efficacy tests, and their penetration, absorption and distribution in tanned leather has been investigated. For protection against moulds, it is possible to use a single active ingredient, but it is recommended that a combination of fungicides is used to improve performance by synergistic effects and by broadening the activity spectrum against moulds (Orlita, 2004). Defects caused by fungi that mostly appear in tanning and post-tanning processes lead to drastic economical losses, and it is important that fungi should be controlled from the initial steps of the manufacturing stages. Bactericides with two different compositions (potassium dimethyl dithiocarbamate and quarternized compounds) and a fungicide (2-thiocyanomethylthio benzothiazole-based), each at a concentration of 0.5%, are used in a study wherein bactericides are added into the main soaking float with or without fungicide (Yapici, 2008). With the use of two antimicrobial agents in the aforesaid concentration, a synergistic effect appears; whereas the number of fungi detected in the main soaking liquor decrease at a considerable level, all of the bacteria are controlled and so no bacterial growth occurred.

Ozone as a Biocide

Among the beamhouse processes, soaking process poses the most significant bacterial threat to the skins and hides. In soaking, various bactericides are used; but these chemicals, which are lethal to bacteria, may also be harmful to the environment and human health. As an alternative to these chemical-based bactericides, the possible use of ozone in the soaking process has been investigated. The effect of ozone applications for various durations has been investigated and the effect is compared to that of a sodium dimethyl dithiocarbamide-based bactericide. It is observed that 15 minutes of ozone application is found to be most effective for bacterial growth prevention (Mutlu *et al.*, 2009).

Detergents

Rehydration during soaking process is accelerated by employing chemicals having wetting effect on the fibre structure. There are three types of detergents; anionic, cationic and nonionic. The nonionic detergents such as alkyl polyethylene oxide have proved to be the most effective especially in removing triglyceride grease, which is difficult to remove in low-temperature soaking conditions since the melting point of the grease is 40°–45°C. The merit of using nonionic detergents is that they do not interact with collagen and hence, they will not interfere with the processing such as fixing and dyeing. Both the cationic and the anionic types have a tendency to get fixed to the hide protein and carried over into subsequent operations. The nonionic detergents also facilitate rehydration of hides as well as penetration of dehairing agents in subsequent processing. The use of nonionic surfactants such as Tween 80 and Triton X-100 in soaking as well as in dehairing operations has been reported (Sivasubramanian *et al.*, 2007). They break lipid–lipid and lipid–protein interactions and thereby dissociate the protein aggregates. Dissociation of protein aggregates possibly open up the collagen fibre structure besides facilitating entry of enzyme or other chemicals used in the subsequent dehairing operation. It is also reported that these nonionic surfactants enhance the protease action by disaggregating them (Bhairi, 1997). Sebaceous grease that comprises components such as sterols, sterol esters, wax esters, phospholipids etc. can be effectively removed with anionic detergents; however, they interact with collagen and affect the subsequent processes.

Additives

Some tanners add a small quantity of sodium sulphide or sodium tetrasulphide to the soak. These chemicals act on the keratin follicle and subsequently, the dehairing process starts. Sodium tetrasulphide assists in the solubilization of the soft globular proteins and helps in the removal of fine hairs. Common salts help in the removal of globular proteins and also prevent microbial growth in the soak solution. If the soaking is carried out in a drum, the liquor ratio is maintained in such a way that the salt washed out from the cured hides is sufficient for the removal of some of the soluble proteins. In case of prolonged soaking systems, the addition of salt permits soaking operation at higher temperatures that decreases the possibility of bacterial damage (Thorstensen, 1985).

Soaking of hides is improved in the presence of organic sulphur compounds at acidic pH. An improved soaking process has been described where well-washed salted calf skins are soaked for 6 h at 25°C with periodic agitation in a bath comprising water (80%), thioglycollic acid (0.07%) and sodium chloride (0.15%) at a pH between 4.2 and 5.0. The soaked hides could be limed in the same bath. Thioglycollic acid pretreatment gives a faster and more uniform dehairing during liming.

For soaking dry salted hides and subsequently, to permit better dehairing, a mixture of an organic thiosulphate, a surfactant and a bactericide in water is recommended. Hahn *et al.* (1981) have reported that dry salted hides treated in water with a soaking formulation composed of sodium (carboxy methylene) thiosulphate, commercial surfactant and benzyl trimethyl ammonium chloride have given cleaner and more easily dehaired hides after liming.

Another method has been developed for soaking using the nitrogen compound like dicyandiamide. The salted ox hides after initial washing for 2 h at 30°C are soaked for 4 h in 100% water containing dicyandiamide. The pH of the soak liquor is maintained between 9.8 and 10.5. After soaking, the hides have shown no sticking of the fibre structure. Dehairing and bating could be carried out later in the same liquor.

The effects of calcium hydroxide concentration, soaking time and temperature of soak on the contents of collagen and non-collagen proteins in shark skin and on the removal of sodium chloride from the salted skins to optimize the soaking process have been studied. Adequate and uniform hydration of the skin with minimum loss of collagen, removal of preservatives and soluble proteins and prevention of bacterial growth are achieved by soaking in liquor containing 1% calcium hydroxide and 0.5% nonionic surfactant both on skin weight basis at 15°–20°C for 4–6 h at a liquor/goods ratio of 3.0: 3.5.

Duration

Salted skins and hides usually require 6–12 h to remove enough salt to ensure proper rehydration of material, whereas dried stocks require 24–36 h or more.

ADVANTAGES OF USING ENZYMES AS SOAKING AIDS

Proteolytic enzymes, especially acid proteases, increase the rate of soaking for dried fur skins. The short soaking time and low pH prevent damage to the hair bulbs and subsequent hair loss. The advantages include loosening of the scud, initiation of the opening of the fibre structure and production of leather with less wrinkled grain when used at an alkaline pH of less than 10.5 (Taylor *et al.*, 1987).

Protease degrades non-structural proteins and enhances rehydration of skins and hides. Besides, it solubilizes and helps in the removal of the inter-fibrillary cementing substance from hides and skins. In addition to proteases, if small amounts of amylases and lipases are added, the glycoproteins as well as lipoproteins are all degraded and removed in the soaking operation itself. This facilitates the subsequent dehairing and bating operations also. It is reported that amylase can be used to remove proteoglycans through hydrolysing the carbohydrate moiety and this results in fibre opening (Thanikaivelan *et al.*, 2002).

Hyaluronic acid (HA) is an anionic polysaccharide formed by repeated combinations of the disaccharides β1-4-D-gluconic acid and β1-3-D-N-acetyl glucosamine, which does not participate covalently in proteoglycan aggregates or with proteins; in this sense, this is unique among GAGs. HA is found in skins and hides as a highly viscous gel, filling a large volume of interfibrillary space and exerting a negative influence upon substance diffusion; consequently, it is important to eliminate it during the soaking process. The elimination of HA, as well as plasma proteins (globulins and albumins), during soaking is favoured by the presence of sodium chloride and by the action of an enzyme with proteolytic activity (Sivasubramanian *et al.*, 2008b). A practically total elimination of these non-collagenous components from salted hides requires soaking for 24 h and significant mechanical action; with short soaking periods, i.e., 4 h, and limited mechanical action, the elimination of the above components is not significant; this situation can only be compensated during the dehairing process (Cantera, 2001b). The compact fibrous structure of the grain layer and the presence of highly keratinous and hydrophobic epidermal layer put at stake the elimination of HA and non-fibrous proteins; for such purposes, the use of enzymatic preparations during soaking facilitates access to the structure and consequently, the elimination of these components and the opening of the collagen fibres occur (Cantera, 2001b). Proteoglycans (PG) such as dermatan sulphate PG (decorin), heparin sulphate PG and chondroitin sulphate PG are not removed during the soaking process in the absence of enzymes as they closely interact with collagen fibrils and they are only eliminated during dehairing operations.

There is a reduction in the content of total hexosamine and uronic acid in soaked skins when compared to raw skins and this is primarily due to removal of some amount of hyaluronic acid. Though soaking of skins in water removes hyaluronic acid to some extent, sulphated PGs that form aggregates with hyaluronates in interfibrillar spaces are not removed. It is reported that hyaluronic acid can be slightly removed in soaking with water and the removal is accelerated in the presence of sodium chloride (Cantera, 2001a). (For more details, refer to 'Mechanism of dehairing' Chapter 16).

Though there are reports on the use of other enzymes such as amidases, hyaluronidases, phospholipases, chondroitinases and lignocellulases in the soaking process, their use has not gained significance due to their low specific activity, compatibility in the presence of other agents, cost, etc. (Addy *et al.*, 2001; Covington, 2009).

RESEARCHES ON ENZYMATIC SOAKING AGENTS

Enzymes from *Aspergillus parasiticus, A.flavus, A.oryzae* and *B.subtilis* have been used alone or in mixture and the resulting leathers are found to be full and supple and show no loose grain. Rokhvarger and Zubin (1971) have suggested the use of carbohydrase from the mould culture *A.awamori* in soaking. This enzyme is used at pH 4.5–5.0 at 30°C for 18–24 h which improves the wetting of hides. Subsequently, the use of rhizopine at 38°C, pH 4.0–5.0 for 8 h to soften woolskins has been advocated (Chebotereva *et al.*, 1975). This treatment effectively removes 50% of the mucopolysaccharides, separated the bundles of collagen fibres and generally softens the fur.

Christner (1992) has reported the use of microbial alkaline protease and alkaline lipase in soaking of hides along with surfactants and NaOH in the pH range of 9.0 to 11.0. Lipases hydrolyse triglyceride grease, thereby not only reducing the hydrophobicity of the fibre structure but also aiding wetting. Lipases can replace the use of anionic detergents in soaking because the latter exhibits strong interaction with collagen, which affects subsequent processes.

Non-salted and preserved sheep skins are wet back with the enzymes of *A.oryzae* in 4–5 h (Toshev and Esaulenko, 1972). Optimum conditions for extracting nitrogenous components and carbohydrates by the use of Bioferm (pancreatic amylase) and sodium bisulphite at pH 5.0 and 35°C are described and it is shown that by reducing the temperature to 25°C, the extraction of monosaccharides is reduced markedly (Esaulenko *et al.*, 1975). Botev *et al.* (1976) have shown the use of bacterial amylase for soaking dried wool lamb skins. Bacterial amylase has strong amylolytic activity, much weaker proteolytic activity and no lipolytic activity. At a concentration of 0.5–0.6 g/L and pH around 5.0 in the presence of 1.5 g/L sodium hydrosulphate, the enzyme activity is increased by about 35%. Hence, bacterial amylase could be used at 1/3rd concentration necessary for amylase derived from animal sources.

The advantages of using mould proteases at pH 5.0 or lower for soaking, dehairing and bating are discussed, and soaking with pepsin and papain at a pH of 3.0–4.5 is described (Monsheimer and Pfleiderer, 1970). Subsequently, use of alkaline proteases of bacterial and fungal origin is recommended and this reduces the need for the liming chemical by 30–60% (Monsheimer and Pfleiderer, 1972; 1974). In a later patent, Monsheimer and Pfleiderer (1981c) have proposed a formula for soaking hides, skins and fur skins using a proteolytic enzyme at a pH of 10.5. After 4 h, the hides are uniformly soaked and show no sticking of the fibre structure, whereas the fur skins are also well-soaked with no evidence of hair slip.

It has been reported by Leberfinger *et al.* (1976) that smooth, clean and properly dehaired hides are obtained after a 2 h treatment of the previously fleshed hides with a mixture of enzymatic surface-active soaking agent and sodium hydrosulphide followed by a 12–13 h treatment with calcium hydroxide. Within another 10 h, the rest of the process including chrome tanning could be concluded so that a 24 h rhythm could be accomplished with preservation of quality standards.

To increase area yield and also to improve the quality, furs have been first treated with water followed by a two-step pickling process in the presence of an enzyme. The soaking and the second step of pickling have taken place in the presence of 0.1–0.3 g/L polygalacturonase while the first step of pickling is in the presence of lactic acid and sulphuric acid.

Soaking of hides in 1% solutions of protogidrolitin (a proteolytic enzyme preparation) or in actinomitset II (a lipolytic–proteolytic enzyme preparation) tends to remove the non-protein constituents, thus facilitating further processing. A typical procedure is to treat washed hides in 1.5 vol (v/w) of solution containing sodium sulphite 0.3%, surfactant 0.3% and enzyme preparation 0.5% at 36°C for 1 h, then to add 3.0% of the enzyme preparation and continue the treatment for 12 h.

According to Moiseeva *et al.* (1980), softening and unhairing of dried hides could be improved using protosubtilin G 10 (alkaline protease) in pre-soaking operation. The optimum pre-soaking conditions are: pH 11.0, temperature 50°C, enzyme 0.75% on hide weight and sodium acetate 10 g/L for 6–8 h. The pH and temperature are important factors. Raising of temperature from 30° to 50°C increases the proteolytic activity by 20%. However, pre-soaking carried out above 50°C will denature the proteins of the raw hides and the quality of the final leather will be of poor quality.

It has been shown that dried furs are soaked in an acidic aqueous bath containing 1% acid proteinases from *Rhizopus rhizopodiformis* and sodium bisulphite at 25°C for about 20 h and the soaked furs are finished in the usual way. However, exact pH of soak bath is required to be maintained to get optimum results.

Three commercial bacterial alkaline protease preparations have been tested for the soaking of salted cow hides. Use of enzyme preparations has resulted in a decrease in soaking time. The relaxation of cow hides soaked in the presence of enzymes increases with prolonged soaking time compared to virtually no changes in the relaxation of hides soaked in pure water or water containing a surfactant. All enzyme preparations are shown to increase significantly the amount of nitrogenous compounds in the liquor after soaking.

BIBLIOGRAPHY

Addy, V.L., Covington A.D., Langridge, D. and Watts, A. (2001). "Microscopy methods to study fat cells. Part 1: Characterization of ovine cutaneous lipids using microscopy." *J. Soc. Leath. Chem. Assoc.* 85, 6–15.

Bhairi S.M. (1997). A *guide to the properties and uses of detergents in biology and biochemistry*. Calbiochem-Novobiochem Corporation, San Diego.

Bienkiewicz, K.J. (1990). "Leather–water: a system." *J. Amer. Leath. Chem. Assoc.* 85: 305–325

Botev, I., Esaulenko, L. and Toshev, T. (1976). "Enzyme soak for dry unsalted lambskins with Bulgarian produced á-amylase." *Kozh-Obuvn Prom-st.* 17, 13.

Bowes, J.H and Kenten, R.H. (1950). "The swelling of collagen in alkaline solutions. Swelling in solutions of sodium hydroxide." *J. Biochem.* 46, 1–8.

Cantera, C.S. (2001 a). "Hair saving unhairing process. Part 1:Epidermis and the characteristics of bovine hair." *J. Soc. Leath. Tech. Chem.* 85, 1–5.

Cantera, C.S. (2001b). "Hair saving unhairing process. Part 3: Cementing substances and the basement membrane." *J. Soc. Leath. Tech. Chem.* 85, 93–99.

Christner, J. (1992). "Enzymatic soaking method." US Patent No. 5089414.

Chebotareva, L.G., Rokhvarger, O.D., Zubin, A.M. and Afonskaya, N.S. (1975). "Composition for softening of hides." *Kozh-Obuvn Prom-st.* 17, 26.

Covington, A.D. (2009). "Curing and preservation of hides and skins." In: *Tanning chemistry: The science of leather.* RSC Publishing, Cambridge, UK. pp. 72–93.

Dutta, S.S. (1985). *An Introduction to the Principles of Leather Manufacture.* Indian Leather Technologists Association, Calcutta, p.108.

Esaulenko, L., Toshev, T. and Papazyan, L. (1975). Change in enzymic activity of bioferm preparation." *Kozh-Obuvn Prom-st.* 16, 9.

Hahn, E., Lach, D. and Streicher, R. (1981). "Auxiliary, and process, for the soaking of hides." German Patent No. 2938078.

Leberfinger, R., Landbeck, F. and Matschkal, H. (1976). "Determination of the sulphide content in unhaired hides." *Leder.* 27, 160–165.

Moiseeva, L.V., Shestakova, I.S. and Belyaeva, Z.A. (1980). "Study of new enzyme preparation of alkaline proteinase". *Kozh-Obuvn. Prom-st.* 22, 19.

Monsheimer, R. and Pfleiderer, E. (1970). "Treatment of hides and skins." East German Patent No.1800891.

Monsheimer, R. and Pfleiderer, E. (1972). "Treatment of hides, skins etc." East German Patent No.2059453.

Monsheimer, R and Pfleiderer, E. (1974). "A method for softening of hides and skins." East German Patent No.2308967.

Monsheimer, R and Pfleiderer, E. (1981c). "Soaking method." German Patent No. 2944462.

Mutlu, M.M., Cadivel, B.H., Ozgunay, H., Adiguzeu, A.C. and Sari, O. (2009) "Ozone as a biocide in soaking." *J. Soc. Leath. Trades Chemists.* 93, 18–20.

Orlita, A. (2004). "Microbial biodeterioration of leather and its control: a review." *International Biodeterioration and Biodegradation,* 53, 157–163.

Ozyaral, O. and Birbir, M. (2005). "Examination of the fungal community on salt used in Turkish leather industry." *J. Soc. Leath. Chem. Assoc.* 89, 237–241.

Puvanakrishnan, R. and Dhar, S.C. (1988). CLRI Publications, Adyar, Chennai.

Rokhvarger, O.D. and Zubin, A.M. (1971). "Determination of mucopolysacharides in the skin tissues of hides." *Kozh-Obuvn Prom-st.* 13, 30.

Sarkar, K.T. (1981). *Theory and practice of leather manufacture.* Ajoy Sorkar, Chennai. p.71.

Sivasubramanian, S, Naidu, R.B., Kamini, N.R., Gowthaman, M.K., Ramalingam, S., Kanagaraj, P., Balaram, P., Chandrababu, N.K. and Puvanakrishnan, R. (2007) "Production of a halotolerant protease for industrial applications." Indian patent application 2375/DEL/2007.

Sivasubramanian, S., Manohar, B.M., Rajaram, A. and Puvanakrishnan, R. (2008a). "Mechanism of enzymatic dehairing of skins using a bacterial alkaline protease." *Chemosphere.* 70, 1025–1034.

Sivasubramanian, S., Manohar, B.M. and Puvanakrishnan, R. (2008b). "Ecofriendly lime and sulphide-free enzymatic dehairing of skins and hides using a bacterial alkaline protease." *Chemosphere.* 70, 1015–1024.

Taylor, M.M., Bailey, D.G. and Feairheller, S.H. (1987). "A review of the uses of enzymes in the tannery." *J. Amer. Leath. Chem. Assoc.* 82, 153–165.

Thanikaivelan, P., Rao, J.R., Nair, B.U. and Ramasami, T. (2002). "Zero Discharge Tanning: A Shift from Chemical to Biocatalytic Leather Processing." *Environ. Sci. Technol.* 36, 4187–4194.

Thanikaivelan, P., Rao, J.R., Nair, B.U. and Ramasami, T. (2001). "Approach towards zero discharge tanning: Exploration of NaOH-based opening up method." *J. Amer. Leath. Chem. Assoc.* 96: 222.

Thorstensen, T.C. (1985). Practical Leather Technology, 3rd edn. Robert E. Krieger Publishing Co., Malabar, Florida. pp. 87–88.

Toshev, T.M. and Esaulenko, L. (1972). "Accelerated refreshing of dry, nonsalted, preserved sheepskins with enzymes of *Aspergillus oryzae.*" *Kozh-Obuvn. Prom-st.* 13, 12.

Valeika, V., Balciuniene, J., Beleska, K., Skrodenis, A. and Valeikiene, V. (1996). "Use of NaOH for unhairing hides and the influence of salts on the hide properties." *J. Soc. Leath. Tech. Chem.* 81, 65–69.

Vedaraman, N. and Iyappan, K. (2006). "A process for leather making using saline water." European Patent 1656459.

Yapici, A.N. (2008). "The effect of using a fungicide along with bactericide in the main soaking float on microbial load." *African J. Biotechnol.* 7, 3922–3926.

16

DEHAIRING—CONVENTIONAL AND ENZYMATIC METHODS

INTRODUCTION

The leather industry is a major polluting one with the disposal of tannery effluents, creating technical difficulties. Dehairing is the primary pretanning operation in leather manufacture where the use of saturated lime solution along with toxic sharpeners such as sodium sulphite employed in high concentration result in toxic sludge. No rational approach has been adopted in the use of these chemicals as the proportion varies widely with respect to the region and quality of the hides and skins. This foul odorous effluent, containing organic amines, hydrogen sulphide, protein degradation products and pathogenic microbes, is a potential hazard both to the environment as well as to tannery workers. In addition, hair and wool are valuable by-products of the leather industry and hence, proper collection of hair or wool requires a careful processing technique.

Conventional dehairing processes employing lime and sulphide contribute to 50–60% of total pollution load in terms of biochemical oxygen demand (BOD), chemical oxygen demand (COD), total dissolved solids (TDS) and total suspended solids (TSS) besides a high alkaline effluent of 100% toxicity (Taylor et al., 1987; Marsal et al., 1999). Hydrogen sulphide, emanating from the dehairing process, is proven to be fatal even at concentrations as low as 200 ppm (Hannah and Roth, 1991; Roth et al., 1995). The physiological and toxicological effects of hydrogen sulphide are well-known and the current 8 h time-weighted-average for the Occupational Exposure Standard for hydrogen sulphide is 10 ppm (Davies, 1997). The extensive use of hazardous sulphide would not only result in unfavourable consequences on the environment but also would undermine the efficacy of the effluent treatment plants. Hence, rationalization of the dehairing process by systemic use of enzymes, especially proteases, in place of lime and sulphide becomes an issue of primary importance in leather processing. This ultimately leads to a substantial reduction in effluent load and toxicity in addition to improvement in leather quality (Puvanakrishnan and Dhar, 1986; 1988).

METHODS OF DEHAIRING

The process of dehairing is generally achieved by the following methods viz. (i) by attacking the hair and reducing it to a pulp, and this is called hair burn method and (ii) by destroying or modifying the epidermal tissue surrounding the hair bulb, so that the hair is loosened and removed mechanically. This is known as 'hair saving' method. Hair destruction methods involve the rupture of the disulphide and other bonds which stabilize the hard keratins of the epidermis while hair loosening methods have been observed to involve only the partial or complete destruction or only a softening of the tissues that hold the hair in place.

Six methods of dehairing can be adopted, viz., (i) clipping process (ii) scalding process (iii) chemical process (iv) sweating process (v) enzyme-assisted process and (vi) enzyme only process. Among them, currently, chemical dehairing is being practised widely, followed by enzyme-assisted process.

Clipping Process

This is carried out by clipping the wool of the sheep by a suitable machine when the animal is either alive or dead. Although best grade wool is obtained by clipping, it generally results in a loss of at least 15–20% wool. Wool obtained by clipping is not as long as that obtained by either 'painting' or 'sweating' as the portion left in the skin has got no value.

Scalding Process

Thermal dehairing or scalding process is carried out by the immersion of soaked skins in warm water (55–60°C) for a few minutes which brings about hair loosening without serious damage to the skin itself. Scalding is used for the removal of bristles from pig skins. When skins have been scalded at 58–60°C, a pronounced split in the outer sheath of the root has been observed. This may be caused by thermal contraction of the collagen and elastin structures surrounding the follicle (Merrill, 1956).

A promising method of grain removal from pig skins has been worked out (Wojdasiewicz, 1976). In this method, a limed pig skin is passed very quickly between a pair of rollers, one of them heated to 160–200°C, and the contact time is about 0.05–1.0 sec. By this process, the grain proteins are denatured and skins having soft and uniform surface are obtained. The disadvantages of the scalding process are that it requires exact temperature and time and a slight change in these parameters may damage the skins.

Chemical Process

The principal objectives of the chemical process using lime-sodium suphide are (a) to remove the hair by attacking the epidermis, which forms the outermost sheath in the structure of hides and skins, (b) to dissolve cementing substances and to loosen it from corium, (c) to saponify the skin lipids and to remove them, (d) to split up the fibre bundles and to open up the texture and (e) to swell the adipose tissue to facilitate removal of the adhering flesh.

It is common practice to use sharpeners such as sodium sulphide, dimethylamine, sodium borohydride, etc. in lime liquor and painting compositions. The main function of such additives is to help the easy removal of hair. Among them, sodium sulphide has been used as a sharpening agent because of its established efficiency, commercial availability and low cost.

'Paste' and 'pile' is a traditional hair-save method for calf, sheep or long-haired goat skins. It is used in cases where the hair/wool is valuable. This is a chemical process of dehairing of skins wherein the skins are piled after painting is done on the flesh side of washed and soaked skins with a thin paste composed of lime-sulphide mixture. The skins are spread out flat, flesh side up and the paint is applied by suitable brushes.

For the conventional process, 10% lime and 2% sodium sulphide are added along with 15–20% water. The paste thus prepared is applied on the flesh side of the skin. The skins are then piled up flesh to flesh and left for 18 h at ambient temperature (28–32°C) covering with a moist gunny bag until the hair becomes loose. Dehairing is done manually using conventional beam-and-knife method (Figure 16.1) or by mechanical means. The dehairing chemicals penetrate the skin from the flesh side and destroy the hair roots. It is then a simple task to scud off the hair that is not in contact with the dehairing chemicals. The hair is practically intact, although partial damage at the end of the hair root may be quite possible. Although the paste method of dehairing of skins does not generate liquid effluent disposal into the effluent stream, it causes significant pollution problems related to their toxicity as they contain high content of sulphide and lime (Sivasubramanian *et al.*, 2008a). The inherent disadvantage of this method is that it is time-consuming and involves heavy labour costs especially in a tannery where production capacity is very high.

Dip method is adopted for hides wherein the soaked hides are suspended in dehairing float comprising 100% water and lime-sulphide mixture in a pit or tub. The lime and sulphide employed for this process ranges between 10–15% and 2–5% respectively. The proportion of lime and sulphide usage varies with respect to skins and hides and their quality. Due to their thickness, hides employ higher proportion of lime and sulphide than skins.

The hair/wool recovered by direct liming process has no high value as it is damaged by high alkalinity and strong action of sulphide. When lime and sulphide are used for painting, insoluble calcium proteinates are produced in the fibres, which cannot be removed by mere washing. If the alkaline action is not stopped in time, other decomposition reactions may occur and consequently damage the wool/hair. Some of the dyeing defects of the tannery wool are due to long contact of the wool with the alkali. When processing sheepskins, it is preferable to perform enzymatic dehairing without lime or sulphide because the wool is more valuable than the leather.

Figure 16.1 Conventional dehairing

Other chemical dehairing methods either as a complete or partial replacement of lime and sulphide have been reported. These methods involve the use of chemicals such as hydrogen peroxide, sodium hydroxide, soluble silicates, chlorine dioxide and nickel carbonate. Most of the methods are operated at alkaline pH and hence need an extensive dealkalization step.

The demerits of the oxidative dehairing process using hydrogen peroxide in the presence of sodium hydroxide include hair damage and the resultant pollution load in terms of high COD and suspended solids. In addition, this method always employs high alkaline medium with higher proportion of H_2O_2 (4–12%) and sodium hydroxide (6–9%). When using H_2O_2 for dehairing, disruption of the $-S-S-$ bond occurs through a different mechanism. The oxidative attack of the $-S-S-$ bond is due to the formation of peroxy anion from H_2O_2 due to the highly alkaline medium. The use of sodium hydroxide in combination with sodium sulphide has been reported in chemical dehairing that is free of lime. The use of sodium hydroxide is more effective than calcium hydroxide in fibre opening and swelling of the collagen structure. This is due to (a) higher osmotic effect of Na^+ with respect to Ca^{2+} (b) the capacity of Ca^{2+} in cross linking with the carboxylic sites of the collagenic fibrils to give a more packed structure and (c) quick dissolution of sodium hydroxide in water and rapid swelling at high concentrations. Dehairing methods based on chlorine dioxide and nickel carbonate are not practised because of their toxic nature. The use of soluble alkali silicates such as sodium silicate in the concentration of 4 to 6% has been reported as lime substitute for collagen fibre opening. The resulting pelts show greater cleanliness of grain surface with well-split fibre bundles. This method facilitates exhaustion of subsequently applied auxiliaries such as chrome and fat liquor oils. Further, these auxiliaries are uniformly distributed. However, this method generates high pollution load in terms of COD and suspended solids.

Though reports are available for non-enzymatic dehairing methods, they do not result in commercial application due to toxicity and inefficiency. Bronco *et al.,* (2005) have reported the use of oxidant, hydrogen peroxide, in alkaline medium for dehairing; but, reduction in pollution load, especially COD, is not significant. Sehgal *et al.,* (1996) have developed a non-enzymatic dehairing process by painting method, free of sulphide, using nickel compounds and kaolin; but, the disposal of nickel compounds poses health hazards.

Hence, to evolve suitable methods for effluent disposal and to protect the environment from pollution hazards, several attempts have been made to replace the hair-destroying methods by hair-saving methods like sweating and enzymatic dehairing.

Sweating Process

Fresh or soaked skins are kept either by hanging in the air or in piles inside specially designed chambers with temperature and humidity controls for about 4–5 days until the epidermis and the hair roots are attacked by the enzymes secreted by bacteria resulting in the loosening of hair. The sweat pit or the closed damp room is a structure usually above the ground and has a double wall between which waste bark or other materials are filled to prevent loss of heat by radiation. The temperature of the pit is controlled by means of steam pipes running under the false bottom, and water is sprinkled to provide moisture and for lowering of the temperature if necessary. There are two methods of sweating process, viz., cold sweat and warm sweat. In the former, the temperature is kept around 21°C and in the latter, it is maintained at about 27°C.

The sweating process gives a maximum yield of high-quality wool but the process is slow and difficult to control and may damage the pelt if the enzyme action is not stopped in time. Satisfactory results are obtained in the case of the sweating process, when only the right type of bacteria, which will attack only the epidermis and hair roots without affecting collagen, thrives during the process. In actual practice, it is very difficult to have a rigid control over the process.

Enzyme-assisted Dehairing Process

In this process, dehairing of skins and hides is performed by using enzymes, especially alkaline proteases, having optimal activity at high pH along with only a small proportion of lime and sodium sulphide. When proteins in skin have been chemically modified, protease will accelerate the breakdown and removal.

Based on the soaked weight, enzyme (1–2%), lime (5%), sodium sulphide (0.5–2%) and water (10–20%) are used to make a paste for enzyme-assisted process for skins. For hides, dip method is adopted wherein 2% enzyme, 5–10% lime, 2% sulphide and 100% water are employed.

This process may result in soft leathers with more opening up of fibre structure. This process may eliminate the need for a 'bating' step and this depends on the elastolytic activity of the protease. Enzymes having elastolytic activity provide flat stretchless leather because of the removal of ripples in the skin. Reduction in the use of lime and sulphide contribute to moderate decrease in pollution load in terms of TDS, TSS, BOD and COD when compared to conventional chemical dehairing. Currently, many commercial enzyme preparations are being used for enzyme-assisted dehairing processes; but the use of lime and sulphide is not completely eliminated (Frendrup, 2000). Some of the reported lime- and sulphide-free enzymatic dehairing processes adopt either an enzyme carrier such as sparingly soluble kaolin in significant quantities (Dayanandan *et al.*, 2003) or soluble silicates (Saravanabhavan *et al.*, 2005). This may contribute to a significant rise in COD, TDS and TSS of effluent.

Several enzyme-assisted dehairing methods nowadays use commercial enzymes thereby reducing the quantity of lime or sulphide or both.

Laxman *et al.* (2007) have reported that a protease preparation from a fungal microbe, *Conidiobolus coronatus,* is used along with a reduced quantity of sulphide in dehairing of skins and hides without using lime.

The use of soluble silicates as a lime substitute in the dehairing process has been reported recently (Munz and Sonnleitner, 2005; Saravanabhavan *et al.*, 2005); yet, the reduction in pollution parameters is not that significant. Another report describes an invention related to dehairing and fibre opening process involving commercial protease formulations and silicates, wherein the use of lime and sulphide is eliminated in the enzymatic dehairing step. However, use of silicate salt in dehairing contributes to a rise in total dissolved solids as well as chemical oxygen demand in effluent.

Enzymatic Dehairing Process

Proteolytic enzymes find widespread applications in many industries (Underkofler, 1976). A recent report shows that approximately 50% of the enzymes used as industrial process aids are proteolytic enzymes (Godfrey and Reichelt, 1983). Proteolytic and amylolytic enzymes derived from various sources, viz., microbial, animal and plant sources have been applied individually or in combination to produce effective dehairing of hides and skins. Proteolytic enzymes are more efficient in enzymatic dehairing than amylolytic enzymes.

Use of animal proteases   The use of pancreatic enzymes for depilation was first studied by Rohm (1910) after treating the skins with caustic alkali for swelling. Lindroth (1961) has reported a pancreatic enzyme dehairing system without any alkaline treatment and has compared the leathers produced by enzymatic dehairing with those from lime sulphide dehairing. Pig skins have been dehaired by an active enzyme preparation from pork pancreas and hair loosening is attained within 6 h when the enzyme treatment is preceded by an alkaline swelling with lime and activation with ammonium sulphates and sodium sulphite (Studniarski *et al.*, 1975). Pancreatin has been utilized for the depilation of pig skins in a buffered medium of pH 8.0–9.0 at 37°C using a system containing a pancreatic enzyme preparation, ammonium sulphate and sodium bisulphite. After dehairing and removal of skins, the dehairing bath is regenerated and used repeatedly several times.

Role of autolytic enzymes   It has been reported that skins and hides are dehaired by the autolytic enzymes present in the skin at a pH range of 7.0–8.5. Leathers manufactured from autolytic enzyme-dehaired goat skins are found to be deficient in fullness and smoothness of grain as compared to conventional leathers. These properties have been improved by giving alkaline treatment to the enzyme-dehaired skins. The dehairing of skin or hide is also reported to be caused by the autolytic action of a proteolytic enzyme present in the skin or hide and the optimum conditions of dehairing are found to be incubation at 37°C in slightly alkaline pH with a relative humidity of about 100% (Sivaparvathy *et al.*, 1974a;1974b).

Use of plant proteases   The latex of madar plants (*Calotropis gigantea*), which contains rich amount of a potent proteolytic enzyme 'Calotropain', has a simultaneous action of dehairing and bating of hides and skins in 24 h. Burton *et al.* (1953) have shown that fresh calf skin could be dehaired by using pectinases as well as diastase. However, Gillespie (1953) has observed that satisfactory dehairing could not be achieved by using commercial diastase and amylase. The findings of Burton *et al.* (1953) have been partly confirmed by Bose *et al.* (1955) who have reported on the dehairing actions of amylase from germinated "Ragi" (*Eleusine coracana*) and commercial diastase. Bose *et al.* (1960) have also studied the optimum conditions for the hydrolysis of skin mucoids by "ragi" amylase and they have described a process of simultaneous dehairing and bating using the same enzyme.

The powdered leaves and barks of the jawasee shrub, which contains rich amounts of a proteolytic enzyme, are used for dehairing hides and skins in the Indian states of Gujarat, Rajasthan and Madhya Pradesh. The leathers made out of the dehaired pelts using jawasee process of dehairing have been found to possess certain characteristic properties. A process for the manufacture of grain garment leather using jawasee protease has also been described (Yeshodha *et al.*, 1978).

A pineapple proteinase is found to have maximum activity at pH 3.5–4.5, and good hide dehairing effects are obtained at pH 5.5 in the temperature range of 30–42°C. A two-stage dehairing by liming process is proposed using the proteinases isolated from the fruit of *Adenopus breviflorus*. Complete dehairing is found to occur within 6 h at 40°C and pH 7.5 and a further treatment with a 5% mixture of $CaCl_2 : KOH : Mg(OH)_2$ (11:8:1) for 10 h at 32°C and pH 2.0, would be enough for opening up the collagen fibres for subsequent tanning (Adewoye and Lollar, 1983). Subsequently, Adewoye and Bangaruswamy (1984) have developed a single unit process for dehairing of hides using the neutral proteinases from the fruit *Adenopus breviflorus*. There is a complete hair recovery, the fruit material does not pollute the environment and no H_2S is evolved in the beamhouse. Rose *et al.* (2007) have reported the use of animal and plant enzymes for total elimination of lime and sulphide in the dehairing process. This process eliminates a separate bating step. However, proteases from microbes are more preferred over plant and animal sources since production from the latter sources are not readily inducible and scalable.

MICROBIAL PROTEASES

The inability of the plant and animal proteases to meet current world demands has led to an increased interest in microbial proteases. Microorganisms represent an excellent source of enzymes owing to their broad biochemical diversity and their susceptibility to genetic manipulation. Microbial proteases account for approximately 40% of the total worldwide enzyme sales (Godfrey and West, 1996). Proteases from microbial sources are preferred to enzymes from plant and animal sources since they possess almost all the characteristics desired for their biotechnological applications. Commercial production of proteases from bacterial and fungal sources has gained much importance because of their extracellular production, submerged cultivation, high yield and short duration of production and easy recovery of the enzyme. Yeast proteases are mainly intracellular in nature and therefore, these enzymes have not gained significant commercial interest.

Proteases from Bacteria

Most commercial proteases, mainly neutral and alkaline, are produced by organisms belonging to the genus *Bacillus*. Bacterial neutral proteases are active in a narrow pH range (pH 5.0 to 8.0) and have relatively low thermotolerance. Due to their intermediate rate of reaction, neutral proteases generate less bitterness in hydrolysed food proteins than do the animal proteinases and hence are valuable for use in the food industry. Some of the neutral proteases belong to the metalloprotease type and require divalent metal ions for their activity, while others are serine proteinases, which are not affected by chelating agents. Bacterial alkaline proteases are characterized by their high activity at alkaline pH, e.g., pH 10.0, and their broad substrate specificity.

Although a wide range of microbes are known to produce proteases, major industrial enzymes with increasing market potential are derived from *Bacillus* group of bacteria due to their higher capacity and activity. Being a fast growing and non-pathogenic GRAS (Generally Regarded As Safe) microbe, *Bacillus subtilis* strain is used in the production of protease and is beneficial in terms of scalability, high yield and commercial exploitation. Proteolytic enzymes derived from a large number of *Bacillus* sp. and *Streptomyces* sp. have been used in dehairing of hides and skins.

A lime and sulphide-free process of dehairing has been developed for the manufacture of suede from sheep skins using protease from *B. subtilis*. It is reported that use of protease from *Bacillus subtilis* strain avoids the use of sulphide but employs lime for pH adjustment (Macedo *et al.*, 2005). Schlosser *et al.* (1986) have reported a method of depilation in an acid medium containing *Lactobacillus* culture. In dehairing, the hair loosening is effected at pH 10.0 using fungal or bacterial enzymes, the treatment period being approximately 12–16 h, followed by hair removal using mechanical means. The treatment period can be substantially reduced if the enzyme solution is fed in from the flesh side under pressure. This method suffers from the disadvantage of solubilization of collagen. Ibrahim *et al.*, (2011) have shown that an alkaline serine protease from *B. pumilus* has the potential for dehairing of skins. An alkaline protease, with pH stability between 8.0 and 9.0 and isolated from *Bacillus cereus,* has been found to be effective for the dehairing of goat skins in leather processing (Shakilanishi *et al.*, 2011). An alkaline protease from *Bacillus cereus* isolated from protein-rich soil sample is found to have maximum activity at 30°C and pH 8.0. The serine type protease is found to be effective for dehairing of goat skins (Sundararajan *et al.*, 2011).

Enzymatic hair loosening processes play a role wherever high-quality hair, wool or bristles are to be recovered. Although many reports are available for enzymatic dehairing, either free of lime or sulphide or both, none of these methods have found commercial application in tanneries, and sustained research efforts are underway to find a way for commercial acceptance.

Proteases from Fungi

Fungi elaborate a wider variety of enzymes than do bacteria. For example, *Aspergillus oryzae* produces acid, neutral, and alkaline proteases. The fungal proteases are active over a wide pH range (pH 4.0 to 11.0) and exhibit broad substrate specificity. However, they have a lower reaction rate as well as heat tolerance than do the bacterial enzymes. Fungal enzymes can be conveniently produced in a solid-state fermentation process. Fungal acid proteases have an optimal pH between 4.0 and 4.5 and are stable between pH 2.5 and 6.0.

Fungal neutral proteases are metalloproteases that are active at pH 7.0 and are inhibited by chelating agents.

The protease from *A. flavus* was earlier used for dehairing, and later it was reported that simultaneous dehairing and bating is possible with the protease of *A. flavus*. Clarizyme, an alkaline serine protease produced by *A. flavus,* is reported to have dehairing effects on skins and hides in the absence of sulphide. *A. flavus* grows rapidly on wheat bran and produces large amounts of extracellular proteases. Dehairing enzymes which avoid the use of lime and sulphide are produced by microorganisms such as *Aspergillus flavus* (Malathy and Chakraborty, 1991), *Rhizopus oryzae* (Pal *et al.*, 1996) and *Bacillus* sp. (Raju *et al.*, 1996). However, these enzymes are applied by painting method and use of these enzymes are used only for skins and not for hides. It is also reported that the enzyme preparation from cultures of *A. oryzae, A. parasiticus, A. fumigatus, A. effusus, A. ochraceus, A. wentii,* and *P. griseofulvum* exhibit marked depilatory activity on sheep skins.

Proteases from Viruses

Viral proteases have gained importance due to their functional involvement in the processing of proteins of viruses that cause certain fatal diseases such as AIDS and cancer. Serine, aspartic, and cysteine peptidases are found in various viruses (Rawlings and Barrett, 1993). All of the virus-encoded peptidases are endopeptidases; there are no metallopeptidases.

Although proteases are widespread in nature, microbes serve as a preferred source of these enzymes because of their rapid growth, the limited space required for their cultivation, and the ease with which they can be genetically manipulated to generate new enzymes with altered properties that are desirable for various applications.

TYPES OF PROTEASES

According to the Nomenclature Committee of the International Union of Biochemistry and Molecular Biology, proteases are classified in subgroup 4 of group 3 (hydrolases). However, proteases do not comply easily with the general system of enzyme nomenclature due to their huge diversity of action and structure. Currently, proteases are classified on the basis of three major criteria: (i) type of reaction catalysed, (ii) chemical nature of the catalytic site, and (iii) evolutionary relationship with reference to structure (Barrett, 1994).

Proteases are grossly subdivided into two major groups, i.e., 'exopeptidases' and 'endopeptidases', depending on their site of action. Exopeptidases cleave the peptide bond proximal to the amino or carboxy termini of the substrate, whereas endopeptidases cleave peptide bonds distant from the termini of the substrate.

Exopeptidases   The exopeptidases act only near the ends of polypeptide chains. Based on their site of action at the N or C terminus, they are classified as amino- and carboxypeptidases, respectively.

Aminopeptidases   Aminopeptidases act at a free N-terminus of the polypeptide chain and liberate a single amino acid residue, a dipeptide, or a tripeptide. They are known to remove the N-terminal methionine that may be found in heterologously expressed proteins but not in many naturally occurring mature proteins. Aminopeptidases occur in a wide variety of microbial species including bacteria and fungi (Watson, 1976). In general, aminopeptidases are intracellular enzymes, but there has been a single report on an extracellular aminopeptidase produced by *A. oryzae* (Labbe *et al.*, 1974).

Carboxypeptidases   The carboxypeptidases act at C-terminals of the polypeptide chain and liberate a single amino acid or a dipeptide. Carboxypeptidases can be divided into three major groups, serine carboxypeptidases, metallocarboxypeptidases, and cysteine carboxypeptidases, based on the nature of the amino acid residues at the active site of the enzymes.

Endopeptidases   Endopeptidases are characterized by their preferential action at the peptide bonds in the inner regions of the polypeptide chain away from the N and C termini. The presence of the free amino or carboxyl group has a negative influence on enzyme activity. The endopeptidases are divided into four subgroups based on their catalytic mechanism, (i) serine proteases, (ii) aspartic proteases, (iii) cysteine proteases, and (iv) metalloproteases.

Serine proteases   Serine proteases are characterized by the presence of a serine group in their active site. They are numerous and widespread among viruses, bacteria, and eukaryotes, suggesting that they are vital to the organisms. Serine proteases are found in the exopeptidase, endopeptidase, oligopeptidase, and omega peptidase groups. The isoelectric points of serine proteases are generally between pH 4.0 and 6.0.

Serine alkaline proteases that are active at highly alkaline pH represent the largest subgroup of serine proteases. Although alkaline serine proteases are produced by several bacteria, subtilisins produced by *Bacillus* spp. are the best known. Subtilisins of *Bacillus* origin represent the second largest family of serine proteases. The active-site conformation of subtilisins is similar to that of trypsin and chymotrypsin despite the dissimilarity in their overall molecular arrangements.

Aspartic proteases   Aspartic acid proteases, commonly known as acidic proteases, are the endopeptidases that depend on aspartic acid residues for their catalytic activity. Most aspartic proteases show maximal activity at low pH (pH 3.0 to 4.0) and have isoelectric points in the range of pH 3.0 to 4.5. Their molecular masses are in the range of 30 to 45 kDa. Microbial aspartic proteases can be broadly divided into two groups, (i) pepsin-like enzymes produced by *Aspergillus*, *Penicillium*, *Rhizopus*, and *Neurospora* and (ii) rennin-like enzymes produced by *Endothia* and *Mucor* spp.

Cysteine/thiol proteases   Cysteine proteases occur in both prokaryotes and eukaryotes. About 20 families of cysteine proteases have been recognized. The activity of all cysteine proteases depends on a catalytic dyad consisting of cysteine and histidine. Generally, cysteine proteases are active only in the presence of reducing agents such as HCN or cysteine. Based on their side chain specificity, they are broadly divided into four groups: (i) papain-like, (ii) trypsin-like with preference for cleavage at the arginine residue, (iii) specific to glutamic acid, and (iv) others. Papain is the best-known cysteine protease.

Metalloproteases Metalloproteases are the most diverse of the catalytic type of proteases. They are characterized by the requirement for a divalent metal ion for their activity. They include enzymes from a variety of origins such as collagenases from higher organisms, hemorrhagic toxins from snake venoms, and thermolysin from bacteria. Based on the specificity of their action, metalloproteases can be divided into four groups, (i) neutral, (ii) alkaline, (iii) *Myxobacter* I, and (iv) *Myxobacter* II. The neutral proteases show specificity for hydrophobic amino acids, while the alkaline proteases possess a very broad specificity. *Myxobacter* protease I is specific for small amino acid residues on either side of the cleavage bond, whereas *Myxobacter* protease II is specific for lysine residue on the amino side of the peptide bond. All of them are inhibited by chelating agents such as EDTA but not by sulphydryl agents or DFP.

Collagenase is another important metalloprotease, whose action is very specific; i.e., it acts only on collagen and gelatin and not on any of the other usual protein substrates. Elastase produced by *Pseudomonas aeruginosa* is another important member of the neutral metalloprotease family.

METHODS FOR THE APPLICATION OF ENZYMES

For using the protease preparation for dehairing of hides and skins by paint (Figure 16.2) or dip method, it is necessary to standardize the optimal conditions such as pH, enzyme concentration and duration. The optimal treatment for the production of dehaired pelts from goat skins and cow hides by dip method are: 6–18 h, 1% enzyme and 100% float at pH 7.5–9.0 for goat skins and 2% enzyme and 100% float at pH 7.5–9.0 for cow hides. Sodium carbonate can be used for pH adjustment if the dehairing protease is active in the alkaline range.

A spraying technique (Figure 16.3) can also be adopted for the application of enzyme concentrate in dehairing. The advantages of this method over the paint and dip methods are: (a) even concentrated solutions can be sprayed, (b) easier penetration of enzymes is achieved when the spray is forcefully applied on the flesh side and (c) this method is free of effluent and the process is complete with 100% removal of hair from the skin surface. A high-speed penetrator under pressure has also been employed for instantaneous penetration of enzymes and the advantages are (i) reduction in the time of application as well as in dehairing time and (ii) uniform application.

Figure 16.2   Application of enzymes by paint method

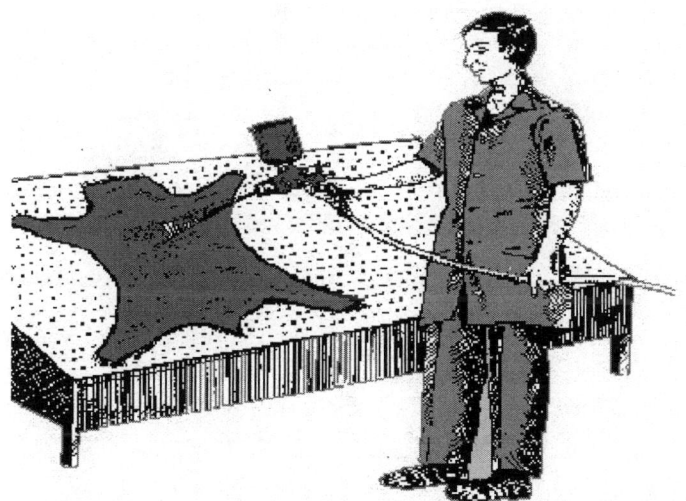

Figure 16.3   Application of enzymes by spraying technique

IMMOBILIZED ENZYMES FOR DEHAIRING

Immobilized enzymes have been successfully used in many industries. The greatest advantage of immobilized enzymes is their reuse. For the first time, a new immobilized pancreatic enzyme product has been prepared by the immobilization of activated pancreatic enzymes on a cheap and indigenously available support using a bifunctional agent (Puvanakrishnan *et al.*, 1980). A novel method of dehairing of sheep skins using an immobilized pancreatic enzyme product for the manufacture of grain garment leather has been developed and the efficient reuse (thrice) of the immobilized enzyme product in dehairing operation has been demonstrated (Puvanakrishnan *et al.*, 1981).

ROLE OF SHAVING IN DEHAIRING

El Baba *et al.* (2000) have demonstrated that removal of hair from cattle hides by cutting with electric clippers would reduce the COD and suspended solids to nearly half the level when compared to conventional hair-burn dehairing. The hair cutting allows a reduction in the sulphide usage, which can be complimented by including alkaline protease, to accelerate the breakdown of the degraded hair.

ROLE OF ALKALINE TREATMENT IN ENZYMATIC DEHAIRING

An alkaline treatment is necessary at some stage (in connection with enzymatic depilation) to solubilize the cementing protein for facilitating the removal of hair. Though enzymatic dehairing is found to cause the splitting up of the fibre bundles, subsequent alkaline treatment is required to further open up the weave pattern (Andrews and Dempsey, 1967). Hannigen *et al.* (1968) have observed that it is desirable to swell the hides with lime after the enzyme treatment to prevent 'drawn grain'. It has been reported that it is important to give the hides a second dehairing with small quantities of lime or sulphide to remove the fine hairs and that 'bating' may not be necessary after enzymatic dehairing (Herfeld and Schubert, 1969). The development of a stable alkaline protease in lime and other alkaline media and achievement of the simultaneous dehairing and swelling in a single step using that enzyme have also been reported (Yates, 1972).

Using an enzyme preparation from pork pancreas, hair loosening has been attained in pig skins within 6 h when the enzyme treatment is preceded by alkaline swelling with $Ca(OH)_2$ and activation with ammonium sulphate and sodium sulphite (Studniarski *et al.*, 1976).

A chemical enzymatic method of liming and dehairing yields cattle hides with higher contraction (shrinkage) temperature and degree of collagen crystallinity than the classical Na_2S-$Ca(OH)_2$ procedure. This method involves soaking hides in water containing $Ca(OH)_2$ 1.2 parts, NH_4OH 2.0 parts, molasses 0.8 parts and Na_2SO_4 2 parts for 1 h at 24–26°C, then adding 1.8% $Ca(OH)_2$, more water and soaking for 20 h. The hides are then dehaired with proteolytic enzymes. The wastes from this process contain no toxic impurities.

CHALLENGES IN THE DEVELOPMENT OF DEHAIRING ENZYMES

There are several factors impeding the implementation of enzymatic dehairing process and they are outlined below:

1. The removal of residual fine hairs remains the greatest obstacle to the development of hair-saving enzymatic process (Paul *et al.*, 2001).
2. Dehairing enzyme imparts damage to the structure of collagen and grain surface of the skins and hides.
3. Incomplete opening up of collagen fibre bundles.
4. Inadequate plumpness due to absence of osmotic pressure-driven swelling results in lower degree of fibre splitting and fiber swelling when compared to chemical dehairing.
5. The cost of enzyme production, scaling up and bioprocessing.
6. Identification and assessment of the suitability, specificity and efficacy of enzyme preparation to process different raw materials such as skins and hides without damaging the collagen
7. Enzyme action is influenced by many factors, viz., pH, temperature, substrate specificity, etc.
8. Enzymatic and enzyme-assisted dehairing processes require careful controls as they are shown to attack the dermal collagen, leave residual hairs, and damage fine fibres in the grain matrix, thus diminishing the quality of leather.

In order to use enzymes in leather processing, it is imperative to know the characteristics of a particular enzyme, its nature and the specificity of substrates on which the enzyme will act and finally the end results which are to be achieved.

An ideal dehairing enzyme should yield

(a) dehaired pelts completely free of fine hairs and epidermis
(b) the pelts should appear cleaner and whiter without discolouration than the pelts of chemical dehairing process due to elimination of sulphide in the process.
(c) The grain surface of enzyme-dehaired pelt should feel smooth and even without scud and surface damage.
(d) The process should yield significant quantity of intact hair of good quality and free of damage that could be a value-added saleable by-product
(e) Opening up of collagen fibre structure should be complete, regular and even in dermis and corium.
(f) Physicochemical properties such as tensile strength, colour, softness, etc. of the leathers produced by enzymatic process should be equivalent to or better than that obtained by chemical processing.
(g) There should not be any perceivable damage to the collagen fibres in enzyme dehaired pelts.
(h) Dehairing should be achieved in short duration without a high degree of control as it obviates the problem of collagen damage by prolonged protease action.

MERITS OF ENZYMATIC DEHAIRING

Recovery of Hair, A Valuable by-Product

Figure 16.4 shows that the recovery of intact hair is possible in enzymatic dehairing whereas chemical and enzyme-assisted dehairing processes impart damage to the hair. Microscopic analysis has confirmed that the hair recovered from experimental dehairing (enzyme only) is intact in both skins and hides and the hair is not damaged (Sivasubramanian *et al.*, 2008a). Recovered intact hair from skins and hides is used in the manufacture of felt, organic fertilizer and poultry feedstuff. Also, the recovered hair when subjected to hydrolysis by thermal, biological and chemical means has an array of applications ranging from biogas generation, regeneration of keratins, recovery of melanin for the preparation of sun-tan lotions and cosmetics, hair conditioner, amino acid production in pharmaceutics, retanning agent as well as chrome exhaustion in leather manufacture and in the manufacture of synthetic products such as nylon (Cantera and Buljan, 1997).

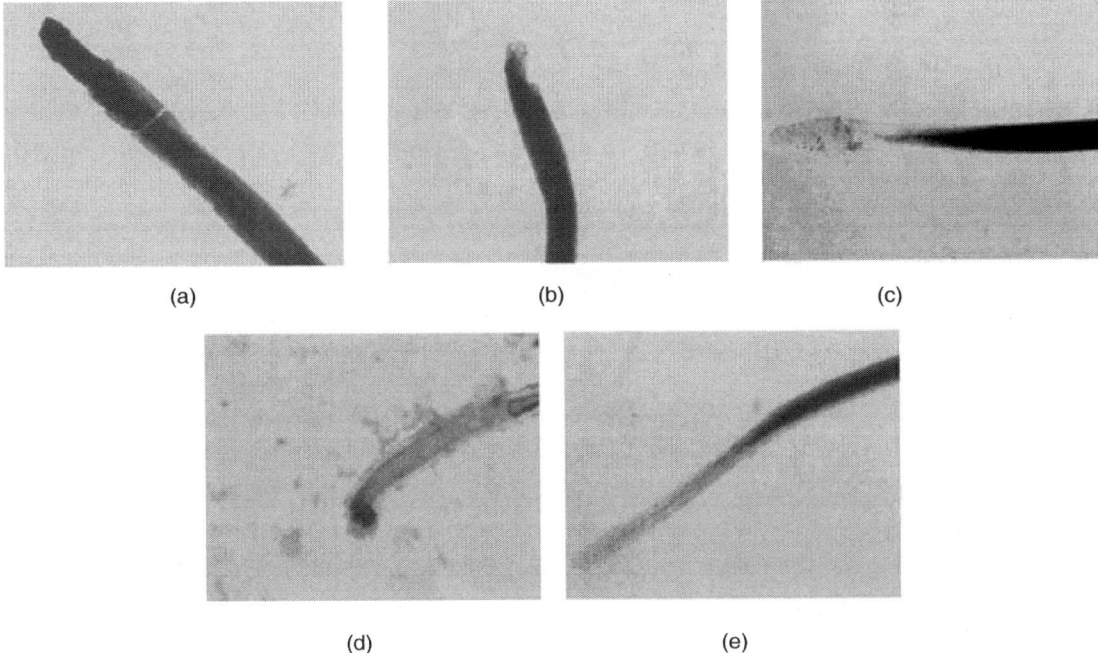

(a)   (b)   (c)

(d)   (e)

Figure 16.4 Microscopic structure of the hair from goatskins (a–c) and cow hides (d, e) from conventional chemical (a), enzyme-assisted (b, d) and enzymatic dehairing (c, e). (×150). (*See* Plate 2.1).

Enzymatic dehairing helps in easy handling of the pelts avoiding discomfort to tannery workers, minimizes the problem of effluent disposal in the tanneries and simplifies the pretanning processes by cutting down time and chemical costs in processes such as liming, deliming and bating. Enzymatic dehairing not only increases the surface area of the leather when compared to conventional dehairing but also provides leather with cohesive original and natural structure of hide or skin along with an increase in strength properties.

Fibre Opening

The grain surface appears to be smooth and even and also, complete absence of hair is shown in enzymatic dehairing against chemical and enzyme-assisted dehairing (Figure 16.5 a–c). Opening up of collagen fibre structure is more complete, regular and even in dermis and corium of enzymatically dehaired pelts than chemical and enzyme-assisted processes, while the fibre structure is compact and moderately opens up in the conventional and enzyme-assisted systems respectively. There is no apparent damage to the collagen fibres in enzyme-dehaired pelts (Figure 16.5 d–f).

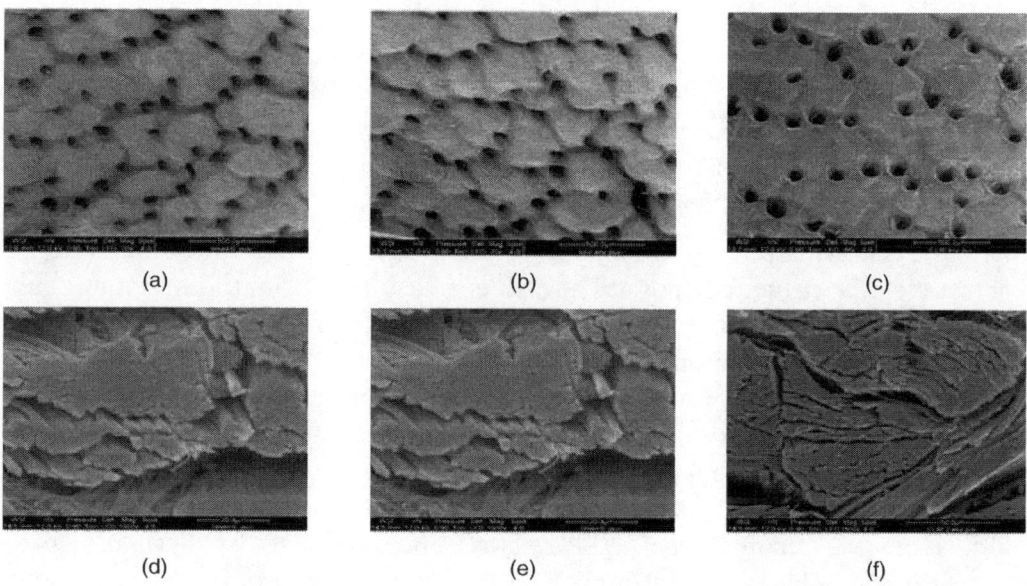

(a)   (b)   (c)

(d)   (e)   (f)

Figure 16.5   SEM showing grain surface of dehaired pelts from (a) conventional chemical, (b) enzyme-assisted and (c) enzymatic dehairing of cow hides (80×); cross sectional veiw of dehaired pelts from (d) conventional, (e) enzyme-assisted and (f) enzymatic dehairing of cow hides (1500×) (Sivasubramanian *et al.*, 2008a). (*See* Plate 2.2).

Physicochemical, Objective and Softness Assessment of Dyed Crusts Obtained by Chemical, Enzyme-assisted and Enzymatic Methods

Physical characteristics of the dyed crusts obtained by chemical, enzyme-assisted and enzymatic methods are shown to be in agreement with the standards and there are no significant difference between them (Sivasubramanian *et al.*, 2008a). Semi-quantitative assessment of leather quality was made through a measurement of softness, by means of compression index (CI) values deduced from 'compression measurement data' of dyed crusts obtained by enzymatic dehairing of skins and hides. CI values are higher than that of chemical and enzyme-assisted systems indicating an increased softness and quality. Softness of leathers is indirectly related to opening up of fiber bundles (Thanikaivelan *et al.*, 2002; Lokanadam *et al.*, 1989) and it also depends on the extent in filling.

Chemical analysis of the dyed crusts obtained from control and experimental dehairing systems exhibits a rather similar pattern (Sivasubramanian *et al.*, 2008a). Dyed crusts of enzymatic treatment have high content of hide substance because the collagen content is not affected by enzyme treatment. Finished leathers show quality on par with chemical treatment with regard to properties such as grain flatness, fullness, softness, grain tightness, dyeing characteristics and general appearance.

Environmental Benefits

In the beamhouse, the conventional dehairing process is responsible for most pollution: 84% of the BOD, 75% COD, 92% of the suspended solids and 100% of the toxicity. Hence, it is a necessity to opt for a greener mode of dehairing especially using enzymes that can diminish the eco-toxicological parameters.

Enzymatic dehairing process has a definite advantage over the lime–sulphide system in terms of trouble-free effluent disposal. The effluent, resulting from the enzymatic dehairing methods, is found to be much better adaptable to biological purification, mainly because of its lower pH value. Even though enzymatic dehairing might have been followed by alkaline swelling, the observed total amount of wastes are much lower than in normal liming. In the liming process, the effluent problems are principally due to the large quantity of total solids arising from lime and sodium sulphide. TDS and TSS of enzymatic dehairing effluent are greatly reduced (by about 85% for both skins and hides) compared to dehairing with lime and sulphide (Sivasubramanian *et al.*, 2008a). This is mainly due to the elimination of sludge-forming, sparingly soluble lime in the process. Pulped hair and keratinous substances contribute to high BOD and COD in the effluent from lime–sulphide process. The extent of reduction in terms of BOD and COD in the enzymatic dehairing effluent is observed as 78% and 84% for skins and 85% and 90% for hides respectively. The proportional reduction of BOD and COD is more pronounced in the effluent of enzymatic dehairing of hides than that of skins due to the absence of pulped hair in the effluent of normally pressed hides.

Pretanning processes generally account for 70–80% of the total COD of effluent from all leather-making processes (Marsal *et al.*, 1999). About 75% of the organic waste from a tannery is from pretanning processes and 70% of this waste is from hair rich in nitrogen (Kamini *et al.*, 1999). Elimination of hair-destructive sulphide in the process leads to the complete recovery of intact hair which results in huge reduction in COD.

Studies on the effluent from enzymatic dehairing of skins and hides devoid of lime and sulphide show that there is a great reduction not only in the TSS, TDS, BOD and COD but also in the toxicity due to complete elimination of sulphide. Contrary to chemical and enzyme-assisted dehairing, the enzymatic process does not generate high alkaline effluent as lime is eliminated in the process. It is noteworthy that the pH of the effluent from the enzymatic process is near neutral, indicating that it is less toxic. In addition, treatment of an effluent with neutral pH will be less expensive compared to that having a high alkaline pH.

The effluent derived from enzymatic dehairing of skins and hides contain a complex mixture of degraded noncollagenous proteins, proteoglycans, insignificant amount of enzyme, dislodged intact hair and epidermal keratinous substances. The effluent from conventional and enzyme-assisted processes contain the above constituents besides reacted and unreacted products of lime and sulphide addition, hair in pulped form due to its degradation by destructive sulphide in the alkali bath.

MECHANISM OF DEHAIRING

In enzymatic dehairing, hair loosening commences with an attack on the outermost sheath and is continued by the swelling and breakdown of the inner root sheath and parts of the hair that are not keratinized (Kuntzel and Stirtz, 1958). The process of enzymatic dehairing by means of proteolytic and amylolytic enzymes has been suggested to depend essentially on the hydrolysis and removal of the skin mucoids (Burton *et al.*, 1953; Reed, 1953; Bose *et al.*, 1955). When the skin mucoids are removed, the cohesion of the cementing of the protein fibres is reduced and the entire structure gets loosened (Reed, 1953). Money and Scroggie (1971) have reported the rupture of lysosomal membranes at acidic pH and the liberation of lysosomal enzymes which cause hair loosening. Since proteins and mucopolysaccharides are found to be affected in lysosomal dehairing, it is presumed that cathepsin D, which degrades both proteins and mucoproteins, might play an important role in this type of dehairing.

Observations from earlier studies show that enzymes enter into skins only from the flesh side when the dehairing process is carried out at room temperature. When the skins are treated with enzymes in warm alkaline condition, enzymes pass through both the grain and flesh surface, the effect being more prominent in the latter case. The enzymes entering through both the surfaces have different dehairing action, and these dehairing actions also depend on the state of the raw skins, the pretreatments and the dehairing methods. Enzymes which enter through the flesh side are important in maintaining the quality of the leather. In the early stages of dehairing, the amount as well as the penetration of enzymes entering from the flesh side is greater than those entering from the grain side.

In another study on the relation between effective concentrate and temperature of enzymatic depilation, it has been shown that dehairing temperature and the amount of enzyme are decreased with increasing mechanical action, viz., vibration and speed of drum rotation. However, loss of collagen is found to increase with increasing mechanical action.

The mechanism of traditional hair burn dehairing by the combined actions of lime and sulfide has been known for a long time (O'Flaherty *et al.*, 1978). Disulphide link in cystine, the major stabilizing bond in keratin, is subjected to nucleophilic attack by sulfide ions. Lime, being an alkali, contributes to swelling of skins by removing the electrical charge from the basic groups in collagen and changing the dimensions of its structure (Menderes *et al.*, 2000). Lime also opens up the collagen fibre bundles that are abundantly distributed in the dermis and the underlying corium layer of the skin by removing much of the charged glycosaminoglycans (GAG) from proteoglycans (PG) through β-carbonyl elimination (Anderson *et al.*, 1965; Alexander *et al.*, 1986). In conventional dehairing process, lime cleaves the covalent linkage between the GAG and protein core and thereby releases GAG. But, the liberation of GAG and other oligosaccharides, dissociation of PG aggregates, opening up of collagen fibre structure and enzyme action upon the noncollagenous components of the various layers of skin vary in enzymatic dehairing process as against chemical process (Sivasubramanian *et al.*, 2008b).

When skins are treated with lime and a reducing thiol agent such as sodium sulphide in liming process, hard and soft keratins of the epidermal layer undergo modification and degradation. So, the structural integrity of the epidermal layer is greatly affected, thus facilitating its removal from the skin (El Baba *et al.*, 2000). The dermis, which is closely and tightly associated with the epidermis through an amorphous basement membrane, consists of glycoproteins, heparan sulphate PG and non-fibrous collagens of type IV and type VII (Stanley *et al.*, 1982). Basement membrane, which is fragile and sensitive to chemical, enzymatic and mechanical treatment, is consequently removed along with the epidermis on liming, creates a smooth and clean surface of the grain layer (Cantera, 2001a). Proteases do not directly act upon epidermal components and hair; instead, it causes depilation by degrading these components of basement membrane at the dermal–epidermal junction. Disruption of this region leads to the simultaneous detachment of both basement membrane and epidermis from the dermal layer.

Upper dermis, also known as grain layer, is rich in soluble proteins, type I collagen, cementing substances such as hyaluronic acid and dermatan sulphate PG and elastic fibres apart from fibroblasts. When the hair bulb is subjected to protease action, hair is dislodged from the surrounding follicles, as the sac of the hair bulb is less in fibrous keratin (El Baba *et al.*, 2000). Lower dermis, the corium layer, also contains the above components except elastin and is enriched with entangled fibres and thick fibre bundles containing type I collagen fibrils. These collagen fibrils are encircled by a sheath of fine reticular fibrils together with cementing substances filling in the interfibrillary spaces. Collagen of both the grain and corium is extensively converted to leather-making substance. Hence, elimination of noncollagenous components from the skin is necessary. Adhering to the corium is flesh, the layer usually preferred by enzymes for penetration. The cementing substance in the dermis has a marked influence on collagen stability by making the fibres more compact. They exert a negative influence upon diffusion of substances. To achieve more enzyme diffusion and to get access onto the proteins of hair bulbs and follicles, cementing substances have to be removed from the dermis and corium. PG constituents are greatly reduced in the dehairing process of both lime–Na$_2$S and the enzyme systems. But, the removal rate is higher in enzymatic dehairing than lime–sulphide dehairing. In order to access interfibrillar and interfibre contents of the corium, the enzymes have to penetrate through the flesh layer, adjoining the corium. Once the enzyme establishes its contact with the fibre structure of the corium, it progressively removes PG constituents. Penetration of the enzyme through the much thinner epidermis is rather difficult than through the thick flesh layer (Yates, 1972b) as the epidermis is highly keratinized, pigmented, scaly and hydrophobic in nature, impermeable to any solution containing high molecular weight proteins and unaffected by proteolytic enzymes except keratinases (Cantera, 2001b).

Uronic acid and hexosamine are the components of both sulphated and non-sulphated GAG and their ratio is nearly 1:1. They present as disaccharide repeats in both sulphated and non-sulphated GAGs. These components are progressively removed during the course of the dehairing process in both lime–Na$_2$S and enzymatic systems. This indicates the elimination of interfibrillary substances such as sulphated PGs and non-sulphated GAG such as hyaluronic acid. Since hyaluronic acid is a non-sulphated GAG, it is not susceptible to proteolytic attack. However, it interacts with PG non-covalently through linker proteins, forming high molecular weight aggregates (Pearson *et al.*, 1983). Integrity of these PG–hyaluronate aggregates can be affected by attacking the protein core of PGs. Preserving the content of collagen without causing damage to its cohesive structure is a prerequisite for application of enzymes in leather processing. The protease preparation can have elastolytic activity but not collagenolytic activity and it should be safely used in dehairing experiments without the need for a high degree of control. Remarkable reduction of all PG components except collagen is achieved in the enzymatic dehairing process signifying that their elimination is crucial for opening up of the collagen fibrous structure. This, in turn, assists the diffusion of enzyme through collagen layers towards epidermis, consequently dislodging the hair without any damage from its follicle.

Histochemical Studies

Histochemical studies with Masson's trichrome staining show that in the dehaired pelts produced using the protease, the noncollagenous components are completely removed and the collagen network is fully opened up (Figure 16.6a–c). Gomori's staining of reticular fibres shows that they are less abundant in lime–Na$_2$S and enzyme-dehaired pelts when compared to soaked skins and this is probably due to removal of some of these fibers during the process (Figure 16.6d–f). Verhoeff's staining shows the absence of elastic fibers in enzyme dehaired pelts indicates that the enzymatic treatment degrades the elastin network (Figure 16.6g–i).

Lime and sulphide do not have significant action upon elastic fibres but they are removed later in a step called bating wherein trypsinlike protease is employed. Removal of elastin in the enzymatic process imparts softness and more flexibility on the final leather and this feature is beneficial for producing certain types of leather such as upholstery and gloving leather, where softness and drape are important properties (Alexander *et al.*, 1986). Leathers that require stiffness, stretch or resilience, will not benefit from this bating step. A major advantage of the enzymatic dehairing process is that the bating step in leather processing can be avoided. Histological studies indicate that the protease removes elastin completely and the reticulin probably at a lower rate from skins. Alcian blue-PAS (Figure 16.7a–c) and Alcian blue-neutral red (Figure 16.7d–f) staining methods have shown the elimination of noncollagenous components such as glycoproteins and mucoid substances (PAS–reactive), PGs and hyaluronic acid in dehaired pelts of the enzymatic system. However, the presence of PAS-reactive substances such as glycoproteins and mucoid substances around hair follicles and glands are observed in the lime–sulphide dehaired pelts indicating the incomplete removal of these substances. This is probably due to resistance of these substances to lime–sulfide treatment. As these staining methods are not specific for collagen, fibres of collagen are not visible prominently. Glycoproteins and mucopolysaccharides are essential for holding the hair follicle with outer root sheath present in basement membrane, and removal of these substances results in the complete dislodging of hair in enzyme-dehaired pelts.

Immunohistochemical studies on the removal of decorin, a principal cementing substance of collagen fibrils of dermis and corium, reveals that the PG is removed partially in dehaired pelts of lime–Na$_2$S and completely in enzymatic process respectively, subsequently opening up the collagen fibre bundles (Figure 16.8a–c). This further substantiates the fact that the removal of decorin is a prerequisite for opening up of collagen fiber bundles.

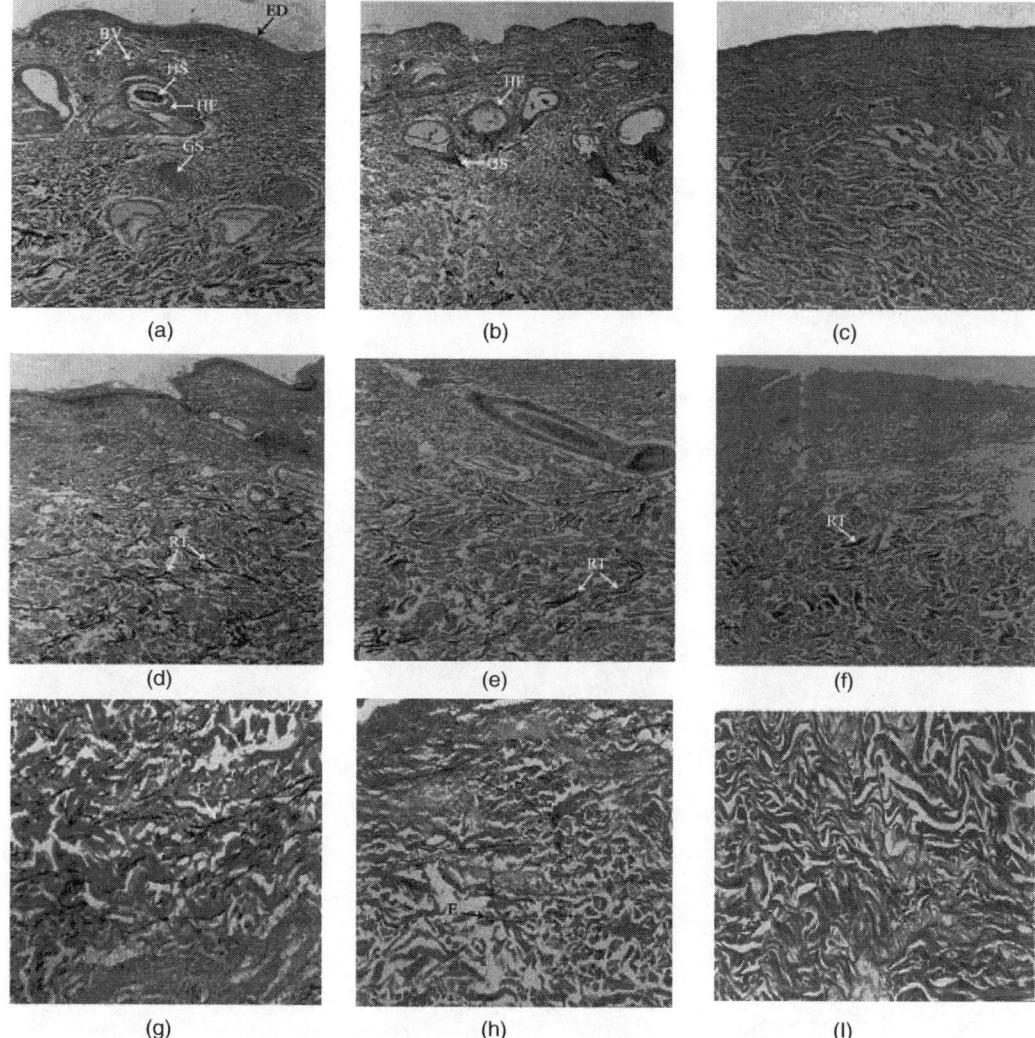

Figure 16.6 Masson's trichrome staining of sections from (a) soaked skin and dehaired pelts of (b) lime–sulfide and (c) enzymatic processes (×175); Gomori staining of sections from (d) soaked skin and dehaired pelts of (e) lime–sulphide and (f) enzymatic processes (×175); Verhoeff's staining of sections from (g) soaked skin and dehaired pelts of (h) lime–sulphide and (i) enzymatic processes (×520). ED—epidermis; BV—blood vessel; GS—glandular structure; HS—hair shaft; HF—hair follicle; RT—reticulin and E—elastin. (*See* Plate 3.1).

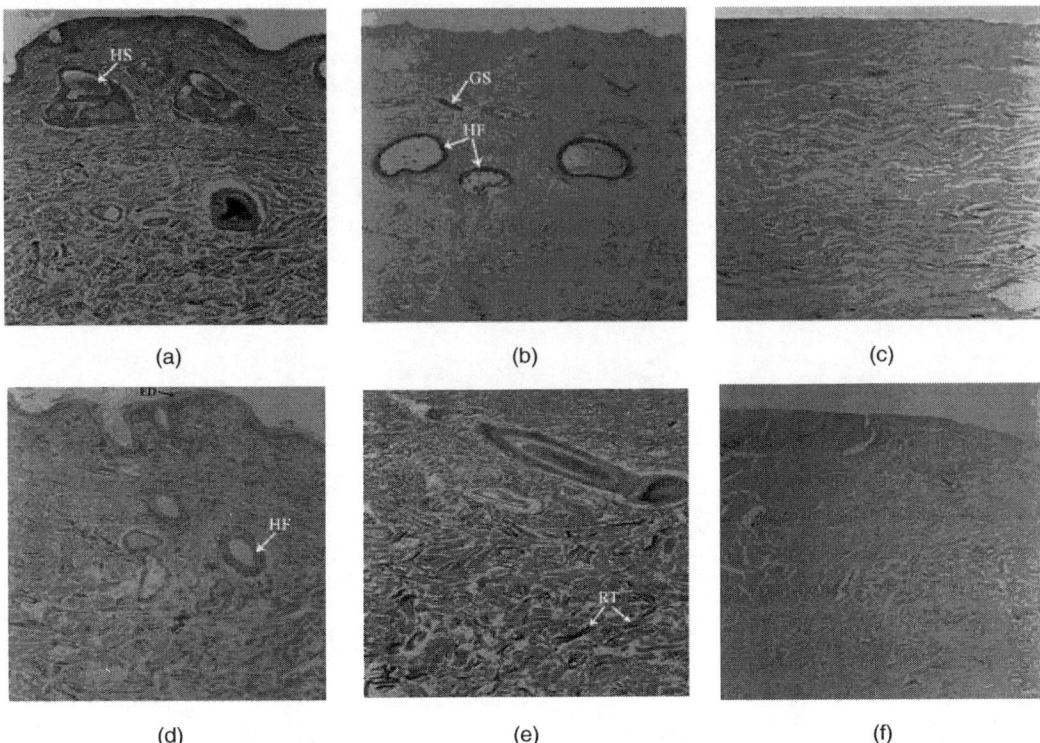

(a) (b) (c)

(d) (e) (f)

Figure 16.7 PAS–Alcian blue staining of sections from (a) soaked skin and dehaired pelts of (b) lime–sulfide and (c) enzymatic processes (x175); Alcian blue neutral red staining of sections from (d) soaked skin and dehaired pelts of (e) lime–sulphide and (f) enzymatic processes (×175). ED—epidermis; GS—glandular structure; HS—hair shaft and HF—hair follicle. (*See* Plate 4.1).

Although hydrodynamic, colligative and other biological properties of PGs are attributed to GAG chains, importance of the protein core is exemplified by its interaction with high molecular weight hyaluronate, thereby forming high molecular weight PG aggregates in interfibrillar spaces. The PGs show a high affinity for native collagen I molecules, with a dissociation constant in the order of 10^{-8}–10^{-9} M (Schonherr *et al.*, 1995a, 1995b; Svensson *et al.*, 1995). While only dermatan sulphate chain can play a minor role in its interaction with collagen, high-affinity binding depends on the core protein (Hedbom and Heinegard, 1993), especially a stretch of 40 amino acid residues of the central portion of core protein (Blaschke *et al.*, 1996). Protease increasingly degrades the protein core of decorin and releases O-linked GAG side chains and N-linked oligosaccharides along with peptide counterparts. This further leads to a drop in the viscosity of these aggregates, weakening of hyaluronate–decorin interactions with collagen fibrils and simultaneous dissociation of PG aggregates from interfibrillar spaces. This can possibly result in the opening up of fibre bundles in the dermis and corium layers of skin, thus favouring the penetration of enzyme. The biochemical action of lime and proteases upon the cementing substances is thus quite different from each other.

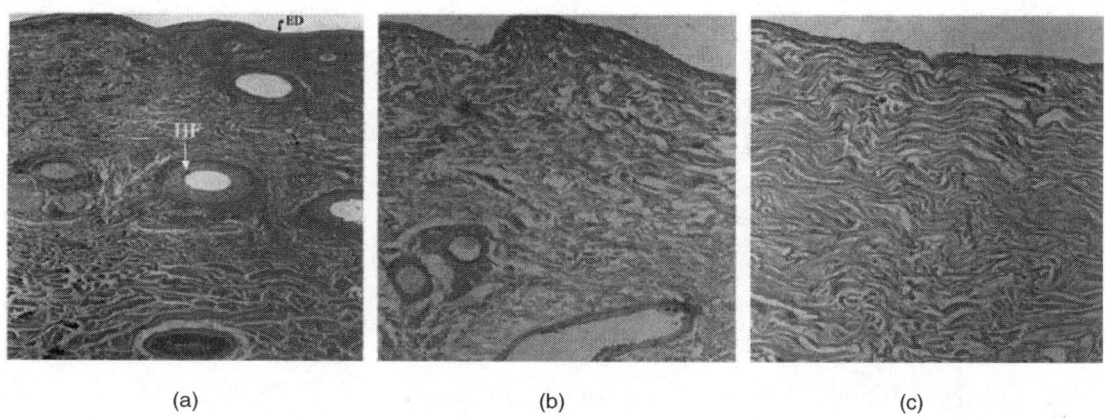

(a)   (b)   (c)

Figure 16.8 Immunohistochemistry (×175). Sections from (a) soaked skin and dehaired pelts of (b) lime—sulphide and (c) enzymatic processes. ED—epidermis; HF—hair follicle. (*See* Plate 4.2).

BIBLIOGRAPHY

Adewoye, R.O. and Bangaruswamy, S. (1984). "A single unit process for dehairing of hides using the neutral protease from the fruit of *Adenopus breviflorus.*" *Leder.* 35, 78.

Adewoye, R.O. and Lollar, R.M. (1983). "*Adenopus breviflorus* protease. A potential depilatory agent for the tanning industry." *J. Leath. Res.* 1, 1.

Alexander, K.T.W., Haines, B.M. and Walker, M.P., (1986). "Influence of proteoglycan removal on opening-up in the Beamhouse." *J. Am. Leather Chem. Assoc.* 81, 85–100.

Anderson, B., Hoffman, P. and Meyer, K. (1965). "The O-serine linkage in peptides of chondroitin 4- or 6-sulfate." *J. Biol. Chem.* 240, 156–167.

Andrews, R.S. and Dempsey, M. (1967). "Some investigations into methods of unhairing II – Experiments on enzyme unhairing." *J. Soc. Leath. Trad. Chem.* 51, 247–258.

Barrett, A. J. (1994). "Classification of peptidases." *Meth. Enzymol.* 244:1.

Blaschke, U.K., Hedbom, E. and Bruckner, P. (1996). "Distinct isoforms of chicken decorin contain either one or two dermatan sulfate chains." *J. Biol. Chem.* 271, 30347–30353.

Bose, S.M., Madhavakrishna, W. and Das, B.M. (1955). "Mechanism of unhairing skins and hides by means of certain proteolytic or amylolytic enzymes." *J. Am. Leath. Chem. Assoc.* 50, 192.

Bose, S.M., Madhavakrishna, W. and Das, B.M (1960). "A process of unhairing skins and hides by enzymes from germinated ragi (*Eleusine corocana*) for the manufacture of leather." *Bull. Cent. Leath. Res. Inst.* 6,590.

Bronco, S., Castiello, D., Delia, G., Salvadori, M., Seggiani, M. and Vitolo, S. (2005). "Oxidative Unhairing with Hydrogen Peroxide: Development of an Industrial Scale Process for High-Quality Upper Leather." *J. Am. Leath Chem. Assoc.* 100, 45–53.

Burton, D., Reed, R. and Flint, F.O. (1953)."The unhairing of hides and skins without lime and sulphide: the use of mucolytic enzymes." *J. Soc. Leath. Trad. Chem,* 37, 82–87.

Cantera, C.S. (2001b). "Hair-saving unhairing process Part 3. Cementing substances and the basement membrane." *J. Soc. Leath. Technol. Chem.* 85, 93–99.

Cantera, C.S. (2001a). "Hair-saving unhairing process Part 1. Epidermis and the characteristics of bovine hair." *J. Soc. Leath. Technol. Chem.* 85, 1–5.

Cantera, C.S. and Buljan, J., (1997). "Hair—a new raw material—Overview". *World Leather.* 10, 51–56.

Davies, R.M. (1997). "Setting of consent limits for tanning industry trade effluents." *J. Soc. Leath. Technol. Chem.* 81, 32–36.

Dayanandan, A., Kanagaraj, J., Sounderraj, L., Govindaraju, R. and Rajkumar, G.S.(2003). "Application of an alkaline protease in leather processing: An ecofriendly approach." *J. Clean. Prod.* 11, 533–536.

El Baba, H.A.M., Covington, A.D. and Davighi, D. (2000). "The effects of hair shaving on unhairing reactions Part 2: A new mechanism of unhairing." *J. Soc. Leath. Technol. Chem.* 84, 48– 53.

Frendrup, W. (2000). "Hair-Save Unhairing Methods in Leather Processing." UNIDO Report. pp. 1–37.

Gillespie, J.M. (1953). "The depilation of sheep skins with enzymes." *J. Soc. Leath. Trad. Chem.* 37, 344–353

Godfrey, T. and West, S. (1996). *Industrial Enzymology,* 2nd edn. Macmillan Publishers Inc, New York. p. 3.

Godfrey, T. and Reichelt, J. (1983). *Industrial Enzymology: The application of enzymes in industry.* Nature Press, N.Y.

Hannah, R.S., Roth, S.H., 1991. "Chronic exposure to low concentrations of hydrogen sulfide produces abnormal growth in developing cerebral Purkinje cells". *Neurosci. Lett.* 122, 225–228.

Hannigan, M.V., Happich, M.L., Jones, H.W., Windus, W. and Naghski, J. (1968). "Sole leather from enzyme-unhaired hides." *J. Amer. Leath. Chem. Assoc.* 63, 522.

Hedbom, E. and Heinegard, D. (1993). "Binding of fibromodulin and decorin to separate sites on fibrillar collagens". *J. Biol. Chem.* 268, 27307–27312.

Heidemann, E. (1982) "Newer developments in the chemistry and structure of collagenous connective tissues and their impact on leather manufacture." *J. Soc. Leath. Technol. Chem.* 66, 21–29.

Herfeld, H., and Schubert, B. (1969). "The influence of swelling and plumpness of animal hides in the liming process on the properties of leather". *J. Am. Leath. Chem. Assoc.* 64, 198–226.

Ibrahim Syed, K., Muniyandi, J. and Karutha Pandian, S. (2011) "Purification and Characterization of Manganese-Dependent Alkaline Serine Protease from *Bacillus pumilus* TMS55." *J. Microbiol. Biotechnol.* 21, 20–27.

Kamini, N.R., Hemachander, C., Mala, J.G.S. and Puvanakrishnan, R. (1999). "Microbial enzyme technology as an alternative to conventional chemicals in leather industry." *Curr. Sci.* 77, 80–86.

Kuntzel, A. and Stirtz, T. (1958). "Hair root studies." *J. Am. Leath. Chem. Ass.* 43, 445.

Labbe, J.P., Rebegrotte, P., and Turpine, M. (1974). "Demonstrating extracellular leucine aminopeptidase (EC 3.4.1.1) of *Aspergillus oryzae* (IP 410): leucine aminopeptidase 2 fraction." C R Acad Sci (Paris), 278D:2699.

Laxman, R.S. (2007). "Process for the preparation of alkaline protease." US Patent 7186546 B2.

Lindroth, E.G., (1961). "A study of a pancreatic enzyme unhairing system." *J. Amer. Leath. Chem. Assoc.* 56, 31.

Lokanadam, B., Subramaniam, V. and Nayar, R.C. (1989). "Compressibility measurement and the objective assessment of softness of light leathers." *J. Soc. Leath. Technol. Chem.* 73, 115–119.

Macedo, A.J., Silva, W.O.B., Gava, R., Driemeier, D., Henriques, J.A.P., Termignoni, C., (2005). "Novel Keratinase from *Bacillus subtilis* S14 Exhibiting Remarkable Dehairing Capabilities." *Appl. Environ. Microb.* 71, 594–596.

Malathy, S. and Chakraborty, R. (1991). "Production of Alkaline Protease by a New *Aspergillus flavus* Isolate under Solid-Substrate Fermentation Conditions for Use as a Depilation Agent." *Appl. Environ. Microbiol.* 57, 712–716.

Marsal, A., Cot, J., Boza, E.G., Celma, P.J., Manich, A.M., (1999). "Oxidizing unhairing process with hair recovery. Part I. Experiments on the prior hair immunization". *J. Soc. Leath. Technol. Chem.* 83, 310–315.

Menderes, O., Covington, A.D., Waite, E.R., Collins, M.J., (2000). "The mechanism and effects of collagen amide group hydrolysis during liming." *J. Soc. Leath. Technol. Chem.* 83, 107–110.

Merrill, H.B. (1956). " *The Chemistry and Technology of Leather* : Vol. I. F. O'Flahertv. W. T. Roddv. R. M. Lollar. (Eds.). Reinhold. - I. New York.

Money, C.A. and Scroggie, J.G. (1971) "Lysosomal hair loosening of hides." *J. Soc. Leath. Trad. Chem.,* 55,333

Munz, K.H. and Sonnleitner, R., (2005). "Application of Soluble Silicates in Leather Production." *J. Amer. Leather Chem. Ass.* 100, 66–75.

O'Flaherty, F., Roddy, W.T., Lollar, R.M., 1978. In: The Chemistry and Technology of Leather, vol. 1. Krieger publishing company, Malabar, FL, pp. 229–322.

Pal, S., Banerjee, R., Bhattacharya, B.C., Chakraborty, R. (1996). "Application of a proteolytic enzyme in tanneries as a depilating agent." *J. Amer. Leather Chem. Ass.* 91, 59–63.

Paul, R.G., Mohamed, I., Davighi, D., Covington, A.D., Addy, V.L. (2001). "The use of neutral protease in enzymatic unhairing." *J. Amer. Leather Chem. Ass.* 96, 180–185.

Pearson, C.H., Winterbottom, N., Fackre, D.S., Scott, P.G. (1983). "The NH_2-terminal amino acid sequence of bovine skin proteodermatan sulfate." *J. Biol. Chem.* 258, 15101–15104.

Puvanakrishnan, R. and Bose, S.M. (1984) "Immobilization of pepsin on sand: Preparation, characterization and application." *Indian. J. Biochem. Biophys.,* 21, 323-326.

Puvanakrishnan, R., Bose, S.M. and Dhar, S.C., (1981). "Comparative studies on enzymatic unhairing using immobilized pancreatic enzyme product and CLRI enzyme depliant "M" in the manufacture of grain garment leather." *Leath. Sci.* 28, 32–38.

Puvanakrishnan, R., Bose, S.M. and Dhar, S.C., (1980). "A process for the production of immobilized pancreatic enzyme product for use in leather manufacture." Indian patent No.151088.

Puvanakrishnan, R. and Dhar, S.C., (1986). "Recent advances in the enzymatic depilation of hides and skins." *Leather Sci.* 33, 177–191.

Puvanakrishnan, R., and Dhar, S.C., (1988). Enzyme technology in beamhouse practice. Enzymes in Dehairing. NICLAI Publication, Chennai. pp. 92–120.

Raju, A.A., Chandrababu, N.K., Samivelu, N., Rose, C., Rao, N.M. (1996). "Eco-friendly enzymatic dehairing using extracellular proteases from a bacillus species isolate." *J. Am. Leather Chem. Ass.* 91, 115-119.

Rawlings, N. D., and A. J. Barrett. (1993). "Evolutionary families of peptidases." *Biochem. J.* 290, 205.

Reed, R. (1953). "The significance of the mucoids of hides and skins in the lime yard." *J. Soc. Leath. Trad. Chem.* 37, 75–81.

Rohm, O. (1910). "Dehairing and cleaning of skins and hides." German Patent No.268, 873.

Rose, C., Suguna, L., Rajini, R., Samivelu, N., Rathinasamy, V., Kuttalam, I., Ramasami, T. (2007). "Process for lime and sulfide free unhairing of skins or hides using animal and/or plant enzymes." US Patent No. 7198647

Roth, S.H., Skrajny, B., Reiffenstein, R.J. (1995). "Alteration of the morphology and neurochemistry of the developing mammalian nervous system by hydrogen sulphide." *Clin. Exp. Pharmacol. Physiol.* 22, 379–380.

Saravanabhavan, S., Thanikaivelan, P., Rao, J.R., Nair, B.U. (2005). "Silicate Enhanced Enzymatic Dehairing: A New Lime-Sulfide-Free Process for Cowhides." *Environ. Sci. Technol.* 39, 3776–3783.

Schlosser, L., Keller, W., Hein, A., Heidemann, E. (1986). "The utilisation of a Lactobacillus culture in the beamhouse." *J. Soc. Leath. Technol. Chem.* 70, 163–168.

Schonherr, E., Hausser, H., Beavens, L., Kresse, H. (1995b). "Decorin-Type I Collagen Interaction. Presence of separate core protein-binding domains." *J. Biol. Chem.* 270, 8877–8883.

Schonherr, E., Witsch-Prehm, P., Harrach, B., Robenek, H., Rauterberg, J., Kresse, H. (1995a). "Interaction of Biglycan with Type I Collagen." *J. Biol. Chem.* 270, 2776–2783.

Sehgal, P.K., Ramamurthy,G., Muralidharan, C., Gupta, K.B. (1996). "Unhairing of Goat Skins by an Alternative Non-Enzymatic and Sulphide Free Process." *J. Soc. Leath. Technol. Chem.* 80, 91–92.

Shakilanishi, S., Chandrababu, N.K., and Shanthi, C (2011) "Alkaline protease from *Bacillus cereus* VITSN04: Potential application as a dehairing agent." *J. Biosci. Bioeng.* 111, 128.

Sivaparvathy, M., Nandy, S.C.and Dhar, S.C. (1974a). "Autolytic enzyme action on skin." *Leath. Sci.* 21, 183–191.

Sivaparvathy, M., Nandy, S.C.and Dhar, S.C., (1974b). "Effect of various factors in the autolytic enzymatic unhairing of skin." *Leath. Sci.* 21, 297–304.

Sivasubramanian, S., Manohar, B.M. and Puvanakrishnan, R. (2008b). "Mechanism of enzymatic dehairing of skins using a bacterial alkaline protease." *Chemosphere.* 70, 1025-1034.

Sivasubramanian, S., Manohar, B.M., Rajaram, A. and Puvanakrishnan, R. (2008a). "Ecofriendly lime and sulfide free enzymatic dehairing of skins and hides using a bacterial alkaline protease." *Chemosphere.* 70, 1015–1024.

Stanley, J.R., Woodley, D.T., Katz, S.I., Martin, G.R. (1982). "The structure and function of basement membrane." *J. Invest. Dermatol.* 79, 69–72.

Studniarski, K., Sagala, J. and Jonczyk, W. (1975) "Enzymatic unhairing of pig skins." *Leder.* 26, 202.

Studniarski, K., Sagala, J. and Jonczyk, W. (1976) "Determination of the molecular weight of tanning materials from dicyandiamide in DMF-DMSO solution. Use of gel permeation chromatography (GPC) and a flame ionization detector (FID)." *Hikaku Kagaku*, 22, 94.

Sundararajan, S., Kannan, C.N. and Chittibabu, S. (2011) "Alkaline protease from *Bacillus cereus* VITSNO4" Potential application as a dehairing agent." *J. Biosci. Bioeng.* 111, 128–133.

Svensson, L., Heinegard, D., Oldberg, A. (1995). "Decorin binding sites for collagen type I are mainly located in leucine-rich repeats 4-5." *J. Biol. Chem.* 270, 20712–20716.

Taylor, M.M., Bailey, D.G., Feairheller, S.H. (1987). "A review of the uses of enzymes in the tannery." *J. Am. Leather Chem. Assoc.* 82, 153–165.

Thanikaivelan, P., Rao, J.R., Nair, B.U., Ramasami, T. (2002). "Zero discharge tanning: A shift from chemical to biocatalytic leather processing." *Environ. Sci. Technol.* 36, 4187–4194.

Underkofler, L.A. (1976). *Industrial Microbiology.* Miller, B.M. and Litsky, W.(eds.). p.128, McGraw-Hill Book. Co., N.Y.

Watson, R. R. (1976). "Substrate specificities of aminopeptidases: a specific method for microbial differentiation." *Methods Microbiol.* 9:1–14.

Wojdasiewicz, W., Szumowka, K. and Morawiec, R. (1976). "Method for obtaining nubuck leather surfaces by thermal correction of grain." U.S Patent No.3951593

Yates, J.R. (1972). "Studies in depilation. Part X. the mechanism of the enzyme depilation process." *J. Soc. Leath. Technol. Chem.* 56, 158–177.

Yeshodha, K., Dhar, S.C. and Santappa, M. (1978). "Studies on the degreasing of skins using a microbial lipase." *Leath. Sci.,* 25, 36.

17

BATING—STATE OF ART

INTRODUCTION

Bating is the only step in leather processing where enzymatic processes cannot be substituted by chemical processes. The process of bating gives certain desired characteristics in the finished leather. Bating has been practised in different ways over the centuries. The bating process as practised today is by far more pleasant, safer and less offensive in odour as compared to the earlier process using the infusion of dog dung or manure. Modern bating procedures employ pancreatic enzymes or proteolytic enzymes of bacterial origin. Enzyme bate is one of the essential auxiliaries and its use is absolutely essential for the manufacture of leathers such as gloves, shrunken grain softie, nappa, garments and glace kid. The concept of softening hides by treating them in a warm infusion of animal dung has been termed as 'bating' and the product used for such process is known as 'bate'. From the ancient times up to the early part of the 20th century, dungs of dogs, pigeon and hen have been used for this purpose. Strictly speaking, 'bating' refers only to the use of hen or pigeon manure. When dog dung is used, the process has been known as 'puering'.

The first patent on bating agent used a mixture of pancreatic extract and ammonium salts. Wood's (1912) classical investigations opened up a new chapter in the field of bating and puering. He showed that bating was brought about by the action of enzymes secreted by the bacteria present in the dungs. He also demonstrated the presence of protease, amylase and lipase in the warm infusion of dungs. Recently, bacterial organisms have become the major source of bating enzymes that replaces the unpleasant and odour-forming bates and its associated practices of older versions.

OBJECT OF BATING

The main object of bating is to remove some of the non-leather forming proteinous materials such as albumin, globulin and mucoids or cementing substances from hides and skins and to allow the splitting up of collagen fibres so as to help in the penetration of tanning materials and other processing chemicals, thereby giving the finished leather the desired feel, softness, pliability and other characteristic properties. After liming, the skins are usually in swollen condition. The swelling in the pelt fibres must be reduced so that they will not be resilient when tanned. Bating brings about this reduction by removing the lime. But, bating is much more than a mere deliming. Goatskins, which have only been thoroughly delimed, yield a very inferior product.

Bating also brings about the following other effects in the pelts:

(a) produces a silky grain

(b) makes the pelts slippery, non-elastic and flaccid

(c) loosens the scud or dirt, i.e., short and fine hairs, remnants of hairs at follicle site, grease and lime soaps, dark coloured melanin pigments and traces of epidermis, which are easily removed by scudding

(d) loosens remains of hypodermic tissue so that they can easily be removed by scraping

(e) increases the degree of stretch.

METHOD OF BATING

Deliming and bating, the subsequent steps in the processing of pelts after liming, are really two separate operations although they are usually carried out in one step and often overlap each other. Two alternative procedures are being adopted for bating of pelts. One method is to wash pelts and then to carry out a partial deliming. In the latter stages of deliming, the temperature is raised and the bate is added. The other method is to add the bate immediately after washing of the limed pelts is completed and to rely on ammonium salts in the bate for deliming. Almost all commercial bates invariably contain a good proportion of ammonium salts.

EFFECT OF DIFFERENT FACTORS ON BATING

Temperature The most suitable temperature for bating is around 37°C. At higher temperature, the activity is lost, pelt is damaged and a large amount of hide substance is dissolved due to gelatinization. At lower temperatures, the enzyme activity is reduced since bates derived from microbes and trypsin exhibit optimal enzymatic activity around 35–40°C. Warming of the bate liquor serves two purposes:

1. It aids in the removal of decomposed protein matter which has not been washed out with cold water and

2. It provides suitable conditions for the action of the bate enzymes.

Duration   Bating duration varies with the kind of leather produced and the type of raw stock under process. Insufficient bating as well as excessive bating should be avoided. A bating time of approximately 1 h is normally considered to be effective for upper leather manufacture and 3–5 h for glace kid and similar leathers. At times, overnight bating is customary for glace kid, garment leathers and glove leathers. In such cases, where very strong bating action is required, overnight bating can be carried out without the danger of damage to the skin provided precautions are taken to prevent bacterial damage. Fresh bate liquor should be used and pH should be adjusted to 8.0–9.0. Sometimes, common antiseptics such as sodium fluoride, sodium pentachlorophenate or beta naphthol are used for the control of bacterial growth. It is very important that the antiseptic chosen is effective in mildly alkaline solutions and does not inactivate the enzyme. In the manufacture of pigskin glove leather, the relationship of the bating time and the enzyme activity of the bate with the chrome tanning parameters such as chrome absorption and fixation has been critically discussed.

Enzymes take longer duration to penetrate the skin completely, i.e., throughout all the layers of skin. However, most of the non-structural proteins and cementing substances are removed during dehairing or liming operation producing a well-opened-up structure with swelling. The bating step will further remove the above-said materials that are not removed during dehairing or liming operations. Since enzymes have slow penetrating ability, action of the bate enzymes is confined to the grain layer to clean the grain of residual epidermis and damaged hair or keratin and the flesh side. It also avoids the risk of excessive loosening of corium.

Intact collagen is resistant to attack by bating enzyme due to high-order structure, and collagenases are required for degrading them. However, alkali-damaged or modified collagen is susceptible to degradation by proteases that may or may not have collagenolytic activity. If collagen is excessively degraded, damage occurs in the form of grain-enamel loss and/or corium loosening. The duration can be optimized for the bate to ensure the desired effect on the skin or hide subjected to the bating step.

Concentration   The strength of the bating liquor or the quantity of bate to be used is calculated as a percentage on the stock to be bated. Commercial bate preparations contain bate enzymes having varied proteolytic activities, and the tanner can follow the specifications according to the manufacturers. Besides, commercial bate products are not pure enzymes. They may contain a mixture comprising more than one type of enzyme and the types vary in their concentrations. The precise reaction they catalyse may not be known. If the activity is high, the duration to effect the bating step may be reduced. Normally, the quantity varies between 0.5–2.0% on the weight of the limed pelt and for more intensive bating, a higher percentage may be used.

pH   It is well known that the optimal activity of trypsin-type enzyme is at the pH of 8.0-9.0 while the pH of the limed pelt is around 12.5. The pH of the bate liquor may be kept around 8.5 by adding a suitable deliming agent. At the end of the bating process, the pH will be around 7.8–8.0.

The formation of H_2S during deliming and bating of hides, especially at pH 9.0 with a short float has been observed and several methods for decreasing this formation have been suggested (Skrabs, 1976; Schubert, 1976). Substantial amounts of sulphide may be eliminated by additional washing. Also, the sulphide may be oxidized with Na_2O_2 or $MnSO_4$. Under these conditions, drums should not be opened or it is necessary to wear a protective mask.

Use of delimer   The principal materials which a bate contains are a proteolytic enzyme, a carrier for the enzyme like wood flour and a suitable deliming agent such as ammonium chloride or ammonium sulphate or both. The deliming agents are used for the removal of lime and lime salts and to bring down the pH of the pelt to about 8.0 to ensure proper enzyme action. Practical experience shows that ammonium sulphate gives a firmer as well as fuller flank leather and is preferred for the production of shoe upper leather from calf skins and cattle hides. But, if the pelts are not washed properly, insoluble calcium sulphate forms a crust on the grain surface of the pelt and creates problems in the subsequent processes.

On the other hand, if ammonium chloride is used for deliming, readily soluble calcium chloride is formed. That is why it is still preferred by some tanners. However, a higher concentration of calcium chloride has a detrimental effect on collagen. Hence, it is advisable to use both the ammonium salts in suitable proportion to obtain better results. The concentration of the deliming agent should be chosen so that it should have no denaturing or inhibitory action on the enzymes. A concentration of 0.1% ammonium chloride will be sufficient to render enzymes sufficiently active.

Deliming with agents based on naphthalene sulphonic acid can lead to partial or even total enzyme inactivation if hides are treated simultaneously or immediately after addition of delimer with bating agent. It is, therefore, recommended to separate the deliming stage from the bating stage and add the bating agent only 15–20 min. after the addition of deliming agent based on naphthalene sulphonic acid.

Washing prior to bating   The dehaired stock contains lime and loose materials resulting from the action of alkali on the skin proteins. The stock is usually washed in a drum or paddle to remove extraneous matter and also some of lime present within the skin. Inadequate washing may leave excessive lime. Hence, the washing operation should be considered as part of bating and should be carefully controlled.

Lime blast   Large amount of water is used for washing skins during leather processing. It is important that the quality and uniformity of the material are controlled rigidly in all steps. Water containing high proportion of carbonates may cause precipitation of insoluble carbonate of lime on the grain of the skin. This precipitation is known as 'lime blast'. This phenomenon may also occur upon excessive exposure of dehaired stock to air. Limed stock should always be protected against drying-out, between the time it is dehaired and it is entered into the bate paddle. Lime blast is usually detected by the harsh feel of the surface caused by the precipitated carbonate. The water supply should be tested for temporary hardness and water of more than 5° temporary hardness is not desirable in the washing and bating operations. Despite precautions, if lime blast has occurred, a careful wash of the stock with sulphuric acid or hydrochloric acid just before the addition of bate should correct the condition.

Uniformity of stock   In order to get uniform bating effect, it is always better to select uniform stock. If hard and soft skins or skins of different sizes are mixed together and are subjected to bating operations, it is impossible to get uniform bating effect. Hence, it is advisable to select the pelts before bating and to grade them according to the size and nature of the skin.

Mechanical agitation   Certain physical changes take place when the skins undergo bating operation. It is always better to perform bating operation in a paddle for proper regulation of temperature and for controlling deliming as well as enzyme action. Nevertheless, drum bating also has some advantages, viz., (a) economy in handling goods since drum can be used for other operations as well and (b) use of smaller quantity of bate liquors. It is advisable to stop the drum periodically to prevent the development of too much heat and also the drum speed should be kept minimum. Bating in drums is prevalent in European tanneries.

Water–skin ratio   The proportion of water to skin is important in bating. If the water is less, there will be undue "crowding" and the skins will not open up properly. If water is too much, the skin will float resulting in the enzymes getting unnecessarily diluted. It is better to maintain the ratio of skins to water close to 1:4 in the case of paddle bating and 1:3 in the case of drum bating.

Use of old bate liquors   In most of the glace kid tanneries in USA and UK, old bate liquors are also used to get proper bating action within shortest possible time so as to finish the bating operation within the same day.

Effect of crosslinking agents   Modified glutaraldehyde has a good fat-dispersing effect, and its use in bating imparts a beneficial effect on the grain pattern, promoting pickling penetration and chromium dispersion. This facilitates efficient retanning and dyeing.

EXTENT OF BATING

In practice, bating cannot be defined to a formula and must be empirically adjusted to take into account the following variables:

   (a) the type of leather being manufactured
   (b) the liming process which precedes bating and
   (c) the tanning and finishing process, which is to follow.

An experienced tanner, however, knows by practice, the desired effects from the use of a good bate. In the case of glove leather, the main emphasis is to have leather which is soft, with a grain having a pleasing look and of maximum possible stretch. To achieve these objectives, the bating action must be strong enough so that the interfibrillary substances are loosened and partially solubilized for subsequent removal by washing. In the case of upper leather, the function of the bate is merely to clean the grain to give the leather a smooth appearance. Therefore, obviously, the bating in this case will be very light. For sole leather, bating is not required. However, some tanners use weak bate for the manufacture of cycle saddle leather.

SIGNS OF BATING EFFECT

Although the effect of bating operation is due to the action of enzymes present in the bate, in practice, bating cannot be defined and the efficiency of bating must be judged only empirically as well as qualitatively by experience. The following signs appearing on the bated pelts will indicate the effect of proper bating:

  i.   the grain of the bated pelt should feel silky and slippery
  ii. the thumb impression should be retained on the grain when pressed
  iii. the pelt should be perfectly fallen and flaccid
  iv. the pelt should be white, clean and porous
  v.   the scud should be easily removable
  vi. the flesh should come off easily when scratched with finger nail

MECHANISM OF ACTIVATION OF PANCREATIC COMPLEX

The pancreas of pig, buffalo, goat, sheep and cattle have been widely used for the production of bates. A comparatively rich source for the proteolytic enzymes is the big pancreas. The bovine pancreas, however, is available abundantly in India. Throughout the manufacturing process of bate from pancreas, care must be taken for the preservation of the proteolytic enzyme system. The proteolytic enzymes in the pancreas are present in inactive form: chymotrypsin as chymotrypsinogen, trypsin as trypsinogen, and carboxypeptidase as procarboxypeptidase. Hence, for the preparation of enzyme bates, the pancreas is suitably treated to convert the inactive precursor of the enzyme into active form.

Trypsinogen is produced by the pancreas and secreted in the pancreatic juice. The conversion of trypsinogen to trypsin, a serine protease, is by the splitting of the 'strategic bond' between the residues Lys (6) and Isoleu (7) releasing a hexapeptide (Davie and Neurath, 1955). Irrespective of the method of activation, the aim of conversion is to split this bond for the formation of the first detectable active enzyme, β-trypsin. The adjacent four aspartyl acid residues have clearly a shielding effect and once this is broken, deshielding of the catalytic centre takes place and the enzyme becomes active.

Subsequent cleavage of β-trypsin between Lys (131) and Ser (132) leads to the formation of α-trypsin, a two-chain structure held by disulphide bonds. A further cleavage of α-trypsin at the bond between Lys (176) and Asp (177) yields another active form of trypsin, psuedotrypsin (Smith and Shaw, 1969). Trypsin cleaves the peptide bond on the carboxyl side of basic amino acids, lysine and arginine.

Chymotrypsinogen, which has a single polypeptide chain of 245 residues and five intra-chain disulphide crosslinks contributed by five cystine residues, is activated by trypsin. Trypsin removes two dipeptides from positions 14-15 and 147-148 of chymotrypsinogen by hydrolysis to yield active chymotrypsin. Procarboxypeptidase is converted to carboxypeptidase by enterokinase or trypsin. Proelastase is also activated by trypsin. Chymotrypsin cleaves the peptide bond on the carboxyl side of amino acids with aromatic side chains, i.e., phenylalanine, tyrosine and tryptophan.

Fresh bovine pancreas, collected immediately after the slaughtering of the animals, is washed and cleaned in water. They are minced using a meat-mincing machine. The fine pulp collected is used for activation. There are three methods of activation of pancreas.

Autocatalytic activation   Trypsinogen is converted to trypsin by trypsin itself and this process is called autocatalytic activation of trypsinogen. To the finely minced pancreatic pulp, 1 N NH_2SO_4 is added to adjust the pH to 4.5. The pulp is thoroughly mixed and left for conversion of the precursors to their active form over a period of 20 h at a temperature of 25°–30°C. After conversion, the pulp is taken for further processing to manufacture bate. The pH value of the pancreatic pulp is kept below 5.0 since the yield of trypsin is more.

Activation by acid protease   The term 'acid protease' denotes a proteolytic enzyme or group of enzymes, which catalyse the hydrolysis of peptide bonds at the acid pH range. The major source for the isolation of these acid proteases is the culture filtrate of moulds. Several fungal acid proteases from *Aspergillus saitoi*, *Aspergillus niger*, *Aspergillus oryzae* and *Penicillium janthinellum* have the ability to activate pancreatic trypsinogen to trypsin. The mechanism of activation of trypsinogen by acid protease and that of autocatalytic activation are said to be identical, since both mechanisms involve the cleavage of the Lys (6)–Isoleu (7) bond with liberation of the hexapeptide—(valyl-(aspartyl)$_4$–lysine). CLRI has patented a process in which the activation of pancreatic enzymes has been carried out by the use of acid protease from *A. fumigatus*. Maximum activation is obtained under the following conditions:

i.   The pH of activation — 3.1

ii. The temperature of activation — 37°C

iii. Concentration of enzyme — 450 mg for every 50 g of pancreatic pulp; and

iv. The duration of activation — 60 min.

After conversion, the pulp is taken for further processing to manufacture bate.

Activation in presence of calcium ions   When 0.02 M calcium chloride is added to the pulped pancreas, it will inhibit completely the formation of inert protein and the trypsinogen is quantitatively converted to trypsin. The pancreatic pulp is activated using 0.1 M $CaCl_2$ in borate buffer, pH 8.0. The conversion is allowed to take place for 20 h at a temperature of 18°–20°C. After transformation, the pulp is taken for further processing to manufacture bate.

Further processing steps to manufacture bate   The pancreatic pulp, after activation of the inactive precursors, is mixed with wheat bran or wood flour, ammonium sulphate and ammonium chloride in the following ratio : For one unit weight of pancreas, (i) dry wheat bran - 2 unit weights, (ii) commercial ammonium sulphate - 2 unit weights and (iii) commercial ammonium chloride - 1 unit weight. Wheat bran adsorbs the enzyme and makes the product dry. After thorough mixing, the blended mixture is dried in a suitable drier wherein the temperature is maintained below 45°C. After drying, the material is ground to a fine powder in a disintegrator. Wood flour is a lignocellulose that serves as an alternative diluent of the bate. Since there is no affinity between diluent and bate, the enzyme is readily available for reaction. It has the advantage of not contributing to the ammoniacal nitrogen in the effluent, but it contributes to the raise in suspended solids in the effluent.

The use of kaolin, a form of clay, as a bate diluent is less common. It will contribute to an increase in the content of suspended solids in the effluent. Besides, due to strong affinity between bate enzyme and the diluent, there will be a delay after adding the bate to the solution before the enzyme is desorbed from the surface of diluent.

ENZYME BATES

From Animal Pancreas

Enzymes from animal pancreas are the oldest in use for the production of enzyme bates. The pancreatic enzymes required for these bates are obtained from animal pancreas while the fungal enzymes have been extracted from the mould *A. parasiticus* or *A. oryzae*. The final products are based on the blends of the two enzymes. The bates thus developed have been tested on various types of hides and skins and the action of these bates has been compared with those of standard imported bates. A pancreatin product for use in bating has been prepared by treating finely pulped animal pancreas with H_2SO_4 and then with fused $CaCl_2$ and blending the activated pulp with powdered husk of paddy or a similar carrier material, ammonium sulphate and ammonium chloride (Dhar *et al.*, 1978).

From Plants

Rose *et al.* (2007) have reported the use of animal or plant enzymes in lime and sulphide-free dehairing process wherein dehairing and bating steps are carried out in a single step. The process employs proteases obtained from plant tissues and exudates of *Carica, Euphorbia,*

Calotropis and *Plumeria* and the animal proteases are sourced from pancreas. Besides, the process employs mixtures of plant and animal proteases. Monsheimer and Pfleiderer (1976) have described the preparation of tannable pelts from skins and hides wherein concurrent soaking, dehairing, opening up of the fibre structure and bating are effected in a single procedural step by treatment with enzymes in an aqueous bath containing enzymes and a thioglycolic agent. It employs papain, trypsin, and fungal and bacterial proteases for producing the tannable pelt. A recent report shows that when dehairing is carried out using enzyme, bating step may not be necessary (Sivasubramanian *et al.,* 2008).

From Moulds

For the commercial production of enzymes, moulds are grown on suitable medium containing wheat bran and other ingredients. Underkofler and Hickey (1954) have described a process for the manufacture of enzyme bates from mould source. Trabitzch (1966) has reported the use of enzymes from *Aspergillus* species in bating and dehairing. Systematic studies have revealed that *A. parasiticus* is one of the best sources for the production of enzymes.

From Bacteria

Bating properties of several kinds of bacterial proteinases have been studied in comparison with conventional pancreatic enzymes and the bating effect is determined by the liberation of nitrogen before and after the enzymatic treatment of hides using the modified biuret method (Orlita, 1975). A procedure has been developed for bating pig skins using an enzyme preparation from *B. subtilis* and the bated skins have exhibited good physicomechanical properties. Bacterial preparation from *S. rimosus* and *B. licheniformis* have been tested for their bating action and it is found that solubilization of collagen has been less pronounced under the influence of microbial proteases than under the influence of pancreatic proteases (Pokorny *et al.,* 1983). Subtilisin and nagarase are proteases isolated from *B. subtilis*, which preferably act against small polypeptides by non-specific cleavage. Another non-specific enzyme having broad specificity that is used in bating formulation is pronase, which is obtained from *Streptomyces griseus*.

Laboratory-scale experiments have been carried out to test the efficiency of the extracellular protease from *Bacillus cereus* MCM B-326 for bating of buffalo hides. Cattle dung and commercial bate powder (com bate) have been used as controls. Results show that protease-treated pelt is shown to be free from scud and pigments, with clean and fine grain, and is white, smooth and silkier. Histological sections of protease-treated bated pelts show greater opening up of collagen fibres. This study indicates that *Bacillus* protease could be developed as a potential bating agent (Zambare et al., 2010).

MERITS AND DEMERITS OF DIFFERENT SOURCES OF ENZYMES FOR BATE MANUFACTURE

Many workers have tried to find out the comparative efficiency of bates prepared from different raw materials. But, no definite conclusions have been drawn. Eling and Lollar (1953) have pointed out that both microbiological and pancreatic bating preparations have similar action. According to Moore (1952), however, pancreatic enzyme has greater efficiency in bating than the bacterial or fungal enzyme since it contains protease in addition to lipase and amylase. Practical experience shows that pancreatic enzyme is superior to microbial enzyme in imparting the fineness of grain for the manufacture of glace kid. Since pancreas is used for many medicinal preparations, it is rather expensive. Hence, moulds, grown on a cheap medium, can be used as a substitute. If a combination of both mould and pancreatic enzymes in suitable proportions is made, it will be an ideal bate for different types of leather.

The usage of pancreatic or bacterial bates has to be justified because the effects produced by them in leather is different in terms of area, objective as well as softness assessment and the physical properties. This is because of the functional difference between pancreatic and bacterial bates, especially the ability to degrade elastin. Pancreatic and digestive enzymes do not degrade elastin during the dehairing and bating operation whereas bacterial bates often contain elastases. Therefore, a comparision between leathers made by pancreatic and bacterial bates is essential and this provides information about leather character in terms of elastin removal and the absence of elastolytic activity. Microbial bates especially bacterial bates are gaining importance from the the economical viewpoint as these can be manufactured in bulk by fermentation, and scale-up of these enzymes is also highly feasible. Besides, microbes offer enzymes with wide-ranging properties, and the improvement of enzyme function is also achievable by recombinant DNA technology.

ACID BATING

Normally, bating is carried out in the alkaline pH range, since the enzymes present in pancreatic and microbial bates are optimally active at that pH range. Studies have indicated that in comparison with the enzymatic bating process in alkaline or neutral pH range, the pelts treated with proteolytic enzymes in acidic pH range derive special characteristics (Dhar, 1974). The main features of bating in acidic condition are the following:

1. bating operation can be carried out at a temperature between 25°C and 32°C;
2. deliming, bating, pickling and chrome tanning can be carried out in the same drum avoiding the frequent removal of pelts from the drum;
3. reduces the operational cost;
4. manual scudding can be avoided; and
5. the grain remains tight after finishing.

Other important advantages are (i) partially chrome-tanned skins can be treated with acid bate for the manufacture of good-quality garment leathers. Alkaline bate is not suitable for such treatment and (ii) pelts which are limed in the acid condition with an oxidizing agent like chlorine dioxide are subsequently bated using an acid bate (Dutta, 1985).

Acid bating is carried out by two methods (Panneerselvam and Dhar, 1980). In the first method, well-soaked goat-skins, after dehairing by a standard liming process, are delimed in a drum using 1.5% ammonium chloride. The delimed pelts are lightly pickled in a drum to bring the pH of the pelts to 3.5. Bating is carried out in the same drum with 1.5% acid bate for 2 h at a temperature of 30°–32°C. After the period, the pelts are pickled, chrome-tanned and finished as glace kid. In the second method, well-soaked goat skins are limed, delimed and pickled as usual. The pickled pelts are partially tanned using 5% chrome extract. The well-washed blue chrome leathers are bated with 2% acid bate at pH 3.5 in a drum for 2 h at room temperature (30–32°C). The blue chrome leathers after acid bating are fully chrome-tanned using an additional 5% chrome extract. They are then neutralized, fat-liquored, dyed and finished as grain garment leather.

A critical review of the literature has shown that not much work has been done on the production and application of acid bates in leather manufacture. Earlier, certain pepsin-type proteolytic enzymes, which are active at pH 1.2–2.0, have been tried, but no typical action has been observed. The use of fungal and bacterial proteases, whose optimum activity occurs in acidic pH range, in bating of hides and skins has been described by Pfleiderer (1968). A process for the manufacture of an enzyme product using papain for bating of skins and hides in acidic pH range has been patented. An acid proteinase with optimal activity in the pH range of 3.0–4.0 has been prepared from the culture filtrate of *Penicillum janthinellum* and this product has been found to possess good bating property when tested with delimed pelts. Comparative chemical, physical and microscopical studies of glace kid produced by using CLRI Acid Bate, Acid Protease and an alkaline bate have shown that CLRI Acid bate and Acid Protease exhibit good bating efficiency similar to that of a standard alkaline bate (Dhar and Sambasiva Rao, 1974). A systematic study on the chemical, physical and microscopical assessment on the quality of leathers produced using Fungal Acid Bate AF and Unipon ORU has shown that both Fungal Acid Bate AF and Unipon ORU produce equally good glace kid as well as grain garment leathers (Panneerselvam and Dhar, 1980).

For the manufacture of garments from fur skins, bating should be such that it removes soluble proteins and cementing substances present in the natural skins without loosening the wool of the skin. The natural grease or fat present in most of the fur-bearing skins should also be removed completely. To achieve these objectives, CLRI Acid Bate has been used for bating fur skins in acidic condition. Addition of a small quantity of fungal lipases whose optimum activity lies in acidic pH range could remove skin grease quite efficiently.

Recently, delimed cowhides have been subjected to bating with a pancreatic amylase and optionally, a proteinase at an acidic pH range for 20 min. After the period, the pH is adjusted to 5.0 with a non-swelling acid like sulphophthalic acid or naphthalene sulphonic acid. Pickling is also undertaken in the same bath by adjusting the pH to 3.5 with the same non-swelling acid (Monsheimer and Pfleiderer, 1980).

IMMOBILIZED ENZYMES FOR BATING

Immobilized enzymes have been the subject of increasing interest because of the enhanced stability achieved upon immobilization (Klibanov, 1979). The main advantage of immobilized enzymes is their reuse. For the first time, an immobilized pancreatic enzyme product has been prepared by the immobilization of activated pancreatic enzymes on a cheap and indigenously available support using a bifunctional agent (Puvanakrishnan *et al.*, 1983). A bating process for the manufacture of glace kid using the immobilized pancreatic enzyme product has been described in detail and the efficient reuse of the same immobilized enzyme product successively thrice has been demonstrated (Puvanakrishnan *et al.*, 1980).

MECHANISM OF BATING

Many theories have been put forward from time to time to explain the mechanism of bating. None of them, however, could satisfactorily explain all the known effects of bating on the pelt as well as on the finished leather.

Since alkaline condition is essential for the removal of certain denatured proteins present in the skins after dehairing, a proteolytic enzyme must be selected which will be active in the presence of a small amount of alkali. Hence, trypsin type of enzymes have been utilized for the digestion of the denatured protein in alkaline condition.

It is well known that one of the actions of bate is the removal of breakdown products such as grease, epidermal residue and glands, which is brought about by a combination of proteolytic and lipolytic enzymes present in the bate. Bating enzymes produce a collapse of the collagenous fibre structure (Pfannmueller, 1956) and such a collapse may expel the degraded epidermal components or scud from the interior portion of the skin. Further, mechanical agitation during bating loosens and also aids in the removal of scud and surface debris.

The decrease of plumpness in pelt during bating has not yet been fully explained. It is not possible to obtain such an effect by applying ammonium salts alone and this deswelling might be due to the removal of non-collagenous proteins that still persists after dehairing (Bienkiewicz, 1983).

Elastic tissue fibres are found in the epidermal system as a network of branching fibres beneath the grain and also in the arteries of the blood vessel system throughout the skin

(Pfannmueller, 1956). As per the old schools of thought, bating does not change elastin. But, according to Ornes and Roddy (1960), 30–80% of elastin is removed by the use of pancreatic enzyme preparations. Although pancreas contains elastase, it is not definitely known whether elastase is present in other bating preparations containing fungal enzymes. Since elastase and non-collagenolytic activities are the requirements of efficient bate preparations, assessing elastase as well as collagenase activity in bating preparations is necessary.

Kritzinger (1951) has stated that the partial breakdown of the hair muscle (erectorpili) is one of the important functions of bating. The efficiency of bates is related with their power to digest erectorpili muscle and the digestion of hair muscle would be sufficient to prevent raising of the grain in bated pelts.

Electron microscopical studies on the action of proteolytic enzymes on collagen by Borasky and Rogers (1952) have revealed that intact fibrils are resistant to tryptic attack but filamented fibrils are readily digested. They have suggested that the main function of bating is to remove all non-collagenous proteins and degraded collagen fibrils.

During bating of pigskins with α-amylase, it has been shown that α-amylase attacks the carbohydrates, thereby promoting the loosening of fibre structure. This effect is shown by the increase in carbohydrates and hydroxyproline and by the changes in protein–peptide fractions and molecular weight distribution in the thermal decomposition products. α-amylase catalyses the hydrolysis of bonds in the main chains of the collagen molecules as well as the inter- and intramolecular cross linkages and it degrades the collagen to units of 24,000–28,000 kDa molecular weight, which are the basic building units associated with the carbohydrates.

Studies on the penetration of bating enzymes such as trypsin and chymotrypsin in hides have shown that a loose fibre interweave increases bating time and high enzyme dosage aids the penetration power of these enzymes and increases the enzyme concentration in the hide.

BIBLIOGRAPHY

Bienkiewicz, K. (1983). *Physical chemistry of Leather Making.* Robert E.Krieger Pub.Co. p.288.

Borasky, R. and Rogers, J.S. (1952). "Effects of tannery processes on the electronoscopic appearance of bovine hide collagen fibrils." *J.Am.Leath.Chem.Assoc.* 47,312.

Davie, E.W. and Neurath, H. (1955). "Identification of a peptide released during autocatalytic activation of trypsinogen." *J.Biol.Chem.* 212, 515–529

Dhar, S.C, Bose, S.M. and Lakshmibarathi, K. (1978). Indian Patent No.143755

Dhar, S.C. and Sambasiva Rao, P. (1974). "Studies on enzymes for leather manufacture: part V. Comparative assessment of acid and alkaline bates for the manufacture of glace kid." *Leath.Sci.* 21,81–86.

Dhar, S.C. (1974). "Production and application of enzymes in the pretanning process of leather manufacture." *Leath. Sci.* 2, 39–47.

Dutta, S.S. (1985). *Introduction to the Principles of Leather Manufacture,* 3rd edn. Indian Leath.Technol. *Assoc.,* p.194

Eling, R. and Lollar, R.M. (1953). "The effect of microbiological and pancreatic bates on limed skins." *J.Am. Leath.Chem.Assoc.* 48,135.

Klibanov, A.M. (1979). "Enzyme stabilization by immobilization." *Anal. Biochem.* 93, 1–25.

Kritzinger, C.C. (1951). "A new approach to beamhouse processes." *J.Am.Leath.Chem.Assoc.* 46, 350.

Monsheimer, R. and Pfleiderer, E. (1976). "Method for preparing tannable pelts from animal skins and hides." US Patent no. 3966551.

Monsheimer, R. and Pfleiderer, E. (1980). "Fermentative decomposition of depilated stock leather within acidic range by enzyme under floor loosening condition simultaneously." German Patent No.2856320

Moore, H.N. (1952). "Bating studies. II. The comparative bating efficiency of bates of different biological sources." *J.Am.Leath.Chem. Assoc.* 4,110.

Orlita, A. (1975). "Bacterial proteases and their bating effects." *Kozarstvi.* 25,131.

Ornes, C.L. and Roddy, W.T. (1960). "The isolation of elastic tissue of animal skin." *J. Am. Leath. Chem. Assoc.* 55,124.

Panneerselvam, M. and Dhar, S.C. (1980). "A new procedure for bating with fungal acid protease and its effects on finished leathers." *Leath.Sci.* 27, 207–213.

Pfannmueller, J. (1956). *Chemistry and Technology of Leather.* (O'Flaherty, F., Roddy, W.T. and Lollar, R.M.,eds.). Reinhold Pub. Co. pp.359–60.

Pfleiderer, E. (1968). "Application of proteolytic enzyme in acid pH range for soaking, dehairing and bating." *Das Leder,* 19, 301–303.

Pokorny, M., Restek, A and Petravic, J. (1983) "Activity of pancreatic and microbiological enzymes in leather bating." *Kozarstvi,* 33, 224.

Puvanakrishnan, R., Bose, S.M. and Dhar, S.C. (1980). "Comparative studies on bating using immobilized pancreatic enzyme product and pancreatin bate in the manufacture of glace kid." *Leath.Sci.* 27,81–87.

Puvanakrishnan, R., Bose, S.M. and Dhar, S.C. (1983). "A process for the production of immobilized pancreatic enzyme product for use in leather manufacture." Indian Patent no.151008

Roddy, W.T. and Lollar, R.M. (eds.) Reinbold Pub. Co. pp. 359–60.

Rose, C., Suguna, L., Rajini, R., Samivelu, N., Rathinasamy, V., Kuttalam, I., Ramasami, T. (2007). "Process for lime and sulfide free unhairing of skins or hides using animal and/or plant enzymes." US Patent no. 7198647.

Schubert, B. (1976) "Investigations on the reduction of formation of hydrogen sulfide in dliming and bating." *Leder*. 27,157.

Sivasubramanian, S., Manohar, B.M., Rajaram, A. and Puvanakrishnan, R. (2008). "Mechanism of enzymatic dehairing of skins using a bacterial alkaline protease." *Chemosphere*. 70, 1025–1034.

Skrabs, R. (1976). "Hydrogen sulfide hazards in leather production." Leder. 27,153.

Smith, R.L. and Shaw, E. (1969). "A modified bovine trypsin produced by limited autodigestion." *J. Biol. Chem*. 244, 4704.

Trabitzch, H. (1966). "Concerning the possibilities of employing enzymes in the beamhouse – A review." *J. Soc. Leath. Trad. Chem*. 50,382–389.

Underkofler, L.A. and Hickey, R.J. (1954). *Industrial Fermentations*. Vol. II p.109, N. Chemical Publishing Co.

Wood, J.T. (1912). *The puering and drenching of skins*. E & F.N Spon Ltd., London.

Zambare, V., Nilegoankar, S. and Kanekar, P. (2010). "Application of protease from *Bacillus cereus* MCM B-326 as a bating agent in leather processing." *The IIOAB Journal*. 1, 18–21.

18

DEGREASING—ANALYSIS OF DIFFERENT SYSTEMS

INTRODUCTION

The presence of excess amount of natural grease in sheepskins results in a number of defects in the finished leather, viz., fatty spues, uneven dyeing and finishing, waxy patches in alum-dressed leather and pink stains in chrome blues. Removal of natural grease from the interfibrillary spaces facilitates more even penetration of tanning materials, fat liquors, dyes, etc. The grease or fat should be removed during the pretanning operations in order to obtain sufficiently soft and pliable leather for garment manufacture. The process of removal of excess natural fat from hides and skins is known as "degreasing" and it is an essential step in the production of glove and clothing leather.

The normal fat present in hides and skins is removed either as soluble lime soap or, hydrolysis products like fatty acids during liming and bating. The liming process is reported to cause the rupture of fat cells present in the epidermis and saponification of the liberated fatty acids (Yeshodha *et al.*, 1978a). The subsequent scudding operation squeezes out the saponified animal fat to the surface, which can then be scraped off. But, this process removes most of the epidermal lipids, leaving only the corium lipids relatively unaltered. Hence, there is the absolute necessity for special degreasing operations for the processing of greasy hides and skins. The degreasing operations are carried out after bating or after pickling.

The problem of pollution control arising out of the discharge of liquid wastes from tanneries has remained almost as critical today as it was at the establishment of tanneries. It has already been observed that the BOD of the pickling effulent is in the range of 1600–2200 mg/L (Kothandaraman and Aboo, 1977). Since degreasing follows pickling, it is but logical to infer that the degreasing effluents due to solvents and surfactants will add to the existing BOD load of the pickling effluent. Moreover, kerosene, chlorinated hydrocarbons and white spirit used in the degreasing system add to the toxicity of environment and effluents, though systems are available to optimize the recovery of solvents. It is imperative that an alternative technology is worked out to obviate the pollution problem arising out of the degreasing of effluents by the use of enzymes.

Grease present in skins is made up of three main constituents, viz., triglyceride fat, waxes and fatty acids. A diagrammatic view of the distribution of fat in a greasy skin is presented in Figure 18.1. In addition to its presence in the fat glands of the papillary layer, grease is also observed as droplets in the reticular layer, at the junction of papillary and reticular layers and in the subcutaneous adipose tissue. The type and amount of grease varies between species (Table 18.1), breeds, different environments and types of feeding.

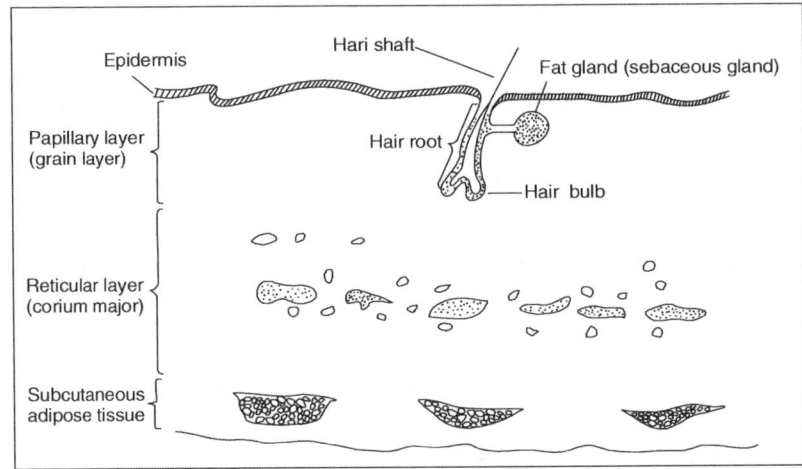

Figure 18.1   Diagrammatic view of the distribution of fat in a greasy skin

TABLE 18.1   *Grease in various hides and skins

Type of skin	Amount of grease (percentage of dry skin weight)
Ox	3–5
Sheep	
New Zealand	20–30
Domestic	20–30
Abyssinian	6–8
Sudan	4–6
Goat	3–5

* (Palmer and Marsde, 1981)

METHODS OF DEGREASING

Solvent Degreasing

Fat solvents like kerosene and trichloroethylene are normally used for degreasing. As these solvents are insoluble in water, they cannot be effectively employed for degreasing wet pelts. They are used along with nonionic emulsifiers which disperse the solvent in water, thus facilitating its entry inside the pelt.

White spirit is the most commonly used solvent under the low flash solvents category. The main disadvantage of low flash solvents is their high volatility and inflammability leading to hazards of toxicity and fire. Hence, application of this type of solvents is impossible using the conventional drum. However, low flash solvents are extremely efficient for wool on sheepskins containing large amounts of grease.

Among high flash solvents, viz., trichloroethylene, perchloroethylene and kerosene, kerosene is the most widely used. Its main advantage is that it can be used in the conventional tannery drum and is much cheaper than any other solvent. Kerosene degreasing is a drum process and it is usual to proceed directly to chrome tanning after degreasing.

Where grease recovery is desired, odourless white spirit is preferred to paraffin, as the latter contains a high boiling fraction which contaminates the recovered grease. Mechanical damage to lambskins can be minimized by the use of kerosene with high aromatic hydrocarbon content. It is also possible to obtain equal degreasing efficiency at 30°C which is 10°C lower than with the normal kerosene.

Before the development of surfactants, for the removal of grease, the skins used to be kept in pickled condition for 3 weeks to allow the moisture content to reduce to about 35–40%. The introduction of surfactants has greatly facilitated solvent degreasing although the amount of moisture remaining in the skin should not exceed 65% because, a stable emulsion of solvent, grease and surfactant may be formed which is exceedingly difficult to remove from the skin and to break for solvent recovery (Briggs,1981).

Mode of application   The solvent (10–20% on pelt weight) together with emulsifier (0.5–1.5% also on pelt weight) is necessary for effective degreasing and the drumming time is kept for about 30–45 min. After degreasing, the pelts are washed in 6.0% (W/V) brine solution for removal of grease (Sarkar, 1981).

Innes and Pankhurst (1952) have first suggested the use of paraffin in the degreasing of woolly skins. From a comparison of white spirit and paraffin for the degreasing of sheepskin, Mitton and Pankhurst (1953) have observed that recovery of grease and solvent is greatly facilitated by the use of white spirit. Studies on the degreasing of rabbit, lamb and Persian lambskins using perchloroethylene and trifluorochloroethane have shown that the latter does not damage the skins during drying after degreasing due to its lower boiling point and lower heat of evaporation (Ries, 1971). It has been reported that extraction with ether produces a pronounced weakening of the stratum corneum, which is reflected in the appearance of cracks and upturned stratum corneum cells (Wolfram *et al.*, 1972).

An improved degreasing method for sheepskin in the tawing process has been developed, in which moisture-free hides have been extracted with a low boiling petroleum fraction at 45°–50°C in the presence of a cationic polyelectrolyte and about 70% of the extraction solvent has been recovered in a decantation stage and recycled to the degreasing operation (Gavend *et al.*, 1975). It has also been shown that the extent of degreasing achieved increases from 15 to 80% when the moisture content of the hides is reduced gradually.

The addition of emulsifiers to solvent system containing perchloroethylene and light petroleum or kerosene has increased their degreasing potential so that the residual fat in sheepskin and the cowhide is about 5% and gives stable emulsions of the fat which is easily separated in effluent control systems. It has also been shown that pickled sheepskins are more effectively degreased in a double greasing system using 16.4% kerosene, 8.6% perchloroethylene and 0.3% nonionic surfactant at 35°C. The process removes 78.9% of the fatty materials.

For the removal of high quantities of inter-fibre fat of sheepskin, the procedure involves drumming in 10% trichloroethylene and 1% surfactant with the pelts in bated condition at pH 7.0–8.0, when they are less swollen and the structure is more accessible (Bienkiewicz, 1983). In the case of New Zealand pickled pelts containing about 25% fat by weight, it is common to degrease by drumming with 40% kerosene and 1% surfactant. The resultant milky emulsion of grease solution in solvent and water is then washed off with 5% brine solution (Sharphouse, 1983).

Dry Degreasing

This process is carried out on the dry leather and it consists of treating the leather with solvents such as white spirit, trichloroethylene or perchloroethylene. The leather is hung on frames in steel tanks or in chambers which are then either filled with the solvent or solvent is sprayed over the skins by rotating arms. A considerable amount of solvent remains in the leather and this is recovered by blowing warm air into the chambers, and taking the vapour to water-cooled condenser system which strips out the solvent for reuse. This process is used in cases where it is undesirable to dry-drum the leather, as in the case of firm vegetable-tanned leather or long-woolled sheepskins where matting or felting of the wool must be avoided (Sharphouse, 1983).

Aqueous Emulsification or Aqueous Degreasing

Surfactant solutions, because of their property of forming micelles, can dissolve and suspend grease and wax, although they are not efficient as simple solvents (Reed, 1966). They are applied at a temperature of about 60°C, so that all the grease is in a mobile liquid state and thus readily emulsifiable. Pickled pelt shrinks in the region of 50°–60°C and hence before this method can be used, the skins must be pretanned suitably with formaldehyde or glutaraldehyde. The tanning agent must not combine with fatty acids, and hence salts of chromium, zirconium, iron or aluminium cannot be used. Having raised the shrinkage temperature to 70–80°C by pretannage, the skins can withstand a temperature of 60°C.

Temperature, pH, concentration of emulsifying agent, neutral salt content etc. are the important controlling factors in aqueous emulsification method of degreasing. Drumming of pelts with 10% cationic or nonionic detergent solution for 1 h at 15–25°C at pH 4–5 is the ideal condition. Addition of sodium chloride increases the efficiency of emulsification and for better results, the pickled stock is depickled to a pH of 4.0–5.0 before the degreasing operation (Dutta, 1985).

It has been shown that cationic and nonionic detergents are slightly absorbed by collagen fibres and that they could remove grease from pelts satisfactorily. Aqueous emulsions of chlorinated hydrocarbons have also been shown for degreasing of hides and furs. A combination of surface-active agents renders satisfactory degreasing of fur skins and leathers without utilizing any solvent. Removal of fat from pigskins by extraction with hot surfactant solutions has been optimized statistically by varying the temperature, concentration of surfactant, duration of the extraction and detergent type, and it has been observed that removal of fats increases with increase in surfactant concentration.

A salt and solvent-free method for sheepskin degreasing using sodium salts of polyphosphoric acids leaves only 3% grease content in leather, and glutaraldehyde acts as a soft tanning agent that helps to release grease which is then removed by surfactants (Brufau, 1985).

SCOURING

For woolled sheepskins meant for garments and headgear, before degreasing, scouring is resorted to immediately after soaking (Briggs, 1981). The objective of scouring is to remove natural fat from the wool and to reduce the natural grease in the skins as far as possible. The efficiency of scouring is improved by working at a temperature of 32–35°C. The scouring agents in general use are: (i) soda ash, sodium silicate or ammonia (1.0–1.6 g/L), (ii) soap, anionic or nonionic detergents and (iii) mixtures of detergents with solvents. A combination of these materials is usually used. Scouring is done in a paddle using 10–30 litres of water per skin. A preliminary run of 30 min with alkali and detergent is followed by addition of emulsified solvent, which is then run for a similar period. Failure to remove the maximum amount of grease at this stage is detrimental if a mineral tannage is used later.

Degreasing by Extraction Under Pressure

The approach of this method is that the grease must be sufficiently fluid to enable it to be expressed when pressure is applied. It is safer to give the pickled pelts a light vegetable tannage before pressing. The lightly tanned pelts are warmed by immersion in warm water and the temperature must be controlled to avoid heat damage to the skins. Pressures up to about 2 tones per sq.inch are usually sufficient. Any calcium soap formed during liming should be concerted into fluid-free fatty acids before applying pressure. Pickling or light vegetable tannage takes care of this aspect. This process takes about 4 hours for completion and is mainly applied in the manufacture of roller leathers.

Ultrasound Method of Degreasing

Ultrasound is a sound wave with a frequency above the human audible range of 16 Hz to 16 kHz. The ultrasound (50 kHz, 2.5 W/cm^2) application in water is relatively trouble-free and quite effective for degreasing. It also improves the texture of pigskin facilitating its further processing (Baldano and Shestakova, 1973).

Use of power ultrasound in aqueous degreasing process has been studied and compared with different degreasing systems. The results indicate that about two-fold increase in fat removal has been observed due to the use of ultrasound as compared to controls (Sivakumar *et al.*, 2009).

Enzymatic Degreasing

The possibility of applying lipolytic enzymes from plant, animal and microbial sources for the degreasing of hides and skins has been suggested by early investigators. Lipases are enzymes which hydrolyse the carboxyl-ester bonds in triglycerides at an oil–water interface (Verger and de Haas, 1976). Hydrolysis of triglycerides by lipase results in the liberation of free fatty acids, diglycerides, substantial amounts of monoglycerides and sometimes free glycerol. Acid lipases obtained from various sources have been found to hydrolyse glycerolesters optimally in the pH range of 4.0–6.0.

The advantages of using enzymes for degreasing are the elimination of solvents, reduction in surfactants and possible recovery of valuable by-products viz., fatty acids. The disadvantages are that the lipase does not remove all types of fats in the same way that solvents do and the process cost is a little escalated (Taylor *et al.*, 1987).

Plant lipases   The well-known acid lipase of the plant source is the castor bean lipase. Madhavakrishna and Bose (1959) have shown the use of castor bean lipase for degreasing sheepskins at a pH range of 4.5–5.0 and best results with respect to the hydrolysis of skin lipids by this enzyme are obtained when the lipid is acted upon under emulsified condition.

Pancreatic lipases   Lubert *et al.* (1949) first showed that pancreatic lipases which are active at the pH range of bating are suitable for degreasing. Studniarski *et al.* (1975) have observed that a dehairing enzyme prepared from pig pancreas is capable of simultaneous degreasing and dehairing and the grease content of the skins is reduced by about 80% thereby obviating further degreasing treatment.

Microbial lipases   Commercial microbial lipases have wide applications in food processing and detergent industry (Hasan *et al.*, 2006), cosmetics (August, 1972) and leather industry (Puvanakrishnan and Dhar, 1986). Nashif and Nelson (1953) have shown the use of bacterial lipases for degreasing. A novel method of degreasing raw pig- and sheepskins based on lipolytic breaking down and emulsification of natural fats present in the raw material by the use of enzyme preparations from *Aspergillus oryzae* and *Aspergillus flavus* has been shown.

A fungal lipase, having a pH optimum between 5.8 and 7.9 and pancreatin bate, have been used for simultaneous bating and degreasing and a process for the manufacture of suede clothing leather has been developed by this method (Yeshodha *et al.*, 1978b).

Leather having good properties is obtained by degreasing with alkaline lipase and the degreasing effect is improved when the alkaline lipase is used in combination with proteinases and pancreatin. Xue (1982) has observed that lipase has got good degreasing effects on the surface but it is less effective for the inside layers. The use of an alkaline lipase at a pH of 9.0–9.3 in the degreasing of pigskin has got the advantages of short degreasing time and high degreasing efficiency (Hu *et al.*, 1983).

A combination of enzyme-compatible surfactants with enzymes has a synergistic effect in both soaking and bating and thus ensures optimum wet degreasing. In comparison, the use of surfactants alone results in a less intensive emulsification and dispersion of the natural fat than that required for obtaining evenly dyed, largely finished aniline leathers. No special process stages are needed for the degreasing, and a level of 60% degreasing is achieved (Pfleiderer, 1983).

In the washing and degreasing of sheepskins with an aqueous solution containing an alkaline reagent, nonionic and anionic surfactants and formaldehyde, the sheepskin weight is decreased and the elasto-plastic properties are improved. Degreasing is improved further by the addition of an enzyme preparation protosubtilin G 10X (0.4–1.5 g/L) along with sodium bicarbonate, surfactants and formaldehyde to the solution, 25–35 min. after the beginning of processing.

Skins of high quality are produced by washing, treatment with pulsed electric current (c.d. 0.01–0.25 A/cm^2, pulse length 5–100 ms, number of pulses 40-4500), degreasing with a granulated detergent containing alkaline protease, washing at pH 3.0–3.5 and drying. Immediately after degreasing, an enzyme treatment is carried out at 20–25°C with 1.0–2.0% 1:1 mixture of protosubtilin G 10X and testipan.

Acid lipases   The main advantages of degreasing in acidic conditions are the following.

i.   Bating, pickling, degreasing and chrome tanning can be carried out in the same drum avoiding the frequent removal of pelts from the drum.

ii.   The operational cost is reduced

iii.   The grain remains tight after finishing.

A microbial acid lipase acting at a pH range of 3.2–3.6 has been reported for the degreasing of woolled sheepskins (Yeshodha *et al.*, 1978a). An acid lipase from *Rhizopus nodosus* acting at a pH of 5.6–5.8 has been observed to be very effective in the degreasing of sheepskins for the manufacture of suede clothing leather (Muthukumaran and Dhar, 1982).

It has been shown that the degreasing of hides is carried out by soaking in an acidic bath containing a proteolytic enzyme (0.01–3%) and a nonionic surfactant (0.2–1.5%) or its mixture with anionic emulsifiers. It has been observed that a combination of proteolytic enzymes and emulsifiers gives optimum results in wet degreasing of sheepskins but a large amount of enzymes and reagents and prolonged treatment times are required (Brufau, 1985).

Mode of application of acid lipase   For degreasing, pickled pelts are kept immersed in an enzyme bath containing1% microbial lipase and 200% water (on pelt weight), adjusted to a pH of 3.6 and left in the same bath overnight at a temperature of 28–32°C. Then the degreased pelts are removed from the bath and subjected to salt wash twice with 100% water and 8% common salt (on pelt weight) for 40 min. The washed pelts are repickled, chrome-tanned and taken for further processing (Yeshodha *et al.*, 1978a).

MERITS AND DEMERITS OF DIFFERENT DEGREASING SYSTEMS

Earlier studies have shown that solvent extraction using kerosene with 1% sulphated amyl oleate is more efficient than aqueous degreasing. Dempsey *et al.* (1954), comparing the kerosene process with pressure degreasing, have concluded that both are capable of removing considerable amount of grease from sheepskins, although the kerosene process results in more complete removal and a more even distribution of the residual grease over the area of the skin. A combination of both processes, kerosene followed by pressure, has produced superior results than when used alone. Douglas (1957) while discussing the economics of various methods of degreasing, has reported that a controlled paraffin-degreasing process using kerosene is cheap and reliable.

A critical appraisal of the data on the physical characteristics and chemical analysis of the suede clothing leathers produced from coarse woolled sheepskins by treatment with the lipases from *Rhizopus nodosus* and kerosene indicates no significant differences (Muthukumaran and Dhar, 1982). This shows that both enzymatic as well as solvent degreasing produce equally good effects. However, Vulliermet *et al.*(1982) has stated that, for the degreasing of delimed and pickled sheepskins, fungal lipase has given unsatisfactory results when compared to the classical solvent method.

Aqueous degreasing agents are efficient and reasonably economical when they are employed at a temperature of 60°C. However, the surfactants used at present are biologically "hard", nonionic surfactants, i.e., not capable of being degraded wholly by lipolytic bacteria, and their large-scale use is likely to cause difficulties in effluent treatment (Palmer and Marsden, 1981). Biologically "soft" surfactants are available but they are expensive. A disadvantage is that formaldehyde or glutaraldehyde pretannage could give rise to a reduction in strength or lack of run in the final leather.

Both the aqueous emulsification and solvent extraction methods involve the use of surface-active agents, which must be carefully selected. Anionic agents, which are readily bound by pickled and even depickled pelts are not generally useful. The choice has to be made between cationic and nonionic agents. The use of nonionic accessory agent which is not adsorbed by the pelt and aids penetration of the solvent by an emulsification mechanism has met with success.

Pressure degreasing is less widely practised on account of its inconvenience. It may be necessary to give the skins a slight vegetable tannage before pressing, to aid the manual loading of the press. Moreover the application of pressure necessitates dry milling of the skins to reopen the fibre structure.

BIBLIOGRAPHY

August, P. (1972). "Lipase containing defatting creams". German Patent no.2064940

Baldano, Z.K. and Shestakova, I.S. (1973). *Mater Nauchn Konf, Vost-Sib Technol Inst, Sekts Khim-Technol.* 11th 1972, 95–97.

Bienkiewicz, K. (1983). *Physical Chemistry of Leather Making.* Robert E.Krieger Publishing Co., U.S.A.

Briggs, P.S. (1981). *Clothing and Gloving leathers.* Tropical Products Institute, London. pp. 67–78.

Brufau, J.G. (1985). Non-pollutants degreasing agents: Several degreasing processes of sheepskins without dissolvents." *Boletin Tecnico AQEIC.* 36, 277–282.

Dempsey, M., Innes, R.F. and Mitton, R.G. (1954). "Comparison of the paraffin and pressure degreasing proceses: The effect on distribution of the residual grease and the properties of the leather." *J. Soc. Leath. Trad. Chem.* 38, 265–272.

Douglas, G.W. (1957) "Oil tannages." *Tanner.* 10, 25–28.

Dutta, S.S. (1985). *An Introduction to the Principles of Leather Manufacture.* Indian Leather Technologists Association, Calcutta.

Gavend, G., Philippe, B. and Rouzieres, J. (1975). "Degreasing in tawing plants.".*Technicuir.* 9, 5.

Hasan, F, Shah, A.A. and Hameed, A. (2006) "Industrial applications of microbial lipases." *Enz. Microb. Technol.* 39, 235–251.

Hu, T., Zhou, Z., Zhong, J., Cao, Q., Huang, C. and Gu, X. (1983). "Use of alkaline lipase for degreasing of pigskin leather." *Pige Keji.* 3,19.

Innes, R.F. and Pankhurst, K.G.A. (1952). "The degreasing of woolskins." *J. Soc. Leath. Trad. Chem.* 36, 358–363.

Kothandaraman, V. and Aboo, K.M. (1977). "Charateristics and treatment of tannery wastes", paper presented at the 12th Tanners'Get together held at Central Leather Research Institute Chennai (1–4 February).

Lubert, D.J., Smith, L.M. and Thornton H.R. (1949). "An extraction-titration method for estimation of bacterial lipase." *Can.J.Res.* 27, 491.

Madhavakrishna, W. and Bose, S.M. (1959). "Studies on enzymic unhairing and degreasing for production of leather." *Bull. Cent. Leath. Res. Inst.* 5, 351.

Mitton, R.G. and Pankhurst, K.G.A. (1953). "A comparison of white spirit and paraffin for degreasing sheep skins for chrome gloving leather". *J. Soc. Leath. Trad. Chem.* 37, 331–339.

Muthukumaran, N. and Dhar, S.C. (1982). "Comaparative studies on the degreasing of skins using acid lipase and solvent with reference to the quality of finished leathers." *Leath.Sci.* 29, 417.

Nashif, S.H. and Nelson, F.E. (1953). "The Lipase of *Pseudomonas fragi.* I. Characterization of the Enzyme." *J.Dairy Sci.* 36, 459.

Palmer, N.W. and Marsden, E.P. (1981). "Glove Leathers." In: *Gloving, clothing and Special Leathers.*" P.S. Bridge. (ed.). Tropical Products Institute, London. pp.43–46.

Pfleiderer, E. (1983). "Analytical problems of bating products for leather." *Leder,* 34,181.

Puvanakrishnan, R. and Dhar, S.C. (1986). "Recent advances in the enzymatic depilation of hides and skins." *Leather Sci.* 33,177.

Reed, R. (1966) *Science for Students for Leather Technology.* Pergamon Press, Oxford.

Ries, H. (1971). "Treatment of pelts in organic solvents." *Das Leder.* 22,17.

Sarkar, K.T. (1981). *Theory and Practice of Leather Manufacture.* Ajoy Sorcar, Chennai. pp.120–121.

Sharphouse, J.H. (1983). *Leather Technicians Handbook*. Leather Producers Association, Northampton. pp.138–143.

Sivakumar, V., Chandrasekaran, F. Swaminathan, G. and Rao, P.G. (2009). "Towards cleaner degreasing method in industries: Ultrasound-assisted aqueous degreasing process in leather making." *J Cleaner Prodn*. 17, 101-104.

Studniarski, K., Sagala, J. and Jonczyk, W. (1975). "Enzymatic unhairing of pig skins." *Leder*. 26, 202.

Taylor, M.M. Bailey, D.G. and Feairheller, S.H. (1987). "A review of the uses of enzymes in the tannery." *J.Am. Leather Chem Assoc*. 82,153–165

Verger, R. and de Haas, G.H. (1976). *Annual Reviews of Physics and Bioengineering*. Vol.5. L.J. Mullins. (ed.). Annual Reviews Inc., Palo Alto. pp. 46–59.

Vulliermet, A., Carre, M.C., Sanejouand, J. and Jullian, C. (1982). "Enzymic degreasing of sheep skins." *Technicuir*. 16, 64.

Wolfram, M.A, Wolejsza, N.F. and Laden, K. (1972) "Biomechanical properties of delipidized stratum corneum." *J. Invest.Dermatol*. 59,421–426.

Xue, J. (1982). "Mechanism of enzymic depilation and its relation to pig skin shoe leather production." *Pige Keji*. 12, 36.

Yeshodha, K., Dhar, S.C. and Santappa, M. (1978a) "Studies on the degreasing of skins using a microbial lipase." *Leather Sci*. 25, 77–86.

Yeshodha, K., Dhar, S.C. and Santappa, M. (1978b). "Fungal lipase and its use in degreasing of sheep skins." *Leath.Sci*. 25, 267–273.

19

RECENT TRENDS IN WASTE MANAGEMENT

INTRODUCTION

The leather industry is one of the most traditional export-oriented industries, earning significant foreign exchange. But, the industry carries the burden of disposal of tannery wastes which cause perennial environmental pollution. Leathers are turned into final products of desirable quality after several steps of chemical processes. The four main processes for conversion of crude skin/hide to finished leather are pretanning or beamhouse process, tanning process, wet-finishing or post-tanning process and finishing process. About 130 different types of chemicals are applied in leather manufacture, depending on the type of raw material used and the end product desired. The salt used for preserving the raw skin/hide discharges huge amount of pollution load in terms of total dissolved solids (TDS) and chlorides. Other polluting chemicals are lime, sodium sulphide, ammonium salts, nitrogen compounds, sulphuric acid, chromium salts, syntans, vegetable tanning materials, solvents, detergents, acids, dyes, fat liquors, resins, etc. Solid wastes generated are suspended solids, settleable solids and gross solids. The pollution load in tannery effluent results in increased chemical oxygen demand (COD) and biological oxygen demand (BOD). Apart from solid and effluent wastes produced, tanning processes also generate gaseous pollutants causing an obnoxious odour. The various pollutants generated by leather processing industries are listed in Table 19.1.

There are about 1600 tanneries located in India with a total processing capacity of 1.5 million tons of hides and skins per year *(Directory of tanneries in India, 1998; Report of All India Survey on raw hides and skins, 2005).* More than 80% of the tanneries implement chrome tanning process and the remaining tanneries adopt vegetable tanning process. The leather industry is under increasing pressure to reduce water consumption and improve the quality of the effluent that is discharged.

Table 19.1 Categorization of pollutants from different operations of leather industry (Bosnic *et al.*, 2000)

Stages of processing	Pollutants generated

Abattoir operations
  1. Slaughter of animals
  2. Curing
	Green fleshings, trimmings and shavings, salt and biocides

Pretanning/beamhouse operations
  1. Soaking and washing
  2. Liming and dehairing
  3. Deliming and bating
  4. Pickling and degreasing
	Water, dirt, blood, manure, urine, salts, disinfectants, detergents, sulphide, lime, pulped hair, limed fleshings, H_2S gas, suspended solids, ammonium salts, pigments, fat residues, protein residues, organic solvents, alkaline scouring agents, organic acids, dissolved solids, ammonium gas

Tanning operations
  1. Chrome tanning
  2. Vegetable tanning
  3. Synthetic tanning
	Chromium, vegetable tannins, syntans, fungicides, wet blue/ vegetable/ syntan shavings and trimmings, buffing dust

Post-tanning/wet finishing operations
  1. Retanning
  2. Dyeing
  3. Fat liquoring
  4. Drying
	Chromium/ vegetable tannins/ syntans, dyes, fat liquors, resins, polymers

Finishing operations
  1. Conditioning
  2. Buffing
  3. Finishing
	Buffing dust, chrome trimmings, organic solvents, solid and liquid finisher residues

For the discharge of effluents to surface water bodies, the Minimum Acceptable Standards (MINAS) are set by the Central Pollution Control Board (CPCB), India (Table 19.2).

Table 19.2   Minimum acceptable standards for tannery effluent discharge

Parameters	Chrome tanning effluent (mg/L)	Vegetable tanning effluent (mg/L)
pH	6 – 8	6 – 8
TSS	100	100
TDS	2100	2100
BOD	30	30
COD	250	250
BOD/COD	-	-
Total Kjeldahl nitrogen	100	-
Ammoniacal nitrogen	50	-
Chloride	1000	1000
Sulphate	1000	-
Sulphide	2	2
Chromium	2	-
Colour	-	-
Phenols	1	-

ENVIRONMENTAL IMPACT OF TANNERY WASTES

A large quantity of chemicals are used for leather processing, but only a small amount is bound to the final product, leather, and the remaining chemicals are discharged as effluents. The diverse pollutants from the tanning industry are adversely affecting human life, agriculture, livestock and aquatic life. The common disorders faced by tannery workers are eye diseases, skin irritation, kidney failure and respiratory and gastrointestinal problems. Chromium, extensively used in the tanning process, is a highly toxic substance and a potent carcinogen. The effluent with a low pH due to use of acids is corrosive to water-carrying systems and this may lead to the dissolution of heavy metals in the wastewater. The high pH in tannery wastewater due to liming chemicals causes scaling in sewers. A large fluctuation in pH exerts stress on water environment which may kill some sensitive species of aquatic plants and animals.

Large quantities of proteins and their degraded products affect the environment by increasing the BOD and suspended solids. Organic matter does not cause direct harm to aquatic environment. But, it has an indirect effect—it deprives the water of its dissolved oxygen content. A total lack of dissolved oxygen as a result of high BOD can kill all natural life in an affected area. Due to sulphide discharged from the dehairing process, hydrogen sulphide is released at a pH value lower than 8.5. This gas has an unpleasant smell even in trace quantities and is highly toxic to many forms of life. Sulphide in public sewer can create problems due to corrosion by sulphuric acid produced as a result of microbial action.

When the suspended solids settle, it covers the natural fauna on which aquatic life depends. This can lead to a localized depletion of oxygen supply in the water. It also reduces light penetration and consequent reduction in photosynthesis due to the increased turbidity of water. Sodium chloride used in the tannery affects freshwater life when its concentration becomes too high. Similarly, sulphate also affects the fresh water inhabitants and in addition, at higher concentration it causes corrosion to concrete structures. The vapours of finishing chemicals like volatile organic compounds can affect the respiratory functioning of workers (Karabay, 2008).

According to the present Environmental Pollution Control Regulations, tanneries are instructed to set up individual or Common Effluent Treatment Plant (CETP) for treatment of liquid waste generated from tanneries. Before sending the effluent to the CETP, it is compulsory to pre-treat the effluent wastewater in individual tanneries mainly to recover and reuse chromium. Generally, a few individual tanneries follow all wastewater treatment steps on site and many tanneries follow only pre-treatment or no treatment at all, sending the effluent to a CETP. However, treatment is absolutely necessary due to the wide range of toxic effects on the environment caused by untreated tannery effluent and sludge.

TREATMENT AND UTILIZATION OF SOLID WASTE FROM LEATHER INDUSTRY

Solid wastes are generated from almost all the stages of leather making. Solid wastes generated from various operations in abattoirs and tanneries can be classified as follows:

  i. wastes from raw hides/skins (green fleshing wastes, trimmings);
  ii. wastes from pretanned hides/skins (skin trimmings, hair residues, fat residues, lime fleshing residues);
 iii. wastes from tanned leather (wet blue and vegetable tanning trimmings and having wastes, buffing dust);
 iv. wastes from dyed and finished leather (trimmings from leather) and sludge generated from effluent treatment plants.

Out of 1000 kg of raw hide, nearly 800 kg is generated as solid wastes in leather processing. Only 200 kg of the raw material is converted into leather. The proportion of solid wastes generated from tannery is as follows:

  ❋ fleshing: 56–60%,

❋ chrome shaving, chrome splits and buffing dust: 35-40%,

❋ skin trimming: 5–7% and

❋ hair: 2–5% (Kanagaraj *et al.,* 2006).

Solid wastes create a major problem for leather industry in terms of their variety and quantity. The variety and quantity of solid wastes depend on animal species, breeding conditions, slaughterhouse practices, conservation conditions, leather process stages, mechanical operations, and chemicals used in processes. The wastes contain variable amount of nutrients, mainly proteins, which can be converted into useful by-products and it provides additional earnings to the leather industry.

Solid wastes can be used as raw material for different industries. In India, studies have been carried out regarding the use of tannery sludge and other solid wastes for making building materials. Central Leather Research Institute (CLRI)–The Netherlands Co-operation (TNO) programme and United Nations Industrial Development Organization (UNIDO) have recommended the suitability of the sludge from the CETP for making bricks. The sludge can also be used as a soil conditioner in agriculture or as a backfilling material to raise soil levels. However, under the existing Indian waste management regulations, this practice is prohibited due to diverse opinion on quality and quantity of chromium content in the sludge. It should be noted that the sludge from tannery effluent treatment normally contains several useful nutrients such as nitrogen, phosphorus, etc., and composting of this sludge could be a worthwhile option, if the chromium content could be kept below the permissible level. But, the most suitable and practical solution for the disposal of unutilized hazardous sludge from CETP and pre-treatment system of tanneries is the controlled/secure landfill (Rajamani *et al.,* 1995).

VALUE-ADDED PRODUCTS FROM SOLID WASTES

Animal products such as skin, wool, horns, hoofs, hair, fleshings, trimmings, etc. are significant waste by-products of slaughterhouse and leather industry. A range of valuable products has been developed from various wastes. These wastes contain large amounts of the structural fibrous proteins, keratin and collagen. A promising alternative is to hydrolyse the waste by acidic or enzymatic hydrolysis to obtain protein hydrolysates containing valuable peptides and amino acids and these hydrolysates may find novel applications (Sundar *et al.,* 2011).

Chemical-Free Wastes

The solid waste can be hydrolysed with the help of proteolytic/lipolytic enzymes and used as a useful by-product. The fresh green fleshings, trimmings and shavings have been found not to inhibit the biological process of composting and can be used as fertilizer. It can also be used for the recovery of grease, meat meal and pet chews by using protease and lipase enzymes.

The raw hide/skin trimmings collected before processing operations, can serve as a source of high-value products like collagen, and they can be used in preparation of collagen-based biomaterials for medical applications (Rose, 2007). The untanned proteinaceous tannery solid waste (animal fleshings) is used as a substrate for acid protease production by *Synergistes* sp. *Synergistes* sp. can be grown to produce 350–420 U/ml of mesophilic acid protease using animal fleshings (Kumar *et al.,* 2008). Slaughterhouse fleshing and tannery fleshings, the major solid waste emanating from the beamhouse, and solid sludge from primary and secondary treatment processes are subjected to biomethanation. These wastes are completely liquefied biologically and the resultant liquefied fleshing is treated in anaerobic reactors to produce bioenergy/biogas. Liquefied tannery fleshing, sludge and cow dung can be used for biogas production (Vasudevan and Ravindran, 2007; Bhirnd, 2007). Untanned waste fleshings, mixed with tannery sludge after digestion, produces methane having a residual solid phase which is suitable for composting. Based on the chromium content, it can be applied directly to agricultural land as a soil conditioner. Animal wastes after alkaline hydrolysis can also be used as soil conditioner (Gousterova *et al.,* 2003).

Biomethanation (or anaerobic digestion) systems are tested and proven processes that have the potential to convert tannery wastes into energy efficiently, and they achieve various goals—prevention/reduction of pollution elimination of uncontrolled methane emissions and odour, conserving the bio-energy potential and production of stabilized residue for use as low-grade fertilizer. Anaerobic digestion is a favorable technological solution to alleviate the problem of pollution as it can be used as a tool for the degradation of a substantial part of the organic matter contained in the sludge and tannery solid wastes and this generates valuable biogas. Digested solid waste is biologically stabilized and can be used in agriculture as fertilizer.

Collagen dispersions have remarkable water-retention capability and this property has led to the research on their applications in environmental remediation, biotechnology, biomedical engineering, etc. The protein hydrolysates from calfskin and blood wastes can be used as nitrogen source in bacterial growth media (Tonkova *et al.,* 2007). Untanned collagen wastes from trimmings are used in the preparation of collagen-based biomaterials for use in clinical applications such as tissue adhesive, vascular grafts, drug delivery matrices, wound dressing and tissue engineering scaffold. Hide collagen is known to have uses in food preparations such as binder, filler, moisturizer, texturizer, edible meat product casings, petfoods, thickening agent, etc. (Mokrejs *et al.,* 2009). The protein-rich hydrolysates from leather industry can be converted to value-added commercial products such as fertilizers, biodegradable polymers and additives for cosmetic industry, etc. The hydrolysate of pigskin and fish bone can be used as cosmetics due to its high water-retention capacity, ability to repair rough skin, lack of odour and absence of harmful effects on skin (Morimura *et al.,* 2002). Use of chromium free protein hydrolysates obtained by chemical and chemical-enzymatic hydrolysis are rich in amino acids and can be used as foliar growth enhancers and biostimulators for fruit and vegetable crops (Lacatus *et al.*, 2009). Biodiesel can be produced from fleshing fat wastes by transesterification or by using immobilized lipase-catalyst for use as a replacement of fossil fuels (Kolomaznik *et al.*, 2009; Shin *et al.*, 2008). Raw hide trimmings stabilized with glutaraldehyde can act as an effective adsorbent for removal of dyes from wastewaters (Fathima *et al.*, 2009). Enzymatic skin hydrolysates and dialdehyde starch are reacted together to produce hydrogels which are used as biodegradable packaging materials (Langmaier *et al.*, 2008).

Hair Wastes

There are a number of promising uses reported for the recovered hair from hair-save processes. These include felt production, keratin hydrolysate, cosmetics and pharmaceutical products. Keratin-containing hair waste is hydrolysed using conc. NaOH or HCl and the hydrolysate thus obtained helps to improve the chrome exhaustion of tanning as well as rechroming bath. Hair hydrolysate has found applications in the pharmaceutical and cosmetics industry as a thermal and acoustic insulation material and as a manure supplement with household and garden refuse. Hair recovered through a hair-save process can be incorporated into existing composting processes, as it is a valuable source of nitrogen and organic carbon. Hair can also be directly used as a slow-release source of organic nitrogen and carbon for fertilizing purposes. Hair from pig skins is a valuable material that is used for brushes and other consumer products. Hair recovered from sheepskins is of longer length and is utilized for the manufacture of low-priced rugs and carpets. The human hair contains a number of amino acids, inclusive of cysteine, cystine and tyrosine which are of commercial value. Similarly, dissolution by reduction and extensive hydrolysis of peptide bonds can be employed for preparing amino acid formulations from animal hair (Rose, 2007). Animal hair, through hydrolytic dissolution and chemical modification, can be used as foam stabilizer for fire extinguishers, cosmetics and animal feeds. Acid, alkali or enzymatic hydrolysis of keratin wastes can be used in formulating keratin-based cosmetics, additives in the preparation of concrete and ceramics, organic fertilizers, etc. (Karthikeyan *et al.*, 2007).

Limed and Delimed Fleshings

Limed fleshings and trimmings are hydrolysed with alkaline protease and the resultant collagen hydrolysate is used for glue preparation, and the protein concentrate can be used as fertilizer in agriculture/horticulture (Serrano *et al.*, 2003). Ravindran *et al.* (2008) have vermicomposted fleshings using earthworm (*Eisenia foetida*) for use as nutrient-enriched compost. Delimed fleshings mixed with sugar and common salt are inoculated with *Pediococcus acidolactici* and on fermentation yield silage which can be used in animal and aqua feeds to supplement arginine, phenylalanine and tyrosine (Bhaskar *et al.*, 2007). Limed fleshing can be mixed with municipal solid waste to produce biogas by anaerobic digestion (Shanmugham and Horan, 2009).

Sludge from lime pits can be used for land filling in low-lying areas as well as in construction of low cost housing. Barks and nuts from vegetable tanning can be used as fuel for boiler and brick kilns. Limed splits, high in nitrogen and low in carbon, will compost readily.

Chrome and Vegetable Tanning Wastes

Collection and safe disposal of chrome-containing solid waste and sludge is a critical issue related to environmental pollution and implementation costs. Fleshing wastes which contribute to a major portion of solid waste can be hydrolysed and the hydrolysate improves the uptake of chromium in chrome tanning and rechroming. The fleshings can also be digested with pancreatic enzymes and the hydrolysate thus obtained is rich in protein and can be used in feed formulation by mixing with other feed ingredients (Kanagaraj *et al.*, 2006).

Chrome-tanned leather, splits and trimmings are treated with H_2O_2, macerated, ground and extracted to obtain glue, gelatin, protein flavour and reconstituted collagen. Wet blue trimmings, shavings, buffing dust and trimmings from crust and finished leather are converted into glue, feed and fertilizers. Chrome shavings treated with magnesium oxide, carbonates and hydroxides under mild alkaline conditions result in gels, adhesives and films of high molecular weight gelable protein fraction. The chrome products are detanned to obtain gelatin and isolation of collagen fibres. Chrome shavings are treated with pepsin to isolate high-quality gelatin and the generated hydrolysate is again treated with alkaline protease to form a protein hydrolysate. The remaining solid after gelatin removal is chemically treated to recover and reuse chromium in tannery. Chrome-tanned wastes are hydrolysed with proteolytic enzymes to produce protein hydrolysate and chrome sludge. The chrome sludges are found suited for pigment in glass making, heat-resistant bricks and alkaline chromate. The hydrolysates are used for concrete mixture, protective coating and plaster binder (Choudhary *et al.*, 2004). Chrome shavings are treated with proteolytic enzymes and the protein hydrolysate obtained after the isolation of protein from chromium mass is a potential feed, fertilizer and additive in cosmetic industry. The hydrolysate from chrome-tanned wastes is used for producing biodegradable plastic particularly useful in agriculture. Chrome shavings treated with pepsin–trypsin and pepsin–alkaline proteases have yielded products such as gelatin, chrome cake and hydrolysate. Chrome shavings are treated with acidic protease–pepsin followed by cross linking with glutaraldehyde and the resulting dry material can be used for leather board. The wet blue shavings are digested into smaller material with alkaline protease and the dried material can be used for casein formulations in leather finishing. Chrome-tanned leather hydrolysate after treatment with glutaraldehyde results in thermo-reversible gels used for glue preparation and thermo-irreversible gels used in encapsulation. *Bacillus subtilis* strain 11 isolated from compost and standard *Bacillus subtilis* ATCC 6633 are used for digestion of solid tannery wastes and the obtained digest may also be used as fertilizer or as a source of nitrogen for microbial populations (Zerdani *et al.,* 2004).

Table 19.3 Utilization of leather industry wastes (Rose, 2007 and Ramalingam, 2011)

Type of wastes	Product	Current Utilization	Value addition
Raw hides/skins trimmings and shavings	Glue, gelatin, hydrolysate, dog chews, hide meal	Industrial gelatin, pet foods and treats, aqua feed formulations	Pharmaceutical-grade gelatin
Fleshings	Glue and gelatin, fleshing meal and hydrolysate	Glue manufacture, aqua and poultry feed formulation	Biodiesel and fertilizers
Hair/wool	Cleaned hair, keratin hydrolysate, amino acids extraction	Rugs and carpets, leather filter, feed, biofertilizer	Keratin-based shampoo
Splits	Cheap leather	Washers, inner lining for leather goods	High split finish leather
Chrome / vegetable/ syntan shaving dust	Pressed leather board, chrome-peptide condensate, hydrolysate	Printing boards, retanning agent, feed/fertilizer additive	Separation of protein for application as fillers and biofertilizers
Tannery sludge		Secure landfill	Highly compressed bricks
Autospray waste		Incineration	Water-based paints
Buffing dust	Insole board, briquettes	Leather boards /land fill/ Incineration	Generation of biogas energy

Grease from the degreasing process can be used as a component of low-grade fat-liquors through a sulphitation process. Various laboratory and industrial trials have demonstrated that chromium-containing leather waste may be thermally treated to produce an ash containing chrome oxide, which can be used by the chromium chemicals manufacturing industry. Enzymatic digestion of shavings results in a high-quality and valuable hydrolysate or gelable protein, along, with a protein-contaminated chromium sludge. The hydrolysate can be used in retanning agents, as foam stabilizers, in the chipwood industry and gypsum industry. The chromium sludge can be reused in a dichromate reduction plant for the production of chromium sulphate. Mixing the sludge with clay and bricketting, and solidification with fly ash and cement would minimize leaching of chromium. The solid wastes can also be utilized in developing new products and the details are shown in Table 19.3.

TREATMENT OF TANNERY EFFLUENTS

When effluent is discharged directly into streams and rivers, it needs to be of good quality as the environment is sensitive and highly susceptible to degeneration or decay. The receiving bodies of water slowly increase in chloride content and hardness. Presence of ammonia, sulphides, tannin and chromium in the water bodies causes pollution and is highly toxic to aquatic lives. Approximately, 50,000 kg of wastewater is produced by processing one ton of raw hide, which generates pollution load of COD (235–250 kg), BOD (100 kg), suspended solids (150 kg), Chromium (5–6 kg) and sulphide (10 kg) (Sundar *et al.*, 2011).

Generally, the tannery effluents are subjected to primary and secondary treatment prior to discharge into the environment. Primary treatment includes physical and chemical processes like screening, equalization, coagulation and flocculation for the removal of suspended solids, sulphide treatment and for precipitation of chromium. Secondary treatment of biological processes by aerobic/anaerobic methods is carried out to remove the organic matter dissolved in effluents (Ganesh and Ramanujam, 2009).

Mechanical and Chemical Treatment

To start with, the raw effluent undergoes mechanical treatment such as screening to remove coarse material. Up to 30–40% of TSS in the raw waste stream can be removed by properly designed screens/grits. Mechanical treatment also includes removal/skimming of fats, grease, oils and gravity settling. After mechanical treatment, physico-chemical treatment is usually carried out, which mainly involves the chrome precipitation and sulphide removal. Physico-chemical pre-treatment supplies an effluent containing easily biodegradable organics for the standard aerobic biological treatment plants. Oxidants, especially hydrogen peroxide (H_2O_2), have been used for many years to reduce the COD and BOD of industrial wastewaters. The other oxidizing agents such as sodium hypochlorite (NaOCl) and calcium hypochlorite (Ca [OCl]$_2$) have been used to reduce COD from tannery wastewater. Electro-Fenton oxidation of wastewater can also be used for the treatment of organic matter in tannery wastewater. Coagulation and flocculation are also part of this treatment for the removal of a substantial percentage of the COD and TSS. In the coagulation-flocculation (CF) process, the commonly used inorganic coagulants are aluminium sulphate (alum), ferric chloride ($FeCl_3$), ferrous sulphate ($FeSO_4$), ferrous chloride ($FeCl_2$) and polyaluminium ferric chloride (PAFC). They have been used to reduce organic load and suspended solids as well as to remove toxic substances such as chromium before biological treatment (Lofrano *et al.*, 2006; Islam *et al.*, 2011). Improved coagulation followed by biological treatment can result in safe effluent to aquatic environment. The effluent coming out from the various upstream units of the leather plant (except chrome tanning) are combined and treated by a two-step pressure-driven membrane process involving nanofiltration and reverse osmosis. It is carried out after a pretreatment consisting of gravity settling and coagulation. The performance criteria of these membrane processes are evaluated in terms of COD, BOD, conductivity, TSS and the permeate flux of the treated effluent. It is observed that TDS, BOD and COD values of the finally treated effluent are well within the permissible limits in India (Jain *et al.,* 2006). Effluent from tanneries after mechanical and physico-chemical treatment is easily biodegradable in biological treatment plants.

CETP Method

The simplest method of tannery effluent treatment followed in CETP is described in Figure 19.1. In the first step, liquid wastes generated from all the processing operations (except chrome tanning) are combined in equalization and settling tanks. The pretanning liquors with the soluble organic matter flocculate and the flocs are removed by settling. Liming, pickling and dehairing liquors are mixed and coagulated by ferric chloride, whereas chrome liquors are neutralized by sodium carbonate. After separate sedimentation of pretanning and tanning liquors, all the effluents are mixed and transferred into the treatment plant. The tannery effluents contain a great amount of soluble or colloidal matter. In equalization tanks, colloidal matter flocculates naturally, and if it is not effective, coagulants can be added for enhancing flocculation. It has been found that alum sulphate is the most suitable coagulant for vegetable tanning effluents. From the economy point of view, alum is often replaced by ferrous sulphate. In recent years, organic polycoagulants at low doses have been used with good results. The settled sludge is voluminous and contains about 1–3% of insoluble solids, of which about 50% is organic and the rest is inorganic in nature. A substantial reduction in the concentration of chromium compounds is achieved when the chrome waste is mixed with the beamhouse liquor.

The chrome effluent after mixing with beamhouse liquors generally has Cr^{3+} content below 30 mg/L. The precipitated chromium hydroxide is highly hydrated and the flocs settle down very slowly. Mixing with fresh lime slurry and addition of polycoagulant decreases the Cr^{3+} content by 5–10%. After the primary treatment, the effluent is being treated by biological processes such as activated sludge process (ASP) and upflow anaerobic sludge blanket (UASB) which are commonly followed in CETPs in India.

The method of biological treatment of tannery effluents was introduced in 1940. The older effluent treatment methodologies consisted mainly of mixing acid and alkaline wastes followed by sedimentation or lagooning. Nowadays, treatment of tannery effluents is normally completed by the biological treatment process after mechanical and chemical pretreatment. Among the various treatment methods, biological treatment of wastewater is found to be effective whereas a combination of physical, chemical and biological methods is found to be better than biological treatment alone. During primary treatment, chromium, TDS and TSS in chemically treated wastewater will be reduced to 40–50% of their initial concentration, which facilitates proper biological treatment. The BOD reduction by the activated sludge method is between 80–90% while the chemical pretreatment reduces the BOD by 60 percent. A combination of both chemical pretreatment and the activated sludge method reduces the BOD content by 95%. The effect of biological treatment depends on several factors, e.g., pH, content of sulphide, chromium, vegetable and synthetic tannins, conductivity, etc.

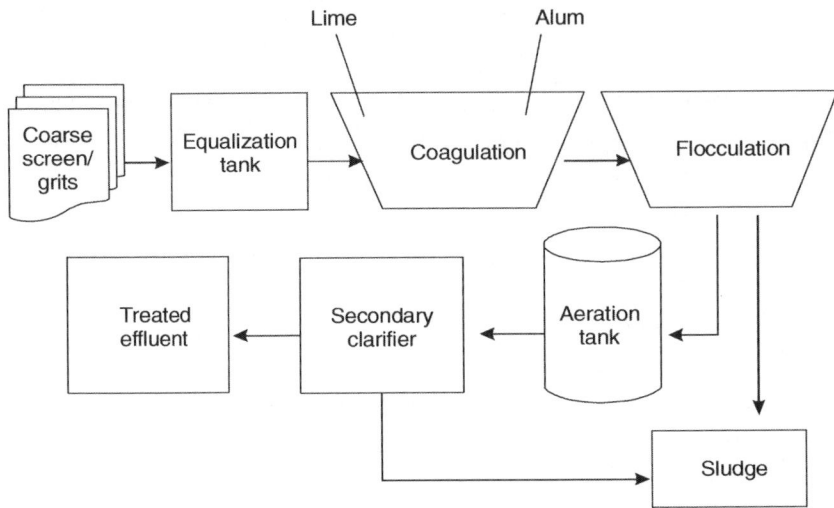

Figure 19.1 Effluent treatment method being adopted in CETP

Biological Treatment

The various biological treatment methods presently used are activated sludge process, sequence batch reactor, trickling filter, aerated lagoons, rotating biological contactor, fluidized bed reactor, oxidation pond, anaerobic lagoons, upflow anaerobic sludge blanket process, anaerobic fixed biofilm reactors, membrane bioreactors, phytoremediation, etc. The most commonly used processes for biological treatment of tannery effluent in CETPs in India are ASP and UASB. Technology is being constantly upgraded to achieve cost-effective end-of-pipe solutions.

Aerobic Process

Activated sludge is a process for treating industrial wastewaters using air and a biological floc composed of bacteria and protozoans. It facilitates oxidizing carbonaceous matter, biological matter, nitrogenous matter and nitrification. The process involves air or oxygen being introduced into a mixture of screened effluent combined with organisms to develop a biological floc which reduces the organic content of the effluent. The feasibility of ASP for the treatment of tannery wastewater has been evaluated by adjusting the parameters related to growth of microorganisms and substrate utilization. A bench-scale model comprising an aeration tank and final clarifier is used, which is operated continuously for 267 days. The results of the study demonstrate that an efficiency of more than 90% and 80% reduction in respective BOD and COD load are obtained keeping an aeration time of 12 h (Haydar *et al.*, 2007). A CETP-based ASP, employed for effluent treatment, is found to improve significant reduction in BOD and COD. In addition, a reduction in bacterial counts especially for pathogens such as *Escherichia coli, Vibrio* sp. and *Pseudomonas* sp. to an extent of 98.4%, 87.5% and 96.2% respectively is observed (Pramod *et al.*, 2010). Sequencing batch reactor (SBR) can be used for handling all wastewaters commonly treated by conventional activated sludge process. This method is advantageous as it results in reduced operating costs, improved nitrogen removal and less sludge formation. Laboratory-scale aerobic SBRs, after a primary treatment, fed with tannery wastewater and operated for 12 h show a reduction of 85–93% BOD, 80–82% COD, 78–80% total Kjeldahl N and 83–99% ammoniacal N (Ganesh and Ramanujam, 2007). Four different concentrations of untreated tannery effluents (25, 50, 75 and 100%) are bioremediated using *Aspergillus niger* and *Bacillus cereus* and the experiment is carried out for 72 h. At all the concentrations of tannery effluent, *A. niger* is shown to reduce BOD and COD significantly followed by *B. cereus* (Mohamed *et al.*, 2011). Two fungal cultures, *Aspergillus niger* and *Penicillium* sp., isolated from tannery environment are used for the biodegradation of effluent wastes such as soak liquor, vegetable tannin (especially myrobalan) liquor and effluent dye solution. This has resulted in reduction of 82–94% COD for soak liquor, 77–92% for vegetable tannin liquor and 92–95% for dye liquor respectively (Kanagaraj *et al.*, 2011).

The highly saline tannery wastewater (1–10% w/v) cannot be effectively treated by conventional biological treatment. Salt-tolerant microbes can adapt to these saline conditions and degrade the organics in saline wastewater. Four salt-tolerant bacterial strains, viz., *Pseudomonas aeruginosa* isolated from tannery saline wastewater, *Bacillus flexus* isolated from marine soil, *Exiguobacterium homiense* isolated from salt-lake saline liquor and *Staphylococcus aureus* isolated from seawater are taken individually or in mixed consortia for treating saline tannery waste water with varying salt concentrations (2% to 10%). Salt-tolerant mixed bacterial consortia show appreciable biodegradation at all saline concentrations with 80% COD reduction at 8% salinity level. Hence, it is suggested that the consortia could be used as suitable working cultures for tannery saline wastewater treatment (Sivaprakasam *et al.*, 2008).

The pre-denitrification – nitrification process is found to be efficient for simultaneous removal of nitrogen and organic substrates from tannery wastewaters. Normally, Cr^{3+} is not shown to inhibit performance operations. A pilot wastewater treatment plant consisting of pre-denitrification–nitrification process is constructed and operated for 6 months, and it is observed that total nitrogen and COD at 98% and NH_4 nitrogen at 95% are removed (Durai and Rajasimman, 2011).

There are a number of methods available for the removal of Cr^{6+} from industrial wastewaters. The conventional treatment techniques used for removal of Cr^{6+} ions from wastewaters include chemical precipitation, oxidation–reduction, ion exchange, electrolysis, etc. Regarding chromium recovery/reuse from tannery wastewater, two methods are currently in vogue. The first is by precipitation of chromium as hydroxide [$Cr\,(OH)_3$)], followed by filtration and subsequent dissolution in a known quantity of sulphuric acid to form basic chromic sulphate which is further reused in the tanning process. The second one is by direct recycling, involving filtration of the waste liquor followed by chemical replenishment. Chromium removal is also possible by biological means using microbes. Microorganisms such as algae, bacteria, fungi and yeasts have proved to be potential metal biosorbents. There are several fungal and bacterial strains such as nonpathogenic *Trichoderma* sp., *A. fumigatus* and species of *Pseudomonas* which are used for chromium remediation. Biosorption studies show a significant chromium uptake almost 100% by *Trichoderma* filamentous fungi. *Aspergillus niger* MTCC 2594 biomass is found to be effective in the biosorption of Cr^{3+} at 83% after 48 h and Cr^{6+} at 79% after 36 h from spent chrome liquor. Thus, it is shown that fungal species can be a good cost-effective biosorbent for removal of Cr^{6+} from industrial effluents (Mala *et al.*, 2006).

In recent years, cyanobacteria have been used as a bioadsorbent for the removal of certain heavy metals. The biomaterial, *Tridax procumbens* (Asteraceae – Compositae), is used in the preparation of activated carbon. It is shown that the heavy metal removal efficiency of the activated carbon of *Tridax* biomaterial is highly significant. The percent removal of Cr^{6+} in synthetic wastewater is 97% with an effective dose of 5 g of bioadsorbent while the percent removal of Cr^{6+} in tannery wastewater is 98%. This method can be recommended for the removal of high concentration of Cr^{6+} from tannery wastewater (Singanan and Singanan, 2007). The cyanobacteria, *Spirulina fusiformis,* can remove large quantities of chromium (93–99%) depending on the concentration of chromium in the growth media. Similarly, a few species of cyanobacteria such as *Oscillatoria* sp., *Nostoc* sp., *Gloeocapsa* sp. and *Synechococcus* sp. are used for bioremediation of chromium-containing effluents (Pandi *et al.,* 2009).

Tadesse *et al.* (2006) have experimented with horizontal settling tanks and subsequent Advanced Integrated Wastewater Pond System (AIWPS) of reactors for checking the removal efficiency of chromium from effluent. After one day retention time, 58–95% of Cr^{3+} was removed in the settling tank at pH 8.0 which is optimum for Cr^{3+} precipitation. When indigenous sulphur-oxidizing *Acidithiobacillus thiooxidans* was used as inoculum in a 2-L bubble column bioreactor with aeration for 5 days at 30°C, 99.7% of chromium was leached from tannery sludge. Hence, *A. thiooxidans* can be used effectively in chromium bioremediation (Wang *et al.,* 2007).

Pentachlorophenol is a toxic compound mainly used as a preservative in leather industry; but it creates problems during handling and also during biological effluent treatment. A few bacterial strains viz., *Pseudomonas aeruginosa, Arthrobacter* sp. *Proteus* sp. and *Bacillus* sp., isolated from tannery effluent sludge have been tested for their activity in the presence of sodium pentachlorophenate (Na-PCP) as the only source of carbon and energy. It has been observed that the *Arthrobacter* species has the highest PCP degrading potential (55%) within 30 days followed by *Pseudomonas* sp. (47%), *Bacillus* sp. (44%) and *Proteus* sp. (38%). The treatment of a tannery wastewater is performed at the laboratory scale using the ascomycetous fungus, *Botryosphaeria rhodina* MAMB-05. The wastewater samples collected in the retanning and dyeing steps are inoculated with *B. rhodina* MAMB-05 and it is observed that COD and total organic carbon (TOC) are reduced by about 91 and 93% respectively (Hasegawa *et al.*, 2011).

Use of Immobilized Cells

Immobilized cells as biofilms, beads or inert supports have been found to be most effective in designing bioreactors for heavy metal degradation. Chromium remediation studies have been carried out with a variety of organisms such as *Pseudomonas, Aeromonas, Bacillus, Micrococcus* and *Microbacterium* and out of these, *Pseudomonas* species is shown to be the most efficient. Fungi are being exploited for heavy metal degradation because of the presence of the metallothioneins (MT) and phytochelatins which are small cysteine-rich polypeptides that can bind metals. The chromium remediation ability of *Bacillus subtilis, Pseudomonas aeruginosa* and *Saccharomyces cerevisiae* in consortia and in their immobilized forms has been studied and their efficiencies are compared. The best activity is observed by *S. cerevisiae* and *P. aeruginosa* consortia, followed by immobilized beads of *S. cerevisiae* alone and in consortia with *B. subtilis* (Benazir *et al.*, 2010). Similarly, Cr^{6+} reduction has also been reported by *B. coagulans* and *B. circulans*. Immobilized cells appear to have greater potential in controlling particle size, efficiency of regeneration, easy separation of biomass and effluent recirculation, promoting at the same time, high biomass loading, minimal clogging and slower depletion of nutrient source. It has also been reported that immobilized cells have been found to be the most effective in designing small and large-scale bioreactors for heavy metal degradation. This study is the basis for implementation of advanced technologies like exo-polymer and bioreactor technology for rapid and effective removal of chromium from polluted water bodies. *In situ* bioremediation with biostimulation and bioaugmentation may prove to be highly efficient in chromium remediation.

The biosorption of Cr^{3+} by live and immobilized algal cells of *Spirulina platensis* has been investigated for simulated chrome liquor in the concentration range of 100–4500 ppm. Both live and dried biomass are very good biosorbents as they could remove high amounts of chromium from tannery wastewater. Polyurethane foam and sodium alginate are used as entrapment agents and their performances are compared. *Spirulina platensis* is an excellent biosorbent and the sorption could be accomplished with high yield even at high chromium concentrations up to 4600 mg/L, the maximum value encountered in any tannery environment (Shashirekha *et al.,* 2008). Algae, *Spirogyra condensata* and *Rhizoclonium hieroglyphicum* are employed for chromium removal from tannery effluent in immobilized and free form. It is reported that *S. condensata* and *R. hieroglyphicum* have the maximum uptake capacity of about 14 mg of Cr^{3+}/g and 11.81 mg of Cr^{3+}/g of algae respectively. Immobilized algae give better Cr recovery than native algal cells (Onyancha *et al.,* 2008).

Immobilized algae have been investigated for their potential for the accumulation of waste materials in effluent especially for the uptake of nitrogen, phosphorus and heavy metals. Immobilized cells are found to accumulate more metals than free cells. It is reported that *Chlorella vulgaris, Scendesmus* sp. *Phormidium laminosum, Scendesmus obliquus, Chlorella pyrenoidosa, Scendesmus intermedius* Chod. and *Nanochloris* are effective in the removal of ammonia and phosphorus at 80–90% and 88–100% respectively. *Chlorella vulgaris* cells entrapped in alginate are found to be a good system to remove a wide range of heavy metals. *Aulosira fertilissima, Chlorella ellipsoida, Scendesmus quadricauda, Navicula canalis, Scendesmus acutus* and *Chlorella vulgaris* are reported to accumulate chromium and other metals (Hameed and Ebrahim, 2007).

Dye Removal from Effluent

Dye released into the waste stream, if not properly treated, could exert great impact on the environment. There are chemical, physical, and biological methods for treating the dyes. Currently, the main methods of dye-containing wastewater treatment are by physical and chemical oxidation means. The oxidation agents used are chlorine, hydrogen peroxide, H_2O_2 with ferrous sulphate (Fentons reagent), ozone and UV irradiation. The physicochemical processes used for treatment of dye from effluents are adsorption, precipitation, chemical oxidation, photodegradation, membrane filtration, ion exchange resins, use of activated carbon, electrofloatation, etc. (Ahmad *et al.,* 2002). Conventional physical and chemical treatment methods of dye degradation are not so effective.

A wide variety of microorganisms are capable of degrading dyes. Pure bacterial strains such as *Pseudomonas luteola, Aeromonas hydrophila, Bacillus subtilis, Pseudomonas* sp. *Proteus mirabilis* and *Bacillus* sp. decolorize dyes under anaerobic condition (Prasad and Rao, 2010). Bioremediation/biodegradation is the current trend as well as a promising technology for dye removal. Aerobic and anaerobic biological treatment methods involving bacteria (*Bacillus subtilis, Pseudomonas* sp. *Aeromonas hydrophilia, Acetobacter liquefaciens*), fungi (*Phanerochaete chrysosporium, Aspergillus niger, Rhizomucor pussilus, Geotrichum candidum, Thelephora* sp., etc.) and algae (*Chlorella* and *Oscillatoria*) are the most commonly employed methods of bioremediation.

Another alternative for the removal of heavy metals and dyes from wastewater and involving aquatic plants has gained considerable interest. Some freshwater macrophytes including *Potamogeton lucens, Salvinia hergozi, Eichhornia crassipes, Myriophyllum spicatum, Cabomba* sp., *and Cratophyllum demersum* have been investigated for their potential for heavy metal and colour removal from dyes. Their mechanisms of metal and colour removal by biosorption can be classified as extracellular accumulation/precipitation, cell-surface sorption/precipitation, and intracellular accumulation. These mechanisms can result from complexation, metal chelation, ion exchange, adsorption and microprecipitation. Water hyacinth performs well in dilution of wastewater with freshwater at 1:3 and 1:2 ratios while plants are normally unable to survive even for a few days. The treatment results show reduction of 41.63% BOD, 35% total nitrogen and 37.60% TDS. It also reduces COD from 412.9 mg/L to 171.2 mg/L and conductivity from 23.5 μ mhos to 16.84 μ mhos and increases dissolved oxygen (DO) by 35% (Shah *et al.*, 2010).

Anaerobic Process

The anaerobic treatment of tannery wastewater has utilized technologies based on lagoons, contact filter, UASB reactor and high-rate biomethanation. UASB process is the most widely used for anaerobic wastewater treatment (Figure 19.2). High-concentration wastewater is introduced from the bottom of a modified UASB reactor through the inflow distributor. The wastewater passes upwards through the anaerobic sludge bed where the microorganisms contact with wastewater substrates. The sludge bed is composed of microorganisms that naturally form granules with high sedimentation velocity and thus resist wash-out from the system. The upward motion of released gas bubbles causes hydraulic turbulence that provides reactor mixing without any mechanical agitation. At the top of the reactor, the water phase is separated from the sludge solids and gas in a three-phase separator.

Treated water is discharged as effluent while biogas is collected at the top of the reactor and sludge solids settle down at the bottom. There is a major constraint in using the UASB system for treatment of tannery wastewater. The presence of high sulphate content results in the generation of methane contaminated with H_2S which restricts the use of biogas as a source of energy. An empowered system has been developed by CLRI wherein the sulphate is completely removed as elemental sulphur and the treatment of tannery wastewater is carried out in the conventional UASB reactor thereby generating energy. The system ensures reduction of COD and sulphate TDS to 60% and 90% respectively.

The closed anaerobic bioreactors offer a possibility to recover sulphate and bioenergy from methane. Anaerobic treatments result in only 50 to 60% of BOD and COD reduction. Post-anaerobic treatment technologies are utilized to reduce the BOD and COD loads to permissible limits. These include microbial degradation in aerobic conditions through the use of aerators. A new post-treatment technology, viz., Chemo-Autotrophic Activated Carbon Oxidation (CAACO) system has been developed and adopted for treatment of anaerobically treated tannery waste. It is an integrated system of biological and chemical oxidation carried out in a single hybrid reactor packed with activated carbon-based redox resins. Anaerobic treatment in the reactor is followed by wet-air oxidation using activated carbon which contains chemoautotrophic bacteria in immobilized state. Oxygen required for the oxidation of organics is supplied in stoichiometric quantities as air from the bottom of the reactor. This treatment eliminates about 92% of suspended solids, 98% BOD, 85% COD, 100% sulphide and 100% odour with high degree of performance consistency. The pH of the treated effluents is very close to neutral. This technology has been applied in many tanneries, in multiple textile units, Sago units and for treating domestic wastewater.

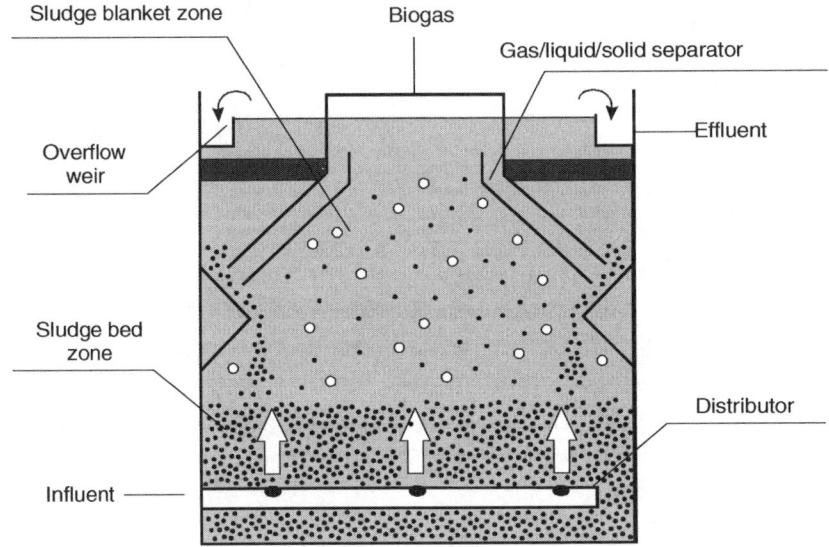

Figure 19.2 UASB reactor

Membrane bioreactor (MBR) systems essentially consist of a combination of membrane and biological reactor systems. These systems are the emerging technologies, currently developed for a variety of advanced wastewater treatment processes. Removal of TDS is a major environmental concern in the tanneries. Low cost and compact membrane bioreactor processes have been developed and successfully applied to tannery effluent treatment. Reduction of COD (95%), BOD (100%) and a total removal of suspended solids has been achieved. The high quality effluent after the membrane bioreactor treatment is further purified using reverse osmosis, which facilitates salt-free water recycling. Membrane bioreactors and reverse osmosis together demonstrate an 80% water recovery.

Phytoremediation

Phytoremediation technology is an emerging technology, which is considered for the removal of chromium in contaminated systems because of its cost-effectiveness, aesthetic advantages and long-term applicability. *Pterocarpus indicus* and *Jatropha curcas* are able to remediate Cr^{6+}-polluted soil containing less than 90 mg kg^{-1}. At a given concentration of Cr^{6+} and to achieve 50% Cr^{6+} removal, *Jatropha* is more effective than *Pterocarpus* and no significant effect on plant dry matter as well as aerial plant parts is observed. Compost amendment is preferable when Jatropha is planted in Cr^{6+}-polluted soil since *Jatropha* could remediate higher concentration than *Pterocarpus*. Further study is needed to investigate compost amendment into Cr^{6+}-polluted soil that brings about adverse effect on plant dry matter of *Pterocarpus*.

Reed Bed Technology

Reed beds are self-contained, artificially engineered, wetland ecosystems. In simple terms, a reed bed is a hole in the ground, lined with an impermeable liner, filled with one or more solid media (*e.g.*, soil) and planted with a sufficiently robust reed species. The system is then fed with effluents and drained by gravity. Reed beds are a cost-effective method of wastewater treatment. The systems are robust and well-proven, requiring only a fraction of the maintenance of traditional methods of treatment. Reed bed fits with the landscape and significantly reduces the investment and operational costs compared to conventional biological treatment. A further advantage is that reed beds do not produce surplus sludge and are free from odour. Scholz *et al.* (2007) have reported that after physical and chemical treatment, the soluble fractions of the effluent are treated in a sequence of seven horizontal reed beds and they have recorded a significant reduction of COD (80%). The reed bed of the halophytic plant, *Phragmites australis,* has been found to have a potential to remove chromium from the diluted tannery wastewater (1:1 dilution). Overall removal of about 99% chromium could be achieved providing reasonably purified water. The results also indicate that there is a significant accumulation of total Cr reaching 80% on the surface strata of the planted system. *Phragmites australis* accumulates significantly high amount of Cr in the roots. A horizontal subsurface flow reed bed, *Phragmites australis,* is also operated at different NH_4-N concentrations between 10–30 mg/L by adding NH_4Cl into the tannery effluent. Almost complete removal of NH_4-N (99%) and over 40% COD removal is obtained at 20 mg/L of NH_4-N after 7 days (Kucuk *et al.,* 2003).

There are reports that emergent hydrophytes such as *Typha, Phragmites, Scirpus, Leersia, Juncus* and *Spartina* have been used in phytoremediation of effluents to reduce the levels of heavy metals. Such hyperaccumulator plants can be exploited for treatment of heavy metals containing wastes. The metal bioaccumulation capability of a common anchored hydrophyte, *Typha angustifolia* L., has been studied in a greenhouse trial. Different concentrations of tannery sludge are prepared in water, and plants of *Typha angustifolia* are exposed to the sludge for 30, 60 and 90 days. The plants are found to absorb significant amounts of the heavy metals like chromium, copper and zinc from tannery sludge. The percentage reduction in all metals is significant while cadmium and lead are found to be totally absent. A significant reduction in the levels of sodium chloride and chlorides and COD is observed. A maximum reduction of 62% for Na, 42% for Cr, 38% for Cu and 36% for Zn is observed in 30% sludge after 90 days exposure to *T. angustifolia*. *T. angustifolia* is suitable for the decontamination of most of the harmful metals from tannery sludge (Bareen and Khilji, 2008).

Tannery wastewater, municipal solid waste and cow dung are used for manure preparation by means of pit composting, bacterial composting and vermicomposting. After completion of different types of composting, the levels of nitrogen, phosphorus, potassium, calcium and magnesium are increased in each type of compost/manure. In vermicomposting, tannery effluent is used instead of water to maintain moisture. Earthworm species, *Eisenia foetida*, can tolerate the tannery wastewater, and the biomanagement of tannery waste water is successfully achieved and compost is used for plants. Vegetable crop species, *Pisum sativum* and *Trigonella fenugrecum*, show normal growth with the use of all three types of composts (Nalawade *et al.*, 2009).

CONTROL OF AIR EMISSION

Leather manufacture processes discharge considerable emissions into the atmosphere. Rehydration of salted hides and skins generally emit odour of volatile fatty and amino acids due to the biological decomposition. During liming, dehairing, deliming and tanning processes, volatile amines, hydrogen sulphide and ammonia are generated which result in rotten egg and pungent odour. Phenolics are emitted into air during processing of hides in the post-tanning and finishing operations and particulate emissions may occur during shaving, drying, and buffing. Putrefaction of hides due to poor preservation also generates H_2S. New guidelines are being introduced in many countries for odour control. Proper preservation and hide storage can be a solution to odour control. Use of peroxides and elimination of sulphides can completely eliminate foul smell. Various gaseous emissions generated from tannery can be controlled by control devices such as thermal oxidizers to reduce volatile organic compounds and hydrogen sulphide emissions. Particulate emissions during shaving, drying, and buffing are controlled by dust collectors or scrubbers. Biofilters can also be suggested to control odour from tannery.

CLEANER TECHNOLOGY FOR WASTE REDUCTION

Waste reduction is essential to meet the demands for reduced pollution load from tanneries. It is being increasingly recognized that end-of-pipe treatments are not the ultimate strategy for waste management. The implementation of cleaner production processes and pollution prevention measures can provide both economic and environmental benefits. The idea is to control pollution through reduction in the amount of waste generation and effluent discharge. The ideal condition will be to achieve zero or near-zero waste water discharge. These also include cost-effectiveness through recovery, reuse and recycle of waste materials and also profitable use of resources. However, the applicability of these technologies varies from tannery to tannery due to the variety of raw material, processing conditions and the type of finished leather. Waste reduction can be in terms of:

i.   Reduction in water usage by developing technologies that minimize the quantity of water needed during the tanning process and recycling the wastewater

ii. Reducing the amount of salt used by adopting alternative ecofriendly preservation methods such as biocides or radiation or chilling

iii. Reduction in the use of chemicals such as lime, sulphide, etc.

iv. Recycling of dehairing and liming wastewater

v.   Using alternative biosafe enzyme technology for pre-tanning processes

vi. Use of immobilized enzymes in the beamhouse process

vii. Adopting low-chrome methodology or reusing of chrome or direct recycling of chrome tanning float or recycling of chrome after precipitation

viii. Using alternative vegetable tanning methods and

ix. Reduction of volatile organic compounds (VOC) by using aqueous finishes for finishing coatings.

The use of cleaner process technologies to achieve these ends is of great advantage to tanners. It has been demonstrated that a tannery with a production capacity of 2000 kg of hides/skins per day might potentially save ₹ 1.4 million per month by adopting clean process technologies (Ayalew, 2005).

Reduction in Chemical and Water Consumption

There are various methodologies to reduce the total salt load in wastewater. Salt consumption could be minimized if the cooling facility below 4°C is made available for preservation of hides and skins for shorter duration. Fleshing and trimming could be practised in the slaughterhouse. In curing, the use of salt is reduced by mixing additives like sodium carbonate and magnesium oxide for better preservation. Such skins preserved with low salt will not require desalting, and soak liquor will contain less salt. Reusing of pickle liquor and direct chrome liquor recycling reduces the use of sodium chloride. Another way for significant salt reduction is low-salt pickling with polysulphonic acids. Dry salting can also minimize the use of salt for preservation of hides. Preservatives such as isothiazolone products, potassium dimethyl dithiocarbamate, sodium chlorite, benzalkonium chloride, sodium fluoride and boric acid could also be used. Some of these have also been found useful for soaking, pickling and wet blue preservation (Sivakumar *et al.*, 2005; Ramasami *et al.*, 2007). The dusted salts may be reused for pickling after dissolution in water and clarification or filtration. Alternatively, the recovered salts could be used for a number of applications including foundry casting (in the mould), hypochlorite production and defrosting of roads. The salt obtained from mechanical desalting and solar evaporation could be used for curing and pickling or can be disposed off into the sea.

Dusted salt can be reused for curing but a preliminary heat treatment is required to reduce bacterial impact and to limit the presence of organic matter in recovered salt. Fleshing of green hides after soaking is a cleaner alternative over fleshing after liming. Up to 40% of sodium sulphide and 50% of lime can be saved by the direct recycling of the liming float. When tanning and pickling floats are separated, they result in a saving of about 80% of salt and 20–25% of formic or sulphuric acid. Salt concentration in pickling floats can also be reduced by using non-swelling agents. Splitting of limed pelt is a cleaner technology than chrome-tanned splitting as it reduces the amount of chromium used and gives off waste that can be easily used for the production of gelatin (Ramasami *et al.*, 2007).

Reduction of Pollutants

Mechanical desalting by hand shaking, mechanical brushes or a drum type shaker can remove up to 10% of salt added to the hides. This can be reused for pickling after dissolution and removal of solids. Desalting of salted hides could result in a reduction of up to 15% salt loads. A reduction of up to 15% of TDS has also been observed due to use of enzyme-based dehairing processes and better quality lime. Segregating and reusing pickling and chrome-tanning liquors also reduce the TDS by 10% in composite tannery wastewaters. Solvent recovery, extraction of brines and commercial use of recovered grease have been advocated as a cleaner process technology for degreasing. Cleaner processes have resulted in the reduction of effluent loads in composite wastewaters from about *600 to 400 kg/ton* of raw material. Use of enzyme-based technologies for dehairing shows a 50–60% reduction in the sodium sulphide loads. The reduction in sulphides has also demonstrated a potential ability to save at least 8–10% of the cost of end-of-pipe treatment. The use of ammonium salts in deliming is responsible for the generation of about 40% ammoniacal nitrogen. Various nitrogen-free deliming technologies such as use of organic cyclic esters, CO_2 deliming in combination with little amount of organic acids and insertion of H_2O_2 before CO_2 reduce the formation of H_2S (Christner, 2011).

Reduction in BOD and COD

Mechanical desalting, use of enzyme-assisted low sulphide dehairing and cleaner chrome-tanning have resulted in at least 30–40% reduction in the BOD and COD loads per ton of leather produced. Recovery of hair either when it is separated during the liming or at the end of the hair-saving process and reutilization as a nitrogen source may in itself additionally bring down the COD loads by about 15–20% in the mixed effluents and a reduction of 25–30% in total nitrogen.

Recovery and Reuse of Chromium

Tanneries are a major source of chromium pollution as chrome tanning is being followed in more than 85% commercial tanneries. As per an estimate in 1998, 45,000 tons of basic chrome sulphate are used out of which 18,000 tons are released into the wastewater streams. Cr^{3+} is generally used and unused chrome is the major pollutant. The maximum tolerance of total Cr for public water supply has been fixed at 0.05 mg/L as per Indian standards. The environmental protection agency has formulated the maximum permissible levels of Cr^{6+} into water bodies at 50 $\mu g/dm^3$ and in drinking water as 3 $\mu g/dm^3$ and that of Cr^{3+} as 100 $\mu g/dm^3$. Recovery, recycling and reuse of chromium salts are the better way to reduce the chromium toxicity in tannery effluent. Chromium recovery can be achieved either by a direct recycling of the spent liquor by adding make-up quantities. Recovery of chromium by alkaline precipitation and consequent dissolution in sulphuric acid and reuse of the resultant solution is an efficient methodology. The highly reactive alkali may give a voluminous sludge. Lime may also give calcium sulphate which makes the reuse of chromium difficult. Using MgO as alkali is considered more appropriate for small as well as large tanneries because of ease of operation and low investment costs. Commercial polyelectrolytes are also used for facilitating chrome precipitation. The technology is capable of recovering 98 to 99% of chrome (Belay, 2010). Application of nanofiltration and reverse osmosis in combination can provide better recovery of unreacted chromium from highly concentrated spent tanning effluent. Low-cost biosorbents such as saw dust, coir pith, chitosan and vermiculite show the highest sorption capacity for chromium and hence can be used for chromium removal (Sumathi *et al.*, 2005).

Chrome-free Tanning as an Alternative

Synthetic organic tanning agents used alone or in combination with metallic cations can be considered as a substitute for chromium. Vegetable tanning has also got a high pollution potential because of the low biodegradability. Recovery of vegetable tannin floats has been carried out by ultrafiltration in many European countries and the recovered tannins can be used in the tanning process. Vegetable and aluminium tanning can also produce chrome-free leather. Cleaner production technologies during post-tanning operations include avoidance of chromium during retanning and not using polluting dyestuff. Benzidine dyes and halogenated oils in fat liquors are strongly advocated. The finishing processes should ideally work on water-based finishes and pigments (Belay, 2010).

Reduction in Water Consumption

About 40 tons of water is being used to process 1 ton of hide which implies that there is a need to conserve water. This can be achieved by optimization through low float processing and batch type washing instead of rinsing and combining processes and this methodology may give a reduction of about 30% in water consumption. The recycle of soaking, liming, dehairing, pickling and chrome-tanning liquors can also reduce the overall water consumption by 20–40%. Treated effluents can be also used after filtration through membrane filters and this too further reduces the water consumption.

Use of Microbial Enzymes

Enzymes have good potential to be exploited as an environmentally friendly option in the leather processing industry. Microbial enzyme technology can be used not only for producing better quality leather with minimized pollution impact, but also for the treatment of tannery wastes or for by-product utilization. In order to overcome the hazards caused by the tannery effluents, use of enzymes as a viable alternative has been resorted to in pre-tanning operations such as soaking, dehairing, bating and degreasing. Proteases and lipases are mainly used in leather and waste processing (Choudhary *et al.,* 2004; Sivasubramanian *et al.,* 2008). The main advantages of the use of enzymes are specificity, stereospecificity, activity under mild conditions, possibility of producing 'natural' products, nonpollutants, and biodegradability.

BIBLIOGRAPHY

Ahmad, A.L., Harris, W.A., Syafiie and Seng, O.B. (2002). "Removal of dye from wastewater of textile Industry using membrane technology." *J. Teknol.* 36(F): 31–44.

Ayalew, A. (2005). "Cleaner Production Options for Solid Waste Management in the Leather Industry." In: Thesis submitted to Department of Chemical Engineering, School of Graduate Studies, Addis Ababa University, Addis Ababa, pp. 1–80.

Belay, A.A. (2010). "Impacts of chromium from tannery effluent and evaluation of alternative treatment options." *J. Environ. Protect.* 1: 53–58.

Benazir, J.F., Suganthi, R., Rajvel, D., Pooja, M.P. and Mathithumilan, B. (2010). "Bioremediation of chromium in tannery effluent by microbial consortia." *African J. Biotechnol.* 9 (21): 3140–3143.

Bhaskar, N., Sakhare, P.Z., Suresh, P.V., Gowda, L.R. and Mahendrakar, N.S. (2007). "Biostabilization and preparation of protein hydrolysates from delimed leather fleshings." *J. Sci. Ind. Res.* 66:1054–1063.

Bhirnd, R.G. (2007). "Leather solid waste biomethanisation." In: Benign Environmental and Sustainable Technologies, LERIG. pp. 27–32.

Bosnic, M., Buljan, J. and Daniels, R.P. (2000). "Pollutants in tannery effluents." In: Regional Programme for Pollution Control in the Tanning Industry in South-East Asia, UNIDO, pp. 1–26.

Chakraborty, R. (2003). "Ecofriendly solid waste management in Leather Industry." *Environ. Biotechnol. Newsl.* Univ Kalyani. 3: 2.

Choudhary, R.B., Jana, A.K. and Jha, M.K. (2004). "Enzyme technology applications in leather processing." *Indian J. Chem. Technol..* 11: 659–671.

Christner, J. (2011). "Compliance-a business oppurtunity." In: Connecting leather research and industry through chemicals. LERIG. pp. 60–67.

Directory of tanneries in India (1998). CLRI Publication.

Durai, G. and Rajasimman, M. (2011). "Biological treatment of tannery wastewater – A review." *J. Envtl. Sci. and Technol.* 4(1): 1–17.

Fathima, N.N., Aravindhan, R., Rao, J.R., and Nair B.U. (2009). "Utilization of organically stabilized proteinous solid waste for the treatment of coloured wastewater." *J. Chem. Technol. Biotechnol.* 4: 1338–1343.

Firdaus-e-Bareen and Sheza Khilji. (2008). "Bioaccumulation of metals from tannery sludge by *Typha angustifolia*." *L. African J Biotechnol.* 7 (18): 3314–3320.

Ganesh, R. and Ramanujam, R.A. (2007). "Sequencing batch reactor treatment of tannery wastewater – performance studies and model based evaluation." International conference on cleaner technology and environment management, Pondicherry Engineering College, Pondicherry, India, pp. 428-433.

Ganesh, R. and Ramanujam, R.A. (2009). "Biological waste management of leather tannery effluents in India: Current options and future research needs." *Int. J. Envtl. Engg.* 1(2): 165–186.

Gousterova, A., Nustorova, M., Goshev, I., Christov, P., Braikova, P., Tishinov, K., Haertle, T., and Nedkov, P. (2003). Alkaline hydrolysate of waste sheep wool aimed as fertilizer. *Biotechnol. Equip.* 17: 140–145.

Hameed, M.S.A. and Ebrahim, O.H. (2007). "Biotechnological potential uses of immobilized algae." *Int. J. Agri. Biol.* 183–192.

Hasegawa, M.C., Barbosa, A.M. and Takashima, K. (2011). "Biotreatment of industrial tannery wastewater using *Botryosphaeria rhodina*." *J. Serb. Chem. Soc.* 76 (3): 439–446.

Haydar, S., Aziz. J.A. and Ahmad, M.S. (2007). "Biological treatment of tannery wastewater using activated sludge process." *Pak. J. Engg. Appl. Sci.* 1: 61–66.

Islam, K.M.N., Misbahuzzama. K. Kamruzzaman. A. Majumder and Chakrabarty. M. (2011). "Efficiency of different coagulant combination for the treatment of tannery effluents: A case study of Bangladesh." *African J. Environ. Sci. Technol.* 5(6): 409–419.

Jain, S.K., Purkait, M. K., De, S. and Bhattacharya, P.K. (2006). "Treatment of leather plant effluent by membrane separation processes." *Separat. Sci. Technol.* 41: 3329–3348.

Kanagaraj, J., Senthilvelan, M. and Mandal, A.B. (2011). "Biodegradation of tannery wastes using fungal cultures of *Aspergillus niger* and *Penicillium* sp." In: Connecting Leather Research and Industry through Chemicals. LERIG. pp. 55–59.

Kanagaraj, J., Velappan, K.C., Chandra Babu, N.K. and Sadulla, S. (2006). "Solid wastes generation in the leather industry and its utilization for cleaner environment - A review." *J. Sci. Ind. Res.* 65: 541–548.

Karabay, S. (2008). Waste management in leather industry. Thesis submitted to the Graduate School of Natural and Applied Science, Dokuz Eylul University, Turkey.pp 1-118.

Karthikeyan, R., Balaji, S. and Sehgal, P.K. (2007). "Industrial applications of keratins–a review." *J Sci. Ind. Res.* 66:710–715.

Kolomaznik, K., Barinova, M. and Furst, T. (2009). "Possibility of using tannery waste for biodiesel production." *J. Am. Leather. Chem. Assoc.* 104:177–182.

Kowalski, Z. and Walawska, B. (2001). "Utilization of tannery wastes for the production of sodium chromate (VI)." *Ind. Eng. Chem. Res.* 40: 826–832.

Kucuk, O.S., Sengul, F. and Kapdan, I.K. (2003). "Removal of ammonia from tannery effluents in a reed bed constructed wetland." *Water Sci. Technol.* 48(11–12):179–86.

Kumar, A.G., Nagesh, N., Prabhakar, T.G. and Sekaran, G. (2008). "Purification of extracellular acid protease and analysis of fermentation metabolites by *Synergistes* sp. utilizing proteinaceous solid waste from tanneries." *Biores. Technol.* 99: 2364–2372.

Lacatus, V., Ionita, A, Gaidau, C., Niculescu, M., Popescu, M., Acsinte, D. and Filipescu, L. (2009). "Field test for foliar nutritive products formulated with the leather protein hydrolysates." International leather engineering symposium held in Turkey during April 29–May 1.

Langmaier, F., Mladek, M., Mokrejs, P. and Kolomaznik, K. (2008). "Biodegradable packing materials based on waste collagen hydrolysate cured with dialdehyde starch." *J Therm. Anal. Cal.* 93:547–552.

Lofrano, G., Belgiorno, V., Gallo, M. Raimo, A. and Meriç, S. (2006). "Toxicity reduction in leather tanning wastewater by improved coagulation flocculation process." *Global NEST J.* 8(2): 151–158.

Mala, J.G.S., Nair, B.U. and Puvanakrishnan, R. (2006). "Bioaccumulation and biosorption of chromium by *Aspergillus niger* MTCC 2594." *J. Gen. Appl. Microbiol.* 52: 179–186.

Mangkoedihardjo, S. Ratnawati, R. and Alfianti, N. (2008). "Phytoremediation of hexavalent Chromium polluted soil using *Pterocarpus indicus* and *Jatropha curcas* L." *World App. Sci. J.* 4 (3): 338–342.

Mohamed, A.M., Sekar, P. and John, G. (2011). "Efficacy of microbes in bioremediation of tannery effluent." *Int. J. Curr. Res.* 33(4): 324–0326.

Mokrejs, P., Langmaier, F., Mladek, M., Janacova, D., Kolomaznik, K. and Vasek, V. (2009). "Extraction of collagen and gelatin from meat industry by-products for food and nonfood uses." *Waste Manage. Res.* 27: 31–37.

Morimura, S., Nagata, H., Uemura, Y., Fahmi, A., Shigematsu, T. and Kida, K. (2002). "Development of an effective process for utilization of collagen." *Process Chem.* 37:1403–1412.

Nalawade, P.M., Kamble, J.R, Late, A.M, Solunke, K.R. and Mule, M.B. (2009). "Studies on integrated use of tannery wastewater, municipal solid waste and fly ash amended compost on vegetable growth." *Int. J. Agri Sci.* 1(2): 55–58.

Onyancha, D., Mavura, W., Ngila, J.C., Ongoma, P. and Chacha, J. (2008). "Studies of chromium removal from tannery wastewaters by algae biosorbents, *Spirogyra condensata* and *Rhizocolonium hieroglyphicum*." *J. Hazard. Mater.* 158: 605-614.

Ozgunay, H., Colak, S., Mutlu, M.M. and Akyuz, F. (2007). "Characterization of Leather Industry Wastes." *Polish J. of Environ. Stud.* 16(6): 867-873.

Pandit, M., Shashirekha, V., and Swamy, M. (2009). "Bioabsorpbin of Chromium from retan chrome liquor by cyanobacteria." *Microbiol. Res.* 164:420–428.

Pramod, W., Ramteke, P.W., Awasthi, S., Srinath, T. and Joseph,B. (2010). "Efficiency assessment of Common Effluent Treatment Plant (CETP) treating tannery effluents." *Environ. Monitor. Assessment.* 169 (1–4): 125–131.

Prasad, A.S.A and Rao, K.V.B. (2010). "Physico chemical characterization of textile effluent and screening for dye decolorizing bacteria." *Global J. Biotech. Biochem..* 5(2): 80–86.

Rajamani, S., Ramasami,T., Raghavan, K.V., Langerwerf, J.S.A., Van Groenstijn, J.W. and Mulder, A. (1995). "Implementation and dissemination of the results of the TNO-CLRI Co-operation Programme." XXII IULTCS Congress, Germany.

Ramalingam , B. (2011). "Value addition in leather industry." In: Connecting Leather Research and Industry through Chemicals. LERIG. pp. 55–59.

Ramasami, T., Babu, N.K.C., Muralidharan, C., Rao, J.R., Saravanan, P. and Money, C.A. (2007). "Salinity reduction in tannery effluent." In: Benign Environmental and Sustainable Technologies, LERIG. pp. 20–26.

Ravindran, B., Dinesh, S.L., Kennedy, J. and Sekaran, G. (2008). "Vermicomposting of solid waste generated from leather industries using Epigeic earthworm *Eisenia foetida*." *Appl. Biochem. Biotechnol.* 151:480–488.

Report of All Indian Survey on raw hides and s---kins. Vol 1 (2005). CLRI Publications.

Rose, C. (2007). "Management of residuals - Byproducts recovery." In: Benign Environmental and Sustainable Technologies, LERIG. pp. 33–38.

Scholz, W., Lapoulle, A., Cruickshank, D. and Gigante, J.M. (2007). "Tannery effluent tratement using reed bed and nanofiltration technology." *Leather. Int.* 1–2.

Serrano, A.R., Maldonado, V.M. and Kosters, K. (2003). "Characterization of waste materials in tanneries for better ecological uses." *J. Am. Leather Chem. Assoc.* 98:43–48.

Shah,R.F., Kumawat, D.M., Singh, N. and Wani, K.A. (2010). "Water Hyacinth (*Eichhornia crassipes*) as a remediation tool for dye-effluent pollution." *Int. J. Sci. Nat.* 1(2): 172–178.

Shanmugham, P. and Horan, N.J. (2009). "Optimising the biogas production from leather fleshing waste by co-digestion with MSW." *Bioresource Technol.* 100:4117–4120.

Shashirekha, V., Sridharan, M.R. and Swamy, M. (2008). "Biosorption of trivalent chromium by free and immobilized blue green algae: Kinetics and equilibrium studies." *Journal of Environmental Science and Health*. Part A, 43:4, 390–40.

Shin, S.B., Min, B.W., Yang, S.H., Park, M.S., Kim, H.S., Kim, B.H., Baik, D.H. (2008). "Production of Biodiesel from fleshing scrap using immobilized lipase–catalyst." *J. Korean Soc. Appl. Biol. Chem.* 51:177–182.

Singanan, M., and Singanan, A.V.V. (2007). "Studies on the removal of hexavalent chromium from industrial wastewater by using biomaterials." *Electronic J. Environ. Agri. Food Chem.* 6 (11): 2557–2564.

Sivakumar, V., Sundar, V.J., Rangasamy, T., Muralidharan, C. and Swaminathan, G. (2005). "Management of total dissolved solids in tanning process through improved techniques." *J. Cleaner Produc.* 13: 699–703.

Sivaprakasam, S., Mahadevan, S., Sekar, S. and Rajakumar, S. (2008). "Biological treatment of tannery wastewater by using salt-tolerant bacterial strains." *Microbial Cell Factories.* 7:15.

Sivasubramaniam, S., Murali Manohar, B., Rajaram, A. and Puvanakrishnan, R. (2008). "Ecofriendly lime and sulphide free enzymatic dehairing of skins and hides using a bacterial alkaline protease." *Chemosphere.* 70: 1015–1024.

Sumathi, K.M.S., Mahimairaja, S. and Naidu, R. (2005). "Use of low-cost biological wastes and vermiculite for removal of chromium from tannery effluent." *Bioresource Technology.* 96: 309–316.

Sundar, V. J., Gnanamani, A., Muralidharan, C., Chandrababu, N.K., and Mandal, A.B. (2011). "Recovery and utilization of proteinaceous wastes of leather making: a review." *Rev. Environ. Sci. Biotechnol.* 10:151–163.

Tadesse, I., Isoaho, S.A., Green, F.B. and Puhakka, J.A. (2006). "Lime enhanced chromium removal in advanced integrated wastewater pond system." *Bioresource Technology.* 97(4): 529–534.

Tewari, P.C., Andrabi, S. Z. A., Chaudhary, C.B. and Shukla, S. (2011). "Screening of Pentachlorophenol (PCP) Degrading Bacterial strains Isolated from the Tannery Effluent Sludge of Kanpur, India." *Int. J. Sci. Tech.* 6: 77–84.

Tonkova, E. V., Nustorova, M. and Gushterova, A. (2007). "New protein hydrolysates from collagen wastes used as peptone for bacterial growth." *Curr. Microbiol.* 54:54–57.

Vankar, P.S. and Bajpai, D. (2008). "Phyto-remediation of chrome-VI of tannery effluent by *Trichoderma* species." *Desalination.* 222: 255–262.

Vasudevan, N. and Ravindran, A.D. (2007). "Biotechnological process for the treatment of fleshing from tannery industries for methane generation." *Curr. Sci.* 93:1492–1494.

Wang, Y.S., Pan, Z.Y., Lang, J.M., Xu, J.M. and Zheng, Y.G. (2007). "Bioleaching of chromium from tannery sludge by indigenous, Acidithiobacillus thiooxidans." *J. Hazard. Mater.* 147: 319–324.

Zerdani, I., Faid, M. and Malki, A. (2004). "Digestion of solid tannery wastes by strains of *Bacillus* sp isolated from compost in Morocco." *Int. J. Agri. Biol.* 758–761.

20

PROTOCOLS FOR
ENZYME EVALUATION

INTRODUCTION

Enzymes such as proteases, lipases, amylases, etc have multifarious applications. However, their origin, mechanism and specificity widely differ among themselves, and different assay methods are proposed for quantifying their activities for applications in different industries. Although various methods are available for the determination of enzyme activity, these methods have their own limitations and hence cannot be applied universally. Some of these methods are also known to have certain drawbacks with regard to accurate quantitative estimation. In the case of soaking, dehairing, bating and degreasing enzymes used in leather processing, almost every assay method is comparative and such assay of enzyme activity does not necessarily give an accurate prediction of relative performance in these actual operations. Correlation of the degree of enzymatic action in various steps with assayed enzyme activity is, to a great extent, subjective. As a result, the manufacturers select their own assay method which is convenient or reproducible. A method, once chosen, is then used for batch-to-batch control of activity of an enzyme preparation. Its validity depends on the assumption that the nature and proportions of the component enzymes are constant. Problems arise if the products consist of more than one enzyme and the proportions of these vary slightly from one batch to the other. The problem is not confined to enzymes of a particular step alone but also applies to other preparations which may be developed in future.

It is the catalysing activity of enzymes which an assay attempts to measure and not the amount of enzyme present. Activity is assessed by measuring the degree to which a given substrate is acted upon. This is usually achieved by measuring the concentration of breakdown products when an enzyme digests a substrate for a given period. As described earlier, the apparent catalytic activity of enzymes depends upon temperature, pH, concentration of enzyme, substrate concentration and the specificity of each enzyme for a particular substrate. Other factors such as storage conditions and the presence of activators or inhibitors during incubation also have their effects on enzyme activity.

Several assay methods have been proposed with the aim of finding a convenient, reliable and inexpensive method which gives the best possible indication of activity. Casein is regarded as an ideal substrate for proteases since it can be produced to a high degree of purity and is, therefore, uniform in composition. Casein is generally used as a consistent substrate for the comparison of proteolytic activities since all proteolytic preparations exhibit their action on this substrate though they vary in their degree of hydrolysis. It is also questionable whether the results of enzyme assay using casein as substrate would convey the actual effect of enzymes on the hides and skins since these materials have proteins that are structurally completely different from casein. The earlier methods have used casein or denatured hemoglobin as substrate and the enzyme with highest activity has been considered for treating the hides and skins. Gelatin, a denatured collagen, is an equally ideal substrate since it is chemically similar to hide protein-collagen and is also available in pure form. But, gelatin does not contain tyrosine and it cannot be utilized in colorimetric methods, where the colour development is due to the presence of aromatic amino acids such as tyrosine, tryptophan, etc. Moreover, proteases including collagenases act on gelatin and there is a need to distinguish the collagenolytic activity of the enzyme preparation.

Considering the suitability of the various protein substrates, a number of colorimetric methods have been recommended for the estimation of proteolytic activity of enzymes and the enzymatic formulations. The enzymes which are available in the market or the ones being prepared in various laboratories need to be assessed for the optimal conditions of pH and temperature and the substrate on which they act. Based on this knowledge, the suitability to use it on collagen matrix which should not be degraded and the particular step in the leather processing where its functions are optimal can be worked out. A better understanding of the enzymes with specificity to various substrates will enable their effective use in various steps in leather processing and also in other industries. In addition, it is also necessary to ensure that the preparation does not contain collagenases, which may damage the skin matrix. Hence, protocols for the assay of protease, lipase, amylase, elastase, keratinase and collagenase are essential and they are described in this chapter. In addition, procedures for determination of protein content and of hydroxyproline are illustrated.

ESTIMATION OF PROTEIN AND SPECIFIC ACTIVITY OF ENZYME PREPARATIONS

To determine the specific activity of an enzyme preparation, estimation of protein content is necessary. The specific activity of enzyme is expressed as the unit of enzyme activity per mg of protein under the standard assay conditions. Three methods are available which are widely used to determine the protein content of enzyme formulations or preparations.

i. Lowry Method (Lowry *et al.*, 1951)

In the presence of alkali, protein molecules form a complex with copper ions, and amino acids containing phenolic hydroxyl groups such as tyrosine and tryptophan present in the Cu–protein complex react with Folin–Ciocalteau phenol reagent to give blue colour due to the reduction of phosphomolybdate. The intensity of the colour is proportional to the concentration of protein. This method is having sensitive limit to 10 µg of protein, and the sensitivity is moderately constant from one protein to another. When the Folin reagent which is a mixture of sodium tungstate, molybdate and phosphate, together with a copper sulphate solution, is added to a protein solution, a blue–purple colour is produced which can be quantified by its absorbance at 660 nm. The method is based on both the Biuret reaction, where the peptide bonds of proteins react with Cu^{2+} under alkaline conditions producing Cu^+, which reacts with the Folin reagent while in the Folin-Ciocalteau reaction, it involves the reduction of phosphomolybdotungstate to hetero-polymolybdenum blue by the copper-catalysed oxidation of aromatic amino acids. The resultant strong blue colour is therefore partly dependent on the tyrosine and tryptophan content of the protein sample.

Reagents

1. *Reagent A* 2% (w/v) sodium carbonate in 0.1N NaOH 1.35% sodium potassium tartrate

2. *Reagent B* Dissolve 5.0 mg of cupric sulfate in 1.0ml of 1.35% sodium potassium tartrate (w/v), just prior to use.

3. *Alkaline copper reagent (Reagent C)* Prepare freshly, at the time of protein assay, by mixing 50 ml of reagent A with 1 ml of Reagent B (v/v).

4. *Folin-Ciocalteau phenol reagent 1N* Commercially available 2N Folin–phenol reagent is diluted to 1N by adding equal volume of water (v/v).

5. *Bovine serum albumin (BSA)* 12.5 mg of BSA in 50 ml of 0.1N NaOH. The standard solution contains 250 µg/ml of BSA.

Procedure Suitable aliquots of protein solution (0.01 ml) are taken in test tubes. The volume in each tube is made up to 1.0 ml with water. 5.0 ml of alkaline copper reagent is added, mixed and allowed to stand for 10 min. at room temperature (25–30°C). 0.5 ml of 1N Folin–phenol reagent is added to each tube, shaken well and kept for incubation for 20 min. at room temperature (25–30°C). The intensity of the blue color developed is read at 660 nm against a reagent blank containing all the reagents except the protein solution. Amount of protein present is calculated by referring to the standard graph. BSA protein in the range of 20 to 100 µg/ml is used to prepare the standard curve and can be used for estimating the protein content of test samples. The standard graph is drawn by plotting the concentration of the standard solution on the X-axis and the OD on the Y-axis.

$$\text{Protein concentration} = \frac{\text{OD of unknown} \times \text{Standard concentration}}{\text{OD of known}}$$

ii. Bradford Method (Bradford, 1976)

This method is based on the binding of Coomassie Brilliant Blue dye to the protein. At low pH, the free dye has absorption maxima at 470 and 650 nm, but when bound to protein, it has an absorption maximum at 595 nm. The practical advantages of the method are that the reagent is simple to prepare and that the colour develops rapidly and is stable. Although it is sensitive down to 20 µg protein, it is only a relative method, as the amount of dye-binding appears to vary with the content of the basic amino acids, arginine and lysine, in the protein. Coomassie Brilliant Blue (CBB) dye solution is prepared by dissolving 0.01% CBB (G-250) in 50 ml of 95% ethanol, and to this, 100 ml of orthophosphoric acid (85% w/v) is added and made up to 1 L with double-distilled water. This is filtered through Whatman No.1 filter paper and stored in a brown bottle.

Procedure   To 0.1 ml of sample containing 1–10 µg of protein, 1.0 ml of CBB dye is added and mixed well. CBB binds to the aromatic amino acid residues in the protein and gives a blue colored complex which can be measured at 595 nm after 15 min. against dye solution as a reference. The concentration of protein can be estimated from the standard graph using BSA standard as described above.

iii. Kjeldahl Method

This is a general chemical method for determining the nitrogen content of any compound. It is not normally used for the analysis of purified proteins or for monitoring column fractions but, it is frequently used for analysing complex solid samples and microbiological samples for protein content. The sample is digested by boiling with concentrated sulphuric acid in the presence of sodium sulphate (to raise the boiling point) and a copper and/or selenium catalyst. The digestion converts all the organic nitrogen to ammonia, which is trapped as ammonium sulphate. Completion of the digestion stage is generally recognized by the formation of a clear solution. The ammonia is released by the addition of excess sodium hydroxide and removed by steam distillation in a Markham still. It is collected in boric acid and titrated with standard hydrochloric acid using methyl red-methylene blue as indicator. It is possible to carry out the analysis automatically in an autokjeldahl apparatus. Alternatively, a selective ammonium ion electrode may be used to directly determine the content of ammonium ion in the digest. Although Kjeldahl analysis is a precise and reproducible method for the determination of nitrogen, the determination of the protein content of the original sample is complicated by the variation of the nitrogen content of individual proteins and by the presence of nitrogen in contaminants such as DNA. In practice, the nitrogen content of proteins is generally assumed to be 16% by weight.

ASSAY FOR PROTEASE

Anson's Colorimetric Method (Anson, 1938)

The basic principle of this method is to estimate the liberated tyrosine and tryptophan during proteolysis. Tyrosine and tryptophan react with phenol reagent giving a blue colour which is read at 660 nm using a spectrophotometer. Though the phenol reagent develops blue colour with both the amino acids, and the total colour produced by both the amino acids is measured by this method, the results are expressed in terms of tyrosine only.

Standard curve   1% solution of pure tyrosine in 0.2N HCl is prepared and serially diluted to form different strengths ranging from 50 to 500 µg. To each aliquot of 0.5 ml of tyrosine solution, add 2.0 ml of water, 5.0 ml of 0.5N sodium hydroxide solution and 1.5 ml of Folin–Phenol reagent. After 2 minutes of shaking, the blue colour developed is read at 660 nm in the spectrophotometer against water blank. The values thus obtained are plotted against different concentrations of tyrosine.

Reagents required

  i. *Casein solution*   Add 2.0 g of casein (fat-free) to 50 ml of water; keep it in boiling water bath for 10 min. till a clear solution is formed. Add 50 ml of buffer of required pH (7.0–11.0).
  ii. *Hemoglobin solution*   Add 2.0 g of hemoglobin to 50 ml of water. Stir well. Add 50ml of buffer of required pH (1.5–7.0).
  iii. 5% Trichloroacetic acid (TCA)
  iv. 0.5N Sodium hydroxide
  v. Folin–phenol reagent (twice-diluted)

Procedure   Add 1 g of enzymatic formulation to 100 ml of water and keep at 30°C for 15 min. with occasional stirring. After the period, collect the enzyme extract after filtration. Add 2.5 ml of enzyme extract and 2.5 ml of water to 10 ml of casein or hemoglobin solution. Incubate at 45°C for 30 min. The reaction is stopped by adding 30 ml of TCA. Warm the mixture on a boiling water bath for 2–3 min. Cool to room temperature (25° to 30°C) and filter. To 0.5 ml of filtrate, add 2.0 ml of water, 5.0 ml of sodium hydroxide solution and 1.5 ml of Folin–Phenol reagent. After shaking for 2 min., read the blue colour in a spectrophotometer (660 nm). Run a control in the same manner except for the addition of TCA solution before the enzyme addition. Subtract control values from the experimental, and the net values are taken for the calculation of enzyme unit. One unit of enzyme activity is defined as the amount of enzyme preparation which liberates one microgram of tyrosine from casein or hemoglobin at the required pH at 45°C in 30 min.

Of the many methods tried, Anson's modified colorimetric method is found to be one of the best routine quantitative assays of protease activity. This method gives very accurate and reproducible data and is capable of estimating accurately even very low contents of tyrosine and hence very low proteolytic activity.

Casein Digestion Method

This assay can be used for testing the activity of alkaline protease by using casein as a substrate. The small peptides, which are produced by proteolytic digestion, are measured in the filtrate of trichloroacetic acid by means of their absorption at 280 nm since the aromatic amino acids such as tryptophan, tyrosine and phenylalanine have their absorbance maximum around 280 nm. Since casein is soluble in alkaline buffer solutions of pH above 8.0, this method cannot be used to evaluate acid and neutral proteases.

Reagents

1% Casein solution (Hammarsten casein, Biochemical assay grade).

Carbonate buffer of 0.1 M; pH 10.0.

5% Trichloroacetic acid.

Procedure 1.0 ml of casein solution and 0.9 ml of buffer are added to each test tube and warmed at 37°C in a water bath. Prior to mixing, 0.1 ml of suitably diluted enzyme solution is added. After 10 min., the reaction is stopped by addition of 3.0 ml of 5% trichloroacetic acid. All these tubes are kept for 15 min. at room temperature (25° to 30°C). Then, the precipitate is filtered through Whatman No.1 filter paper. The optical density of the filtrate at 280 nm is determined using UV- Visible spectrophotometer. For the blank, TCA is added before the enzyme solution. All determinants and blanks are run in duplicates, and averages are used. It is important to adjust the enzyme quantities in the assay so that OD value at 280 nm is below 1.0. Above that value, the linearity between absorbance and tyrosine concentration no longer holds good. Amount of tyrosine liberation is estimated from the standard graph. One unit of enzyme activity is defined as the amount of enzyme releasing one mg tyrosine for 10 min at 37°C and expressed as 1 U/ml /10 min.

Hemoglobin or Egg Albumin Digestion Method

This method is suitable for evaluation of acid and neutral proteases in the pH range of 3.0–7.0 and not suitable for alkaline proteases. The principle is similar to the casein digestion method except for the buffer used for dissolving the substrate. As the protease activity is expressed in terms of tyrosine liberation, these substrates are not ideal for analysing or comparing the activity of various proteases over a wide pH range.

Azocasein Assay (Brock *et al.,* 1982)

Proteolytic activity is determined by azocasein assay. 100 µl of enzyme is added to 2.0 ml of 0.8% azocasein (Sigma) in 0.05 M Tris-HCl, 0.5 mM $CaCl_2$, pH 7.5, and incubated at

37°C for 15 min. The reaction is stopped by adding trichloroacetic acid (10%, 1.0 ml), followed by centrifugation and measurement of absorbance of the clear supernatant at 450 nm. One unit of activity is the amount of enzyme that causes an increase in optical density of 1 unit/min. at 450 nm.

Azocoll Assay (Chavira, 1984)

Protease digests Azocoll, an insoluble protein substrate impregnated with azo dye and releases colour to the solution. Intensity of colour developed is quantitated spectrophotometrically at 530 nm as protease activity.

Procedure 10 mg of azocoll is added to 2.0 ml of Tris-HCl buffer of pH 8.0 in test tube. To this, 0.1 ml of suitably diluted enzyme is added and kept for shaking for 10 min at 40°C followed by filtration using Whatman filter paper. The absorbance of the filtrate is then measured at 530 nm relative to a blank to which no enzyme is added. One unit is defined as the amount of enzyme required to increase the OD at 530 nm by 0.01 absorbance unit.

Trypsin Assay (Rovery *et al.*, 1953; Schwert and Takenaka, 1955)

Trypsin hydrolyses the ester linkage of N-Benzoyl-L-arginine ethyl ester (BAEE) resulting in increase in absorbance at 247 nm. One BAEE unit of trypsin is the amount of enzyme causing an increase in the absorbance by 0.003/min at 25°C, pH 7.6 and 247 nm.

Reagents

1. *Buffer* 0.1 M KCl, 0.05 M $CaCl_2$, 0.01 M Tris, pH 7.75.
2. *Substrate solution* Dissolve 350 mg of BAEE in a final volume of 100 ml of buffer. The final solution contains 0.01M BAEE and is stored at 4°C.
3. *Enzyme standard solution* It is prepared in 0.01 M HCl at 4°C at a concentration of approximately 10 mg/ml and dialyzed overnight at 4°C against 0.001 M HCl. Insoluble material is removed by centrifugation at 10,000×g for 10 min. The precise concentration is established by measuring absorbance at 280 nm after 1:10 dilution with 0.001 M HCl. (Σ = 15.4 M^{-1} cm^{-1})

Procedure Substrate solution (3.0 ml) is measured into each cuvette of the spectrophotometer. Balance the absorbance of the two cuvettes at 247 nm. Add 0.1 ml of enzyme solution (containing 0.45 µg of trypsin) to the sample cuvette and mix well. The increase in absorbance is measured at 247 nm.

Note: The rate of increase of absorbance should be linear up to an absorbance of 0.24 (65% hydrolysis) and this rate is directly proportional to the concentration of the enzyme.

Chymotrypsin Assay (Schwert and Takenaka, 1955; Hummel, 1959)

Chymotrypsin hydrolyses the ester linkage of N-benzoyl-L-tyrosine ethyl ester (BTEE) causing an increase in the absorbance which is measured spectrophotometrically at 256 nm.

Reagents

1. 0.08 M Tris-HCl buffer, pH 7.8 containing 0.1 M $CaCl_2$.
2. *BTEE substrate*   0.00107 M BTEE in 50% w/w methanol (63 ml absolute methanol added to 50 ml reagent-grade water).
3. *Enzyme*   Dissolve enzyme at 1mg/ml in 0.001M HCl. Dilute in 0.001N HCl to 10–30 µg/ml for assay. Dialyse overnight at 4°C against 0.001 M HCl. Insoluble material is removed by centrifugation at 10,000 xg for 10 min.

Procedure

Pipette the following solutions into cuvettes:

> 0.08M Tris-HCl with 0.1M $CaCl_2$ 1.5 ml
>
> 0.00107 M BTEE   1.4 ml

Equilibrate at 25°C for 4–5 min to achieve temperature equilibrium and record blank rate, if any. Add 0.1 ml of appropriately diluted enzyme and record the increase in absorbance at 256 nm for 4–5 min. Calculate A_{256}/ min from the initial linear portion of the curve.

Peptidase Activity (Strongin *et al.*, 1978)

Assay of peptidase activity is done using benzoyl-DL-arginine *p*-nitroanilide (BAPNA) as the substrate.

1.9 ml of substrate prepared in Tris-HCl buffer, pH 8.0 is mixed with 0.1 ml of suitably diluted enzyme and the mixture is incubated at 30°C for 30 min. Then, 2.0 ml of 5% (w/v) citric acid is added to stop the reaction. The yellow colour developed is measured at 410 nm. One unit of the enzyme activity is defined as the amount of enzyme that hydrolyses 1 µmole of substrate per min. under the assay conditions.

Hide Powder Azure (HPA) Method

It has been proposed that tanners evaluating enzymatic formulations conduct small-scale trials using their own delimed stock as a substrate. The principle applied in Hide Powder Azure (HPA) assay technique is that it employs hide powder covalently dyed with a blue dye as the substrate (Lamb, 1982; Himelbloom and Hosni, 1985; Mozersky, 1992). Digestion of degraded collagen by the enzymes leads to the release of dye into solution. Spectrophotometric measurement of the dye released provides a measure of the degree of digestion achieved.

Reagents Required

100 mg of HPA is suspended in 20 ml of 0.1 M Tris, (trishydroxymethylmethylamine, 12.1 g/L) adjusted to pH 8.5 with hydrochloric acid. The insoluble substrate is transferred to 50 ml screw-capped test tube with 5.0 ml portions of buffer. After tightly capping the tube, it is placed in the test tube rack of the orbital shaker and prewarmed for 15 min at 40°C while oscillating at 200 rpm.

Procedure   0.1 ml aliquot of suitably diluted enzyme is added to the screw cap tube containing substrate solution. After 10 min of incubation, 2.0 ml of 5% TCA is added and the tubes are left at room temperature for 15 min. Then, the precipitate is filtered through Whatman No. 1 filter paper. The optical density of the filtrate at 595 nm is determined using spectrophotometer. The control is run in a similar manner where the enzyme is added after the addition of 5% TCA. One unit of enzyme activity is defined as the amount of enzyme that solubilizes blue dye to increase the absorbance at 595 nm of 1.0 O.D in 10 min.

ASSAY FOR ELASTASE

Elastase Assay using Synthetic Substrate (Bieth *et al.*, 1974)

Elastase hydrolyses N-succinyl-L-Ala-L-Ala-L-Ala-p-nitroanilide (Suc-Ala$_3$-NA) resulting in the increase in the absorbance at 410 nm which is measured spectrophotometrically. One unit hydrolyses 1 µM of Suc-Ala$_3$-NA per min at 25°C and pH 8.0 under the specified conditions.

Reagents

1. 0.1 M Tris buffer, pH 8.0
2. 0.0044 M Suc-Ala$_3$-NA substrate dissolved in Tris buffer, pH 8.0 (2 mg / ml)
3. 0.15 M NaCl
4. *Enzyme*:   Prepare a 1 mg/ml solution in Tris buffer. Immediately before use, dilute further to obtain a rate of 0.02–0.04 A/min.

$$\text{Protein (mg)/ml} = A_{280} \times 0.54$$

Procedure

Pipette into each cuvette the following solutions:

Tris buffer   2.7 ml

Enzyme   0.1 ml

Mix and equilibrate in the spectrophotometer for 4–5 min. To the test cuvette, add 0.2 ml of substrate, mix and record the increase in A_{410} for 3–5 min. Calculate absorbance/min. from the linear portion of the curve.

Elastase Assay using Elastin–Congo Red Substrate (Cahan et al., 2001)

Elastolytic activity is determined with elastin-Congo Red (Sigma) as the substrate. Reaction suspensions containing 10 mg of elastin-Congo Red in 2.0 ml of 0.05 M Tris-HCl, 0.5 mM $CaCl_2$, pH 7.5, and 100 µl of enzyme, are incubated with shaking at 37°C for 1 h. The reaction is stopped by adding 0.1 ml of 0.1 M EDTA, followed by centrifugation and measurement of absorbance at 495 nm of the clear supernatants. One unit of activity is the amount of enzyme that causes an increase in 1.0 OD at 495 nm per h.

Elastase Assay using Elastin–Orcein Substrate (Sachar et al., 1995)

The elastase activity is determined by using elastin–orcein as substrate. This method involves the measurement of the amount of elastin digested and the method has the advantage that orcein is a stain specific for elastin, so that traces of collagen or other protein contaminants in the substrate do not affect the validity of the assay.

Procedure 20 mg of elastin–orcein substrate is placed in each of the test tubes and 20 ml of Tris-HCl buffer, pH 8.0, and 0.5 ml of suitably diluted enzyme are added. The test tubes are placed in the shaking water bath at 37°C for 30 min. Then, the tubes are removed and 2.5 ml of 0.5 M phosphate buffer of pH 7.0 is added. The mixture is centrifuged, filtered and read at 578 nm. One unit of elastolytic activity is defined as the amount of enzyme digesting 1 mg of elastin–orcein per hour.

Keratinase Assay (Santos et al., 1996)

The keratinase activity is determined using keratin-azure as substrate and the liberation of the azo dye is monitored at 595 nm in a spectrophotometer.

Reagents

1. Crude enzyme
2. Keratin-azure (Sigma)

Procedure The reaction medium contains 2.0 ml of crude enzyme and 4 mg of keratin-azure. The reaction is carried out at 42°C with constant agitation and stopped by centrifugation of the reaction medium. Absorbance is measured in the clear supernatant at 595 nm in a spectrophotometer. As a control, enzyme samples kept for boiling for 15 min are added to the substrate and are incubated for the same time period. The keratinolytic activities of the samples are calculated by subtracting the absorbance recorded for the boiled samples. All assays are done in triplicate, and the data presented are mean values of the triplicate assays.

ASSAY FOR COLLAGENASE

Collagenase Assay using
Collagen as Substrate–Method - I (Etherington, 1972)

Bovine tendon collagen is suspended at 10 mg/ml in a solution of 0.35% acetic acid in 25% (v/v) glycerol. The collagen is dispersed using a polytron homogenizer. This fine suspension with a pH of approximately 3.5 remains stable for at least 2 weeks. Collagenase is assayed at pH 3.5 using 15 ml conical centrifuge tubes. Each tube contains 0.4 ml of collagen suspension 0.6 ml of 0.2 M sodium formate buffer containing 10 mM cysteine and 0.2 ml of the enzyme sample. Tubes are incubated at 37°C for 3 h or 16 h depending on the enzyme activity. At the end of the incubation period, the tubes are centrifuged at 2900 x g for 10 min. to remove the residual collagen. Blank readings are obtained from assay mixtures that have been prepared on ice and centrifuged without prior incubation.

The amount of collagen degraded is calculated from the concentration of hydroxyproline in solution, based on the hydroxyproline content (14% w/w) in collagen. A sample (0.2 ml) of the supernatant is hydrolysed in the presence of 6N HCl after evacuating the samples. The hydrolysis is performed at 110°C for 16 h. After evaporating the acid, the hydrolysate is made up to a known volume, and hydroxyproline is estimated (Woessner, 1961).

Estimation of Hydroxyproline (Woessner, 1961)

Hydroxyproline liberated from hydrolysed collagen is oxidized by chloramine T to pyrrole. The reaction is arrested by the addition of perchloric acid and the resulting product reacts with p-dimethylamino benzaldehyde (PDAB) to give a coloured complex which is read at 557 nm.

Reagents

1. *Chloramine T* A 0.05 M solution is prepared freshly by dissolving 1.14 g chloramine T in 20 ml of water. 30 ml of methyl cellosolve and 50 ml buffer are then added and the solution is kept in a glass-stoppered flask.
2. *Buffer* 50 g of citric acid monohydrate, 12 ml of glacial acetic acid, 120 g of sodium acetate trihydrate and 34 g of sodium hydroxide are made up to a final volume of 1 L in distilled water. The pH is carefully adjusted to 6.0 and the buffer is stored at 4°C.
3. Methyl cellosolve (Ethylene glycol monoethyl ether)
4. *Perchloric acid* 3.15 M solution is prepared by diluting 27 ml of 70% perchloric acid (A.R. grade) to 100 ml with water.
5. PDAB (*p*-dimethylamino benzaldehyde) A 20% solution is prepared shortly before use by adding methyl cellosolve to 20 g of PDAB (reagent grade) to give a final volume of 100 ml.

6. *Hydroxyproline standard*   A stock solution is prepared by dissolving 10 mg of L-hydroxyproline in 100 ml of 0.001N HCl. Standards are prepared by diluting the stock with water to obtain concentrations of 2–10 µg/ml.

Procedure   2.0 ml portions containing 2–10 µg hydroxyproline are placed in test tubes. A series of standards are prepared containing 0–10 mg hydroxyproline in a total volume of 2.0 ml. Hydroxyproline oxidation is initiated by adding 1.0 ml of chloramine T to each tube. The contents of the tube are mixed by shaking a few times and allowed to stand for 20 min. at room temperature (25–30°C). The chloramine T is then deactivated by adding 1.0 ml of perchloric acid to each tube. The contents are mixed and allowed to stand for 5 min. Finally, 1.0 ml of PDAB solution is added and the mixture is shaken well. The tubes are placed in a 60°C water bath for 20 min., and then cooled in tap water for 5 min. The color developed is read at 557 nm using a spectrophotometer.

Collagenase Assay using Collagen as Substrate–Method-II (Van Wart and Steinbrink, 1981)

Assay for collagenase activity using bovine tendon collagen as substrate is carried out at 35°C. 100 mg of collagen is dissolved in 980 µl of 50 mM Tricine, 0.4 M NaCl, 10 mM $CaCl_2$, pH 7.5. The assay is initiated by adding 20 µl of enzyme. At various time intervals, 100 µl of aliquots of the solution are withdrawn and added to microcentrifuge tubes containing 50 µl of 2N HCl and 50 µl of 1% phosphotungstic acid to simultaneously quench the reaction and precipitate the undigested collagen. After centrifugation, 100 µl of supernatant is withdrawn from each tube and hydroxyproline is measured as described in Method-I.

Collagenase Assay using N-(3-(2-Furoyl) acryloyl)-Leu-Gly-Pro-Ala, (FALGPA) as Substrate–Method III (Van Wart and Steinbrink, 1981)

Assay with FALGPA is carried out spectrophotometrically by continuously monitoring the decrease in absorbance of substrate after addition of enzyme. The wavelength varies from 324–345 nm, depending on FALGPA concentration, so that an initial absorbance of about 0.5–1.0 is obtained. For most routine assays, a FALGPA concentration of 0.05 M in 50mM Tricine, 0.4 M NaCl, 10 mM $CaCl_2$, pH 7.5 is employed and hydrolysis is monitored in a 1 cm cuvette at 324 nm after addition of 20 µl of collagenase to give a final enzyme concentration of 4 µg/ml. At this wavelength, the background absorbance is 0.695 and the change in absorbance on full hydrolysis is 0.125. Initial velocities are calculated from the slope of the absorbance change during the first 10% of hydrolysis and converted into units of microkatals (µmol / sec) dividing the absorbance change corresponding to full hydrolysis and multiplying by the substrate concentration. Specific activities in units of microkatal/kg are calculated by dividing these rates by the amount of enzyme present.

LIPASE ASSAY—TITRIMETRIC METHOD

The basic reaction catalysed by lipase is the release of fatty acids from triglycerides and hence the quantification of the fatty acids released is an estimate of lipase activity.

Chemicals

Olive oil
Polyvinyl alcohol (PVA)
0.1 M Phosphate buffer; pH 6.5
0.05N Sodium hydroxide
0.05 M Hydrochloric acid
0.05 M Sodium carbonate
Acetone
Ethanol
Phenolphthalein

Preparation of substrate emulsion

* Add 2.0 g of PVA to 100 ml of distilled water and boil with stirring till the PVA dissolves completely.
* Add 25 ml of olive oil to 75 ml of 2% PVA and sonicate at 60% amplitude, 9.0 sec pulse for 3 min.

Standardization of NaOH

* Titrate 10 ml of 0.1N Na_2CO_3 against 0.05N HCl in a burette using methyl orange as indicator.
* Calculate the normality of HCl.
* Titrate 10 ml of 0.05 N NaOH with the standardized 0.05N HCl using phenolphthalein as indicator.
* Calculate the normality of NaOH.

Procedure

1. Add 5.0 ml of the substrate emulsion and 4.0 ml of 0.1 M phosphate buffer (pH 6.5) to a test tube and mix well.
2. Incubate the reaction mixture at 37°C for 10 min.
3. Add 1.0 ml of the enzyme solution and incubate at 37°C for 20 min.
4. Add 20 ml of 1:1 acetone–alcohol to the reaction mixture to stop the reaction.
5. Prepare a control using the inactivated enzyme.
6. Estimate the released fatty acids by titration against 0.05 N sodium hydroxide using phenolphthalein as indicator.

One unit of the lipase activity is defined as the amount of enzyme releasing 1 μmol of free fatty acid in 1 min under standard assay conditions.

$$\text{Lipase activity} = \frac{\text{Volume of NaOH} \times \text{Normality of NaOH} \times 1000}{\text{Time of incubation}}$$

Esterase Activity (Strongin *et al.*, 1978)

Most of the protease as well as lipase preparations exhibit esterase activity. Hence, the nature of hydrolytic activity of these enzyme preparations must be identified and quantified to ensure the required activity for specific applications.

Reagents

1. *Buffer*   0.1 M KCl, 0.05M $CaCl_2$, 0.01 M Tris, pH 7.8.
2. *Substrate solution*   Dissolve 350 mg of BAEE in a final volume of 100 ml of buffer. The final solution contains 0.01 M BAEE and is stored at 4°C.

Procedure   The spectrophotometric determination of esterase activity employs BAEE as substrate. 0.2 ml substrate solution and 0.5 ml buffer are taken in each of reference and sample cuvettes (1 cm) in a spectrophotometer and the absorbance at 254 nm is adjusted to zero. 0.5 ml buffer and 0.5 ml of suitably diluted enzyme solution are added to the reference and sample cuvettes, respectively. The activity is determined by monitoring the increase in OD at 254 nm.

AMYLASE ASSAY

Amylases are enzymes that break down starch, glycogen and few carbohydrates containing glucose backbone chain. Amylases are produced by a variety of microbes, including bacteria and fungi. They secrete amylase extracellularly for digesting insoluble carbohydrates so as to uptake soluble end products such as glucose, a prime substrate for their growth. Although many microorganisms produce this enzyme, the ones most commonly used for their industrial production are *Bacillus subtilis, Bacillus licheniformis, Bacillus amyloliquefaciens, Aspergillus niger,* etc. Amylases are classified based on how they break down carbohydrates.

* *α-amylase*   Reduces the viscosity of starch by breaking down the bonds at random, therefore producing chains of glucose of varied sizes. They catalyse the hydrolysis of α-1,4 glucosidic linkages in polysaccharides of three or more α-1,4-linked D-glucose units to produce maltose and larger oligosaccharides.

* *β-amylase*   Breaks the glucose–glucose bonds by removing two glucose units at a time, thereby producing maltose.

* *Amyloglucosidase (AMG)*   Breaks successive bonds from the non-reducing end of the straight chain, producing glucose

Many microbial amylases usually contain a mixture of these amylases and all amylases can be assayed based on the liberation of glucose that is spectrophotometrically quantitated. The method described below is the modification of an amylase assay reported by Bernfeld (1951).

Reagents

* *DNS reagent*   1.0 g of 3, 5-dinitrosalicylic acid (DNS) is dissolved in 50 ml of reagent-grade water and 30 g sodium potassium tartrate tetrahydrate is added slowly. 20 ml of 2N, NaOH is added to the above solution and diluted to a final volume of 100 ml with reagent-grade water.
* 1% starch solution in 50 mM sodium phosphate buffer (pH 7.0).

Assay procedure

* 1.0 ml of culture enzyme is pipetted into a test tube.
* 1.0 ml of 1% soluble starch prepared in sodium phosphate buffer (pH 7.0) and preincubated in a water bath at 40°C for 30 min. is then added.
* A blank is prepared consisting of 2.0 ml of the enzyme extract that has been boiled for 20 min (boiling inactivates the enzyme), added to the starch solution and treated with the same reagent as the experimental tubes.
* The reaction is stopped by adding 2.0 ml of DNS reagent and heated at 95°C for 5 minutes.
* 10 ml of distilled water is added to the reaction mixture to cool.
* Colour intensity is determined at 540 nm using a spectrophotometer.

A concentration glucose calibration curve is used to convert color to reducing sugar equivalent. Enzyme activity may be defined as the amount of glucose in µM produced per ml in the reaction mixture per unit time.

BIBLIOGRAPHY

Anson, M.,L. (1938). "The estimation of pepsin, trypsin, papain, and cathepsin with hemoglobin." *J.Gen. Physiol.* 22, 79–89.

Bernfeld, P. (1951). *Advances Enzymology*–Vol. 12, ed. FF. Nord. Interscience Publications, New York.

Bieth, J., Spiess, B. and Wermuth, C.G. (1974). "The Synthesis and Analytical Use of a Highly Sensitive and Convenient Substrate of Elastase." *Biochem. Med.* 11, 350–357.

Bradford, M. M. (1976). "A rapid and sensitive method for the quantitation of microgram quantities of protein utilizing the principle of protein-dye binding." *Anal Biochem.* 7, 248– 254.

Brock, F.M., Frosberg, C.W. and Buchanan-Smith, J.G. (1982). "Proteolytic activity of rumen microorganisms and effects of proteinase inhibitors." *Appl. Environ. Microbiol.* 44, 561–569.

Cahan, R., Axelrad, I., Safrin, M., Ohman, D.E., and Kessler, E. (2001). "A secreted aminopeptidase of *Pseudomonas aeruginosa*. Identification, primary structure, and relationship to other aminopeptidases." *J.Biol.Chem.* 276: 43645–43652.

Chavira, R., Burnett, J.J. and Hageman, J. (1984). "Assaying proteinases with azocoll." *Anal. Biochem.* 136, 446–450.

Etherington, D.J. (1972). "The nature of the collagenolytic cathepsin of rat liver and its distribution in other rat tissues." *Biochem J.* 127, 685–692.

Himelbloom, B.H. and Hosni, M. (1985). "Optimization of the hide powder azure assay for quantitating the protease of *Pseudomonas fluorescens*." *J. Microbiol. Methods.* 4, 59–66.

Hummel, B.C. (1959). "A modified spectrophotometric determination of chymotrypsin, trypsin, and thrombin." *Can. J. Biochem. Physiol.* 37,1393–1399.

Lamb, N.C.J. (1982). "Bates and related enzyme products: A review of the approaches to laboratory estimation of their activity." *J.Soc. Leath. Technol. Chem.* 66, 110–113.

Lowry, O.H., Rosebrough, N.J., Farr, A.L. and Randall, R.J. (1951) "Protein measurement with the Folin phenol reagent." *J. Biol. Chem.* 193, 265–275.

Mozersky, M. and David, G.B. (1992). "Hide powder azure and azocoll as substrates for assay of the proteolytic activity of bate." *J. Am. Leath. Chem. Asso.* 87, 287–295.

Rovery, N., Fabre, C. and Desnuelle, P. (1953). "Study of the activation of bovine chymotrypsinogen and trypsinogen by determination of the N-terminal residues in these proteins and their corresponding enzymes." *Biochim. Biophys. Acta.* 12, 547–559.

Santos, R.M.D.B., Firmino. A.A., de Sá C.M. and Felix,C.R. (1996). "Keratinolytic activity of *Aspergillus fumigatus* F resenius." *Curr. Microbiol.* 33, 364–370.

Schwert, G.W/.and Takenaka, Y. (1955). "A spectrophotometric determination of trypsin and chymotrypsin." *Biochim Biophys Acta.*16, 570–575.

Strongin, A.Y.A,, Tzotova, L.S., Alramov, Z.T, Gorodetsky, D.I., Ermakova, L.M. and Stepanov, V.M. (1978). "Intracellular serine protease of *Bacillus subtilis*: sequence homology with extracellular subtilisins." *J. Bacteriol.* 13, 1401–1411.

Van Wart, H.E. and Steinbrink, D.R. (1981). "A continuous spectrophotometric assay for *Clostridium histolyticum* collagenase." *Anal. Biochem.*113, 356–365.

Woessner, J.F. (1961). "The determination of hydroxyproline in tissue and protein samples containing small proportions of this imino acid." *Arch. Biochem. Biophys.* 93, 440–447.

21

WHAT IS AHEAD

Microbial technology has made profound impact on medical and industrial applications. Microbial enzymes are currently used in several industrial processes and new innovative applications are constantly added. The capability of enzymes as effective catalysts even under mild reactive conditions would result in significant savings in energy and water and this will be a great boon to both industry and environment. This focus on 'What is ahead' will provide new vistas on the opportunities and challenges that would define the future of microbial technology applications with more emphasis on leather industry.

In general, SSF is a well-adapted and cheaper process than SmF for the production of a wide range of bioproducts such as enzymes, organic acids, aromatic compounds, antibiotics, biopesticides, etc. In many bioproductions, the quantity of products obtained by SSF is many folds higher than those obtained by SmF. However, some of the problems associated with SSF are designs for upscaling, control of operations, fermentation variables, impurities during downstream processing and contamination problems. Therefore, if SSF variables are well-controlled and the purity of the product is defined, SSF may be a more efficient process.

Microbial technology in leather industry is a quadruped with i) microbial enzymes for use in soaking, dehairing, bating and degreasing, ii) consortium of microbes to treat different pollutants in the effluent, iii) biodegradable bactericides and fungicides (biocides) and plant-based antimicrobial formulations for microbial control in the preservation of hides and skins and iv) recombinant DNA technology for improvement of microorganisms as its four legs. Sustained research and development could be taken up in the abovesaid areas.

Leather industry, under pressure from environmentalists and regulatory authorities, has started deriving eco-friendly processes to combat pollution. It is well known that conventional procedures using chemicals are causing significant pollution in terms of both solid and liquid waste. Waste reduction is absolutely necessary to minimize the pollution load from tanneries. It is a well-known fact that end-of-pipe treatments are not the ultimate solution for waste management. Implementation of cleaner production processes and pollution prevention measures can provide both cost and environmental benefits. However, the applicability of these technologies varies from tannery to tannery due to the varying nature of raw material, processing conditions and the types of finished leather.

Waste reduction can be achieved by some of the following methodologies:

1. optimization of water usage in the pretanning and tanning processes.
2. use of biodegradable bactericides and fungicides (biocides) and plant-based antimicrobial formulations for microbial control in soaking process replacing the conventional salt used for preservation.
3. reduction or complete elimination of chemicals such as lime, sulphide. etc.
4. use of enzyme technology in beamhouse practice.
5. use of immobilized enzymes in the beamhouse processes.
6. recycling of spent water from the beamhouse processes.
7. optimization of the chrome tanning process by using less chrome salts or reuse of chrome liquor or recycling of chrome salts after its precipitation/resolubilization.
8. reduction of volatile organic compounds by using aqueous finishes in the finishing yard
9. use of a consortium of microbes for treating the tannery effluents.

Of the total discharge from a tannery, the effluents coming out of the beamhouse or pretanning processes are causing greater concern due to their large volume and toxic nature of pollutants. The treatment and proper disposal of tannery effluents can be partially solved by rationalizing certain pretanning processes, using enzymes and enzymatic formulations.

Conventionally, curing procedures use common salt. Removal of salt from the effluent is a major problem since it causes very high C.O.D. The disposal of accumulated salt in tanneries is another cause for worry. A major viable option is the development of biocides, derived naturally or chemically. In addition, sustained research efforts should be taken up to develop plant-based antimicrobial formulations which are ecofriendly and are termed "green" technologies.

With the introduction of more rapid processing methods in modern tanneries, use of enzymes in the soak for rapid wetting is an alternative viable option. In addition, enzymes used at this stage could ensure softer leather with cleaner grain.

The development of dehairing methods that will reduce or eliminate the use of lime and sulphide is given first priority by tanners. Whether enzymatic methods can fulfil this need depends primarily on their cost and the speed at which they can be made to work within the time frame. Use of proteases which could effectively remove hair in the same duration as done by sulphide has to be worked out. Even then, current cost levels are prohibitive. To combat this cost problem, development of immobilized proteases, because of their inherent capacity for reuse, is one of the potential alternatives.

As in any other enzyme process, enzymatic dehairing has certain inherent limitations. Factors such as time as well as temperature, concentration of enzyme and pH requirements are to be considered and controlled. Each batch of raw stock is different and processes subsequent to dehairing require further modification. Through sustained research, many of the problems can be eliminated or minimized. With the growing public awareness and the stringent measures enforced by regulatory authorities to control the effulent problem, enzymatic dehairing would replace the traditional dehairing systems in the near future.

Development of keratinases is one of the potential areas for further research. When proteases are used for dehairing, hair is recovered from the roots. But, still, in some cases, short hair is not completely removed. To obviate this problem, immediately after dehairing using proteases, skins/hides are to be given a wash and then, a treatment with keratinases. Again, the duration and pH for use of keratinase are also to be standardized.

In the manufacture of bates, proper standards are not being maintained and it is imperative that the standards should be based on the enzyme activity of the product. With the advancements in fermentation technology, there is a good deal of potential for the development of microbial bates.

In addition to the development of the conventional alkaline bates, more emphasis should be given on the use of acid bates because of the many inherent advantages associated with the latter. Pancreatic enzymes or plant enzymes having pH optimum of 6.0 and above are not effective for acid bating. But, fungal enzymes may be a good source for the production of acid bates.

With regard to assay method for the estimation of proteolytic activity of enzymatic formulations, it is questionable whether the results of enzyme assay using casein as substrate would convey the actual effect of enzymes on the hides and skins. Correlations of the degree of bating with assayed enzyme activity are still, to a large extent, subjective. Problems also arise if the product consists of more than one enzyme and the proportions of these vary even slightly from batch to batch. It would be more beneficial if research institutions, tanners and producers could get together in worldwide collaboration and agree to a common method for the evaluation of enzymatic formulations used in beamhouse processes.

Solvent degreasing is invariably used in many tanneries because of its effectiveness and easy handling. Suitable systems are available for the recovery of solvent, brine and grease. With the renewed emphasis on genetic engineering and the advancements in fermentation technology, there is a lot of potential for microbial lipases to be utilized in the degreasing of skins. Their prohibitive cost, coupled with the inherent toxicological problems and potent pollution hazards associated with their use, has created urgency in replacing petroleum products (solvents) with microbial lipases in degreasing processes. For enzymatic degreasing, a combination of enzymes might be necessary because it is not just the breakdown of the grease that is needed but also the release of the grease from within the hide. Another area of interest is the development of microbial acid lipases, which possess many inherent advantages over alkaline lipases. Use of lipases may be more effective in degreasing if they are used in combination with proteases, and intensive research needs to be done in this area also.

Even though immobilized enzymes and whole cells find a variety of applications in other industries, intensive efforts have not been made for their successful utility in dehairing and degreasing operations. Immobilized lipases and immobilized whole cell systems secreting potent lipases are the suitable alternatives to the conventional microbial lipases. No data has so far been published on the use of immobilized lipases for degreasing. Hence, future researches are to be directed towards these areas.

Offal treatment by enzymes seems to have the potential as an economical method of waste treatment to generate by-products which could provide useful raw materials to the chemical industry. Research in this area would be quite timely. The potential exists not only for the treatment of the fleshings but also for the treatment of the offal from other stages of processing.

Development of technologies for proper waste treatment, utilization of wastes into value-added products and recovery/reuse of chemicals reduce environmental pollution caused by the leather industry. Various advanced biological treatment methods provide a practical and economical solution for waste disposal.

Persistent screening for new microorganisms and their enzymes will open simple routes for synthetic processes and consequently, new ways to solve environmental and pollution problems. Extensive beamhouse operations of the future will depend more on a combination of chemical and enzymatic systems, and enzymes will definitely play a predominant role in the pretanning processes of leather manufacture.

GLOSSARY

Abatement   Reduction in the quantum or toxicity of waste generated by process modification and recycling.

Abzyme   A monoclonal antibody with catalytic activity.

Actinomycete   A group of G+C-rich aerobic gram-positive bacteria which form branching hyphae and asexual spores.

Activated Sludge Process   An aerobic secondary wastewater treatment process using sludge/solid matter/sediment composed of actively growing microorganisms that readily use dissolved organic substrates and transform them into organic matter, additional microbial cells and carbon dioxide.

Activation energy   The energy that must be supplied for a chemical reaction to be initiated; it is usually denoted as E_a and given in units of kilojoules per mole.

Active site   The region of an enzyme molecule that contains the substrate-binding site for converting the substrate into product.

Aerobe   Organism that grows only in the presence of atmospheric oxygen.

Affinity chromatography   A column chromatographic technique which separates proteins on the basis of a reversible interaction between a protein (or group of proteins) and a specific ligand coupled to a chromatographic matrix.

Agarose gel electrophoresis   A type of electrophoresis that uses a matrix of highly purified agar to separate nucleic acid fragments based on size.

Albumins   A group of globular proteins, soluble in water but form insoluble coagulants when heated.

Algae   Unicellular or multicellular phototrophic eukaryotic microorganisms. Blue-green algae are not true algae and they belong to a group of bacteria called cyanobacteria.

Allele   An alternative form of a gene that is located at a specific position on a specific chromosome.

Allosteric enzyme   An enzyme whose active site can be altered by binding with an effector at a different site.

Amino acids   The basic building blocks of proteins, consisting of the basic formula $R-CH(NH_2)-COOH$, where "R" is the side chain which defines the amino acid.

Ampholytes   Molecules that contain both acidic and basic groups (are known as amphoteric) and will exist mostly as zwitterions in a certain range of pH.

Amylase   Enzyme that degrades starch, glycogen and other polysaccharides.

Anaerobe   Organism that grows only in the absence of atmospheric oxygen.

Anaerobic digestion (anaerobic composting) A method of composting or microbial fermentation of organic matter which produces methane and carbon dioxide without the requirement of oxygen for reaction; a sewage treatment process.

Anion A negatively charged ion, which moves towards anode.

Antibiotics Natural substances secreted by microbes or their respective semi-synthetic forms that destroy or inhibit the growth of microorganisms, particularly pathogenic bacteria.

Anticodon A sequence of three bases on the transfer RNA that pair with the bases in the corresponding codon on the messenger RNA.

Archaea Evolutionarily distinct group of prokaryotes consisting of isoprenoid glycerol diether or diglycerol tetraether lipids in their membranes and archaeal rRNA, e.g., methanogens, most extreme halophiles, acidophiles and hyperthermophiles.

Autotrophs Organisms which use carbon dioxide as the sole carbon source.

Bacteriophage vector Bacteriophages are viruses that attack bacteria. Several bacteriophages are used as cloning vectors e.g. λ (lambda) and M13 phage.

Bacteriophages Virus that infect bacteria; they are obligate intracellular parasites which usually cause lysis of the host cell (lytic). Alternatively, they remain as prophage by integrating their nucleic acid with host genome by recombination without causing lysis (lysogenic).

Base The purine or pyrimidine component of a nucleic acid chain.

Basement membrane Intermediate layer of skin attaching the epidermal layer to the underlying dermis or corium.

Bate Natural material or a formulation containing proteolytic enzyme and chemical aids, which by enzymatic action removes some of the non-leather -forming proteinous materials such as albumin, globulin and mucoids and also elastin and reticulin.

Bating Treatment of dehaired and limed pelt with a bate.

Beam Semi-convex wooden structure, fixed at an angle to the ground, over which suitably prepared hides or skins are placed for dehairing, fleshing or scudding.

Beamhouse Section of the tannery where hides or skins are processed by the pretanning operations of soaking, dehairing, fleshing, deliming and pickling for subsequent tanning processes.

Binary fission The division of one cell into two by the formation of a septum. It is the most common form of cell division in bacteria.

Binomial nomenclature A system of having two names, genus and specific epithet, for each organism.

Bioaugmentation The addition of microorganisms to the environment that can metabolize and grow on specific organic compounds.

Biodegradation The breakdown of organic substances by microorganisms or their enzymes.

Biological oxygen demand (BOD) The amount of molecular oxygen required by aerobic microbes during oxidation of organic substances in wastewater. The BOD test measures the oxygen consumed (in mg/L) over 5 days at 20°C.

Biomethane A mixture of carbon dioxide and methane which is the by-product of the bacterial decomposition of vegetable and animal wastes.

Biopesticide Pesticides obtained from natural sources such as microorganisms, plants, etc. for the control of pests, diseases and weeds. The advantage of using biopesticides is that they are enviro-friendly and target-specific.

Bioremediation The process by which living organisms act to degrade or transform hazardous organic contaminants.

Biosensor A device for the detection of an analyte, which combines a biological component with a physico-chemical detector component. It consists of 3 parts—the sensitive biological element, the transducer and the detector element.

Biosorption A physiochemical process that occurs naturally in certain biomass which allows it to passively concentrate and bind contaminants onto its cellular structure.

Biostimulation A process that increases activity of microbes capable of degrading biodegradable contaminants.

Biotechnology Term to describe the application of bioscience for benefiting humankind. Encompasses a wide range of disciplines, procedures and processes but is often mistakenly thought to refer exclusively to the industrial-scale use of genetically modified organisms.

Bleaching Deprivation of colour from a coloured material, through exposure to the sun and weather or oxidizing chemicals, in such a way as to remove, or lighten its colour to the maximum or to become colourless, pale or white.

Blunt ends DNA termini without overhanging 3′ or 5′ ends.

Bright-field microscope A microscope that illuminates the specimen directly with bright light and forms a dark image on a brighter background.

Bromelin A proteolytic enzyme found in pineapples used to treat inflammation.

Budding A type of asexual reproduction beginning as a protuberance from the parent cell, and which grows and detaches itself to form a smaller daughter cell (e.g., yeast).

Buffing Abrade or grind a leather surface, especially the grain surface, by a moving band of abrasive paper or cloth.

C:N ratio It represents the relative proportion of the two elements, i.e., ratio of the mass of carbon to the mass of nitrogen in a nutritional substrate.

Cation A positively charged ion, which moves towards cathode.

cDNA Complementary DNA synthesized from a messenger RNA template in a reaction catalysed by reverse transcriptase.

Cell wall A rigid outer coating found in many plant, fungal, and bacterial cells, which accounts for its ability to withstand mechanical stress or abrupt changes in osmotic pressure. Cell wall encloses the cell membrane and cytoplasmic contents and is composed of a variety of polysaccharide-based components such as peptidoglycan (bacteria), chitin (fungi), and cellulose (plants, algae and fungi).

Centrifugation Method of separating components by spinning at high speed.

The g-force causes materials to pellet or move through the centrifugation medium. Uses include spinning down whole cells, cell debris, precipitated nucleic acids, or other components. Also used for separating macromolecules under gradient centrifugation.

Chemical oxygen demand (COD) The amount of oxygen required in milligrams per litre to oxidize both organic and oxidizable inorganic compounds.

Chemolithotrophic autotrophs Microorganisms that oxidize reduced inorganic compounds to derive both energy and electrons; CO_2 is their carbon source. They are also called "chemolithoautotrophs".

Chemoorganotrophic heterotrophs Organisms that use organic compounds as sources of energy, hydrogen, electrons, and carbon for biosynthesis. They are also called "chemoheterotrophs".

Chemotrophs Organisms that obtain energy from the oxidation of chemical compounds.

Chlamydospore An asexually produced, thick-walled resting spore formed by some fungi.

Chondroitin sulphate A sulphated glycosaminoglycan (GAG) composed of a chain of alternating sugars (N-acetylgalactosamine and glucuronic acid). It is usually found attached to proteins as part of a proteoglycan. Chondroitin sulphate is an important structural component of cartilage, a major component of extracellular matrix, and is important in maintaining the structural integrity of the tissue.

Chromatin The condensed nucleoprotein (complex of positively charged histones and negatively charged nucleic acid) fibres of eukaryotic chromosomes.

Chromatography A procedure to separate and/or to analyse complex mixtures. Segregation is usually carried out on paper or in glass or metal columns with the help of different solvents. The paper or glass columns contain porous solids with functional groups that have limited affinities for the molecules being separated.

Chrome leather Leather tanned either solely with basic chromium salts or with chromium salts together with small amounts of auxillary tanning agent called "synthetic tanning agent" or "syntans" used merely to assist the chrome tanning process, and not in excessive amounts to alter the essential chrome-tanned character of the leather.

Chrome liquor Solution of trivalent chromium salts, mainly basic sulphates, prepared by the reduction of acidified potassium/sodium dichromate solution, by sulphur dioxide or by an organic substance, such as glucose, used for chrome tanning.

Chrome tanning Treatment of hides and skins with chrome liquor for conversion into leather.

Cilia Threadlike appendages extending from the surface of some protozoa that beat rhythmically to propel them; cilia are membrane-bound cylinders with a complex internal array of microtubules, usually in a 9+2 pattern.

Cleaner production Processes so designed as to generate minimal or zero waste during production.

Clone A group of genetically identical cells or organisms arising from a single ancestral cell; all members of the clone have identical genetic composition. Also, the term is used to describe the generation of recombinants.

Cloning vector   A self-replicating entity to which foreign DNA can be covalently attached for multiplication (amplification) in host cell.

Coagulants   Chemicals which cause very fine particles to clump together into larger particles resulting in sedimentation. This makes it easier to separate the solids from the water by settling, skimming, draining, or filtering.

Coagulation   The agglomeration of colloidal or suspended matter brought about by the addition of coagulants to the liquid, manually or by mechanical means.

Codon   A nucleotide triplet (sequence of three nucleotides) in a messenger RNA molecule, which specifies a specific amino acid, or a translational start or stop.

Coenocytic   Refers to a multinucleate cell or hypha formed by repeated nuclear divisions.

Coenzyme   An organic molecule that associates with enzymes and influences their activity.

Cofactor   A non-protein component essential for the normal catalytic activity of an enzyme. It may be an organic molecule or inorganic ions.

Cohesive ends   Those ends (termini) of DNA molecules that have short sticky complementary sequences that can join together two DNA molecules. Often generated by restriction enzymes.

Collagen   Most abundant fibrous protein constituent produced by fibroblasts of tendons, cartilage, skin, bone and other connective tissues of animals especially mammals. Collagen is approximately 300 nm long and 1.5 nm in diameter, made up of three polypeptide strands (called alpha chains) of about 1050 amino acids long, each possessing the conformation of a left-handed helix.

Collagen fibril   Composed of delicate molecules of tropocollagen aggregated in linear, regular and staggered arrays. It is the fundamental structural element of connective tissue with about 100 nm thickness.

Collagen fibre   An individual fibre composed of collagen fibrils and usually arranged in branching bundles of indefinite length and thickness.

Collagenase   Enzyme that hydrolyses collagen.

Competitive inhibition   A type of enzyme inhibition, where binding of the inhibitor to the active site on enzyme, prevent the binding of substrate with enzyme.

Complementary base sequence   The nucleic acids that are related to a given sequence of nucleic acids, by the rules of base-pairing.

Composting   The decomposition of organic matter by microorganisms.

Conjugation   The form of gene transfer and recombination in bacteria that requires direct cell-to-cell contact.

Corium   The central layer of the hide or skin remaining after the removal of epidermis, hair and flesh (flesh side), and which is converted into leather. This layer amounts to approximately 75–90% of the total thickness of a hide or skin.

Covalent bond   Relatively strong molecular bond in which the electronic configuration of the constituent atoms is fulfilled by sharing electrons.

Curing salt   Common salt of appropriate composition and grain size for salting hides and skins. It may contain denaturants or additives to improve its preservative properties.

Curing The process which essentially consists of bringing about varying degrees of dehydration to the hide or skin either by simple drying or by salting, the curing salt itself acting as an additional inhibitor of putrefaction. Since only a small percentage of hides are sent to processing within 24 hours of flaying, a great majority of hides have to be preserved or 'cured' either for transport or storage.

Cytoplasm The contents enclosed by the plasma (or cytoplasmic) membrane, excluding the nucleus.

Dalton A unit of mass equivalent to the mass of a hydrogen atom (1.66×10^{-24} g)

Decolourizing Deprive, bleach or remove the colour from a material, by chemical treatment, by sunlight or by weathering.

Degeneracy More than one codon specifying the same amino acid.

Degreasing To remove grease by either enzymatic or chemical (using solvents or surfactants) method.

Dehairing Removal of hair using chemicals or enzyme or both.

Deliming Removal of calcium hydroxide from hides and skins by washing or treatment with acid or acidic salts leading to a pelt with neutral pH.

Denaturation The disruption of the native structure of nucleic acid or protein molecule by heat, chemical treatment or changing pH conditions.

Denaturing gel An agarose or acrylamide gel run under suitable conditions which destroy secondary or tertiary structure of protein or RNA.

Denitrification Reduction of nitrate to nitrite and then to nitrogen gas.

Deoxyribonucleotide Nucleotides which are the building blocks of DNA.

Depilation Removal of hair or wool from hides or skins by enzymatic or chemical methods.

Dermatan sulphate A heterogeneous glycosaminoglycan (GAG), formed by repeating disaccharide units consisting of 4-linked glucuronic acid and 1,3-linked N-acetyl galactosamine; carbohydrate moiety of proteoglycans.

Desizing Removal of starch using amylases from threads of fabrics to prevent damage during weaving; traditionally, desizing is performed using chemicals.

Dialysis The movement of molecules by diffusion from high concentration to low concentration through a semi-permeable membrane.

Diploid cell A cell that contains two sets of chromosomes (2N).

Disposal The final handling of solid waste, following collection, processing, or incineration. Disposal most often means placement of wastes in a dump or a landfill.

Dissolved oxygen (DO) Measure of oxygen dissolved in water expressed in milligrams per litre.

Dissolved solids The solids present in sewage or effluent that cannot be removed by regular filtration methods.

DNA (Deoxyribonucleic acid) A polydeoxyribonucleotide (usually double-stranded) in which the sugar is deoxyribose; the main repository of genetic information in all cells and most viruses;.

DNA cloning The propagation of individual segments of DNA as clones.

DNA ligase An enzyme (usually from the T4 bacteriophage) which catalyses the formation of a phosphodiester bond between two adjacent bases of double-stranded DNA.

DNA Ligation The joining together of linear DNA fragments by the enzyme DNA ligase that catalyses the formation of a phosphodiester bond between a 5' phosphate and a 3' OH group.

DNA polymerase An enzyme that catalyses the formation of $3' \rightarrow 5'$ phosphodiester bonds between deoxyribonucleotide triphosphates.

Double helix DNA structure in which two helically twisted antiparallel polynucleotide strands are held together by hydrogen-bonding and base-stacking.

Downstream processing Refers to the recovery and purification of biosynthetic products after fermentation process.

Effluent The liquid that comes out of a treatment plant after completion of any treatment process.

Elastin Highly extensible and chemically resistant yellow, fibrous protein, occurring in certain vertebrate connective tissues such as a fine network in the reticular dermis (corium) layer and around certain arteries, neck ligaments, etc.

Electrophoresis The movement of charged polymers in an electrical field. A commonly used technique for analysis of mixtures of molecules in solution according to their electrophoretic mobilities.

Emissions The physical or chemical evidence let off into the environment by any process, as a result of its activity.

Endonuclease An enzyme that cleaves bonds within a nucleic acid chain. Endonucleases may be specific for RNA (RNAses) or for single-stranded or double-stranded DNA (DNAses). A restriction enzyme is a type of endonuclease.

Endoplasmic reticulum A system of double membranes in the cytoplasm that is involved in the synthesis of transported proteins. The rough endoplasmic reticulum has ribosomes associated with it while the smooth endoplasmic reticulum does not.

Endospore The differentiated cell formed within the cells of certain Gram-positive bacteria and is extremely resistant to heat and other harmful agents.

Endurance The ability of leather to resist surface damage, such as cracking, when folded, grain outwards, in two directions at right angles to give a sharp corner. Flexing endurance may be tested by a machine such as the Bally flexometer.

Ensiling The process of preserving green fodder for livestock by fermenting the fodder mainly by lactic acid bacteria in air-tight conditions, either by using a storage silo or in plastic wrapping. This method of fodder preservation is called silage.

Enzymatic dehairing Loosening of the hair or wool of a hide or skin by treatment with an enzyme preparation.

Enzyme kinetics The study of the chemical reactions that are catalysed by enzymes, with a focus on their reaction rates.

Enzyme A class of proteins that act as catalysts in biochemical reactions.

***Escherichia coli* (*E. coli*)** A gram-negative bacterium commonly found in the vertebrate intestine. It is the bacterium most frequently used in the study of biochemistry, molecular biology and genetics.

Eukaryote Organism having a unit membrane-bound nucleus and usually other organelles.

Exons A sequence of DNA that codes information for protein synthesis and is transcribed to messenger RNA. Exons are also called coding DNA.

Exonuclease An enzyme which hydrolyses DNA beginning at one end of a strand, releasing nucleotides one at a time (thus, there are 3´ or 5´ exonucleases)

Extensibility Capability of a material to be stretched or distorted without breaking.

Extremophiles Microorganisms that grow under harsh or extreme environmental conditions such as very high temperature, high salt concentration or low pH.

Facultative anaerobes Microorganisms that do not require oxygen for growth, but grow better in its presence.

Fat-liquoring Introducing oil into leather, normally by drumming it with an oil-in-water emulsion, to provide lubrication of the fibres in leather.

Fermentation An energy-yielding process in which an energy substrate is oxidized without an exogenous electron acceptor. Usually organic molecules serve as both electron donor and acceptor. Also, the process by which microbes turn raw materials such as glucose into products such as alcohol.

Fermenter A biological reactor for cultivation of microorganisms.

Fibre Extremely long, fine, pliable, cohesive, natural or manufactured threadlike material. Fibre of collagen, wool, cotton, nylon, etc.

Fibroblast Spindle-shaped connective tissue cell.

Filtration Method of separating solid and liquid components of a suspension by passing through a filter.

Finishing Treatments applied to the tanned hide or skin to give it the desired properties such as bleaching, degreasing, dyeing, retanning, introduction of grease, mechanical treatments applied to the wet or dried leather and the final treatment of the leather surface with pigmented finishes and seasons. In the narrower sense, it is limited to those treatments designed to enhance the aesthetic appearance and/or to give the grain or flesh surface special properties which boost the value of the finished leather.

Flexible Term used to describe pliable and supple leathers.

Flocculation Coalescence of minute particles in a liquid resulting in a disposable precipitate.

Fragmentation A type of asexual reproduction in which a thallus breaks into two or more parts, each of which forms a new thallus.

Gene therapy A technique which includes insertion, deletion or alteration of genes for correcting defective genes responsible for disease development.

Gene A hereditary unit consisting of a sequence of DNA that occupies a specific location on a chromosome and determines a particular characteristic in an organism; a segment of DNA coding for a specific protein.

Genetic code The three-letter code (sequence of nucleotides) that translates nucleic acid sequence into protein sequence, i.e., determines the specific amino acid sequence in the synthesis of proteins. It is the biochemical basis of heredity and nearly universal in all organisms.

Genetic engineering DNA techniques and genetic procedures used to isolate genes from an organism, manipulate them in the laboratory and insert them into another system for expression.

Genome The complete set of genetic information defining a particular animal, plant, organism or virus.

Genotype The genetic characteristics of an organism (distinguished from its observable characteristics or phenotype).

Germline gene therapy The therapeutic process in which the genetic make-up of germ cells (an egg or sperm cell) is altered by the introduction of functional genes, which are integrated into their genomes before fertilization.

Gliding motility A type of motility in which a microbial cell glides along when in contact with a solid surface.

Globulins A group of globular proteins generally insoluble in water and present in blood, eggs, milk, and as a reserve protein in seeds.

Glycoprotein A glycosylated protein.

Glycosaminoglycans Long branched polysaccharides consisting of repeating disaccharide units of hexuronic acid and hexosamine linked to the protein core of proteoglycan either via N or O-linkage. Often, the hexosamine is sulphated except in hyaluronic acid.

Hair Keratinous fibres growing from the skin of most animals, characterized by its stiffness, straightness, melanin pigmentation and special surface pattern of scales.

Half-life The time required for the disappearance of one half of a substance.

Halophile The organism requiring or tolerating a saline environment.

Hazardous waste Waste which can cause or significantly contribute to an increase in serious irreversible incapacitating illness or pose a substantial hazard to human health or the environment because of its quantity, concentration or physical, chemical or infectious characteristics.

Heavy metal Metals of high atomic weight and density whose specific gravity is approximately 5.0 or higher such as chromium, mercury, lead and cadmium that are toxic to living organisms.

Heterotroph An organism that uses reduced, preformed organic molecules as its principal carbon source.

Hide The skin of a fully grown animal such as cow, buffalo, camel, etc.

Holoenzyme The complete enzyme including all subunits often used with reference to RNA and DNA polymerases.

Homogenization Cell disruption by means of mechanical pressure or shear force to release intracellular contents.

Hyaluronic acid (HA) An anionic polysaccharide formed by repeated combination of the disaccharides β–1-4-D-gluconic acid and β–1-3- D-N-acetyl glucosamine, which do not participate covalently in proteoglycan aggregates or with proteins of connective tissues; a unique non-sulphated GAG; HA is found in skins and hides as a highly viscous gel, filling a large volume of interfibrillary space.

Hyperthermophile A prokaryote having a growth temperature optimum of 80°C or higher.

Immobilized enzyme   An enzyme fixed by physical or chemical means to a solid support, e.g., a bead or gel. It could be easily separated and recovered from the products and used again.

Incineration   Controlled high-temperature oxidation of organic compounds. This oxidation generates gas, fumes and ashes.

Induced fit   A change in the shape or conformation of an enzyme that results from the binding of substrate.

Inducers   Molecules that cause an increase in a protein activity when added to cells.

Inhibitor   Molecules that bind to enzymes and decrease their activity.

Inoculum development   The preparation of a population of microorganisms from a dormant stock culture to an active state of growth that is suitable for inoculation in the final production stage.

Introns   A DNA sequence situated between exons and that is removed before translation of messenger RNA and does not function in coding for protein synthesis. Also called as non-coding DNA.

Isoelectric point or pH   The pH at which a protein has no net charge.

Isoelectic focusing   Separation of proteins based on isoelectric pH.

Isomerase   An enzyme that catalyses an intramolecular rearrangement.

Isozymes   Multiple forms of an enzyme that differ from one another in one or more of their properties.

Karyogamy   The fusion in a cell of haploid (N) nuclei to form a diploid (2N).

Keratin   Fibrous scleroprotein present in hair, feathers, hooves and horns.

Kilobase   Symbol kb. A unit used at the molecule level for measuring distances within nucleic acids, chromosomes or genes and is equal to 1000 bases (equivalent to 1000 nucleotides or base pairs).

Landfill   A site where solid waste containing both organic and inorganic material is dumped, covered with soil and allowed for natural decomposition.

Leather   Hide or skin tanned with its original fibrous structure more or less intact and made imputrescible, i.e., to resist bacterial attack as well as subsequent putrefaction.

Ligand   An ion or molecule that donates a pair of electrons to a metal atom or ion to form a coordination complex.

Lime   Calcium oxide or calcium hydroxide, mostly the latter, as used in tanneries.

Liming   Process of treatment of hides and skins with lime solution (saturated calcium hydroxide either alone or with small percentage of sodium sulphide (Na_2S) or sodium hydrosulphide) to loosen the hair, remove non-structural proteins, saponify fatty matter, and open up fibre structure, to facilitate the tanning process.

Lipase   Fat-splitting enzyme that catalyses the hydrolysis of fats or lipids.

Lipopolysaccharide   Complex lipid containing a unique glycolipid found in gram-negative bacteria; often termed as endotoxin.

Lithotroph   An organism that uses reduced inorganic compounds as its electron donor in energy metabolism.

Lyase   Enzymes that catalyse the non-hydrolytic addition or removal of groups which are free during reaction.

Lysis The rupture or physical disintegration of a cell.

Lysogenic virus A virus that can adopt an inactive (Iysogenic) state, in which it maintains its genome within a cell instead of entering the Iytic cycle. The circumstances that determine whether a Iysogenic (temperate) virus will adopt an inactive state or an active Iytic state are often subtle and depend upon the physiologic state of the infected cell.

Lysogeny The state in which a phage genome remains within the bacterial host cell after infection and reproduces along with it rather than taking control of the host and destroying it.

Lysosome An organelle that contains hydrolytic enzymes designed to break down proteins that are targeted at that organelle.

Lysozyme An enzyme occurring naturally in egg white, human tears, saliva, and other body fluids, capable of destroying the cell walls of certain bacteria and thereby acting as a mild antiseptic.

Lytic cycle A virus life cycle that results in the lysis of the host cell yielding progeny virus particles.

Mechanical de-salting Elimination of salt from hides or skins by mechanical equipment.

Membrane bioreactor (MBR) The combination of a membrane process like microfiltration or ultrafiltration with a suspended-growth bioreactor, which is now widely used for municipal and industrial wastewater treatment.

Messenger RNA (mRNA) The template RNA carrying the message for protein synthesis.

Metabolism The sum total of the enzyme-catalysed reactions that occur in a living organism.

Metalloprotease A group of proteases whose functional group in their active site involves a metal.

Methanogenesis Also known as biomethanation, it is the formation of methane by methanogenic microorganisms after digesting organic matter.

Michaelis constant (K_m) The substrate concentration at which an enzyme-catalysed reaction proceeds at one-half of the maximum velocity.

Microaerophile A microorganism that requires low concentration of oxygen for growth, around 2 to 10%, but is damaged by normal atmospheric oxygen levels.

Mixed Liquor Suspended Solid (MLSS) Suspended solids in activated sludge.

Mixed Liquor Volatile Suspended Solid (MLVSS) Volatile part of the suspended solids in activated sludge.

Mutagen An agent, such as a chemical, ultraviolet light, or a radioactive element, that can induce or increase the frequency of mutation in an organism.

Mutagenesis A process that leads to a change in the genetic material that is inherited in subsequent generations.

Mutant An organism that carries an altered gene or a change in its genome.

Mutation The genetically inheritable alteration of a gene or group of genes. May be caused by insertion, deletion, or modification of bases.

Nanofiltration (NF) A process based on separation by membranes which retains particles of between 1 nm and 10 nm.

Native gel An electrophoresis gel run under conditions which do not denature proteins (i.e., in the absence of SDS, urea, 2-mercaptoethanol, etc.).

Nitrification Refers to oxidation of ammonia nitrogen to nitrate nitrogen in wastewater by biological or chemical reactions.

Nitrogenous base An aromatic nitrogen-containing molecule with basic properties. Such bases include purines and pyrimidines.

Noncompetitive inhibitor An inhibitor of enzyme activity whose effect is not reversed by increasing the concentration of substrate molecule.

Nonsense codon A codon which does not code for any amino acid (stop codon)

Nonsense mutation A change in the base sequence that converts a sense codon (one that specifies an amino acid) to one that specifies a stop codon (a nonsense codon).

Nucleic acids Polymers of the ribonucleotides or deoxyribonucleotides.

Nucleoside An organic molecule containing a purine or pyrimidine base and a five-carbon sugar (ribose or deoxyribose).

Nucleotide An organic molecule containing a purine or pyrimidine base, a five-carbon sugar (ribose or deoxyribose), and one or more phosphate groups.

Nucleus The large body embedded in the cytoplasm of all plant and animal cells containing the genetic material. It functions as the control centre of the cell.

Odour control Collection of foul air using a variety of means, and removal of odour using chemical and biological treatments.

Offal The internal organs and parts of organs of a slaughtered animal that cannot be consumed. eg. hair, horns, hoofs, tail, etc.

Okazaki fragment A short segment of single-stranded DNA that is an intermediate in DNA synthesis. They are relatively short DNA fragments synthesized on the lagging strand during DNA replication. In bacteria, Okazaki fragments are 1000–2000 bases in length; in eukaryotes, 100–200 bases in length.

Oligomer A short, single-stranded nucleic acid fragment.

Oligonucleotides Oligonucleotides are relatively short fragments of nucleic acid polymer with defined sequence typically with less than 50 bases in length. The length of the oligonucleotide is usually denoted by "mer".

Oncogene Genes that normally play a role in the growth of cells but, when overexpressed or mutated, can foster the growth of cancer.

Organotrophs Organisms that use reduced organic compounds as their electron source.

Origin A site within a DNA sequence of a chromosome, plasmid, or non-integrated virus at which replication of the DNA is initiated.

Oxidation pond A still water body used for aerobic waste disposal by biodegradation using fast growing aerobic and facultative microbes.

Oxidation A reaction in which the atoms in an element lose electrons and the oxidation number of the element is correspondingly increased.

Two-dimensional (2D) PAGE A technique that combines the principles of isoelectric focusing (IEF) (first dimension), which separates proteins in a mixture according to charge (pI) and the size separation principle of SDS-PAGE (second dimension).

Palindrome A sequence of bases that reads the same in both directions on opposite strands of the DNA duplex (e.g., GAATTC).

Papain A cysteine protease (EC 3.4.22.2) enzyme present in papaya and is also known as papaya proteinase I.

Particulate Also known as particulate matter (PM); fine particles of solid matter suspended in a gas or liquid.

PCR Polymerase chain reaction. A method for amplifying DNA sequences.

Pelt Skin, after it is dehaired or dewoolled, limed, fleshed, bated, and pickled.

Pentose A sugar with five carbon atoms.

Peptide A chain formed by two or more amino acids linked through peptide bonds.

Peptidoglycan The main component of the bacterial cell wall, consisting of a two-dimensional network of heteropolysaccharides (N-Acetyl muramic acid and N-Acetyl glucosamine) running in one direction, cross-linked with polypeptides running in the perpendicular direction.

Phagocytosis The endocytotic process in which a cell encloses large particles in a membrane-delimited phagocytic vacuole or phagosome and engulfs them.

Phasmid vector (phagemid) A type of cloning vector which contains both phage (M13) and plasmid origins of DNA replication. It can grow as a plasmid and also be packaged as ssDNA in viral particles.

Phenotype The observable physical or biochemical characteristics of an organism. It is determined by both genetic make-up and influence of the environmental factors.

Photolithotrophic autotrophs Organisms that use light energy, an inorganic electron source (e.g., H_2O, H_2, H_2S), and CO_2 as a carbon source.

Photoorganotrophic heterotrophs Microorganisms that use light energy and organic electron donors, and also employ simple organic molecules rather than CO_2 as their carbon source.

Phototrophs Organisms that use light as the energy source to drive the electron flow from the electron donors, such as water, hydrogen, or sulphide.

Phytoremediation The use of plants and their associated microorganisms to remove, contain, or degrade environmental contaminants.

Pickling Treatment of pelts in an acid medium (such as a solution of sulphuric acid along with sodium chloride) for long-term preservation or in a preconditioned state for subsequent chrome tanning.

Pinocytosis The endocytotic process in which a cell encloses a small amount of the surrounding liquid and its solutes in tiny pinocytotic vesicles or pinosomes.

Pitch length (or pitch) The number of base pairs per turn of a duplex helix.

Plaque A circular clearing on a lawn (continuous layer) of bacterial or culture cells, resulting from cell lysis by phage.

Plasma membrane The membrane that surrounds the cytoplasm.

Plasmid An extrachromosomal, usually circular, double-stranded DNA which is capable of undergoing replication independent of chromosomes of cells. Engineered plasmids are used extensively as vectors for cloning.

Plasmogamy The fusion of the contents of two cells, including cytoplasm and nuclei.

Polar group A hydrophilic (water-loving) group.

Polyacrylamide gel electrophoresis (PAGE) Technique used to separate proteins and smaller DNA fragments and oligonucleotides by electrophoresis. When run under conditions which denature proteins (i.e., in the presence of 2-mercaptoethanol, SDS, and possibly urea), molecules are separated primarily on the basis of size.

Polyelectrolytes Polymers whose repeating units bear an electrolyte group. These groups will dissociate in aqueous solutions (water), making the polymers charged. Polyelectrolyte properties are thus similar to both electrolytes (salts) and polymers (high molecular weight compounds), and are sometimes called polysalts.

Polynucleotide A chain structure containing nucleotides linked together by phosphodiester bonds.

Polypeptide A linear polymer of amino acids held together by peptide linkages. The polypeptide has an amino-end and a carboxy-terminal end.

Polyribosome (polysome) A complex of an mRNA and two or more ribosomes actively engaged in protein synthesis.

Polysaccharide A linear or branched chain structure containing many sugar molecules linked by glycosidic bonds.

Primary screening A qualitative assay in which a large population of microorganisms is screened either directly or indirectly for a specific type of activity.

Primary structure In proteins, it refers to the amino acid sequence.

Primary waste treatment The mechanical separation of solids, grease, and scum from wastewater.

Primer An oligonucleotide which is complementary to a specific region within a DNA or RNA molecule, and which is used to initiate synthesis of a new strand of complementary DNA at that specific site.

Prion An infectious proteinaceous particle that is the cause of slow diseases like scrapie in sheep.

Prokaryote A unicellular organism which lacks membrane-bound nucleus and cell organelles, and usually contains a single circular DNA molecule.

Promoter That region of the gene that signals RNA polymerase binding and the initiation of transcription.

Prophage The silent phage genome. Some prophages integrate into the host genome; others replicate autonomously. The prophage state is maintained by a phage-encoded repressor.

Prosthetic group A tightly bound non-peptide inorganic or organic component of a protein. Prosthetic groups may be lipids, carbohydrates, metal ions, phosphate groups, etc.

Protease An enzyme that performs proteolysis and is also termed peptidase or proteinase.

Proteoglycan Interfibrillary cementing substance that is located on the surface of collagen fibrils of connective tissues in a highly ordered manner and that ties the adjoining collagen fibrils together through strong electrostatic interactions. It also refers to proteins that are heavily glycosylated and in which the heteropolysaccharide or glycosaminoglycan is usually the major component.

Psychrophiles Organisms that grow at low temperatures (0°C) and show a growth temperature optimum of < 15°C. Not able to grow above 20°C.

Psychrotrophs Organisms that grow at 0°C and above 20°C.

Puering A primitive method of treatment of dehaired skins with solutions of fermented dog dung in a heated infusion. This method has been discarded after the invention of enzymatic bating process.

Pure culture The population of microorganisms composed of a single strain. Such cultures are obtained through selective laboratory procedures and are rarely found in the natural environment.

Purine A heterocyclic ring structure with varying functional groups. The purines, adenine and guanine, are found in both DNA and RNA.

Putrescible Protein-rich organic wastes that are susceptible to decomposition or decay through microbial action.

Pyrimidine A heterocyclic six-membered ring structure. Cytosine and uracil are the main pyrimidines found in RNA while cytosine and thymine are the main pyrimidines in DNA.

Quaternary structure In a protein, the way in which the different folded subunits interact to form the multisubunit protein.

Recombinant DNA DNA that contains genes from different sources that have been combined by the techniques of genetic engineering.

Reed bed Also called reed bed filters (RBF). The process does not rely on microbial degradation as a major process but uses the reactive/adsorptive potential of soil and the uptake by plants. A system successfully used for the reduction of COD.

Renaturation The process of returning a denatured structure to its original native structure, as when two single strands of DNA are reunited to form a regular duplex or when an unfolded polypeptide chain is returned to its normal folded three-dimensional structure.

Replica plating A technique in which an impression of a culture is taken from a master plate and transferred to a fresh plate. The impression can be of bacterial clones or phage plaques.

Replication fork The Y-shaped region of DNA at the site of DNA synthesis

Replicon A genetic element that behaves as an autonomous replicating unit. It can be a plasmid, phage, or bacterial chromosome.

Restriction endonuclease Enzyme which cleaves DNA at specific recognition sequences called restriction sites. They may generate either blunt or sticky ends at the site of cleavage.

Reticular fibre A type of fibre in connective tissue composed of type III collagen.

Reticulin A structural scleroprotein resembling collagen, present in connective tissue as a network of fine fibres, especially around muscle and nerve fibres.

Retrovirus The virus containing single-stranded RNA as its genetic material and producing a complementary DNA by action of the enzyme reverse transcriptase.

Reverse osmosis   A separation process where the solvent molecules are forced by an applied pressure to flow through a semi-permeable membrane in the opposite direction to that dictated by osmotic forces.

Reverse transcriptase   An enzyme that synthesizes DNA from an RNA template, using deoxyribonucleotide triphosphates. The enzyme is usually purified from retroviruses.

Reverse transcription   The process by which double-stranded DNA is formed from single-stranded RNA with the help of reverse transcriptase.

Rho factor   An oligomeric prokaryotic protein, especially that of *E. coli*, that attaches to certain sites on its DNA to assist in termination of transcription.

Ribose   The five-carbon sugar found in RNA.

Ribosomal RNA (rRNA)   The RNA part of the ribosome.

Ribosomes   Small cellular particles made up of ribosomal RNA and protein. They are the sites of protein synthesis.

Ribozyme   A catalytically active RNA.

RNA (ribonucleic acid)   A polynucleotide in which the sugar is ribose.

RNA polymerase   An enzyme that catalyses the synthesis of RNA from ribonucleotide triphosphates, using DNA as a template.

Salting out   A method of separating proteins employing high salt concentrations since proteins are less soluble at high salt concentrations.

Scale up   Process of increasing the quantum of production either in size, number or quantity.

Industrial fermenters are designed to define the parameters for the scale-up studies. Large-scale industrial fermenters are capable of producing the fermentation products as efficient as those produced by small-scale fermenters.

Scouring   Washing/cleaning process for woolled sheep-skins by using hot water containing a detergent to remove the contaminants adhering to wool.

SDS PAGE   This electrophoretic method is based on the separation of proteins according to size; SDS treatment assigns uniform negative charge to proteins.

Secondary screening   Both qualitative and quantitative assays to determine the precise activity of the microorganisms to verify production or degradation of compounds and to evaluate the production potential of the microorganism identified in the primary screening.

Secondary structure   In a protein or a nucleic acid, it refers to any repetitive folded pattern that results from the interaction of the corresponding polymeric chains. In proteins, the most common are á helix and â pleated structures.

Secondary waste treatment   A process involving various types of systems that employ aeration and biological oxidation to decompose dissolved and colloidal organic contaminants.

Secure landfill   A disposal facility designed to absorb and isolate wastes from the environment. This entails burial of the wastes in a landfill that includes clay and/or synthetic liners with provision for leachate and gas collection. The entire set-up is secured with an impermeable cover.

Semiconservative replication Duplication of DNA in which the daughter duplex carries one old strand and one new strand.

Semipermeable The characteristic of allowing only some molecules, usually smaller or uncharged ones, to pass through.

Sequencing batch reactors (SBR) Also called sequential batch reactors, these are industrial processing tanks for the treatment of wastewater. SBRs treat wastewater in batches. Oxygen is bubbled through the waste water to reduce BOD and COD and to make it suitable for discharge into sewers or for use on land.

Settling Tank Or "primary treatment" settling tanks provide an efficient process for the removal of solids that are heavy and will sink to the bottom, as well as materials that float on the surface, such as oil and grease. Settling tanks are designed to hold wastewater for several hours.

Sigma factor A subunit of RNA polymerase that recognizes specific sites on DNA for initiation of RNA synthesis.

Signal sequence A (usually N-terminal) sequence of a protein that directs its processing or localization within the cell.

Site-directed mutagenesis Process by which a defined alteration is made to DNA, and the mutation can be genetically inherited.

Skin Tissue forming the outer covering of the body (both human and animal) with extraordinary rigidness, resilience and elastic stretch along with an ability to transport air or moisture by means of its microporous structure and thermal regulation; also, skins of small animals such as goat, sheep, rabbit etc.,

Slime mould A nonphototrophic eukaryotic microorganism lacking cell walls, which aggregate to form fruiting structures (cellular slime moulds) or simply masses of protoplasm (acellular slime moulds).

Sludge Thick slurry created by falling body and settling at the bottom of settling tanks after secondary treatment.

Soaking Treating hides or skins with water (sometimes along with a disinfectant) to cleanse it from the dirt, dung or clotted blood along with the removal of salt and other soluble matter as well as to rehydrate the skin substrate optimally for subsequent tanning operation.

Sodium sulphide Chemical ($Na_2S \cdot 9H_2O$) normally used in dehairing. When exposed to moist air, it emits hydrogen sulphide, and smells like rotten eggs.

Solid-state fermentation the aerobic microbial fermentation process using solid matrix under controlled conditions in the near absence of free water. It is used for the production of microbial products such as feed, fuel, food, industrial chemicals and pharmaceuticals.

Somatic gene therapy The manipulation of gene expression in somatic cells that will be corrective to the patient but not inherited to the future generations. Somatic cells include all the non-reproductive cells in the human body.

Specific activity Unit(s) of enzyme activity per mg of protein.

Stacking energy The energy of interaction that favours the face-to-face packing of purine and pyrimidine base pairs.

Stale/staling Hide or skin which has undergone putrefactive damage due to delayed curing, or prolonged storage, leading to the development of smell, hair-slip, deterioration of the corium, etc.

Steady state In enzyme-kinetic analysis, the time interval when the rate of reaction is approximately constant with time. The term is also used to describe the state of a living cell where the concentrations of many molecules are approximately constant because of a balancing between their rate of synthesis and breakdown.

Stereoisomers Isomers that are non-superimposable mirror images of each other.

Steroids Compounds that are derivatives of a tetracyclic structure composed of a cyclopentane ring fused to a substituted phenanthrene nucleus.

Stop codon A codon (UAA, UAG, UGA) which terminates translation as they do not code for amino acids.

Structural protein A protein that serves a structural function.

Submerged fermentation A process by which industrial production of antibiotics, enzymes and other substances is carried out by growing the microorganism in submerged culture either in flask or in fermenter to produce the respective product.

Substrate A molecule that is acted upon, and chemically changed, by an enzyme.

Subunit Individual polypeptide chains in a protein.

Supercoiled DNA Super-twisted, covalently closed duplex DNA.

Surfactant Substance introduced into a liquid to increase its surface tension, thus promoting the properties of spreading, wetting and such other properties as foaming.

Svedberg unit (S) The unit used to express the sedimentation constant ($S=10^{-13}$ sec).

The sedimentation constant S is proportional to the rate of sedimentation of a molecule in a given centrifugal field, and is related to the size and shape of the molecule.

Sweating Process for loosening the attachment of the hair or wool of hides or skins by maintaining them under conditions of heat and moisture (wet heat) which promotes the development and action of bacteria on the hair roots and lower epidermal layer.

Syntan Otherwise known as synthetic tannin. Generally, these are prepared as salts of polyphenolic sulphonic acids, or from different simple phenols or from natural phenolic compounds by sulphonation and condensation reactions.

Tallow Fat, or adipose tissue, of cattle, sheep and goats.

Tanning process or tanning Process for converting raw skins/hides to non-putrescible leather.

Tawing A process of treating prepared hide or skin (usually pigskin or goatskin) with aluminum salts and other materials, such as egg yolk, flour, salt, etc. The process increases the leather's pliability, stretchability, softness, and quality.

Telopeptide regions Regions of about 20 amino acids that are not helical and can be found at the c-terminal end of collagen triple helices. They play an important role in holding collagen macromolecules together. They can be removed if collagen is treated with proteases.

Temperate virus Virus which upon infection of a host does not necessarily cause lysis but whose genome may replicate in synchrony with that of the host.

Template A polynucleotide chain which dictates the sequence of the monomers to be added by polymerase in the growing chain.

Tensile strength Measurement of the strength of a material against progressive application of force till the specimen sample of the material shows signs of fracture. The tensile strength is recorded in kg/sq.cm along with the recording of the percentage elongation caused by a specific load, and the percentage elongation at break.

Termination factors Proteins that are exclusively involved in the termination reactions of protein synthesis on the ribosome.

Terpenes A diverse group of lipids made from isoprene precursors.

Tertiary structure With reference to protein or nucleic acid, it is the final folded form of the polymer chain.

Tetramer Structure resulting from the association of four subunits. **Thermophiles:** Organisms whose optimum temperature for growth is between 45° and 85°C.

Thioester An ester of a carboxylic acid with a thiol or mercaptan.

Topoisomerase An enzyme that changes the extent of supercoiling of a DNA duplex.

Total dissolved solids (TDS) Quantity of dry material obtained from the solution in which it is present.

Total Kjeldahl nitrogen (TKN) Determination of all nitrogen excluding ammonium nitrogen in a sample.

Total organic carbon (TOC) The content of carbon present in organic material by measuring the CO_2 after its complete oxidation.

Total solids (TS) The total amount of solids in solution and suspension.

Total suspended solids (TSS) The solid pollutants that either float on the surface of, or are suspended in, wastewater that can be trapped by a filter. TSS can include a wide variety of organic and inorganic material, such as silt, decaying plant and animal matter, industrial wastes, and sewage.

Transamination A biochemical reaction in which an amine group is transferred from an amino acid to keto acid to form a new amino acid and keto acid. The coenzyme required for this reaction is pyridoxal phosphate.

Transcription The copying of a DNA template into a single-stranded RNA molecule.

Transduction Genetic exchange in bacteria that is mediated via phage.

Transfection The process of artificially introducing foreign nucleic acids into the cell.

Transfer RNA (tRNA) A family of low-molecular-weight RNAs (approximately 75–85 bases) that transfer amino acids from the cytoplasm to the template for protein synthesis on the ribosome.

Transferase: An enzyme that catalyses the transfer of a functional group from one molecule to another.

Transformation: The transfer of genetic information or incorporation of new genetic markers into living cells as free DNA.

Transgene: A foreign gene that is introduced into the genome of the other organism naturally or by genetic engineering.

Transgenic: Describing an organism whose genome incorporates and expresses genes from another species.

Transition state: The activated state in which a molecule is best suited to undergo a chemical reaction.

Translation: The process whereby mRNA directs the synthesis of a protein molecule; it is carried out by the ribosome in association with a host of translation initiation, elongation and termination factors.

Transport protein: A protein whose primary function is to transport a substance from one part of the cell to another (organelles), from one cell to another, or from one tissue to another.

Trickling Filter: A filter of natural or synthetic material used to support bacterial growth and provide an aerobic secondary treatment of wastewater. Effluent from the primary clarifier is distributed over a bed of rocks. As the liquid trickles over the rocks, a biological growth on the rocks breaks down the organic matter in the sewage. The effluent is then taken to a clarifier to remove biological matter coming from the filter.

Triplet: A three-nucleotide sequence; a codon.

Trypsin: A proteolytic enzyme that cleaves peptide chains next to the basic amino acids arginine and lysine.

Ultracentrifuge: A high-speed centrifuge that can attain speeds up to 60,000 rpm and centrifugal fields of 500,000 times gravity. It is useful for characterizing and/or separating macromolecules.

Ultrafiltration (UF): A filtration process using membranes which retain particles of between 0.001 micron and 0.1 micron.

Upflow anaerobic sludge blanket (UASB): Normally referred to as UASB reactor, it is a form of anaerobic digester that is used in the treatment of wastewater.

UV irradiation: Electromagnetic radiation with a wavelength shorter than that of visible light (200–390 nm). It causes damage to DNA (mainly by forming pyrimidine dimers).

van der Waals forces: Refers to the combined effect of two types of interactions, one attractive and one repulsive. The attractive forces are due to favourable interactions among the induced instantaneous dipole moments that arise from fluctuations in the electron charge densities of neighbouring nonbonded atoms. Repulsive forces arise when noncovalently bonded atoms come too close together.

Vector: A plasmid, cosmid, bacteriophage, or virus into which a foreign DNA can be inserted. They act as vehicles for transferring DNA from one cell to another.

Vegetable tannin Tanning agent extracted from the barks, fruits, galls, leaves, roots or wood of certain tannin-bearing plants.

Vermicomposting An aerobic composting process for the biological decomposition of solid organic materials using earthworms.

Virion Virus particle, i.e., the viral nucleic acid surrounded by protein coat and in some cases other material.

Viroids Pathogenic agents, mostly of plants, that consist of short (usually circular) RNA molecules.

Virulence The degree of pathogenicity of a parasite or an infectious agent.

Virus A complex of nucleic acid and protein, which can infect and replicate inside a specific host cell to make more virus particles.

Vitamin Organic compounds required by living organisms in relatively small amounts to maintain normal health.

Volatile organic compound (VOC) Large family of carbon-containing compounds, mostly solvents, that may undergo photochemical reactions in the air. Some are toxic. In the leather industry, they are produced from solvent evaporation in the finishing operations.

Water activity (a_w) The vapour pressure of a liquid divided by that of pure water at the same temperature.

Western blot A technique used to detect the presence of a specific antigen (protein) on a nitrocellulose membrane with the help of specific antibodies.

Wet blue Hide, or skin, which has been subjected to chrome-tanning and left wet.

Wet salting The application of dry sodium chloride to the freshly flayed skin or hide.

Wet white Leather, which after tanning with white materials such as aldehydes, aluminium and syntans, is in the wet condition.

Wobble A proposed explanation for base-pairing that is not of the Watson–Crick type and that often occurs between the 3′ base in the codon and the 5′ base in the anticodon.

X-ray crystallography A technique for determining the structure of molecules from the X-ray diffraction patterns that are produced by crystalline arrays of the molecules.

Yeast A unicellular true fungus belonging to the Phylum Ascomycetes, and is widely dispersed in the natural habitat.

Z form A duplex DNA structure in which there is the usual type of hydrogen-bonding between the base pairs but in which the helix formed by the two polynucleotide chains is left-handed rather than right-handed.

Zwitterion A dipolar ion with spatially separated positive and negative charges. For example, most amino acids are zwitterions, having a positive charge on the α-amino group and a negative charge on the α-carboxyl group but no net charge on the overall molecule.

Zymogen An inactive precursor of an enzyme, e.g., trypsin, (the active form), which exists in the inactive form trypsinogen before it is converted to its active form.

INDEX

Made in the USA
Monee, IL
07 July 2026

56548173R00300